On the Frontier

Hugh L. Dryden, for whom the Center was named in 1976. (NASA photo E-4248)

On the Frontier

Experimental Flight at NASA Dryden

Richard P. Hallion
and
Michael H. Gorn

Smithsonian Books
Washington, DC

By special arrangement with the National Aeronautics and Space Administration History Office, this publication is being offered for sale by Smithsonian Books, Washington, D.C. 20013-7012.

Library of Congress Cataloging-in-Publication Data
Hallion, Richard.
On the frontier : experimental flight at NASA Dryden / Richard P. Hallion and Michael H. Gorn.
p. cm.
Includes bibliographical references and index.
ISBN 1-58834-134-8 (alk. paper)
1. Dryden Flight Research Facility—History. I. Gorn, Michael H. II. Title.
TL521.312.H3397 2003
629.1'07'2079488—dc21 2002030484

A paperback reissue (ISBN 978-1-58834-289-8) of the original cloth edition.

British Library Cataloguing-in-Publication Data available.

Manufactured in the United States of America.
15 14 13 12 11 10 5 4 3 2

To My Wife, Christine Gorn
For her Love, Enthusiasm, and Encouragement

and

To Chris Hallion
For Every Possible Reason

Contents

Foreword to the First Edition

A stillness was on the desert. Daylight settled unhurriedly down the hilltops bordering the triangular valley. The indigo sky above and to the west was pierced with the gleam of a solitary planet and a flicker of an occasional second- or third-magnitude star.

The valley bottom was an expanse of flatness. Miles of mirror-smooth clay were marred by neither hummock nor furrow. No tree or bush could be seen on this seemingly endless, waterless lake. No sound from animal or bird punctuated the silence. Wild creatures found little to attract them on the vast empty platter. It was one of nature's quiet hideaways, an outpost of serenity.

There were intruders. On the western shore of this "lake," figures scurried around a strange assemblage. A small shark-sleek craft was being attached to a much larger mother craft. The shark's midsection was banded with ice crystals; puffs of ashen vapor wafted upward and disappeared into the clear sky. The juxtaposition of ancient geology and modern technology, curiously, seemed to fit.

By the mid-twentieth century, the science of aeronautics had grown to substantial maturity. Aircraft were speeding faster and faster and threatening to outrace their own sound. The National Advisory Committee for Aeronautics had a trio of laboratories to study the fundamental problems of flight. They had a wide variety of test facilities and a cadre of bright, able, and dedicated scientists who had performed with remarkable success over the years surrounding the Second World War. For testing of very-high-speed aircraft, however, they needed a new laboratory: a laboratory in the sky.

And so it was that the researchers came to Antelope Valley in California, a valley blessed with clear and uncrowded skies, a sparse population, and Muroc Dry Lake, a natural aerodrome where runway length and direction were, for most practical purposes, nearly limitless.

On the shore of Muroc, NACA established its High-Speed Flight Station and began to challenge of the unknown. The mysteries were perplexing. The search for solutions was tedious, protracted, and often dangerous. The research methods placed men and machines at the boundaries of understanding. On occasion, fine men were lost at those boundaries in the pursuit of knowledge. Their sacrifices will be remembered.

At the dawn of the Space Age, the researchers on the shore of the dry lake were already actively engaged in its planning. After NACA became NASA, their considerable contributions were of substantial significance in the evolution of America's manned spaceflight program.

This book is the story of those researchers and their efforts. Richard Hallion has recorded the history of their flights and captured the spirit of a remarkable and unique institution in the evolution or aerospace progress. He tells of the place, the projects, and, most important, the people. It is a story of men and machines, of success and failure, of time and circumstance.

I had the pleasure of living some of the events recorded here. I take great personal satisfaction in those years, the projects in which I was privileged to participate, and the wonderful and able people I worked with and whose friendship I cherish.

Neil A. Armstrong
October 1983

Foreword to the Second Edition

Dr. Richard P. Hallion's *On the Frontier* was one of the first center histories in the NASA History Series, and it is still one of the best. But it is now almost twenty years old and a lot has happened at Dryden since he wrote the original book. The Dryden Flight Research Facility has regained its status as a full-fledged Center, and it has engaged in a lot of exciting flight research on a variety of vehicles ranging from the X-29 forward-swept-wing research aircraft to several Access-to-Space X-plane programs.

To bring the original edition up to date as of the turn of the century, Dryden Chief Historian Dr. Michael H. Gorn has written a new concluding section to Dr. Hallion's original Chapter 11 and has added Chapters 12 and 13, which cover the last two decades of the century. Together with Dr. J. D. Hunley, Dryden's former Chief Historian, he has also made minor revisions to the earlier chapters to incorporate the literature that has appeared since the chapters were first written and to take advantage of comments on the original chapters by engineers and pilots who were involved in the events Dr. Hallion described.

The result is an updated account that builds on the strengths of the original book. It shows the process of change that has occurred in NASA's flight research since the beginnings of Dryden's history in 1946 and the important work that Dryden and its predecessor organizations have done in partnership with industry, universities, the armed services, and other NASA centers. In many ways, Dryden and NASA remain "on the frontier" of knowledge and development in aeronautics and space. This revised book will help those in and outside of NASA to understand the contributions flight research has made to the Agency and the nation.

Kevin L. Petersen
Director, Dryden Flight Research Center
September 2002

Acknowledgments

For the First Edition

This account of flight research at Hugh L. Dryden Flight Research Center resulted from cooperation between the History Office of the National Aeronautics and Space Administration and the Department of Science and Technology of the Smithsonian Institution's National Air and Space Museum. It would have been impossible to undertake and complete this study without the support and assistance of a large number of persons within NASA, the Smithsonian Institution, and the Departments of Aerospace Engineering and History of the University of Maryland. My debt to all of them is great.

I owe special gratitude to Michael Collins, the former director of the National Air and Space Museum; Melvin B. Zisfein, deputy director; and assistant directors Howard S. Wolko, Donald S. Lopez, and Frederick C. Durant III, together with Dr. Tom D. Crouch, Dr. Paul A. Hanle, Dr. Robert Friedman, and Dr. Richard Hirsh of the curatorial staff. Staff members of the NASM Library, especially Catherine D. Scott and Dominick A. Pisano, were most cooperative in locating obscure reference materials. The staff of the NASA History Office encouraged me at every step. I am especially appreciative of the assistance and cooperation given by Dr. Monte D. Wright, Dr. Frank W. Anderson, Dr. Eugene Emme, Dr. Alex Roland, Lee Saegesser, Leonard Bruno, Carrie Karegeannes, and Nancy Brun. Staff members of the National Archives and Records Administration, especially John Taylor and Jo Ann Williamson, were of great assistance in tracing NACA and NASA record groups. I wish to thank Charles Worman of the Air Force Museum, Carl Berger of the Office of Air Force History, Dr. Lee M. Pearson of the Naval Air Systems Command, and J. Ted Bear of the Air Force Flight Test Center for their assistance. I am grateful to the faculty of the University of Maryland for assistance and wise counsel, especially Professors Alfred Gessow, John D. Anderson, and Jewel B. Barlow of the Department of Aerospace Engineering; and Wayne Cole, Keith Olson, and Walter Rundell of the Department of History. Professor Roger E. Bilstein of the University of Houston at Clear Lake City was most helpful during my research and initial writing, as was Professor Richard E. Thomas, director of the Center for Strategic Technology, Texas A & M University.

It is, of course, to the participants in this history that I owe my greatest debts. With unfailing courtesy, grace, and assistance, the present and former personnel of the Dryden Flight Research Center and their NASA, Air Force, Navy, and industry counterparts welcomed my every inquiry, patiently answered all questions, and assisted in the detailed reconstruction of past events. Through them I learned much of the flight-testing process and the history of the center—and of the character and courage of individuals who test the products of engineering drawing boards. Through Ralph Jackson, director of external affairs for the Dryden Flight Research Center, I had the opportunity to conduct my research in the conducive atmosphere of Dryden during the bustling days of the Space Shuttle. It was a refreshing and novel experience for a historian; as I examined boxes of records a quarter-century old, the rumble of NASA's present experimental aircraft punctuated my research, forcibly reminding me that the history of Dryden is far from over.

Richard P. Hallion
August 1981

For the Second Edition

A little more than 10 years after *On the Frontier* first appeared, Dr. J.D. Hunley inaugurated the NASA Dryden History office, focusing his efforts on the establishment of a historical reference collection. Its creation affected materially the revisions to the first edition. Due to his efforts, the author fashioned the new narrative in the revised edition–the end of Chapter 11, as well as Chapters 12 and 13–mostly from documents cataloged in this collection. Hunley also initiated an oral interviewing project, the recollections of which informed much of the added narrative. Finally, he started an ambitious publications program, this volume being one of many in the series. Hunley devoted himself unstintingly to the second edition of *On the Frontier*. He worked innumerable hours updating and adding new appendices, re-reading the narrative to reduce inadvertent errors, and coordinating a systematic peer review of the entire manuscript. He also acted as valued guide and advisor during the writing of chapters 12 and 13. Although his name does not appear on the title page, he deserves a great measure of credit for this volume.

The historical reference collection founded by Dr. Hunley is staffed by two people of extraordinary abilities: archivist Peter Merlin and volunteer Betty Love. Both contributed immeasurably to the accuracy and completeness of the appendices, as well as to the narrative itself. Ms. Love also took on the daunting task of converting all of the metric numbers in the first edition into English measurements in the second.

The new text in the second edition relied not only on archival sources, but on interviews conducted by the author as well. These included conversations with a number of Dryden employees: former research pilot and chief engineer Bill Dana, X-33 project manager Gary Trippensee, X-34 project engineer John Bosworth, X-34 deputy project manager John McTigue, and X-38 project engineer Christopher Nagy.

The Dryden reference library also proved to be an essential partner in revising and updating this book. Retired Librarian Dennis Ragsdale searched many on-line data bases for articles, conference papers, proceedings, and NASA technical reports pertinent to flight research during the recent period. The NASA web pages likewise provided valuable, up-to-date information, especially NASA Headquarters, the Ames Research Center, the Marshall Space Flight Center, and the Dryden Flight Research Center.

In addition, many members of the DFRC staff read the original and two new chapters and offered invaluable advice and criticism. Some merely re-worded sentences; others suggested new sources to shore up those parts of the narrative involving the present; and a number of project engineers suggested revisions of sections dealing with concepts or vehicles still being developed (and since cancelled). They saved the authors from many errors. The list of contributors includes Betty Love, Peter Merlin, and Bill Albrecht for the Prologue, which has been substantially updated as a result of their knowledgeable comments; Betty Love and Peter Merlin for extended research on Appendix E; Judy Duffield for much of the new data in Appendix R, and Dick Rieder for much new information in Appendix C; Gerald M. Truszynski for Chapter 1; Hubert M. Drake for Chapters 2-3; A. Scott Crossfield and Edwin Saltzman for Chapters 4-5; Johnny Armstrong, Robert Hoey, and Bill Dana for Chapter 6; Calvin R. Jarvis, Bruce Peterson, Wayne Ottinger, and Gene Matranga for Chapter 7; Betty Love, Bob Hoey, and Dale Reed for Chapter 8; Don Mallick and Peter Merlin for Chapter 9; Tom McMurtry and Gary Krier for Chapter 10; and Gordon Fullerton for Chapter 11.

Five persons read much or all of Chapters 12 and/or 13; Dr. Hunley, former Deputy Director of Flight Operations Ed Schneider, former Director of Aerospace Projects Bob Meyer, former Acting Director of Flight Operations Rogers Smith, and one anonymous reader. Additionally, a number of reviewers read Chapters 12 and 13 more selectively, including Roy Bryant, Joe Ayers, Gary Krier, Randy Albertson, Alan Brown, Jennifer Baer-Riedhart, John Bosworth, James Cooper, Dave Lux, Neil Matheny, Lisa Bjarke, Russ Barber, Dean Webb, Lee Duke, Bruce Powers, Alex Sim, Bill Burcham, Trindel Maine, Christopher Nagy, Sheryll Powers, Paul Reukauf, Louis Steers, Ronald (Joe) Wilson, and Robert Navarro. Curtis Peebles of the Dryden History Office read the entire narrative and offered some useful changes. In addition, Dr. Darlene Lister, a highly accomplished editor, rendered the manuscript consistent in style and usage from first page to last.

Three individuals at NASA Dryden merit special notice for their contributions to this book. Senior Photographer Carla Thomas showed customary good humor during the painstaking process of scanning the photos. The multi-talented Steven Lighthill, Chief of Graphics, designed the handsome dust jacket. Lastly, Jay Levine, editor of the Dryden *X-Press,* worked heroically for months on end to render the countless final deletions, additions, and adjustments into consistent and attractive page layout. This project would not have been the same without his labors.

Finally, Smithsonian Books, the publisher of the second edition of *On the Frontier,* has made important contributions to this book. Acquisitions Editor Mark Gatlin has been an enthusiastic champion from the start and offered wise counsel throughout the process. In addition, two individuals in the press's production department—Matt Crosby and Ruth Thomson—helped our collaboration immensely. Dryden Flight Research Center has benefited greatly from our partnership with this distinguished publisher.

Michael H. Gorn
September 2002

Prologue
A Most Exotic Place

Northeast of Los Angeles, beyond the coastal range, lies the Mojave Desert, the southwestern corner of which is called Antelope Valley. This semiarid area produces carrots, onions, fruit, pistachio nuts—and aircraft. The clear weather and vast, unrestricted space have lured the aircraft industry as flowers draw bees. Politicians have pragmatically dubbed it "Aerospace Valley." Its two major communities, Lancaster and Palmdale, cater to the wants and needs of the aerospace community. At Palmdale looms "Air Force Plant 42," where products of Northrop Grumman, Lockheed Martin, and Boeing scoot aloft. Here is the place where the Space Shuttle and the B-1 strategic bomber underwent final assembly. The valley economy would collapse if the aerospace industry declined, and citizens are determined not to let that happen. "Vote your pocketbook! Ketcham = B-1" read one 1976 election poster, and such logic still makes sense to desert residents. Lancaster's economic heart is located at the Air Force Flight Test Center and NASA's Hugh L. Dryden Flight Research Center on the shores of Rogers Dry Lake at Edwards Air Force Base. Lancaster received its name in 1887 from homesick Pennsylvania Amish settlers. In 1950, it had a population of 3,924 and was a sleepy desert community where a shopper could go to a store in the midst of a workday only to find a "Gone Hunting" sign posted on the door. Then came the aerospace boom. A decade later, the population hit 30,000, and at the end of the year 1999 it stood at 131,000. Most Edwards workers, be they Air Force, NASA, or private contractors, live in Lancaster or Palmdale, although increasing numbers of them dwell in other nearby communities.

North of Lancaster the tiny community of Rosamond, home of the former Tropico gold mine, was still a gritty desert town of unadorned houses and mobile homes in the early 1980s. "Welcome to Rosamond-Gateway to Progress," proclaimed a black-and-white sign on Sierra Highway. By century's end, it had become a prosperous community with new houses, schools, a beautiful library, and many new businesses as well as some remnants of the not-so-distant past. Traveling north on State Route 14, turn right at Rosamond Boulevard, and one is soon rolling toward Edwards, running past the smooth baked clay of Rosamond Dry Lake. Ahead, over scrub-covered low hills, stretches the vast parched-silt bed of Rogers Dry Lake.

North of Rosamond and 25 miles above Lancaster, the town of Mojave hugs open desert between brooding Mount Soledad and the Tehachapi range. Mojave was once the terminus for borax-laden mule trains, a brawling, hard-living town. Desert winds sweep across Mojave, sometimes overturning trailers and vans, often closing the roads to truck traffic, usually blowing powdery grit and tumbleweeds across the land. Now the mule trains have been replaced by massive diesel-electric locomotives running north and south with long strings of hoppers and boxcars paralleling the Sierra

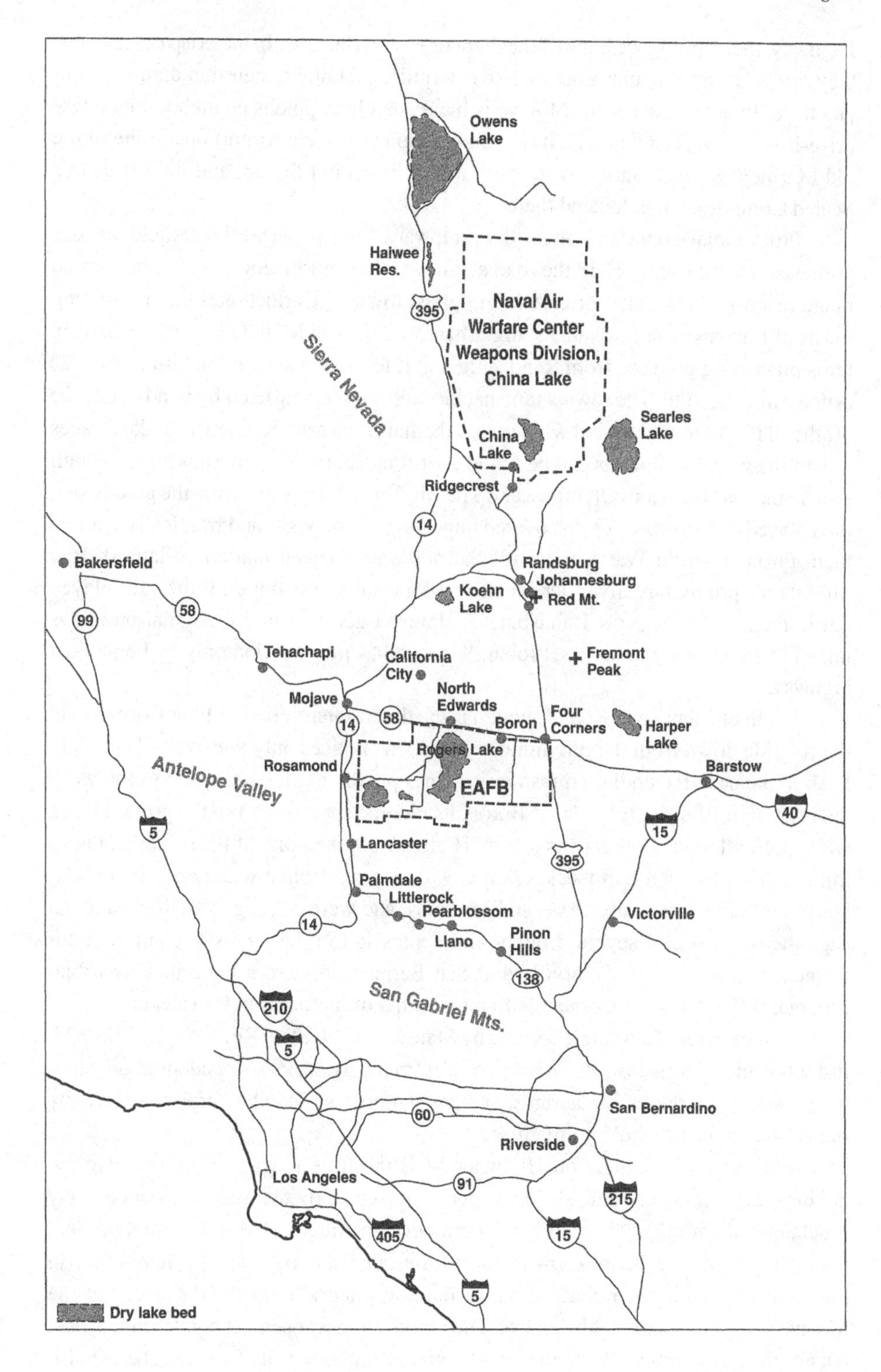

Owens Lake
Haiwee Res.
395
Naval Air Warfare Center Weapons Division, China Lake
Sierra Nevada
China Lake
Searles Lake
Ridgecrest
14
Bakersfield
Randsburg
Johannesburg
Red Mt.
Koehn Lake
58
99
Tehachapi
California City
Fremont Peak
Mojave
North Edwards
14
58
Boron
Four Corners
Harper Lake
Rogers Lake
Rosamond
Barstow
Antelope Valley
EAFB
40
5
15
Lancaster
395
Palmdale
Littlerock
Pearblossom
Victorville
14
Llano
Pinon Hills
138
San Gabriel Mts.
210
5
San Bernardino
60
Riverside
Los Angeles
91
215
405
15
5
Dry lake bed

Highway, then turning west into Tehachapi or east to Barstow. In the crisp desert days, they can be seen from afar, snaking like caterpillars. At night, their thunderous clatter jars the stillness of the desert. Mojave is bleak, barely populous enough to rate a few drive-in restaurants and motels. Its chief attraction is Mojave Airport on the site of the old Marine Corps air station with the National Test Pilot School and industries like Scaled Composites Inc. located there.

From Mojave one can turn northwest toward Tehachapi and Bakersfield, or bear northeast toward Cantil. Here the road again branches: north along the Sierra Nevada range toward China Lake, or east to the ghost town of Garlock and the old mining towns of Randsburg and Johannesburg. Shadowed by Red Mountain, Garlock had six mills processing gold ore from Randsburg, but it fell into disuse at the turn of the 19th century into the 20th. The town's ramshackle buildings are buffeted by winds that race off the El Paso Mountains and whip across the flats of nearby Koehn Dry Lake. Traces of half the world's minerals can be found near Randsburg. Its numerous mines—with such names as Napoleon, Olympus, and The Big Norse—thrived during the gold boom, then played out. Prospectors discovered huge tungsten deposits and frantically worked them through World War I and until the postwar tungsten market collapsed. Red Mountain's prodigious silver deposits caused a resurgent boom until the vein played out in the late 1920s. Now Randsburg is almost a ghost town, its original buildings mixed in among mobile homes. "Joburg" is saved from extinction only by being on a highway.

South of Joburg and Red Mountain is the desert intersection of Four Corners. On the flat ride down from the old mining camps, one notices only the swell of Fremont Peak in the east; the endless transmission lines paralleling the road; and to the west, the FAA air traffic control radar at Boron, its radome sprouting from the ground like a white puffball. Four Corners sits astride Highway 58, the route of the so-called Okies immortalized by John Steinbeck. (Actually, they came from a wider area than Oklahoma, including especially Texas and Missouri, and were coming to California to escape the ravages of a severe drought and depression in the 1930s.) South of Four Corners is the route to Victorville and San Bernardino, across the San Bernardino Mountains. East of Four Corners is Barstow, a major intersection for truckers.

At the center of a triangle formed by State Route 14, State Route 58, and US 395, just a few miles northeast of Mojave, is California City. It boasts residential developments; some businesses, restaurants, and gas stations; a municipal airport; a privately run prison; and an 18-hole golf course.

West of Four Corners on Highway 58 is the little town of Boron, where the double crack-crack! of sonic shocks is heard so frequently that the locals once coyly proclaimed themselves "The Biggest Boom Town in America." Boron's chief product is sodium tetraborate, better known as borax. Introduced into Europe by Marco Polo in the 13th century, borax remained an uncommon mineral until the discovery of the Mojave's deposits in the 19th century. The world's largest open-pit borate mine is just outside the town north of the highway. At night, its high-intensity lights can be seen for miles; by day, dust hangs low in the sky. West of Boron is the little community of

North Edwards; as with Lancaster, most of its citizens are or have been affiliated with Edwards Air Force Base. At North Edwards, the traveler can continue west on Highway 58 to Mojave, or (if authorized) can turn south, onto Rosamond Boulevard, driving down to Edwards, its hangars gleaming in the distance. And here, the visitor sees perhaps the weirdest of sights that the desert has to offer: the 44-square-mile bed of Rogers Dry Lake.

Dry lakes are the flattest of all geological land forms. Rogers Dry Lake is a playa, a pluvial lake, one of 120 such lakes in the western United States. Pluvial lakes are believed to have first appeared in the Pleistocene epoch, about 1.5 million years ago. Glacial activity dropped temperatures and increased precipitation, creating hundreds of pluvial lakes, which fluctuate between wet and dry phases. They appear in arid regions, in the lowest areas of basins, and contain great quantities of sediment. Rogers originally received its water from overflow of rivers in nearby mountains. In time, the water sources disappeared, the lake dried, and the arid Mojave now keeps it that way except for the briefest of periods when rain floods its surface to a depth of a few inches.

The desert winds blow the water (and suspended sediment) back and forth across the lake surface, filling cracks and smoothing the silt. When the water evaporates, the lake is perfectly flat and smooth. Once dry, Rogers is also hard; the water and winds remove dissolved salts from the sediment, which dries to a hard crust—at Rogers, from 7 to 18 inches deep. California has a great number of lakes like Rogers—Rosamond, Mirage, Cuddeback, Harper, Searles, Koehn, China, Ballarat. Rogers, one of the largest dry lakes in the world, is clearly visible to the traveler flying into Los Angeles from the east. In the early part of the 20th century, a silver and gold mining firm established a camp on its shores and named it Rodriguez, after the company's name. Rodriguez eventually became Rogers. The lake, shaped like a lopsided figure eight, is dry for 10 months of each year. During that time its surface can support up to 250 pounds per square inch of pressure. Even the heaviest aircraft can take off from and land on the lake, making Rogers the largest landing field in the world.

Aviation was a long time in coming to Rogers. At first the lake served only as a bed for the Santa Fe Railway and a small camp specializing in the extraction of drilling mud for use in oil wells. In 1910 came the first permanent settlers, Clifford and Effie Corum, and Clifford's brother Ralph. The Corums were determined to start a farm community in the midst of this wasteland; surprisingly, they convinced other settlers to join them. The brothers opened a general store, dug wells for water, and held church services in their home. The Santa Fe Railway's freights always stopped for water. Encouraged, the Corums decided to name the little community after themselves. Here they ran into a snag. Postal authorities objected because California already had a *Coram* township; the similarity in spelling would surely cause confusion. The Corums then suggested Muroc, created by spelling the name backwards, but the Sante Fe Railway objected because of a rail stop named Murdock. The railroad lyrically suggested Dorado, Ophir, Yermo, or Istar. (It is puzzling how many truly desolate desert communities have names connoting beauty, tranquility, and prosperity.) The Corums remained firm. The tiny community became

Muroc; settlers sometimes applied the name to the dry lake as well.

Muroc did not prosper and remained little more than a spot on a map. The 1930s brought Depression and the Okies wending their way along Highway 58, north of the lake. The lake itself gained notoriety as the site of what was supposedly the largest moonshine distillery in southern California; at night, prohibition agents chased liquor runners across the lakebed. Other citizens used the lake to race automobiles. By 1930, designers and pilots already recognized the value of the dry lakes as test sites for new aircraft, and Rogers, together with Harper and Rosamond, became a favorite spot for small aircraft companies to fly their new designs. Here aviatrix and socialite Florence "Pancho" Barnes established a dude ranch and nightclub with a small private airstrip; in future years, Barnes "Fly-Inn" became a popular gathering spot for test pilots and engineers.

The military came to Muroc in 1933, at the behest of Col. Henry H. "Hap" Arnold, commanding officer of the Army's March Field at Riverside, California. Arnold, later to become the Army Air Forces' chief in World War II, needed a desolate site for a bombing and gunnery range. The Navy having denied Arnold the use of the Pacific Ocean, he looked elsewhere. The most logical site for the range was in the vast barren stretches of the Mojave; most of the land around Rogers Dry Lake already belonged to the federal government. In September 1933, a cadre of soldiers from March established a camp on the eastern side of the lake and laid out the gunnery and bombing range. Over the next decade the desert echoed to the snarling throb of brightly painted Boeing P-26A fighters, as well as Northrop A-17 and Martin B-10 bombers, as Army pilots bombed and shot up the desert. The tiny community of Muroc, on the west side of the lake, was not really affected. The community had a couple of live contacts with the "Golden Age" of American aviation, however. In 1927 Charles Lindbergh landed at Muroc, and in 1935 Wiley Post force-landed his *Winnie Mae* on the lake. From the 1940s to the early 1950s, John Northrop's ethereal flying wings were a familiar sight in Muroc skies.

In the summer of 1941, when the Third Reich controlled the European continent and Japan was firming up plans for the Greater East Asian Co-Prosperity Sphere, 13 Army employees operated the bombing and gunnery range. Then on 10 July 140 troops arrived via the railway across the lakebed and staked out a tent camp on the southwestern shore of the lake. The character of operations at Muroc changed. What had been a useful bombing and gunnery site now also became a remote testing site. Here, in 1941, Maj. George V. Holloman experimented with radio-controlled Douglas BT-2 basic trainers in a highly classified project with future implications for pilotless robot weapon development.

On 7 December 1941 the Army's 41st Bomb Group and the 6th Reconnaissance Squadron arrived at Muroc for crew training. That same day, Japanese naval aircraft devastated Pearl Harbor and America was at war. Two days later, four squadrons of Martin B-26 Marauders arrived at Muroc for coastal antisubmarine patrol duty but left in February 1942 for Australia and the bitter New Guinea campaign. The war brought a rapid influx of people, eventually numbering 40,000. The community of Muroc vanished, buried under a tent city erected on the site of what is now South Base. On 23 July the rapidly growing site was designated Muroc Army Air Base. On 8 November 1943 the Army redesignated it Muroc Army Air Field, complete with barracks, sewerage system, control tower, and year-round concrete runway. Engineers built a 650-foot replica of a Japanese *Mogami*-class heavy cruiser on the lake—"Muroc Maru" the pilots dubbed her—and antishipping bomber crews honed their skills. Fighter pilots in P-38s and bomber crews in B-24s and B-25s flew training missions at Muroc before going overseas. Yet Muroc did not remain simply an advanced training base, valuable as this would have been to the war effort; it also became a major research and development center.

Before World War II, the Army's major aeronautical research and development center had been at Wright Field, outside Dayton, Ohio. But Wright was in a heavily residential area; hazardous flight testing of prototypes could endanger the local population. Also the area was too populous to be safe from prying eyes. Anyone could see the latest aircraft going through its paces, just by watching from beyond the airfield's boundary. The Army needed a remote test site. Muroc, a mere 100 miles from the center of the West Coast's aircraft industry but still very isolated from major population centers, was ideal.

The catalyst that caused the big change from Wright to Muroc was the Bell XP-59A program, the United States' first jet aircraft. In April 1941, Hap Arnold had learned of Britain's jet engine research while on an inspection trip to England. In September the Army issued a contract to the Bell Aircraft Corporation for a jet airplane that used a British-developed Whittle engine built by General Electric. The result was the Bell XP-59A Airacomet, a twin-jet single-seat airplane. Obviously, the XP-59A was too secret to test at Wright. Lt. Col. Benjamin W. Chidlaw, one of Arnold's deputies, toured the country looking for a suitable test site. Without question, Muroc was the best possible choice. In 1942, the Materiel Center at Wright Field designated the northwestern end of the dry lake as the Materiel Center Flight Test Site. Subsequently, this site became known as North Base and the training center on the southwestern lake shore became known as South Base. Security, already tight, became vise-like. The XP-59A arrived at North Base in mid-September 1942 and made its initial flight on 1 October. The U.S. entered the jet age, third behind Germany and Great Britain.

Soon the importance of Muroc as a flight test center overshadowed its importance as a training base. North Base conducted its operations strictly separate from South Base, and added its own runway, hangar, and tower facilities. The first tower was a guard shack mounted on two large sawhorses; sometimes it blew over in the desert winds. Known as "Oscar Junior," it had a single Hallicrafter radio connected by a 110-volt extension cord to the operations building, a frame hut. A field telephone, binoculars, and salt tablets for the tower crew completed its equipment. "Oscar Senior," a genuine aircraft control tower with a variety of communications equipment and clear glass sides, entered service in July 1944. By that time, the P-59 was no longer queen of the flightline. Lockheed's XP-80 Shooting Star, dubbed *Lulu-Belle* by Lockheed engineers, had completed its first flight at North Base in January 1944. That same year the Army redesignated North Base as the Muroc Flight Test Base, coequal with its training counterpart to the south. On 15 April 1946, with wartime needs buried in an already fading past, the Army Air Forces ended all training activities at Muroc, designating Muroc solely as a research and development center under the name Muroc Army Air Field. This lasted until 12 February 1948, when it became Muroc Air Force Base following the establishment of the United States Air Force. On 8 December 1949 it was renamed Edwards Air Force Base in honor of Capt. Glen W. Edwards, killed in the crash of a Northrop YB-49 Flying Wing on 5 June 1948.

Muroc-Edwards after World War II remained an important research center. The war had pointed to the importance of such new developments as jet aircraft, and the tempo of wartime research had generated new conceptions of aircraft design such as the swept-wing planform that might prove useful on future military and civil aircraft. The rapid disintegration of the wartime Grand Alliance, underscored by the lowering of the "Iron Curtain" in Central and Eastern Europe, acted as a spur to continued rapid research on projects that might ultimately affect national security. Captured enemy aircraft such as the Heinkel He 162 flew at Muroc in evaluation programs. Work continued on radio-controlled aircraft—two B-17s flew from Hawaii to Muroc under radio guidance in 1946. Weird combination jet-and-piston aircraft such as the Convair XP-81 and Ryan XF2R-1 flew from the lake, as did new bombers such as Douglas's XB-42 and Northrop's graceful though ill-fated XB-35 and YB-49 Flying Wings. But the big news in the fall of 1946 was not the testing of some new aircraft destined for squadron service nor the latest scoop on what the Germans or the Japanese had been up to in the war. Rather it was anticipation of a program of such significance that the whole fabric of aviation might be transformed.

The program revolved around a technological challenge: Could aeronautical science design an aircraft that could fly faster than the speed of sound? Today, what with piloted spaceflight and Mach 2 commercial airline service, such a question seems almost trivial. In 1946, however, that question loomed across the face of aeronautics; highly trained engineers spoke of a mysterious "sound barrier" through which it might be impossible to fly a piloted aircraft. The challenge was not simply a theoretical one that threatened the imagination of designers hunched over drawing boards. Pilots had died as their aircraft approached the speed of sound, died when their aircraft broke up in high-speed dives. The "sound barrier" threatened to deny aeronautical science the high speeds that the jet engine promised, to limit aviation to speeds of about 600 miles per hour.

In September-October 1946, the National Advisory Committee for Aeronautics (NACA), the aeronautical research agency for the United States, sent a small band of engineers and technicians from the Langley Memorial Aeronautical Laboratory at Hampton, Virginia, to Muroc to assist in a supersonic flight research program involving the Bell XS-1 aircraft. This small but slowly growing group became known as the NACA Muroc Flight Test Unit a year later. In October 1947 the XS-1 exceeded the speed of sound in level flight, the first piloted supersonic flight. For the next decade, the NACA group continued to explore the problems and conditions of supersonic flight.

In 1949, the NACA had established the group as the NACA High-Speed Flight Research Station (HSFRS), a division of Langley Laboratory. In 1953, an HSFRS pilot became the first to fly at twice the speed of sound. In 1954, the HSFRS was redesignated the NACA High-Speed Flight Station (HSFS), autonomous from Langley. That summer, the station's 250 employees moved from their shared Air Force quarters to new research facilities located midway between South Base and North Base. Those facilities are still in use, though greatly expanded.

In 1959, after the creation of the National Aeronautics and Space Administration, the High-Speed Flight Station became the NASA Flight Research Center (FRC). The following decade saw the center embark on a strong program of hypersonic research using the North American X-15 aircraft. The X-15, launched over Nevada or eastern California toward Edwards, could streak to Mach 6—six times the speed of sound—over the Nevada and California desert. FRC personnel complemented the X-15 program with a flight test program using lifting-body reentry vehicles—both done in partnership with the Air Force. Center personnel also pursued studies in several space and aeronautics areas. The later 1960s saw a resurgence of interest in advanced supersonic research using such aircraft as the triple-sonic XB-70A and YF-12 Blackbird. In the 1970s the center continued with its lifting-body research in support of the Space Shuttle program, followed by the YF-12 program and such development programs as the F-8 Digital Fly-by-Wire and Supercritical-Wing programs.

On 26 March 1976, NASA renamed the Flight Research Center the Hugh L. Dryden Flight Research Center (DFRC), in honor of an American aerospace pioneer, a man who was fond of saying "the airplane and I grew up together," and who once remarked that "the most important tool in aeronautical research . . . is the human mind." Not a center to remain looking to its past, Dryden looks to the future; less than a year after the Dryden dedication, DFRC undertook the first flight tests of the Rockwell Space Shuttle orbiter *Enterprise*. Five years after dedication, the Space Shuttle *Columbia* landed at Edwards, having completed the first winged reentry of a piloted spacecraft from orbit.

That same year, on 1 October 1981, Dryden consolidated with Ames Research Center at Moffett Field in northern California, to become the Ames-Dryden Flight Research Facility. That arrangement lasted for over a decade, but on 1 March 1994 the facility was again converted to the Dryden Flight Research Center. After a six-month transition period to enable the re-emerging center to institute independent administrative functions, Dryden again assumed full center status on 1 October 1994 with a complement of 465 civil servants and a roughly comparable number of contract employees.

In the period when Dryden was a part of Ames and afterwards, significant flight research continued. Dryden continued to support NASA's space mission with programs that tested the Shuttle drag chute and booster recovery parachutes as well as the Shuttle tires. It also supported a variety of Access-to-Space initiatives such as X-33, X-34, and X-38. It pursued propulsion research in aeronautics with Digital Electronic Engine Control and Highly Integrated Digital Electronic Control that has resulted in improvements to the F100-PW-220 and –229 engines (on the F-15 and F-16 aircraft) as well as more advanced engines. Among a cluster of other projects in the last two decades of the 20th century have been flight research on the Highly Maneuverable Aircraft Technology vehicle; the X-29 forward-swept-wing research vehicle; and a triad of thrust-vectoring projects flying the F-18 High Angle-of-Attack Research Vehicle, the X-31 Enhanced Fighter Maneuverability Demonstrator, and the F-15 Advanced Control Technology for Integrated Vehicles project. All of these efforts promised improvements in areas ranging from composite structures to greater maneuverability to fly at high angles of attack.

Many decry the cost of flight research, the cost in both economic and human terms. They argue for computer simulation and prediction, a turning away from piloted vehicle testing, a turning away from actually building an aircraft and flying it. There is still no better refutation to the hypothesis that flight research is unnecessary than the testimony of NASA Administrator James E. Webb before Congress in 1967:

> Flight testing of new concepts, designs, and systems is fundamental to aeronautics. Laboratory data alone, and theories based on these data, cannot give all the answers . . . . Each time a new aircraft flies, a "moment of truth" arrives for the designer as he discovers whether a group of individually satisfactory elements add together to make a satisfactory whole, or whether their unexpected interactions result in a major deficiency. Flight research plays the essential role in assuring that all the elements of an aircraft can be integrated into a satisfactory system.

At Dryden, flight research is not simply one phase of the center's operation. Rather, it is the center's reason for being. Its fundamental importance has kept Dryden on the frontier of exploration in both aeronautics and space.

Introduction: An Indispensable Role

In the complicated and occasionally hazardous world of aeronautical research, nothing yet devised can take the place of flight itself.[1] In essence, this fact explains the existence of the NASA Dryden Flight Research Center, the subject of this book. Mathematical modeling and analysis yield new theories. Wind tunnels yield approximations of flying properties. Computational fluid dynamics (CFD) creates a virtual world in which software programs replicate many of the factors affecting flight in the real world. These disciplines also make aeronautical discoveries possible in instances where actual flying might be impossible, or at least foolhardy. Still, full-sized aircraft plying the air offer several unmistakable advantages. They give confidence to the decision makers in the airlines, the aircraft manufacturers, the armed forces, airport authorities, and the Federal Aviation Administration to adopt technical innovations. Moreover, the data derived from flying carefully instrumented aircraft often verifies mathematical models, tunnel analysis, and CFD predictions, an important benefit in itself. Yet, in many cases, flight research discovers a spectrum—or perhaps just a point in the flight envelope–entirely unexpected and unpredicted by the other methods of inquiry. The knowledge derived from the clarification of these unanticipated findings has earned flight research–and aeronautics as a whole–some of its biggest rewards.

Milt Thompson, one of NASA's most skilled and well-liked research pilots, experienced a powerful reminder of how sharply actual flight can deviate from predictions, and how it can offer surprises. During the early months of 1963, Thompson and his colleagues at the Flight Research Center (the Hugh L. Dryden Flight Research Center since 1976) prepared to fly a strange-looking aircraft known as a lifting body, a wingless vehicle with an extraordinarily high lift-to-drag ratio. Aerodynamicists at the Ames Aeronautical Laboratory (one of the research centers administered by the National Advisory Committee for Aeronautics, or NACA) first speculated during the 1950s about the practicality of a wingless, conical aircraft with a blunt body and a rounded nose. The scientists believed that such a vehicle would resist overheating at hypersonic speeds (due to the creation of an aerodynamic shock wave), and fly very capably in a high speed regime.

The first flight of a full-scale lifting body–the M2-F1, constructed of nothing more complicated than steel tubing and wood–illustrates the peculiar value of flight research. On March 1, 1963, engineer R. Dale Reed and the Flight Research Center's lifting body group gave the vehicle its first trial in the air. The crew attached a 1,000 foot towline from the gum drop-shaped aircraft to a customized Pontiac Catalina modified for the purpose. As the car sped across a four mile expanse of Rogers Dry Lake on Edwards Air Force Base—accelerating with each run—Milt Thompson raised the nose wheel of the M2-F1 to get a sense of directional control. He liked what he felt. But then he experienced trouble.

[1] For an overview of flight research history see Michael Gorn, *Expanding the Envelope: Flight Research at NACA and NASA* (Lexington, Kentucky: The University Press of Kentucky, 2001), pp. 1-8.

> [O]n the next run, the Pontiac crew accelerated up to 80 knots and stabilized at that speed. Simulation and analysis indicated that I should be able to lift off and fly at 75 knots. I gradually eased the nose up, and as predicted, the M2-F1 became airborne, bouncing from one main wheel to the other. [Unfortunately,] it was uncontrollable in roll. I couldn't keep the wings if you'll pardon the expression, level. We tried a couple more runs, but I still couldn't adequately control the vehicle in the roll axis. The vehicle just bounced from one main landing gear to the other. Dale Reed suggested we install the center fin [on the tail] and try again.
>
> We towed the M2-F1 back to the hangar, installed the center fin, and made another run in an attempt to get airborne. The results were the same. We had major handling deficiencies. During the postflight brief, we discussed the problem and any potential solutions. No one had any good suggestion on how to cure it. *The simulator had not predicted this problem* [and] *we were not certain that the wind tunnel data used to develop the simulator was of the highest quality. None of the wind tunnel models accurately replicated the configuration of the flight vehicle.*[2] [Author's italics.]
>
> An old saying goes, "Flying without feathers is not easy." We were learning that flying without wings wasn't either.

Thus, despite the mathematical analyses that first predicted the airworthiness of the lifting body design, the subsequent wind tunnel analyses, and the resulting simulations, neither Milt Thompson nor Dale Reed anticipated the dangerous roll instability of the M2-F1. Only in the first moments aloft, when the vehicle oscillated from wheel to wheel, did they realize the full magnitude of the challenge posed by the lifting bodies.

* * * * * *

On the Frontier–first published in 1984–recounts many of the surprises, reversals of fortune, indispensable discoveries, and the personal losses that inhabit the landscape of flight research and, specifically, those associated with the Dryden Flight Research Center and its antecedents. When the first contingent from the National Advisory Committee for Aeronautics (the NACA) left the park-like campus of the Langley Memorial Aeronautical Laboratory in Hampton, Virginia, they could not imagine what awaited them in California. There, in the High Desert north of Los Angeles, the government had established the Muroc Army Air Forces Base on the enormous expanse of Rogers Dry Lake bed. Bitterly cold in winter, blistering hot in summer, and barren year round, the open skies and clear weather of Muroc welcomed the pugnacious

[2] Thompson, Milton O., and Curtis Peebles, *Flying without Wings: NASA Lifting Bodies and the Birth of the Space Shuttle* (Washington and London: Smithsonian Institution Press, 1999), p. 70 (block quote).

Walt Williams and four other Langley employees in September 1946. The five ventured across the country because Langley could no longer accommodate the demands of high speed flight. The unobstructed airspace over the Mojave offered better flying opportunities than the more limited flying possible at Hampton. Moreover, in their attempts to understand airflow approaching Mach 1, aerodynamicists at Langley found themselves unable to amass reliable wind tunnel data due to turbulence inside the tunnels. Hence, only one way existed to penetrate the secrets of the speed of sound: by flying highly instrumented aircraft in a series of incremental flights up to, through, and over this mysterious region. Williams and his small band—known then as the NACA Muroc Flight Test Unit—signed up for the desolate outpost mainly to pursue this objective. In so doing, however, they almost accidentally founded perhaps the world's most prominent center for flight research.

During its first fifteen years, the NACA's Muroc outpost staked its reputation on the exploration of the transonic region and beyond, flown in such landmark aircraft as the Bell X-1 and the Douglas D-558 Skyrocket, among others. But from the 1960s onward, NASA flight research took new directions, advancing on two distinct fronts. On the one hand, under the sponsorship of the military and the civilian aeronautics sectors, subsonic flight testing underwent a resurgence not seen since Langley's pursuit of flight research earlier in the twentieth century. Such projects as the supercritical wing, digital fly-by-wire, digital engine control, winglets, propulsion controlled aircraft, and a cluster of high angle of attack investigations crowded Dryden's calendar during the 1970s, 1980s, and 1990s.

At the same time, in the wake of the human spaceflight projects of the 1960s, the center's engineers and scientists became increasingly involved with flight beyond the atmosphere. A distinguished array of space vehicles took to the skies over NASA Dryden. The Lunar Landing Research Vehicle offered the Apollo astronauts a foretaste of the touch down awaiting them on the lunar surface. The X-15 flight research program tested a range of space-related phenomena, including the aerodynamics of hypersonic flight, human physiology under such conditions, the properties of reaction control, aerodynamic heating, and the efficacy of gliding home from space. Dryden researchers also conducted Approach and Landing tests on the Space Shuttle Orbiter, resulting in pivotal modifications of America's "Space Truck." During the lifting body programs, DFRC researchers proved the concept of guided, unpowered reentry from space (reincarnated to some degree during by Dryden experiments on the X-38 Crew Return Vehicle for the International Space Station). Finally, although it never flew, the X-33 demonstrator represented the most recent contribution of flight research to the development of an advanced aerospace plane.

Regardless of whether the research applied to air or space, regardless of the period in which flight research occurred, and regardless of the kinds of vehicles flown, the objective of flight research practiced by the NACA and NASA has remained constant: the collection of reliable data. Once assembled, analyzed, interpreted, and published by NASA, this information became (in most instances) a part of the open literature, available to the aerospace community the world over.

What follows, then, is the story of an institution dedicated to the furtherance of aeronautical knowledge by means of flight research. Its mission and contributions are widely known and well respected. Indeed, a Russian military pilot gave perhaps the most compelling testimony about the center's value. Speaking candidly to a French journalist near Moscow after the end of the Cold War, he observed that his country failed to stay abreast of U. S. aeronautical developments because, in all of its vastness, the U.S.S.R. never had an institution equal to the NASA Dryden Flight Research Center.[3]

[3] The observation of the Russian pilot was told to Michael Gorn by French journalist Frederic Castel and confirmed in an e-mail exchange between Gorn and Castel dated 4 December 2001 and 14 December 2001, DFRC Historical Reference Collection.

I

Exploring the Supersonic Frontier

1

Confronting the Speed of Sound 1944-1948

Since its creation, the NASA Hugh L. Dryden Flight Research Center (as it is now called) has made two major contributions to aviation. The first and most important was its contribution to the early development of supersonic flight technology. The second was its research on the problems of flight out of the atmosphere, including lifting reentry during the return from orbit. Unlike other NASA research centers, Dryden relies almost exclusively on a relatively new kind of research tool—the research airplane, which uses the sky itself for a laboratory. Thus, its research in these two major areas, and many less major ones as well, is bound up in the development and testing of a wide range of specialized jet-and-rocket-propelled research aircraft. Some of these exotic vehicles, such as the X-1, X-15, and the Space Shuttle, have become well-known in their own right, but they all play an integral part in the history of the Dryden center. The history of Dryden—in many ways a microcosm of the history of post-World War II flight research—falls more or less roughly into four chronological phases: the era of the supersonic breakthrough, 1946-1959; the heroic era of piloted spaceflight,1959-1983; a third period overlapping the second, from 1966 to the mid-1990s, in which the emphasis gradually shifted back from space to flight in the atmosphere without neglecting space; and then, beginning in the mid-1990s, a return to a greater emphasis on access to space without neglecting aeronautics.[1] Symbolically, the landing of the Space Shuttle *Columbia* on the sun-baked clay of Rogers Dry Lake in 1981 brought this first phase of piloted spaceflight to a close while reaffirming the importance of the role that Dryden plays in the development of advanced technology for winged vehicles.

Origins

The origins of the Dryden center, "DFRC" as it is known to the world aeronautical community, are inseparable from the story of the postwar assault upon the speed of sound, the infamous and highly touted "sound barrier." By the late 1950s, supersonic flight—flight faster than sound–had become so commonplace that pilots of supersonic planes gave little thought to the cockpit Machmeter when its pointer moved above

[1] These are obviously very broad-brush characterizations, and the dates as well as the descriptions can be argued. For example, as Chapter 13 will show, since the mid-1990s Dryden has done a variety of kinds of flight research including a lot of work with remotely-piloted vehicles in the atmosphere. This continued work with remotely-piloted vehicles dates back to the 1970s.

The Lockheed P-38 Lightning, an early victim of compressibility. (U.S. Army Air Forces photo, copied as NASA photo E95-43116-2)

Mach 1, the speed of sound.[2] Yet a mere decade before, supersonic flight had been a distinct novelty; and two decades before, leading aerodynamicists around the world had debated with great intensity whether supersonic flight was, indeed, possible.

During the 1920s and 1930s, aviation technology had advanced rapidly. In this period, powerful piston engines had been developed. Advances in structural design and a growing appreciation of the need for streamlining an aircraft for high-speed flight enabled creation of high-speed military aircraft by the end of the 1930s that could approach *transonic* speeds. (The transonic region refers to that area between roughly Mach 0.7 and Mach 1.3 where a plane encounters mixed subsonic and supersonic airflow.) Many aircraft had highly undesirable behavior characteristics as they approached high speeds during prolonged dives. The airflow over the wings accelerated and shock waves would form, causing the smooth flow of air around the aircraft to be disturbed and end in a swirling wake of turbulent flow that flailed at the tail section, sometimes inducing structural loads so severe that the tail would be ripped from the craft. Because of the inadequacy of high-speed wind-tunnel design—a shortcoming only overcome by the postwar development by the National Advisory Committee for Aeronautics (NACA) of the so-called "slotted-throat" wind tunnel—the problems of transonic flight, such as compressibility, increased drag and undesirable trim changes, loss of lift, and the onset of "standing" shock waves could not be adequately examined. Many shortcut research solutions were tried, including dropping weighted body shapes from high-flying aircraft and then tracking their descent with radar, firing small rocket-propelled models, and (most useful but also most dangerous) elaborately

[2] The speed of sound varies with altitude, dropping from approximately 760 mph at sea level to about 700 mph at 22,000 feet of altitude and 660 mph at between 36,090 and 65,800 feet. Mach number (after the Austrian physicist Ernst Mach) is the ratio of the speed of an object to the speed of sound.

A North American P-51 Mustang on Rogers Dry Lake after acquisition from the Langley Memorial Aeronautical Laboratory. Among many other things, P-51s were used to gather transonic and supersonic data from air flowing over the wings in dives. (NASA photo E55-2078)

instrumenting a thick-winged P-51 combat aircraft and then diving the fighter in order to obtain transonic and supersonic data. Pending the development of reliable wind-tunnel research methods, however, the best solution seemed to be a new class of research tool: piloted research airplanes powered by jet or rocket engines and capable of attaining high speeds in the relative safety of high altitude, rather than racing toward earth in dangerous dives into the dense lower atmosphere where a plane experiences its greatest structural loadings. The story of supersonic research has not received much attention from historians, though accounts of NACA research work, the development of specific research airplanes, and foreign work in this field do exist.[3]

The establishment of the first research aircraft programs led directly to the creation of the organization that eventually became the Dryden center. The advocates of supersonic research aircraft—notably John Stack of the NACA, Ezra Kotcher of the Army Air Forces (AAF), and Walter Diehl of the Navy—did not realize at first that a special test facility for these aircraft would have to be created. Kotcher and Diehl assumed that the planes would probably pass through the standard service test centers—Muroc and the Naval Air Test Center at Patuxent River, Maryland. Most NACA

[3] See, e.g., John V. Becker, *The High-Speed Frontier: Case Histories of Four NACA Programs, 1920-1950* (Washington, DC: NASA SP-445, 1980); Richard P. Hallion, *Supersonic Flight: Breaking the Sound Barrier and Beyond, The Story of the Bell X-1 and Douglas D-558* (New York: The Macmillan Co., 1972), later republished in a rev. ed. (London and Washington: Brassey's, 1997); Charles Burnet, *Three Centuries to Concorde* (London: Mechanical Engineering Publications Limited, 1979); Jay Miller, *The X Planes: X-1 to X-29* (St. Croix, MN: Specialty Press, 1983); e-mail correspondence, J.D. Hunley to Michael Gorn, relating an interview of De Elroy Beeler by Hunley, 14 July 1999, DFRC Historical Reference Collection. Today (2001), other sources could be added, but generally, no attempt has been made in this revised version of the book to supplement Hallion's sources unless substantially new information has dictated (mostly minor) corrections to his original narrative.

personnel simply assumed that the planes would fly from the NACA's major (and oldest) research laboratory, the Langley Memorial Aeronautical Laboratory at Hampton, Virginia. Several factors worked to change this. First, the NACA did not have the resources to undertake development of such craft on its own; they had to be sponsored by the military services and manufactured by private industry. Second, the aircraft developed had (for their day) hazardous or at least unusual flying characteristics. Some were air-launched from larger airplanes. Others had strange configurations that demanded plenty of room for takeoff and landing. Third, the natural location for such testing—one offering isolation far away from prying eyes, unparalleled year-round flying conditions, and proximity to that hub of the American aircraft industry, the Los Angeles basin—was Muroc, where the AAF had already established its wartime center for advanced aircraft testing.

Two research aircraft programs had begun in 1945: the rocket-propelled and Air Force-sponsored Bell XS-1 and the jet-propelled and Navy-sponsored Douglas D-558. The latter program eventually split into a straight-wing D-558-1 (the Skystreak) and a sweptwing jet-and-rocket propelled D-558-2 (the Skyrocket). Of the XS-1 and D-558, the D-558 came closest to meeting what NACA research airplane advocates—especially Stack—had envisioned, primarily because of its turbojet engine, which enabled the craft to cruise at speeds above Mach 0.8 for over half an hour. The XS-1 (later designated X-1) represented a more radical approach, for at its conception in early 1945, liquid rocket propulsion was regarded—rightly—as unproved, dangerous, and unreliable. Yet the XS-1's rocket engine certainly endowed the craft with much higher potential performance than the contemporary D-558. And the Bell aircraft was also the first of these new research airplanes (which subsequently became known as the postwar "X-series") to be completed. It rolled out of the Bell plant at Buffalo, New York, late in December 1945.[4]

Within a year of the development of the XS-1 and D-558 series, however, the AAF and the NACA began collaborative development of four other research aircraft, adding two more within another three years. All of these were aerodynamic testbeds of one sort or another, or designed to explore the potential benefits or difficulties of some new design configuration. Table l permits comparison of these craft, which constituted the nation's "stable" of transonic and supersonic research aircraft for flight-testing at speeds up to Mach 3. Only one, the XF-92A, bore any relationship to a planned military weapon system (the abortive XP-92 interceptor), though the XF-92A was solely intended for the delta-wing research role it subsequently fulfilled. The Bell X-5 derived from a wartime German research project, the Messerschmitt P 1101, using a generally similar configuration, though its provision for variable in-flight wing-sweeping was uniquely American. The X-4 was greatly influenced by some wartime German research, the contemporary British De Havilland D.H. 108 Swallow, and Northrop's own interest in the tailless or semitailless wing configuration. The Dryden center (under various names) subsequently flew examples of all of these aircraft dur-

[4] The detailed evolution and subsequent employment of the XS-1 and the D-558 series is related in Hallion, *Supersonic Flight.*

ing 1947-1958, with the exception of the ill-fated Bell X-2. The research programs conducted on these aircraft will be discussed later.

The Road to Muroc

In December 1945, the same month that Bell completed the first XS-1, the AAF asked the NACA to supervise all details of the XS-1's data gathering and analysis program. The request was a logical one, since the NACA Langley instrumentation staff had drawn up the instrument requirements for the craft, and it meant that the NACA would have to follow wherever the plane went to fly. The Air Technical Service Command opted to fly the XS-1 first as a glider, air-launched from a modified Boeing B-29A Superfortress, at Pinecastle Field, Orlando, Florida. The Pinecastle trials would enable researchers to assess the craft's low-speed behavior and general handling qualities in much the same way that the Space Shuttle *Enterprise* first flew as a glider at Dryden over 30 years later. Langley Laboratory Director H. J. E. Reid informed NACA headquarters that the Pinecastle tests would determine if the XS-1 could operate from Langley. In fact, the chief of the Air Technical Service Command, Maj. Gen. Franklin O. Carroll, had decided to fly the craft from Muroc, where the AAF had tested its first jet airplanes. Thus, even before the XS-1 first flew, it was evident that the NACA would have to establish a team to accompany the craft, first to Pinecastle and then to Muroc.[5]

And so it fell to the Langley flight-test branch to select a small team under the direction of an engineer to assist the military on the XS-1 trials. Hartley A. Soulé, chief of Langley's Stability Research Division, together with chief NACA test pilot Mel Gough and research airplane advocate John Stack, selected a young but highly experi-

Table 1
Summary of Postwar X-Series Research Aircraft

Aircraft	Configuration	Propulsion	Sponsoring service	Project no.	Year started	First flight	Number built	Number lost	Pilot fatalities	Launch system
Bell XS-1 (X-1)	Straight wing	Rocket engine	USAAF	MX-653	1945	1946	3[a]	1	0	Air drop
Bell X-1 (advanced)	Straight wing	Rocket engine	USAF	MX-984	1948	1953	3[b]	2	0	Air drop
Bell XS-2 (X-2)	Swept wing	Rocket engine	USAAF	MX-743	1945	1955	2	2	2[c]	Air drop
Douglas XS-3 (X-3)	Straight wing (low-aspect ratio)	2 turbojets	USAAF	MX-656	1945	1952	1[d]	0	0	Ground
Northrop XS-4 (X-4)	Swept semi-tailless	2 turbojets	USAAF	MX-810	1946	1948	2	0	0	Ground
Bell X-5	Variable wing sweeping	1 turbojet	USAF	MX-1095	1949	1951	2	1	1	Ground
Convair XF-92A	Delta wing	1 turbojet	USAF	MX-813	1946	1948	1	0	0	Ground
Douglas D-558-1	Straight wing	1 turbojet	USN	Not applicable	1945	1947	3	1	1	Ground
Douglas D-558-2	Swept wing	Jet + rocket[e]	USN	Not applicable	1945	1948	3	0	0	Both[f]

[a]X-1 #2 subsequently became X-1E.
[b]X-1A, X-1B, and X-1D. X-1C canceled.
[c]Plus one launch aircraft crewman.
[d]One canceled when almost complete.
[e]First few with jet; #2 and later #1 modified to all rocket. Designed for mixed jet and rocket propulsion.
[f]All-rocket #2 and #1 air-launched from modified B-29; #3 could operate from ground or with air drop.

[5] Letter, Reid to NACA HQ., 29 Dec. 1945, office files of Robert W. Mulac, NASA LaRC; interview with Walter C. Williams, 13 June 1977, and subsequent comments by Williams to author.

enced and determined engineer for the job: Walter C. Williams. Apparently, Gough—who was Williams's immediate boss during his initial period at Muroc—in particular felt that Williams's forceful personality equipped him to withstand the authoritarian style of Bell Aircraft's Chief of Engineering Robert Stanley, who led the X-1 contractor team at Muroc. Of course, Gough, Soulé, Stack, and Langley's Assistant Chief of Research Floyd Thompson, did not merely send Williams west and hope for the best; they supported him strongly and outfitted him with a handpicked group of engineers and technicians. Much of the NACA's success at Muroc can be attributed to the full backing of these Langley leaders. Williams had worked in the stability and control branch, but just before leaving for California he served in the flight research division under W. Hewitt Phillips, becoming one of the NACA's foremost research airplane advocates with a good background in flight-testing of high-performance aircraft. A New Orleans native, Williams was an inquisitive, take-charge sort of engineer, a man who believed that useful research had to confront actual problems and not be limited to studying theoretical aspects of aeronautical science. At the same time, many of his subordinates at Muroc would find Walt Williams hot-tempered, obstinate, and on occasion all but impossible to work with. Still, he possessed two indispensable obsessions: one for planning, the other for safety. Reflecting two decades later on his role in flight research, Williams summarized his beliefs by stating, "I never bought the philosophy this is a dangerous business, we're going to kill people. I always felt by careful preparation, careful planning in carrying the flight out in a careful manner, you can do some pretty exotic things, like orbiting a man or breaking the sound barrier, without killing people."[6]

With five technicians, Williams journeyed to Pinecastle early in January 1945. The NACA unit used a modified SCR-584 gun-laying radar equipped with a camera to provide accurate flight path data. The saffron-colored XS-1 completed its first glide flight on 19 January 1946, piloted by Bell test pilot Jack Woolams. This and the remaining flights generally went smoothly, but the plane's high sink rate and the problems of keeping the plane in sight amid Florida's frequent clouds added two more votes in favor of the AAF's decision to go to Muroc. In March 1946, the XS-1 went back to Bell for installation of its four-chamber Reaction Motors XLR-11 rocket engine. Over the summer of 1946, NACA Langley prepared to send a larger test support team to Muroc under Williams's direction. On 30 September, Williams and engineer Cloyce E. Matheny arrived, followed by William S. Aiken and preceded on the 15th by two other engineers, Harold H. Youngblood and George P. Minalga. They proceeded to set up an SCR-984 radar tracking system and a telemeter receiving station in addition to checking out all the instrumentation. A second group of six–Joel Baker, Charles M. Forsyth, Beverly P. Brown, John J. Gardner, Warren A. Walls, and Howard Hinman–flew out from Langley, arriving on 9 October. Subsequently this original group would

[6] E-mail correspondence, J. D. Hunley to Michael Gorn, 15 Mar. 1999, relating an interview of De Elroy Beeler by Hunley, DFRC Historical Reference Collection; Hunley to Gorn, 14 July 1999, DFRC Historical Reference Collection; interview of Walter C. Williams by Eugene M. Emme, 25 Mar. 1964, on file in the NASA Historical Reference Collection. See also De Elroy Beeler interview by Michael Gorn, Santa Barbara, CA, 23 Apr. 1999, DFRC Historical Reference Collection.

The NACA's test team at Pinecastle near Orlando, Florida, in 1946, including (from viewer's left to right) Gerald Truszynski, John Householder, Walter C. Williams, Norman Hayes, and Robert Baker. (NASA photo E95-43116-3)

be completed in December with the arrival of two "computers," Roxanah B. Yancey and Isabell K. Martin.[7] The NACA had arrived at Muroc in force.

The team, not surprisingly, was composed primarily of engineers, instrument technicians, telemetry technicians, and computers. Since the 1920s, the NACA had instrumented flight research airplanes to record various kinds of data, but telemetry was a relatively new field. Telemetry involved onboard instrumentation that would measure certain quantities, a transmitter to send a signal from the plane to a ground station, and a receiver to pick up the signal. Active data transmission ("telemetering") had come into its own with the opening of the Panama Canal, which relied extensively on telemetry systems to report on the operation of the canal and its physical environment. Aircraft, missile, and ordnance telemetry development and systems had proliferated during World War II.[8] The XS-1, at the NACA's direction, had a six-channel telemetry installation to transmit airspeed, control surface position, altitude, and normal acceleration to the ground so that, as Walter Williams later explained, "if we lost the airplane, we could at least find out a little about what had happened."[9]

[7] Interview of Williams, and DFRC chronology in the DFRC Historical Reference Collection; telephone interviews of Harold H. Youngblood and William S. Aiken by J.D. Hunley, 3 and 4 Feb. 1997, respectively; comments of Gerald. M. Truszynski on the original version of this chapter.

[8] A good introductory history of telemetery is found in Wilfred J. Mayo-Wells, "The Origins of Space Telemetry," in Eugene M. Emme, ed., *The History of Rocket Technology: Essays on Research, Development, and Utility* (Detroit: Wayne State Univ. Press, 1964), pp. 253-268.

[9] Statement of Walter C. Williams at the History Session, Annual Meeting of the AIAA, San Francisco, CA, 28 Jul. 1965. Copy in the files of the NASA History Office.

In contrast to the telemetry technicians, "computers" were an older institution of the Federal government's scientific establishment. In NACA terminology of 1946, computers were employees who performed laborious and time-consuming mathematical calculations and data reduction from long strips of records generated by onboard aircraft instrumentation. Virtually without exception, computers were women; at least part of the rationale seems to have been the notion that the work was long and tedious, and men were not thought to have the patience to do it. Though equipment changed over the years and most computers eventually found themselves programming and operating electronic computers, as well as doing other data processing tasks, being a computer initially meant long hours with a slide rule, hunched over illuminated light boxes measuring line traces from grainy and obscure strips of oscillograph film. Computers suffered terrible eyestrain, and those who didn't begin by wearing glasses did so after a few years.[10]

The NACA group quickly settled themselves. Walt Williams took an apartment in Palmdale, over 40 miles from Muroc. Single engineers and mechanics lived in "Kerosene Flats," a collection of austere, kerosene-heated Air Force quarters at the town of Muroc, shared with visiting military personnel. Late in 1946, when the Navy Department closed down the Marine air station in the town of Mojave, housing there became available to married NACA personnel. Adding to the Spartan conditions was the attitude of some—but certainly not all—senior AAF base administrators at Muroc, who tended to regard the NACA contingent as visiting contractors rather than partners on a top-level government project. The growing problem with limited housing and workspace (NACA at first had only two small rooms and shared hangar space with the AAF) came to a head in early 1948, triggering action by the NACA headquarters to improve the lot of the Langley contingent.[11]

In early October 1946, the second Bell XS-1, the first destined to make a powered flight, arrived at Muroc. In preparation for its testing, Army technicians had installed two large liquid-oxygen and liquid-nitrogen tanks in the fueling area (the nitrogen was used to pressurize the XS-1's fuel system, for the plane burned liquid oxygen and diluted alcohol) and dug a large loading pit from which the XS-1 could be shackled to the bomb bay of a modified B-29. They also modified a standard Army fuel trailer to function as a mixing tank for the XS-1's diluted alcohol fuel. The Bell test team, headed by project manager Dick Frost and including project test pilot

[10] Interview of Katherine H. Armistead by Richard P. Hallion, 10 Dec. 1976; see also Sheryll Goecke Powers, *Women in Flight Research at NASA Dryden Flight Research Center from 1946 to 1995* (Washington, DC: Monographs in Aerospace History No. 6, 1997). Subsequent interviews by Hallion will not specify that he was the interviewer, as in note 14, whereas interviews by others will do so, as in note 11.

[11] Interview of Williams; NACA High-Speed Flight Station, "10th Anniversary Supersonic Flight," *X-Press* (14 Oct. 1957), pp. 1, 4, 11-14; e-mail correspondence, J.D. Hunley to Michael Gorn, 27 May 1999, relating an interview of De Elroy Beeler by J.D. Hunley, DFRC Historical Reference Collection; De Elroy Beeler, "Muroc in 1947," *X-Press*, 14 Oct. 1957 in Powers, *Women in Flight Research*, p. 33; telephone interview of Don Thompson by Michael Gorn, 11 March 1999, DFRC Historical Reference Collection; interview of Clyde Bailey, Richard Cox, Don Borchers, and Ralph Sparks by Michael Gorn, Palmdale, CA, 30 Mar. 1999, DFRC Historical Reference Collection.

Chalmers "Slick" Goodlin—the previous Bell pilot, Jack Woolams, having been killed in the crash of a racing plane—was ready to fly. Walt Williams's NACA team had its equipment set up, including two SCR-584 radars. The technical people on the lakebed set about to make their mark upon aeronautical science.[12]

Planning the Assault

With the benefit of hindsight, it is easy to be puzzled at all the fuss about transonic flight and the "sound barrier" myth. It is not easy to appreciate just how dangerous the sound barrier seemed to be. By the fall of 1946, most AAF, Bell, and NACA personnel believed that the XS-1 would probably exceed the speed of sound safely, but they could not deny the possibility that it might not. The first group of NACA personnel left Langley just at the time Geoffrey de Havilland died in Great Britain. De Havilland, one of Britain's finest test pilots, had been killed on 27 September 1946 when the tiny De Havilland D.H. 108 Swallow, a tailless aircraft resembling the later American X-4, began violent pitching during a dive to Mach 0.875 while flying at less than 7,600 feet over the Thames estuary. The D.H. 108 had broken up from the severe airloads at lower altitudes, killing the 36-year-old pilot instantly. The accident further reinforced the belief of NACA researchers that all such testing should be undertaken at higher altitudes where the dynamic forces acting on an airplane were less severe.

On 11 October 1946, the XS-1 dropped from its launch aircraft on a seven-minute glide flight, ushering in the era of the rocket-powered research airplane at Muroc. By early December, the craft was ready for powered flights, and on the 9 December Slick Goodlin reached Mach 0.79 at 36,000 feet, still within the scope of contemporary aerodynamic knowledge. Under the terms of the development contract, Bell had to demonstrate that the craft had satisfactory flying qualities up to Mach 0.8; beyond this, the company could not be held responsible for any quirks the plane might exhibit as it approached the speed of sound. By the end of May, both the first and second XS-1s had adequately met the demonstration requirements, having completed 20 powered flights without an accident. The third XS-1 was still at Buffalo awaiting a decision from the AFF on what kind of a fuel-feed system to incorporate in it.

Bell had assumed that when the time came for the actual assault on Mach 1, the company would be called upon to fulfill the mission, using its own test pilots. In fact, however, the AAF and NACA had already decided otherwise. The NACA was to get one XS-1 for its own testing. At NACA headquarters in Washington on 6 February 1947, Col. J. Stanley Holtoner and Col. George Smith with Gus Crowley, NACA's acting director of research, and Hartley A. Soulé, de facto chief of the NACA's research aircraft program, hammered out a joint agreement for the conduct of all research aircraft projects, XS-1 through XS-4. The NACA would furnish its own maintenance and flight crews, as would the AAF. The AAF would also supply spare parts.

[12] R.M. Stanley and R.J. Sandstrom, "Development of the XS-1 Airplane," in *Air Force Supersonic Research Airplane XS-1 Report No. 1,* 9 Jan. 1948, pp. 15-16. Copy in NASA History Office files.

To eliminate wasteful duplication, the AAF would offer to the NACA any available services over which it had control at an AAF base, and the Air Materiel Command (which had replaced the earlier Air Technical Service Command) would provide office space, shelter, housing, and equipment. In recognition of the importance of the growing X-series program, the AAF agreed to assign research airplanes a "I-b" priority, higher than that of tactical aircraft. For its part, the NACA affirmed that it had already placed research airplanes "in the highest priority class of NACA programs." The meeting attendees also agreed that the NACA would "enter research aircraft projects at their initiation, in any case before configurations are fixed."[13] Meanwhile, De Elroy Beeler had joined the Muroc unit from Langley at the suggestion of Floyd Thompson, partly to supervise the XS-1 loads research program, and partly to serve as a counterweight to Williams' somewhat insensitive style of management. In March 1947, Gerald M. Truszynski became project instrumentation engineer. Joseph Vensel, a former NACA Langley test pilot, arrived in April to supervise NACA flight operations. Three months later, the first two NACA pilots arrived for duty at Muroc, Herbert H. Hoover of Langley and Howard C. "Tick" Lilly from Lewis Laboratory.[14]

On 30 June 1947, NACA and Air Materiel Command (AMC) conferees met at Wright Field, Ohio, to discuss the conduct of the XS-1 research program. They agreed to a two-phase program. Using the first XS-1, which had a thin (8 percent thickness/chord ratio) wing planform, the AMC's Flight Test Division would conduct an accelerated test program, with NACA support, to reach Mach 1.1 as quickly as was prudent. The NACA's Muroc team would conduct slower and more detailed research, making thorough examinations of stability and control and flight loads at transonic speeds using the second XS-1 with its thicker (10 percent thickness/chord ratio) wing planform. Bell was out of the supersonic running, though the AMC decided to borrow Dick

A ground engine test of the second Bell XS-1 during the Bell contractor program at Muroc. (NASA photo E95-43116-4)

[13] Letter, J.W. Crowley to Brig. Gen. A.R. Crawford, 19 Feb. 1947, in NASA LaRC files.

[14] Hunley to Gorn, 15 Mar. 1999, DFRC Historical Reference Collection; Interview of Williams, 25 Mar. 1964; interview of Gerald M. Truszynski, 21 May 1971; interview of De E. Beeler, 1 Dec. 1976.

Frost to run a ground school for the AMC test team.[15] Col. Albert Boyd, a highly respected test pilot who directed AMC's Flight Test Division, selected Capt. Charles E. "Chuck" Yeager as project pilot, assisted by Capt. Jack L. Ridley as flight-test engineer and Lt. Robert Hoover as chase and alternate pilot. Yeager, a 24-year-old fighter ace from Hamlin, West Virginia, was a superlative pilot and an intuitive engineer.

On 6 August 1947, the two-pronged NACA/Air Force program got under way with a familiarization glide flight by Yeager in the Air Force XS-1. The take-charge ways of the Air Force jarred the NACA pilots, who were used to the staid and sedate ways of Langley and Lewis. Herb Hoover wrote to Mel Gough at Langley that "this guy Yeager is pretty much of a wild one, but believe he'll be good on the Army ship . . . . On first drop, he did a couple of rolls right after leaving B-29! On third flight, he did a 2-turn spin!" Admiration mixed with shock, not because Yeager's talents went unrecognized—indeed, his exceptional skill as a pilot could not be denied—but because the NACA approach to flight research involved carefully calibrated, data-oriented flying. Yeager and his team nevertheless offered a worthwhile counterpoint to the NACA approach. Spurred by the clear goal of Mach 1, he and his associates brought a sharp sense of purpose—even a feeling of urgency—to the X-1 project. At the same time, Hoover looked upon Williams' work with warm approval. "Williams is doing and has done a fine job," Hoover wrote. "He doesn't lose sight of the fact that a job has to be done."[16] By the end of August, Yeager had completed his first powered flight, reaching Mach 0.85. With Chuck Yeager now fully checked out in the plane, the Air Force and NACA could turn to the series of flights that would, they hoped, take the XS-1 through the speed of sound.

It had become clear that the NACA contingent would be at Muroc for a long, long time. Hugh Latimer Dryden, an internationally known aeronautical scientist, had become the NACA's director of research on 2 September 1947. Among his first actions was a directive informing Walt Williams on 7 September that henceforth the NACA Muroc unit would function as a permanent facility, managed by Langley Laboratory. The group, now 27 strong, would be known as the NACA Muroc Flight Test Unit and would report to Soulé at Langley. Before the end of the month, Dryden and his deputy, Gus Crowley, visited Muroc, where the director of research reaffirmed the NACA's top priority support of transonic flight research.

The NACA Muroc outpost was but the most recent of a series of laboratories and research facilities that Langley had spun off. There had been Ames in California, then the propulsion laboratory at Cleveland, the small Pilotless Aircraft Research Division at Wallops Island, and now the Muroc unit. Langley and its offspring, Ames, had al-

[15] Hartley A. Soulé, Memo for Chief of Research (NACA), "Army proposal for accelerated tests of the XS-1 to a Mach number of 1.1," 21 Jul. 1947, in NASA LaRC files; Charles E. Yeager, "The Operation of the XS-1 Airplane," in *Air Force Supersonic Research Airplane XS-1 Report No. 1,* p. 17; and Walter C. Williams, "Instrumentation, Airspeed Calibration, Tests, Results and Conclusions," in *Air Force Supersonic Research Airplane XS-1 Report No. 1,* pp. 21-22.

[16] Letter, Herbert Hoover to Melvin Gough, 22 Aug. 1947, in LaRC files; Hunley to Gorn, 27 May 1999, DFRC Historical Reference Collection; Hunley to Gorn, 14 July 1999, DFRC Historical Reference Collection.

ways been friendly rivals. The Navy's Walter Diehl, for example, used to play John Stack of Langley and H. Julian Allen of Ames off against one another to get things done, by going to Langley and goading Stack with the latest news about what Allen was doing, and vice versa. Langley engineers unconsciously wanted to show that the parent was still ahead, while Ames engineers smarted under perceived paternalism. Ames's director, Smith J. "Smitty" De France—a skilled engineer, a tireless worker, and a thrifty and sober administrator—suspiciously eyed this Langley offshoot growing in his backyard but remained content to watch what was going on, occasionally sending observers to Muroc to monitor the work of the Langley group on the XS-1.[17] One senior NACA Muroc engineer remembers his first meeting with the strong-willed De France: "Well," boomed Smitty, about to express a fear common to many NACA researchers, "When are you going to blow up the plane, kill the pilot, and go home?"[18] De France later proved very helpful to Williams's band in the desert. And De France was no stranger to flight research or to its hazards; one of the NACA's earliest flight researchers, he had been seriously injured in an aircraft accident at Langley that ended his flying career.

Through the "Sound Barrier"

On the tenth anniversary of the first supersonic flight by a piloted airplane, Walt Williams recollected that as the XS-1 had edged closer and closer to the magic Mach 1 mark on a series of flights, the NACA's engineering staff at Muroc "developed a very lonely feeling as we began to run out of data."[19] The last reliable wind-tunnel data ended at about Mach 0.85; the last useful information from P-51 "wing-flow" dive tests ended at about Mach 0.93. By early October 1947, Chuck Yeager was edging past that, nibbling at the "sonic wall" in the Air Force XS-1, which he had named *Glamorous Glennis*, after his wife. During his flights Yeager worked closely with the NACA engineers, especially Williams.[20] With each succeeding flight to an incrementally higher Mach number, NACA technicians would analyze the telemetry records, pull the onboard instrumentation records (lengths of scratchy oscillograph "traces"), and study the results. Then they would meet with Williams and his chief assistant, De Beeler, and these two would present the results to the Air Force's Yeager and Jack Ridley. The long strips of oscillograph records showed if the plane was losing control effectiveness, if more stabilizer trim was needed, and if lateral (roll), longitudinal (pitch), or directional (yaw) stability was deteriorating.

[17] Interview of Williams; conversation with Walter Diehl, 26 Mar. 1976; interview of Milton Ames, 26 July 1971; Elizabeth A. Muenger, *Searching the Horizon: A History of Ames Research Center, 1940-1976* (Washington, DC: NASA SP-4304, 1985), p. 13.

[18] Confidential source. Others, besides De France, feared the X-1 might fail: see Walter Williams, "The Background," in Powers, *Women in Flight Research*, p. 32.

[19] "10th Anniversary Supersonic Flight," *X-Press* (14 Oct. 1957), p. 3, repr. in Powers, *Women in Flight Research*, p. 32.

[20] On the wall of Walter Williams's office in NASA headquarters later was a photograph of the XS-1 in flight with the inscription "To Walt: The mainspring that made it all possible—Chuck Yeager, Major, USAF."

Early in October, Yeager reached Mach 0.94 and had a nasty surprise: he pulled back on the control column and nothing happened. The plane continued to fly as if he hadn't touched the controls. Wisely he shut down the rocket engine; as the plane decelerated, control effectiveness returned to normal. Williams's engineers later determined that a shock wave had formed on the horizontal stabilizer; as the XS-1 increased its speed, the shock wave had moved rearward, "standing" right along the hinge line of the plane's elevator surfaces (which control pitch) at Mach 0.94, negating their effectiveness. Fortunately the XS-1 had been designed with an adjustable stabilizer, so the NACA/Air Force team decided to control the craft with the conventional elevator up to where it lost its effectiveness, then use the stabilizer "trimmer" for longitudinal (pitch) control as the XS-1 approached the speed of sound.[21]

On 10 October, Yeager again reached an indicated Mach 0.94. During the glide earthwards, frost formed on the inside of the canopy, and despite persistent efforts, Yeager could not scrape it off. Chase pilots Bob Hoover and Dick Frost, flying Lockheed P-80 Shooting Stars, had to talk him down to a blind landing on the lakebed. Gerald Truszynski's technicians removed the oscillograph film and started their analysis, working long into the night. Engineers Hal Goodman and John Mayer compared the data from ground radar tracking with the airplane's internal instrumentation, so that errors in the cockpit Machmeter, induced by airflow changes around the airplane as it approached the speed of sound, could be compensated for. That night, Goodman and Mayer discovered that instead of Mach 0.94, all indications were that the XS-1 had actually reached Mach 0.997 at 40,000 feet; this worked out to approximately 660 miles per hour, infinitesimally close to the speed of sound. Williams had the results by morning and passed them along to the Bell representative, Dick Frost. Both men recognized that all they still needed was a clear-cut case; Williams feared that if too much publicity from the 10 October flight generated overconfidence, the Air Force might storm ahead and wind up losing the aircraft and pilot, with disastrous results for the research aircraft program. Besides, listeners on the ground had not heard the telltale sonic "boom" caused by a plane exceeding the speed of sound, a phenomenon already known to aviation science as a result of German experience with the supersonic V-2 missile. So Williams, Beeler, and other NACA engineers, after telling Yeager and Ridley of the revised results, emphasized the need for a cautious approach to a clear-cut case of supersonic flight. Yeager's enthusiastic reaction surprised no one. "He was really eager to get out there and bust it [Mach 1]," Mayer later recollected. On the other hand, this same Air Force officer who was so outspoken in his wish to conquer Mach 1 had demonstrated such exquisite control over the X-1's rocket thrust that, in keeping with NACA requests, he had come very close to expanding the flight envelope (above Mach 0.88) in increments of a mere Mach 0.02.[22]

[21] Charles E. Yeager, "Flying Jet Aircraft and the Bell XS-1," in Walter J. Boyne and Donald S. Lopez, *The Jet Age: Forty Years of Jet Aviation* (Washington, DC: Smithsonian Institution Press, 1979), pp. 107-108.

[22] Letter, John P. Mayer to Michael Collins, 8 Sept. 1976; copy in author's files; Hunley to Gorn, 14 July 1999, DFRC Historical Reference Collection.

Supersonic flight was achieved 14 October. Preparations for the flight began as the sun peeked over the eastern shore of the dry lake, bathing the desert in a soft orange glow, complementing the saffron XS-1 surrounded by technicians. There was one well-kept secret from all those present except Jack Ridley and Walt Williams—Yeager had two broken ribs, courtesy of a horse that had thrown him over the weekend. Stoically, Yeager had had the ribs taped by a civilian doctor to avoid being grounded by a military one. He confided in Ridley and Williams, however, and Ridley had cut the pilot a short length of broom handle to help him lock the plane's entrance hatch in place! The B-29 launch crew knew of the fall but not of the broken ribs, and they presented him, in jest, with glasses, a rope, and a carrot. That morning, after preflighting the aircraft, Yeager met with Williams and Beeler; they stressed caution, warning the young test pilot not to exceed an indicated Mach number of 0.96 unless absolutely certain, from the behavior of the plane, that he could do so safely.[23]

Technicians winched and locked the *Glamorous Glennis* snugly into the bomb bay of its B-29, then filled its tanks with 311 gallons of supercold liquid oxygen and 293 gallons of diluted ethyl alcohol fuel. At Yeager's suggestion, crew chief Jack Russell rubbed the rocket plane's windshield with Drene shampoo, an old fighter pilot's trick to prevent frost from forming on a canopy at high altitude. Finally, all was ready. The NACA team was standing by the telemetry gear and twin SCR-584 radars. The launch crew and test pilot entered the silver-and-black B-29, and soon its four engines were clattering noisily. At two minutes past ten o'clock, the Superfortress taxied away from its hardstand, the saffron XS-1 clasped tightly underneath, received takeoff clearance, and roared down the runway to the east. At about 5,000 feet, Yeager squirmed through the tiny entrance hatch of the XS-1, in acute pain from his broken ribs. As the B-29 continued to climb, Yeager readied *Glamorous Glennis* for flight. Two P-80 chase planes accompanied the B-29, one escorting the bomber to observe the launch, and the other about 10 miles ahead of the B-29 to join the XS-1 after it completed its rocket-propelled excursion through Mach 1. A minute before launch, Jack Ridley raised Chuck Yeager on the intercom and asked, "You all set?" "Hell, yes, let's get it over with," Yeager replied. At 10:26 a.m., at a pressure altitude of 20,000 feet, *Glamorous Glennis* was launched into the skies over the Mojave Desert.[24]

As the XS-1 dropped earthwards, Yeager briefly checked rocket engine operation by firing the four chambers of the XLR-11 engine, shutting down two and climbing away to altitude on the remaining two, pulling away from one P-80. He fired the other two chambers and under a full 26,800 newtons (6,000 pounds) of thrust, accelerated for altitude, the XS-1 streaming a cone of fire with bright yellow shock diamonds outlined in the exhausts from the rocket chambers. Further behind, a broad white con-

[23] Interview of Williams and subsequent conversations; William R. Lundgren, *Across the High Frontier: The Story of a Test Pilot: Major Charles E. Yeager, USAF* (New York: William Morrow & Co., 1955), pp. 224-235; Hallion, *Supersonic Flight*, pp. 107-108; Tom Wolfe, *The Right Stuff* (New York: Farrar, Straus, Giroux, 1979), p. 57. Wolfe is not reliable for details on the flight itself, however.

[24] Interview of Williams; Lundgren, *Across the High Frontier*, pp. 236-240; Hallion, *Supersonic Flight*, p. 108.

Key members of the XS-1 test team (viewer's left to right): Joseph Vensel, head of operations for the Muroc Flight Test Unit; Gerald Truszynski, head of instrumentation; Capt. Charles "Chuck" Yeager, USAF pilot; Walter Williams, head of the NACA Muroc Flight Test Unit; Maj. Jackie Ridley, USAF pilot and project engineer; and De E. Beeler, head of engineers. (NASA photo E95-43116-3)

trail formed a long spearpoint with the little research airplane at its apex. Second by second the XS-1 was growing lighter, its engine gulping propellants, and the thrust-to-weight ratio rose higher and higher. The plane passed Mach 0.8 and streaked on to Mach 0.9. Above Mach 0.93, the adjustable stabilizer provided adequate longitudinal (pitch) control. He shut down two chambers briefly while he assessed his situation. He may have reflected briefly that his broken ribs created a far from ideal situation in which to attempt a deed many thought impossible, an attempt that recently cost Geoffrey DeHavilland his life. Yeager's courage in pursuit of Mach 1 could never be doubted.

Although his own odds were in doubt, the plane's signs were good. Confident that *Glamorous Glennis* could exceed Mach 1, Yeager leveled off and fired one of the two shutdown cylinders. Now very light from the amount of propellants that had already been consumed, the XS-1 shot ahead. At about Mach 0.98 indicated, the needle on the Machmeter fluctuated, then jumped off the scale, leading Yeager to believe the plane was flying at about Mach 1.05. In fact, postflight data analysis indicated the XS-1 had reached Mach 1.06 at approximately 43,000 feet, an airspeed of 700 miles per hour. The Machmeter jump—a hallmark of supersonic flight since—registered the passage of the bow shock wave across the nose as the plane went supersonic. And on the ground, observers heard the characteristic double crack of a sonic boom. Inside the XS-1's instrumentation compartment, the oscillograph recorded the static and impact air pressure traces' sudden jump on a strip of film, irrefutable proof that the airplane had indeed flown faster than the speed of sound. It remained faster than Mach 1 for a little over 20 seconds. Then Yeager decelerated back through the now-crumbled sonic wall. Fully 30 percent of the craft's propellants remained when Yeager shut down the switches and began the long, cold glide back to earth. There would be time enough to probe further beyond the speed of sound. Fourteen minutes after launch, the rocket plane's wheels brushed the baked clay of the dry lakebed. The dreaded "sound barrier" was a thing of the past.[25]

Shortly after the plane landed, as Yeager shambled off to get some well-earned sleep, Walt Williams placed a long-distance phone call to Gus Crowley and Hartley Soulé. "We did it today," he said. The message required no explanation. At Muroc, the project team planned a party at Pancho's Fly-Inn that night, but two hours after the flight, word came from NACA headquarters that the accomplishment and future flight tests were to be regarded as Top Secret. Dryden, Crowley, and Soulé wanted to be certain that the XS-1 had really gone supersonic. They had all of the craft's records sent back to Langley for examination, which understandably annoyed Williams and his staff of professionals at the lakebed.[26]

Despite a leaked account of the first supersonic flight by the trade journal *Aviation Week* in December 1947, the Air Force and NACA did not formally reveal Yeager's accomplishment until 15 June 1948 when Gen. Hoyt Vandenberg, chief of staff of the Air Force, and Hugh Dryden of the NACA confirmed that the XS-1 had repeatedly exceeded the speed of sound, flown by military and NACA test pilots. The announcement triggered a flood of honors and awards, including the prestigious 1947 Robert J. Collier Trophy shared by John Stack for the NACA, Chuck Yeager for the Air Force, and Larry Bell for the American aircraft industry. By the end of 1947, the XS-1 had flown to over 925 miles per hour—Mach 1.35—twice as fast as a wartime P-51 Mustang. The XS-1's success encouraged the Air Force to order four advanced versions from Bell, of which three were eventually completed (the X-1A, X-1B, and X-1D). The Air Force phase of the two-pronged assault on the speed of sound had clearly been a success; the service reached the maximum flight speed of the XS-1 on 26 March 1948 with a flight by Chuck Yeager to Mach 1.45 (957 miles per hour).

At the same time, the NACA was turning its efforts away from support of the Air Force program to flying its own XS-1, the thicker-winged second aircraft. Herb Hoover had completed his first glide flight in the airplane on 21 October, a week after Yeager's accomplishment. Embarrassingly, during the landing, Hoover touched down hard upon the nosewheel, collapsing it and necessitating repairs that kept the craft grounded until mid-December.[27] On 16 December, he checked out the craft at subsonic speeds. Though

[25] Details of flight are from Yeager's pilot report, which at the time Hallion originally wrote this chapter, was on loan from the Historical Office, Air Force Flight Test Center and on exhibit in the Flight Testing Gallery of the National Air and Space Museum, Smithsonian Institution, Washington, DC. Today, the report is conveniently repr. in *Supersonic Symposium: The Men of Mach 1*, ed. with an introduction by James O. Young, Air Force Flight Test Center History Office, an AFSC Special Study, Sept. 1990, p. 221. See also Hunley to Gorn, 27 May 1999, DFRC Historical Reference Collection.

[26] Interview of Williams and subsequent conversations.

[27] Because of a peculiar handling characteristic during its landing flare, the XS-1 series was prone to land hard, overstressing the nosewheel. Nosewheel collapses plagued the Bell, NACA, and Air Force programs on all XS-1s, including the advanced models procured later.

pleased with its flying qualities, he recognized that the brief amount of flight time at high speed imposed by the rapid consumption of rocket propellants reduced the amount of information that could be acquired from each flight. "It's going to take a long flight program with a lot of flights," Hoover pessimistically but accurately concluded in his flight report.[28]

Another and more critical problem was workload. By early 1948, the NACA unit was ministering to three airplanes: the Air Force and NACA XS-1s and the second Douglas D-558-1 Skystreak, which the NACA had received at Muroc for testing by its pilots. Workload posed a serious problem for the instrumentation staff, since the NACA believed in thoroughly instrumenting and calibrating its research airplanes. In one case, three instrument technicians with the Muroc unit put in over 250 hours of overtime between 10 November and 13 December 1947. Williams placed the XS-1 project ahead of every other research activity.[29]

There was no longer any doubt that the XS-1 could safely exceed the speed of sound, but the NACA test team did wonder what differences might stem from the thicker wing on its airplane. Drag would certainly increase; there might be other undesirable traits as well. So Hoover approached the now-punctured "sound barrier" cautiously. Following a series of proving flights to increasing Mach numbers, Hoover made his first high-speed run on 4 March 1948, when he reached Mach 0.943 at 40,000 feet. Six days later he flew to Mach 1.065, slightly over 700 miles per hour, becoming the first NACA pilot and the first civilian to fly faster than sound; subsequently, he

The Bell XS-1 #1, which completed the world's first piloted supersonic flight on 14 October 1947. (Air Force photo by Lt. Robert A. Hoover, copied as NASA photo EC72-3431)

[28] Hoover pilot report, 16 Dec. 1947.

[29] Interview of Williams; letter, Hoover to Gough, 17 Sept. 1947, in NASA LaRC files; ltr., Edmond C. Buckley to Hartley A. Soulé, 22 Jan. 1948.

The NACA Muroc contingent in October 1947 in front of the NACA XS-1 and the B-29 mothership. (NASA photo EC95-43116-6)

received the Air Medal from President Harry Truman for the feat. On the last day of the month, Howard Lilly became the second NACA pilot to "break the Mach," and the NACA had now firmly joined the growing supersonic club.

The Muroc engineering staff immediately set to reducing the accumulated data from the XS-1 program and generated ten formal NACA research memoranda on the airplane's handling qualities, flight loads, stability and control characteristics, and pressure distribution surveys.[30] The XS-1 tests by Hoover and Lilly—and subsequent ones by Robert Champine and John Griffith, who arrived at Muroc in late 1948 and late 1949, respectively—generated significant aeronautical information. The NACA continued flying the craft in the vicinity of Mach 0.90, for it was interested in investigating the exact conditions of flight at velocities around the speed of sound and in acquiring data that could be used for correlation with ground-based wind-tunnel data. The engineers were especially intrigued by the pronounced increase in controllability that the adjustable stabilizer provided the XS-1 at transonic speeds; that work constituted a pioneering effort in the development of the "all-moving" horizontal tail surfaces that later appeared on such first generation supersonic jet fighters as the F-100. As a result of XS-1 research, Soulé could write in late 1949 that "the power-driven adjustable stabilizer has already become standard equipment in new transonic-speed tactical airplane designs.[31] NACA XS-1 testing also indicated, with shocking impact, just how

[30] XS-1 progress reports submitted by Hubert M. Drake and Hal R. Goodman to Hartley A. Soulé for 9 Dec. 1947, 12 Mar. 1948, 29 Mar. 1948, and 13 Apr. 1948. Chronology of rocket research aircraft flights prepared by Robert W. Mulac of the NASA LaRC; copy on file with the NASA History Office.

[31] Hartley A. Soulé, "Review and Status of Research Airplane Program" (summary report of Research Airplane Projects Panel, 1949), p. 11; copy in author's files.

much drag thick-wing sections added at transonic speeds. The NACA XS-1, with its 10 percent thickness/chord-ratio wing, had 30 percent more overall drag at transonic speeds than did the thinner-wing Air Force XS-1. Thick wing sections simply imposed unacceptable penalties for transonic and supersonic airplane design.

There were serendipitous benefits from XS-1 research as well. The extensive calibration of airspeed measurement systems in the XS-1, together with the results of ground radar tracking, provided a data base for building advanced airspeed measurement systems for high-speed airplanes. In a short period of time, then, the NACA XS-1 effort made notable contributions to aviation science, complementing the Air Force effort with *Glamorous Glennis* and justifying the hopes of the planners of the XS-1 joint program. The supersonic assault had been a success.[32]

[32] Ibid., p. 12.

2

Pioneer Days at Muroc 1948-1950

Muroc Air Force Base in early 1948 was not only remote; but also uninviting to many people coming to the mostly brown desert from the greener areas of the country. In December 1947, the NACA's work there came to a standstill as personnel scrambled away to celebrate the holidays in sections of the country that were more appealing to them. Indeed, one reason for the impressive amount of work that got done might have been sociological: there was little else to do. Even by automobile, a trip to Los Angeles was a chore. Without one, the remaining choice was the afternoon Stage Lines bus that left Muroc for Los Angeles at about 5 p.m. The voyager had to spend the night in Los Angeles and take another bus back the next evening.[33] Word about the discomforts of Muroc soon spread within the NACA labs, making recruitment difficult. Other events soon exacerbated this situation.

NACA Muroc: Unwelcome Tenant or Valued Partner?

Over the summer of 1947, when Langley decided to establish the test team at Muroc permanently as the Muroc Flight Test Unit, personnel officers journeyed to Muroc to ask the workers if they wished to stay on as regular staff, thus losing their $3 and $4 per diem as employees on temporary duty. Those who chose to leave were paid for the return to Hampton, Virginia.[34] During the first year of its existence, the Muroc unit experienced a high turnover in personnel. Workers quickly split into two groups: those who adjusted to and even enjoyed desert life; and those who could not stand the heat, dust, and grit for more than a few weeks. Many stayed because of job satisfactions not readily apparent. They believed they were participating in a program of great national importance that would radically alter the future development of aviation; they considered it both a great responsibility and an honor to have been selected to work on the program. Nevertheless, by January 1948 the morale of the NACA unit at Muroc had begun to slip; the long days and nights of work were taking their toll. The Muroc base administration bore some responsibility for this state of affairs.

In January 1948, Edmond C. Buckley, the chief of Langley's instrument division, visited Muroc and was unpleasantly surprised by what he saw. His critical—

[33] Interview of Armistead. See also the interview of Don Thompson and those of Bailey, Cox, Borchers, and Sparks by Michael Gorn, 11 Mar. and 30 Mar. 1999, respectively, DFRC Historical Reference Collection (both of which also apply to notes 34 to 39).

[34] Interview of Beeler.

perhaps excessively critical—memo to Hartley Soulé was read by Langley Laboratory Director Henry J. E. Reid, who thought that Buckley might have exaggerated the situation. Reid journeyed to Muroc and confirmed at least some of Buckley's complaints. Buckley's memo first discussed the housing situation. The junior married professional staff quarters were acceptable by Langley standards. The senior married professional staff quarters, however, were inferior to what a comparable couple could expect at Langley. Buckley described the married quarters for mechanics as "the equivalent of emergency wartime living conditions."[35] Quarters for the single engineers or mechanics were, in his judgment, "all Grade A fire traps." The Air Force vigilantly made certain that occupants of base housing did not have unauthorized furnishings in their domiciles. In one case, an Air Force inspector discovered "an illegal broken-down chair" in one of the NACA quarters and "left the quarters in such a fury that the door came off the hinges and fell on him." Lavatory facilities were mostly communal, and locked doors were not permitted because of fire hazards.

Buckley disliked the Muroc officers' mess as well, deriding its uncleanliness and citing recent cases of dysentery. Some NACA personnel chose to eat in the GI mess at North Base. Buckley ate there and wrote sarcastically that "the sad lot of the European DPs [displaced persons] came to my mind."

Work areas consisted of open hangars—cold in December and January—which lacked darkroom facilities. With both XS-1 aircraft flying and the NACA's Douglas Skystreak in its checkout stage, the chances for leave were poor; most workers were putting in large amounts of overtime. For amusement, Buckley concluded, "one has the choice of working or going to bed to keep warm. Reading or writing in your quarters is impracticable because of facilities and temperature." As for social life, Buckley made the injudicious observation that "Muroc should be staffed with misogynists." The staff at Muroc was obviously understrength, but Buckley thought he could persuade only three new workers to leave Langley for the desert.

Walt Williams badly needed better living and working conditions for his staff. He wanted a hangar for the exclusive use of the NACA, office areas collocated with the hangar, and men's and women's dormitories. The NACA already shared the east main hangar with the Air Force, but it had inadequate office, shop, and stock space, and the electrical system was incompatible with the NACA's instrumentation requirements. Blowing dust seeped into the work area, compromising satisfactory instrumentation work. Williams wished to move into another hangar and construct offices from wood and sheetrock with suitable electrical and plumbing installations. He was willing to get the NACA to furnish all the materials and labor if the Air Force would just approve the construction, though it was simpler for the Air Force to furnish the materials through the Air Materiel Command, with the NACA furnishing the labor.[36] Further, Williams wanted the service to turn over a building, "T-83," to the NACA for use as a

[35] This and subsequent quotations from Buckley are from his letter to Hartley A. Soulé, 22 Jan. 1948.

[36] Letter, Walter C. Williams to Col. Signa A. Gilkey, 9 Mar. 1948.

dormitory. But all his plans hinged on winning the cooperation of the base administration.

During Henry Reid's visit, the Langley chief had met with the commanding officer of Muroc Air Force Base, mentioning, as Reid later noted, that the NACA needed information on the cost of Muroc quarters "in order that we might ask Congress for money to construct some for our employees." The base commander, Col. Signa Gilkey, was unimpressed, and Reid later wrote that his request "was like waving a red flag, as Colonel Gilkey made it very clear that he did not want other activities spending money for permanent installations at Muroc." Colonel Gilkey was also opposed to turning over Air Force buildings to the NACA. In his notes on the trip, Reid stated, "In general, the living conditions and the attitude of the commanding officer are such as to be demoralizing to everyone. . . . My contacts with the commanding officer lead me to believe that one of the things he is afraid of is that if contractors and the NACA are allowed to fix up quarters and improve their situation, they must allow the Navy the same privilege, and eventually control of the base would be lost." Reid concluded that the only real solution would come when NACA personnel had housing available in Lancaster, 35 miles away, "where civilians can live a normal civilian life."[37]

In truth, the situation was more complicated than Reid thought. Col. Gilkey had drawn up an ambitious master plan for the expansion of Muroc, wherein the base would expand to take in nearby Rosamond Dry Lake to the west, reroute the railroad tracks that bisected Muroc's dry lake and limited its landing area, and add a 15,000-

(Then) Brig. Gen. Albert Boyd and Walter Williams examine a model of the Northrop X-4 research aircraft. (NASA photo E95-43116-7)

[37] H. J. E. Reid, memo for files, 18 Mar. 1948, NASA LaRC files.

foot runway and new building and housing areas. All of this would cost approximately $120 million, and the new facility would include schools and shopping areas. (Eventually, all of this did come to pass, and it is fair to say that Col. Gilkey was the architect of the modern Air Force Flight Test Center complex, a tribute to his foresight.) But he feared that complying with the NACA's requests for improvements to existing structures would delay implementation of the master plan.

The NACA quickly solved its Muroc difficulties to its satisfaction. After his return to Langley, Reid took up the matter of the Muroc staff with Soulé, Crowley, Dryden, and others, and the matter eventually went to Jerome Hunsaker and the NACA Main Committee itself. Air Force committee members quickly supported the plans for the Muroc unit, and in April, Williams received title to the long-sought hangar and access to Air Force materials. Within the NACA structure, the Ames laboratory was directed to support the Muroc effort, and in May NACA personnel, including model makers and technicians from Ames, began work on lean-to offices along the sides of the newly acquired hangar. Construction was completed on the shops and offices in November 1948, and the men's and women's dormitories were finished the next spring. Muroc was still not a bed of roses, but at least conditions became more tolerable.[38]

In retrospect, the brief disagreement between the local Air Force administration and the NACA was a passing incident that failed to mar the otherwise excellent cooperation (if mixed with friendly rivalry) that existed between the NACA and the Air Force at Muroc. It was a remarkably similar pattern to that of the Army/NACA relations over the Langley laboratory in 1918-1920.[39] Certainly on the operating level there were no interagency problems, just lots of teamwork and sweat. Any lingering difficulties disappeared in September 1949 when a new commanding officer arrived on base, Brig. Gen. Albert Boyd, a man known throughout the service as "the test pilots' test pilot." One might have expected two strong-willed and dynamic individuals such as Boyd and Williams to strike sparks, but this was not the case. A strong bond of friendship, respect, and cooperation formed between these two, and the Boyd/Williams relationship soon proved fruitful for both the NACA and the Air Force.[40]

NACA Muroc Loses Howard Lilly

There was yet another unhappy episode in the spring of 1948, this one truly tragic: Howard Lilly was killed on a research flight in an NACA Skystreak on 3 May 1948.

Understandably, the NACA Skystreak program had played second fiddle to the XS-1 throughout late 1947 and into 1948. Not only was it recognized that the plane

[38] Interview of Williams; letter and attachment, Smith J. De France to J. W. Crowley, 5 May 1948, NASA ARC files.

[39] Details of the earlier squabble may be found in Alex Roland's history of the NACA, *Model Research: The National Advisory Committee for Aeronautics, 1915-1958* (Washington, DC: NASA SP-4103, 1985), pp. 81-83.

[40] Interview of Williams; Jean R. Hailey, "Maj. Gen. Albert Boyd, 69, Dies; Father of Modern Flight Testing," *The Washington Post*, 22 Sept. 1976; Hunley to Gorn, 14 July 1999, DFRC Historical Reference Collection.

Test pilot Eugene May (viewer's left) and research pilot Howard Lilly at Muroc with a Douglas D-558-1 Skystreak. (NASA photo E95-43116-8)

could not compete with the XS-1 in terms of maximum speed capability (even while diving, the Skystreak eventually touched Mach 1 only once), but it also had extensive requirements for instrumentation that delayed its flight readiness. NACA technicians worked on this craft when they did not have anything to do on the XS-1 and were able to make two familiarization flights in the Skystreak possible by the end of 1947.

The Skystreak had the same general aerodynamic configuration as the XS-1—a straight wing and tail, both thinner than conventional design practice. Here the resemblance ended, for the D-558-1 (as it was designated) took off from the ground under its own power, propelled by a General Electric TG-180 turbojet engine. Douglas had built three of the D-558-1 Skystreaks, which preceded the firm's three D-558-2 Skyrockets. An agreement among all the agencies concerned affirmed the planned delegation of responsibilities on the Skystreak program: Douglas would fly the first Skystreak in a series of company tests; the NACA would get the second and third Skystreaks, maintain them, fly them with fuel and oil from the Air Force, and perform major aeronautical research; the Navy would accept responsibility for engine overhaul and replacement; and Douglas would perform the major maintenance and modification work, drawing upon Navy funding. This arrangement, confirmed for the NACA by a Navy memo on 4 November 1947, was followed until the retirement of the Skyrockets a decade later.[41]

Though the second Skystreak had earlier set a world's airspeed record of 650 miles per hour, the NACA Muroc unit quickly discovered that the craft was something of a jinx. The landing gear often failed to lock fully in the retracted position; on one

[41] Letter, Chief of Navy Bureau of Aeronautics to NACA HQ and Bureau of Aeronautics representative, Douglas Aircraft Corp., 4 Nov. 1947, Aer-AC-25, in historical files of Naval Air Systems Command.

flight, Lilly had to land hurriedly after the cockpit filled with dense smoke from a small electrical fire. Ground trackers watching the scarlet-colored airplane through tracking photo-theodolites discovered that the aesthetically pleasing plane was very difficult to see against the dark blue desert sky, so the fuselage was painted glossy white to facilitate optical tracking—aside from the high-temperature Blackbirds and X-15, all NASA research aircraft since have been white or other light colors. On 29 April 1948, Lilly reached Mach 0.88 in the plane, part of a planned flight program investigating directional stability at transonic speeds.[42]

On 3 May, Lilly took off from Muroc at noon; as the maintenance staff had come to expect, he had to land because the balky landing gear once more failed to lock properly. Minor adjustments took up most of the afternoon, and it was not until late that Lilly tried again. With a lowering sun already casting lengthening shadows, the ground crew readied the plane for flight. Lilly ran up the engine, the TG-180 emitting a rising wail, and started his takeoff, the jet finally lifting off after a run of about a mile. Witnesses saw the landing gear fold up into the plane; the Skystreak accelerated. Then, somewhere within the jet engine's compressor section, strain became too great and some component failed. In the whirling compressor, such a failure had all the catastrophic impact of the flywheel of a huge steam engine coming apart. Whole sections of the compressor housing and blades slashed through the engine casing and through the fuselage skin. Some pieces cut the main fuel lines and severed the craft's control lines as well. Lilly had no control over the plane, its tail section erupted in flames. Today, in the era of "zero-zero" (zero altitude/zero airspeed) ejection seats, he might have had a chance. But all the ailing Skystreak had was a jettisonable nose section so the pilot could abandon the Skystreak at high altitude. Witnesses saw the jet, low over the dry lake, shed a large section of fuselage skin, followed by a gout of flame-streaked smoke. Horrified, they watched the Skystreak wallow along for a few seconds before sickly slipping into a left yaw and roll, diving into the lakebed, and exploding. Howard Clifton "Tick" Lilly, a five-year NACA veteran and the third pilot to fly faster than sound, became the first NACA test pilot killed in the line of duty.[43]

Lilly's death deeply affected the Muroc staff. It was still a small group, no more than 40, and the gregarious Lilly, with his West Virginia twang, had been a close friend of many. The accident especially shocked the safety-conscious Williams. Langley Laboratory, the administrative headquarters for Muroc, established an accident board chaired by veteran NACA pilot Mel Gough. The board reached the conclusion that disintegration of the engine compressor section had severed critical control and fuel lines. Both Williams and the accident board urged that all future research aircraft have the latest model engines, incorporating all up-to-date engine modifications and changes (the unfortunate Skystreak had had an early model TG-180, not up to standard as compared with later TG-180s on other aircraft). They also insisted that all research aircraft incorporate armor plating around the engine in the vicinity of control lines, fuel and hydrau-

[42] D-558-1 progress reports by William H. Barlow for 8 Dec. 1947, 18 Jan. 1948, 2 Feb. 1948, 13 Feb. 1948, 2 Mar. 1948, 29 Mar. 1948; Mulac flight chronology.

[43] Details from M. N. Gough, A. Young, and H. A. Goett, "Aircraft Accident Investigation Report Douglas D-558-1 Airplane, Bu. No. 37971, Muroc Air Force Base, Muroc, California, May 3, 1948" (Hampton: NACA LMAL, 1948).

lic lines, and fuel tanks. Subsequently, the NACA was most uncompromising at contractors' "mock-up" inspections when the question of protecting planes from disintegrating engines came up. Lilly had given his life, but he would be remembered: visitors to the Dryden Flight Research Center drive down Lilly Avenue from Rosamond Boulevard. And in Building 4825, where they get badges to enter the Center, there hangs a portrait of this promising and sorely missed research pilot.

Exactly two weeks after Howard Lilly died at Muroc, Ames laboratory test pilot Ryland Carter perished when a P-51H Mustang broke up during a dive. These two accidents, coming after years of a safe research record, caused certain persons to suggest that the NACA use contract test pilots for flight research, offering—as private industry did—bonuses for hazardous aircraft-testing. Hugh Dryden called a meeting at headquarters to thrash out an answer. Herb Hoover represented the Langley-Muroc group, and Larry Clousing, another NACA test pilot with a distinguished flying record, represented Ames. Hoover and Clousing, as well as the other NACA pilots present, were adamant that NACA pilots fly NACA research aircraft. Dryden concurred and rejected any further consideration of using non-NACA pilots on NACA research aircraft projects.[44]

The death of Lilly caused a temporary shutdown of NACA flight operations at Muroc. Herb Hoover had returned to Langley after Lilly had checked out in both the NACA XS-1 and the ill-fated NACA Skystreak. Now Hoover returned briefly to the desert to train a replacement pilot, Robert A. Champine, who had considerable flying experience with the sweptwing Bell L-39 research airplane, a background that made him particularly well qualified for the upcoming NACA program at Muroc on the sweptwing Douglas Skyrocket. Sweptwing airplanes had tricky behavior at low speeds and during abrupt maneuvering flight. Champine completed his first flight at Muroc on 23 November 1948, when he checked out in the NACA XS-1. Hoover returned to Langley for good in December.[45] NACA Muroc was back in the air.[46]

"X-Series" Administration

The hiatus in flight operations at Muroc caused by Lilly's crash did not mean that development of the research aircraft program was similarly slowed. Any impartial observer of NACA affairs in mid-1948 would have recognized how the scope of the research airplane program had changed. Originally conceived for the XS-1 and D-558, the program had expanded to embrace an XS-2 for sweptwing Mach 3 research, an XS-3 for sustained Mach 2 turbojet research, an XS-4 for transonic research on the

[44] Interview of Ames; interview of Williams and subsequent conversations; Edwin P. Hartman, *Adventures in Research: A History of Ames Research Center, 1940-1965* (Washington, DC: NASA SP-4302, 1970), pp. 168-169.

[45] Hoover himself perished in the crash of a B-45 test plane near Langley on 14 August 1952 when the plane broke up in midair.

[46] Interview of Robert A. Champine by Richard P. Hallion, 11 Nov. 1971.

tailless configuration, and the XF-92A for delta-wing research at transonic speeds.[47] Another batch of advanced X-1s had been ordered, and one more projected vehicle, a variable wing-sweep design that eventually became the X-5, was being discussed by Bell, the Air Force, and the NACA. Each of the NACA laboratories was busily at work on phases of the research aircraft program: Ames and Langley were doing wind-tunnel research on configurations, Lewis was following up with engine work on turbojets, and the Pilotless Aircraft Research Division (PARD) at Wallops Island was firing off models of proposed research aircraft.[48] The research aircraft program involved extensive dealings with outside parties: the military services financed the development of the aircraft and their engines, and private contractors manufactured them. Already the load of paperwork and administrative chores had justified the appointment of an administrative officer, Marion Kent, to the Muroc unit in April 1948. Increasingly Hugh Dryden came to believe that the NACA research airplane effort required a central point of focus for coordination and communication. Eventually, this led to the creation of the NACA Research Airplane Projects Panel (RAPP).

Since 1945, Hartley A. Soulé had been acting as the NACA's chief of research airplane projects and activities, and his duties had dramatically increased. On 9 August 1948, Dryden recognized those increased responsibilities by making Soulé a member

The NACA's X-1 (formerly XS-1) research aircraft. (NASA photo E95-43116-9)

[47] "XS" became simply "X" after 11 June 1948 as a result of a change in Air Force aircraft designation policy. "X" is used subsequently throughout the text.

[48] The extensive and important activities at Wallops in relation to the X-series research airplane program and the DFRC are reviewed in Joseph Adams Shortal, *A New Dimension: Wallops Island Flight Test Range: The First Fifteen Years* (Washington, DC: NASA Reference Publication 1028, 1978). The Wallops station made two important contributions to the X-series program early in its development: discovering a so-called aileron reversal problem on the Bell X-2 and discovering a directional instability problem on the X-3. Both discoveries forced needed redesign of the craft.

of his staff as NACA's research airplane projects leader. The laboratories were told that "the research airplane program involves all laboratories as well as the Muroc Unit, and the program coordination is therefore a function of NACA headquarters."[49] Soulé would report to Dryden's deputy, Gus Crowley, on research airplane matters. Soulé wasted little time in expanding upon the project leader concept. Desiring to improve interlaboratory communications and relationships on the research aircraft, Soulé sent a memo to NACA headquarters at the end of the month recommending the establishment of a special research airplane panel, with a representative from NACA headquarters, the chief of the Muroc Flight Test Unit, and each laboratory. The panel would "effect proper coordination of the interests of the three laboratories in Muroc projects [including] the status of the research airplane projects at or proposed for Muroc, the current position of supporting investigations at each of the laboratories, and technical problems relating to each project."[50] On 2 September 1948, the plan was approved and Soulé was appointed chairman.[51] Over the next two weeks, Soulé notified each lab and Walt Williams of the panel's creation, receiving in return their concurrence in the decision and nomination of representatives to the panel.[52] Establishing the RAPP under Soulé's leadership codified an existing administrative relationship by giving Soulé's actions the trappings of a formal bureaucratic structure. The action demonstrated that the program had grown to such size that it was no longer possible to manage or monitor it on a laboratory level. It required management directly from NACA headquarters—though, wisely, Dryden selected Soulé, the former Langley manager, for the position. With Dryden and Crowley at the helm, and Soulé next, the research airplane program had the unequivocal support of the NACA's highest echelons.

Further, the creation of the RAPP gave the NACA better and more streamlined coordination of the laboratories' activities on research aircraft projects. The RAPP fit smoothly into the NACA's lifestyle; since its inception, the NACA had been governed by the Main Committee, with its fields of research overseen by specialized committees or panels. Every year, until the panel was abolished on the eve of the X-15's flight program, Williams submitted a detailed annual report to the panel, outlining the research programs at the dry lake and the programs being planned. The panel was mostly a formality as far as Williams was concerned. In most cases, he won easy endorsement of his plans from the panel at its annual meetings, usually held early in February. Through RAPP participation, the laboratories learned some of the operating problems facing the Muroc unit, and the RAPP played a crucial role in sorting out some of the difficulties in the X-series development programs.

Creation of the RAPP also marked implicit recognition of the usefulness of public image. The research airplane program was the NACA's most visible symbol of

[49] Letter, Crowley to NACA laboratories, 9 Aug. 1948, NASA LaRC files.

[50] Letter, Soulé to NACA HQ., 30 Aug. 1948, NASA LaRC files.

[51] Letter, Ira Abbot to Soulé, 2 Sept. 1948, NASA LaRC files.

[52] Letter, Soulé to Williams, 9 Sept. 1948, NASA LaRC files.

postwar research. Wanting to keep its image as a far-seeing, up-to-date scientific organization (an image tarnished by its prewar failure to pursue turbojet propulsion), the NACA found that the glamorous research aircraft gave NACA's public image a badly needed shot in the arm. Participation in the research aircraft effort had begun somewhat casually. According to the memoirs of one engineer, "It took form gradually, manipulated and developed in innumerable lunchroom conversations and other contacts."[53] But by mid-1948, the program had assumed such stature that it provided some of the NACA's strongest cards whenever Jerome Hunsaker or Hugh Dryden took the NACA's budget to Congress for approval. The RAPP helped by drawing greater attention to the NACA's commitment to the research aircraft effort. Interestingly, it imposed few administrative or bureaucratic chores on program administrators at Muroc and the NACA laboratories. NACA's traditional pattern of delegated authority for project management minimized paperwork and meetings. While the Wallops model-rocket testing program and the Muroc effort were exceptions—requiring interlaboratory ties within the NACA and ties to outside organizations as well—management, in the words of engineer John Becker, remained "delightfully simple, direct, unobtrusive, and inexpensive."[54]

The same climate that helped create the RAPP and elevate Hartley A. Soulé to the position of research airplane projects leader generated the next change in the status of the Muroc unit itself. It was time to raise it organizationally from the level of a detached unit under the direction of a remote parent (Langley Laboratory) to that of a semiautonomous NACA "station," only one notch below a "laboratory." Langley had designated the Muroc Flight Test Unit as its permanent appendage in September 1947. Certainly, by mid-1949, NACA administrators could foresee a continuing need for the Muroc facility for at least a decade since the NACA planned to participate in the X-1, X-2, X-3, X-4, D-558-1, D-558-2, advanced X-1, and XF-92A programs, as well as consultant on some of the Air Force projects being tested at Muroc. Other projects, such as the gestating X-5, were in the discussion stage. By now, the value of having a specialized locus within the NACA for flight research of high-performance aircraft was also readily apparent. Muroc offered unsurpassed year-round flying conditions, permitting maximum utilization of research aircraft. It was also the flight-testing center of the Air Force, the service that played the major role in financing and supporting the postwar X-series. For the NACA Muroc Flight Test Unit to fulfill its growing responsibilities in testing and research on these aircraft, it would have to expand. Already growth was rapid. At the time of Yeager's flight, the unit had 27 workers. A little over a year later, in January 1949, it had 60. In January 1950, the number had doubled again, to 132. Through fiscal year 1949, Langley Laboratory had had funded the Muroc unit. However, in August 1949, at the launch of fiscal year 1950, Muroc appeared for the first time as a line item on its own: The NACA's budget, approved by Congress on

[53] Becker, *The High-Speed Frontier*, p. 90.

[54] Ibid., p. 118.

24 August 1949 included $685,072 for the NACA Muroc unit. (By comparison, Langley received over $16 million.) On 14 November 1949, the Muroc unit was redesignated the NACA High-Speed Flight Research Station (HSFRS), a title more accurately reflecting the broad scope of flight research contemplated for it than its previous one.[55]

Expanding Upon the Sonic Breakthrough

The year 1949 was important to the NACA Muroc installation in several ways; there were, of course, the changes in the administration of the field site, reflected in its new title. But 1949 held particular importance as the year that the Muroc unit fully resumed research flying, after the death of Howard Lilly and the loss of the NACA Skystreak. The year also saw active involvement with three new research airplanes: a replacement Skystreak, the Northrop X-4, and the sweptwing D-558-2 Skyrocket. By 1950, the Skystreak and Skyrocket had added significantly to the transonic aerodynamic information acquired by the two X-1s, and the X-4 program was causing Williams and the NACA staff innumerable headaches, as will be seen.

Although in retrospect there proved to be little reason to build both the Skystreak and the Bell XS-1, it would have taken a gambler to predict that outcome before the sonic barrier was breached. The major reason for the Skystreak was that it could cruise for an extended time above Mach 0.8, freeing the XS-1 for Mach 0.9 and higher, thus complementing the research program on the rocket airplane. But this was a justification after the fact; when the Skystreak was first proposed, it was competing for the same mission as the rival XS-1. And unlike the AAF-sponsored XS-1, the Skystreak was a Navy-sponsored program. John Stack, the NACA's leading research airplane advocate, saw the Skystreak as much more in line with what the NACA wished a transonic research airplane to be—jet-propelled and relatively conventional in concept. The Navy, for its part, hoped that the D-558-1 Skystreak would lead to a military fighter derivative. The XS-1, in fact, was almost single-handedly the result of AAF research airplane advocate Ezra Kotcher and his unrelenting efforts to develop a Mach 1.2 rocket-propelled craft. As John Becker has stated, it is ironic that Stack and the NACA eventually shared the Collier Trophy for the achievements of the research airplane they least favored, the XS-1.[56] Nevertheless, one should not minimize the importance of the Skystreak to the NACA's flight research effort: from 1948 through 1952, it was the nation's most sophisticated straight-wing turbojet-powered research airplane for transonic flight-testing.

The NACA resumed its research with the Skystreak in early 1949. By the end of the year, Soulé was writing that the data from the X-1 and the D-558-1 were affording "very complete coverage of design information for high-speed straight-wing airplanes

[55] DFRC chronology, NASA Dryden Historical Reference Collection; NACA, *Thirty-Sixth Annual Report of the NACA: 1950* (Washington, DC: NACA, 1951), p. 68; Interview of Williams and subsequent conversations.

[56] Becker, *The High-Speed Frontier*, pp. 92-93.

from takeoff to the transonic speed ranges."[57] Despite the airplane's sleek appearance, tests of the Skystreak quashed hopes by the Navy and Douglas that they might spin off a tactical fighter. As it neared Mach 1, the Skystreak's handling qualities deteriorated rapidly. The force a pilot had to exert on its control wheel for longitudinal trim increased some six times—from 5 pounds to 30 pounds—between Mach 0.82 and 0.87. It tended to wallow about the sky at transonic speeds, certainly not an efficient weapon platform for service use.[58]

The Skystreak did make one major contribution to aeronautical engineering practice, a contribution indicative of the relatively easygoing and freewheeling managerial style at Muroc in the late 1940s. Langley's John Stack had concluded that adding little metal tabs or vanes (called vortex generators) in a row running in a spanwise direction (wingtip to wingtip) on the top and bottom of a wing might act to stabilize the position of shock waves on the wing, reducing undesirable trim changes and raising the so-called "limiting" Mach number of the plane. He called Walt Williams, who installed the tabs on the Skystreak by simply gluing them to the wing since the Skystreak had a fuel-filled ("wet") wing that prohibited riveting. The row of generators indeed worked, raising the plane's maximum controllable speed by 0.05 Mach, a significant increase. Industry quickly applied the results to new aircraft such as the Boeing B-47 Stratojet medium bomber. Vortex generators subsequently appeared on many other aircraft as well. Williams was criticized in certain administrative circles for not securing prior approval from NACA headquarters; However, Williams thought it was more impor-

The NACA's Douglas Skystreak cruises high over the Antelope Valley during a transonic research flight. (NASA photo E-713)

[57] Soulé, "Review and Status of Research Airplane Program," pp. 13-14.

[58] Douglas Aircraft Company, "Summary Report U.S. Navy Transonic Research Project Douglas Model D-558," Rept. E.S. 15879, 31 Aug. 1959, pp. 22, 50-51, in the files of the historian, Naval Air Systems Command.

tant to secure results quickly and expeditiously rather than tie a project up while seeking bureaucratic approvals. One need only compare the rapid implementation of the vortex generator idea with the 1970s winglet research program to appreciate the simplicity and directness of the earlier approach. The Skystreak completed its last vortex generator research flight in June 1950, when it reached Mach 0.99, the limits of its performance. Though not retired until 1953, the Skystreak had reached its zenith.[59]

It fell to the sweptwing Douglas Skyrocket to explore another interesting aerodynamic situation, this one a potentially dangerous instability predicted by wind-tunnel tests—the pitch-up phenomenon that plagued early sweptwing aircraft designs.

The Skyrocket seemed an unlikely choice for a successful research airplane, given its early history. Designed for both a jet and a rocket engine—the rocket for high-speed boost—and to take off from the ground, the Skyrocket appeared in early 1948 as a graceful sweptwing design having only its jet engine installed because the planned rocket propulsion system was well behind schedule. The first Skyrocket (the D-558-2, as it was known) flew at Muroc on 4 February 1948, piloted by company test pilot John Martin. The NACA received the second one, built at the end of the year. Following installation of an instrumentation package, the plane completed its first NACA research flight on 24 May 1949, piloted by Bob Champine. The Skyrocket was not viewed favorably by the NACA Muroc unit. Without its planned rocket engine, the Skyrocket lacked the necessary thrust for really meaningful transonic research. As Walt Williams later recalled, "We had to get off the ground before the temperature reached 80° F. You'd struggle to get to 24,000 feet, using almost all your fuel for the climb, and then you had to dive to get to 0.9 Mach. Flight endurance was thirty minutes or less."[60] Nevertheless, the NACA hoped that flight-testing of the craft would complement earlier low-speed work at Langley with the L-39 as well as a companion effort at Ames laboratory with a specially instrumented North American F-86A jet fighter. Happily, the NACA's expectations for the Skyrocket were met, and the D-558-2 program joined the X-1 as one of the two most successful of the early research airplane programs.

The Skyrocket's first brush with pitch-up came on 8 August 1949, when test pilot Champine banked into a tight 4 g turn at the modest speed of Mach 0.6. Suddenly, without warning, the Skyrocket nosed upward violently, its gravity force recorder indicating that the structure had sustained a momentary 6 g loading. Shaken, Champine applied down elevator, regained control, and landed. Wind-tunnel studies had indicated that sweptwing airplanes might experience the pitch-up phenomenon during "accelerated" maneuvers such as high g turns because of changes in the lifting characteristics of the wing and a decrease in effectiveness of the horizontal tail, particularly if the plane's flight attitude "blanketed" the tail from the oncoming airflow. Champine's flight gave NACA aerodynamicists the first opportunity to study data taken during an

[59] Interview of Williams and subsequent conversations; "DRB" [Donald R. Bellman], *Status of Research with the D-558-1 Airplane*, 14 Sept. 1950, pp. 3-4.

[60] Williams's comments to author after reviewing earlier draft; in author's notes.

actual pitch-up excursion in flight, as well as a new appreciation of the seriousness of the problem. (During takeoff and landing, for example, pitch-up might stall a sweptwing airplane and plunge it into the ground before the pilot had a chance to recover; at high speeds, the danger of pitch-up might unduly restrict the maneuvering performance of sweptwing jet fighters.) Subsequently NACA Muroc pilot John Griffith had an even more serious encounter with pitch-up during a similar 4 g turn. He tried to fight the maneuver by forcing the nose down, but the Skyrocket's tail effectiveness was low, and the plane began rolling and yawing before spasmodically snap-rolling. Griffith recovered handily, but later in the same flight, while he was performing an approach to stall with the craft's wing flaps and landing gear extended, the Skyrocket abruptly pitched up as its tail speed dropped below 130 miles per hour; again Griffith tried to fight it, and this time the plane rolled into a spin, dropping about 6,900 feet before Griffith was able to return it to level flight. NACA Muroc discontinued the Skyrocket's pitch-up program in 1950, when the NACA and the Navy sent the craft back to Douglas to be modified exclusively for rocket propulsion and air-launch from a mother airplane, like the Bell X-1. Nevertheless, the NACA realized that it had encountered a serious aerodynamic problem, and the 1949 pitch-up studies presaged a much more thorough investigation during 1951-1953 using another Skyrocket, about which more will be said below.[61]

In contrast to the productive work on the Skystreak and Skyrocket, the Northrop X-4 program caused the NACA a great deal of concern during 1948-1950. The X-4 was a small twinjet airplane having a swept wing but no horizontal tail surfaces. Instead, it relied on combined elevator and aileron control surfaces called "elevons" for its control in pitch and roll. It was similar in general configuration to Britain's ill-fated De Havilland D.H. 108 Swallow, which had crashed in 1946, and the NACA suspected (rightly) that the X-4 might suffer from the same stability-and-control problems as the D.H. 108—especially a dangerous pitching oscillation as it neared the speed of sound. Some engineers within the Air Force and Northrop hoped that the X-4 might offer a reasonable configuration for high-speed flight. The NACA was under no such illusions, though engineers thought the craft might prove very useful for dynamic-stability studies and studies of varying an airplane's lift-to-drag ratio so as to understand better the behavior and handling qualities of airplanes having extremely low lift-to-drag ratios. Much of this latter work benefited the later X-15 program.

The NACA had hoped to receive in December 1948 one of the two X-4 airplanes being built, but because of manufacturing delays, the first airplane only completed its maiden flight that month. The contractor program on the X-4 did not go smoothly; the plane was, in pilot's parlance, a maintenance "dog," far worse than the Skystreak that had killed Lilly. Northrop's test pilot completed only three flights in the plane in six months. Much against its will, the NACA Muroc technical staff found itself increasingly involved with the plane. It should not have been involved until its own aircraft

[61] Interview of Champine; D-558-2 progress reports for 15 Aug. 1949 and 14 Nov. 1949; Mulac D-558-2 flight chronology.

The NACA's Douglas D-558-2 Skyrocket aircraft. (NASA photo E49-200)

arrived for testing, but Northrop needed help and drew upon the Muroc unit for analysis of flight-test data. Normally, Williams would not have objected, but he had his hands full with the NACA X-1, Skystreak, and Skyrocket. He simply lacked the manpower to perform data reduction and even engineering duties in support of a contractor's program. Adding insult to injury, Northrop alleged that its delays stemmed from the NACA's slowness in working up data from the flights! In response to a puzzled inquiry from Hartley Soulé, Williams sent back a blistering memo castigating Northrop's operating procedures and mechanical problems, concluding that "the airplane is a difficult machine to operate and the research information to be gained is of small value for the work involved."[62] The NACA, Williams promised, would do what it could to support the Northrop program. He added, "We, however, have better use for these people, and as has been stated before, the sooner we drop the project the better off we will be."[63]

The X-4 did have some NACA friends, especially Smitty De France of Ames, who wanted to use the aircraft as a dynamic-stability research vehicle in support of some Ames research. Headquarters had already planned to rotate certain engineers through the Muroc site to familiarize other laboratories with the work being done in the desert, and Ames detailed a staff engineer, Melvin Sadoff, to the Muroc station as X-4 project engineer. Eventually, nearly two years behind schedule, the NACA received the second X-4, the one built for research; the first airplane made only 10 flights before being grounded and becoming a source of spare parts for the second. Completing its first NACA mission in November 1950, the second X-4 soon proved a valuable

[62] Letter, Williams to Soulé, 16 May 1949, NASA LaRC files.

[63] Ibid.

Jack Russell, head of the NACA High-Speed Flight Station's rocket shop, prepares to do pressurization tests on the XLR-11 rocket engine. The console provided the readings for the test of the rocket engine systems.

research tool for dynamic stability research, largely because it was a much more reliable craft than its predecessor. Delays such as those with the X-4 were not uncommon in first generation research airplanes. With the exception of the first X-1s and the Skystreaks, all subsequent programs experienced delays—primarily, it appears, from contractors underestimating the work required to develop specialized research airplanes.[64]

The End of the Beginning

In November 1949, the Air Force had offered *Glamorous Glennis*, the original X-1, to the NACA as a research airplane. But the NACA was already so committed to advanced research aircraft that Soulé was not about to accept a well-worn though historic hand-me-down. He recommended instead that the X-1 be sent to the Smithsonian Institution. Following its last flight—fittingly enough, piloted by Chuck Yeager—on 12 May 1950, it was sent there.

[64] Letter, Williams to Soulé, 12 Oct. 1949; Mulac X-4 flight chronology.

The retirement of the X-1 marked the end of the first tentative phase of supersonic research, the first nibbling away at the speed of sound, the first cautious edging beyond Mach 1. The next phase would come with the detailed examination of transonic flight by such craft as the X-3, X-4, X-5, XF-92A, and D-558-2; as well as with the continuation of frontier-pushing to Mach 2 and 3 with the advanced X-1s, the all-rocket D-558-2, and the X-2.

By 1950, the NACA was readying two large "slotted-throat" tunnels for transonic research, one having a 7.9 feet test section useful to Mach 1.15, and the second having a 15.8 feet test section capable of Mach 1.08. There still was a small "gray area" just around the speed of sound, beyond about Mach 0.98. The absence of ground-based research facilities for transonic testing that led to the early X-series aircraft had been overcome in rapid order largely because the X-series provided a research focus and an urgency that stimulated development of new methods of ground research and new tools such as the slotted-throat tunnel. Because of the forcing function that the X-series imposed upon the development of ground research methods and tools, the principal accomplishments of the early X-series (the X-1 and the D-558-1) lay less in their providing unique new information than in their validating the utility of new laboratory research techniques by providing "real-world" comparison data taken from flight research.[65]

Finally and most importantly, though, there was an undeniable psychological benefit coming from the first supersonic flights of these first research aircraft, a benefit aptly summarized by one program participant: "The most basic value was the liberation of researchers and aircraft designers from their fears and inhibitions relative to the 'sonic barrier.' The awesome transonic zone had been reduced to ordinary proportions, and aeronautical engineers could now proceed with the design of supersonic aircraft with confidence."[66]

[65] Becker, *The High-Speed Frontier*, pp. 98-113, 183; Hallion, *Supersonic Flight*, pp. 193-195; interview of John Becker, 12 Nov. 1971.

[66] Becker, *The High-Speed Frontier*, p. 95.

3

Testing the Shapes of Planes to Come 1950-1956

On 27 January 1950, the Air Force held a special dedication ceremony at Muroc, renaming the desert facility Edwards Air Force Base in honor of test pilot Glen Edwards, who had died in a test flight from the site in 1948. The ceremony symbolized the increasing emphasis that the Air Force was placing upon flight-testing, an emphasis that led to the designation of Edwards in 1951 as the Air Force Flight Test Center (AFFTC) with responsibility for testing aircraft, operating other test facilities, and providing support and services for contractors and other government agencies, such as the NACA. The 1950s, old-timers recall, were the golden age for Edwards, a decade of unparalleled expansion when new speed and altitude records were set almost monthly and the boom of igniting rocket engines punctuated conversations, giving the AFFTC its own distinct and exciting character. The Korean War stimulated expansion at Edwards. Air Force expenditures for the base leapt in fiscal years from $3.5 million in 1950 to $28.7 million in 1955, and then to $82.3 million in 1960. Personnel grew from 3,938 to 8,278 in that decade. The base expanded from about 305 square miles in 1952 to over 468 square miles by mid-1955, making it the largest flight-test center in the world.[67]

In the nine years after 1950, the NACA station at Edwards worked at an intensive level. The unit concluded its major role in the supersonic breakthrough (fittingly enough, it was a NACA pilot who first exceeded Mach 2), tested and evaluated a wide range of vehicles having new configuration concepts for high-speed flight, supported the development of military service aircraft, and undertook theoretical studies that eventually prepared the way for the hypersonic X-15 of the following decade. The station's growth mirrored that of the Air Force installation, though on a smaller scale. The number of employees grew from 132 in January 1950 to 332 in December 1959, and in those nine fiscal years its budget rose from $685 thousand for 1950 to $3.28 million for 1959. (A year later, reflecting the X-15 drive, its budget jumped to $6.99 million, rising to $32.97 million by 1968.) During the 1950s, the NACA Edwards installation gained complete autonomy from Langley. After the NACA became NASA in 1958, the NACA High-Speed Flight Station was redesignated the NASA Flight Research Center (FRC) on 27 September 1959, making it equal administratively with other NASA centers.[68]

[67] AFFTC History Office, *History of the Air Force Flight Test Center, July1-Dec. 31, 1965* (Edwards AFB: AFFTC, 1966), pp. 17, 22-23, and 30, in the files of the AFFTC History Office.

[68] Jane Van Nimmen and Leonard C. Bruno, with Robert L. Rosholt, *NASA Historical Data Book, 1958-1968*, Vol. 1: *NASA Resources* (Washington, DC: NASA SP-4012, 1976), pp. 271-78.

NACA pilots plan a research flight (viewer's left to right): Joe Vensel, Scott Crossfield, Joe Walker, and Walter P. Jones. (NASA photo E95-43116-10)

Autonomy Arrives

The cutting of the umbilical with Langley was not surprising. Since 1946, the Muroc(later, Edwards) facility had moved steadily and surely away from the parent. Although the work of the two centers was complementary, Langley's aeronautical thinking was dominated by the wind tunnel, while the thinking at Edwards was dominated by the research airplane. Since the High-Speed Flight Research Station already reported directly to headquarters through Hartley Soulé as research airplane projects leader, there was little except nostalgia to keep it allied firmly with Langley. However, an autonomous center required all the trappings of a major research facility: it must have good quarters, research areas, and workspace; an independent administration; and a fiscal organization defensible before outside agencies. Assisting the hopes of those who sought autonomy was the situation with the Air Force. With the adoption of the Edwards master plan, the Air Force had committed itself to moving from its old South Base to a new location midway between South Base and North Base. The NACA would have to move as well, so why not take advantage of the situation and move into a full-blown research facility? In August 1951, Congress approved $4 million for construction of new NACA laboratory facilities at Edwards, supplementing the $919,281 granted for salaries and expenses for 1952. The Air Force issued a lease to the NACA for less than a square mile on the northwestern shore of the dry lake, and construction started on the NACA station in early February 1953. One large complex would have hangar space to house the research airplanes, shop and instrumentation facilities, and offices.[69]

By early 1954, the new site was nearing completion. On 17 March 1954, NACA headquarters designated it an autonomous unit effective 1 July 1954, with the title NACA High-Speed Flight Station (HSFS). The transition to autonomy involved a lot more than just a change in title; the HSFS did not play Minerva to Langley's Zeus. Every facet of its administration and operation had to be accounted for, expanded upon, and separated from Langley. This included budget, management (already autonomous except in name), safety, establishment of its own library, preparation of a procedures manual, appointment of a legal officer, appointment of a procurement officer, selection of a color code for correspondence, design of its letterhead, transmission of NACA general directives and policy letters, issuance of a code letter for use in HSFS reports, appointment of a defense materials officer, and transmission of a complete set of NACA reports to the HSFC. All of this took weeks to do. Finally, on 26 June 1954, the NACA group of people moved from their make-do offices and hangar space on South Base to the nucleus of the present Dryden FRC facilities. They were on their own.[70]

Groundbreaking for the new NACA High-Speed Flight Research Station facilities, 27 January 1953 (viewer's left to right): Gerald Truszynski, Joseph Vensel, Walter Williams, Marion Kent, and California state official Arthur Samet. (NASA photo E-908)

[69] NACA, *Thirty-Seventh Annual Report of the NACA*: *1951* (Washington, DC: NACA, 1952), pp. 53-54; Walter C. Williams, "A Brief History of the High-Speed Flight Station," July 1956, p. 6, copy in the files of the NASA History Office.

[70] Letter, Soulé to NACA HQ, 19 Apr. 1954; letter, E. H. Chamberlin to NACA Langley Laboratory, 5 Apr. 1954; Williams, "A Brief History of the HSFS," p. 6.

The initial facilities of the NACA High-Speed Flight Station, completed in 1954 at a cost of $3.8 million. This is still the core of the Dryden Flight Research Center, although the initial building in the center of this photo has had many additions and numerous other facilities have been added over the years. (NASA photo E-1613)

By 1954, the NACA station at Edwards already was a research facility with strong in-house technical capabilities; likewise, the fundamental organization of the station was well established, a basic arrangement followed for many years. There were four branches—later termed "divisions" and then, under NASA, "directorates." These were administration, research, operations, and instrumentation.

Administration, of course, meant Walt Williams and his staff. He did not have a deputy, although in his absence De E. Beeler often assumed the role of acting chief. Williams's managerial style emphasized minimal paperwork, informal communication and decision-making, rigorous attention to time and cost schedules and, above all, an unwavering commitment to safety. His great flexibility in structuring management defies placing Williams within any of the standard industrial-organization schools of management such as "Theory X," "Theory Y," "MBO," etc. His style of management most closely mirrored the "gamesman" approach but without the "gamesman's" frequently cynical view of his role within an institution. Williams was without question a highly effective administrator, as was his successor Paul F. Bikle, a man who reflected many of the same attributes.

Research was supervised by Beeler, an intense, hard-driving individualist. Research involved the station's mathematicians, engineers, and physicists. This branch did the work on aircraft stability and control, flutter and vibration, loads, structures, performance, and other special research, including design conceptualization of advanced aerospace vehicles. This group supervised the flight research portion of the aircraft flight-testing programs. Eventually the research branch got into aircraft simulation as well, for flight-planning, pilot-training, systems analysis, and performance prediction.

Operations was the fiefdom of Joe Vensel, a veteran NACA test pilot who ran the pilots office, supervised flight operations and maintenance of the aircraft, and helped plan and monitor the flight programs. Crusty but fatherly, Vensel ruled with an iron hand. Somewhat deaf from his years in open-cockpit biplanes, Vensel had the habit of turning off his hearing aid and going to sleep if a meeting became boring. The research pilots under him maintained close liaison with the engineers in the research branch.

Instrumentation was the responsibility of Gerald M. Truszynski, who established a reputation for thoroughness that helped make him a senior NASA administrator a decade later. This branch undertook the instrumentation and calibration of the various research airplanes and provided flight-tracking and data acquisition services. Although the HSFS occasionally did its own instrument fabrication, it generally relied upon Edmond Buckley's instrumentation group at Langley for development. The NACA at Edwards still relied on a pair of old SCR-584 radars, though it was obvious that as the capabilities of the X-series advanced to Mach 3 and beyond, so would the need for a specially instrumented high-speed flight corridor with several data-linked tracking stations. This would come to pass with the establishment of the X-15 High Range.

In a broad sense, the research aircraft program involved the cooperation of three parties—industry, the military services, and the NACA. This was reflected in the way testing took place at Edwards. The testing process closely followed the military pattern of airplane acquisition and testing, except that the NACA added another aspect all its own. First, a contractor would build a research airplane to military specifications, usually derived in conjunction with NACA; this was particularly true for the rocket-research aircraft and the D-558 series. Standard practice called for the contractor to deliver the first aircraft built to Edwards for so-called Phase I testing. This involved the contractor's own pilots demonstrating that the airplane had generally satisfactory handling qualities and conformed to the contract. Then, the contractor would usually deliver the aircraft to the Air Force, with a second craft going to the NACA for detailed research investigations. The NACA HSFS would generally provide data acquisition and analysis support to the contractor and the Air Force on their programs. Despite the oft-heard claim that the military and contractor programs were "scientific research," more often than not, especially on the rocket-propelled aircraft, the programs were little more than contractor verification of the plane's low-speed flying qualities, followed by repeated attempts by the contractor and, later, the service, to set new speed and altitude records. Such flying was always viewed with disfavor by the NACA because it seemed an unnecessary risk of expensive research tools. The De Havilland D.H. 108 lost in Britain, for example, had been destroyed during a practice speed run for a planned airspeed record flight attempt. Aside from seeking records, service research tended to emphasize pragmatic military values rather than the niceties of aerodynamic and propulsion studies; as has been mentioned, the Navy closely watched the Skystreak program to see if it could spawn a tactical airplane. The Air Force evaluated the Bell X-5 variable-sweep aircraft to see if it could be modified into a cheap fighter for NATO and other foreign countries. As part of the Air Force's

NACA headquarters officials inspect the High-Speed Flight Station's new facilities in 1954, with some local staff (viewer's left to right, front row): Jerome Hunsaker, Walt Williams, Hugh Dryden; (middle row) Scott Crossfield, Joe Vensel, John Victory; (back row) Marion Kent, De Beeler, Gerald Truszynski. (NASA photo E95-43116-11)

Cook-Craigie acquisition plan (discussed in Chapter 5), NACA laboratories around the country received current generation military aircraft for flight-testing in support of the military research and development programs on these aircraft. The NACA facility at Edwards tested many of these aircraft as well.

Generally speaking, then, flight-testing at the Edward's NACA facility during the 1950s involved research on the X-series aircraft and research support on various military aircraft programs. The X-series itself broke down into two major subcategories of aircraft: configuration explorers—aircraft having unique and unusual design shapes requiring verification or refutation, such as the X-3, X-4, X-5, XF-92A, and, to a lesser extent, the D-558-2 Skyrocket—and supersonic aerodynamic research vehicles having rocket propulsion and being air-launched from modified Boeing B-29 or B-50 bombers—such as the advanced X-1s, the all-rocket D-558-2 Skyrocket, and the Bell X-2. The configuration testbeds were rarely flown beyond Mach 1, because most were simply transonic in performance. The X-3, for instance, was a planned Mach 2 configuration testbed that failed to fly anywhere near that mark because the manufacturer had to use less powerful engines than originally intended. The third D-558-2 Skyrocket, which retained both jet and rocket propulsion, is included in the configuration group because of its extensive sweptwing pitch-up investigations undertaken during the 1950s. The rocket-propelled supersonic research aircraft, on the other hand, were the aircraft that first exceeded Mach 2 and 3.

In the mid-1950s, the research aircraft program continued to expand. Hartley Soulé and other program officials could see three broad streams: the early rocket research airplanes and configuration explorers, a hypersonic research vehicle that soon became the X-15, and, beyond, a true winged orbital spacecraft (termed a "boost-glider") known as "Dyna-Soar" (for *Dyna*mic *Soar*ing). These roughly sequential streams or "rounds" caused Soulé to dub the early rocket research aircraft and configuration explorers "Round One." The X-15 became "Round Two," and the Dyna-Soar became "Round Three." This after-the-fact classification quickly passed into the NACA's official records and nomenclature.[71]

The NACA's Configuration Explorers

The High-Speed Flight Research Station's research on new aircraft really involved studies of aerodynamic, stability-and-control, and handling qualities on five basic configurations: the sweptwing, the semitailless, the delta wing, the variable-sweep wing, and the low-aspect-ratio (AR) thin wing.

Configuration	Aircraft	Speed Range
Sweptwing	Douglas D-558-2 #3	Mach 1.0
Semitailless	Northrop X-4 #2	Mach 0.9
Delta wing	Convair XF-92A	Mach 0.9+
Variable-sweep	Bell X-5 #1	Mach 0.9+
Low-AR thin wing	Douglas X-3	Mach 0.95+

All of these aircraft were also associated with particular aerodynamic research or dynamic-stability problems as well:

Aircraft	Research Problem
Douglas D-558-2	Sweptwing pitch-up during maneuvering.
Northrop X-4 #2	Pitching oscillation of increasing severity approaching Mach 0.95.
Convair XF-92A	Delta pitch-up during maneuvering.
Bell X-5 #1	Unacceptable stall-spin behavior: sweptwing pitch-up during maneuvering.

[71] Williams's comments to Hallion.

Douglas X-3	Coupled motion instability during abrupt rolling maneuvers.

Each of these problems was a major concern to an aircraft industry undertaking the design of new combat aircraft vastly different in configuration and speed potential from those of only five years before, and to the Air Force and Navy, whose pilots might have to fly and fight in these new designs. Thus, any detailed understanding of these difficulties would be welcomed as a significant contribution.

With the exception of the Douglas X-3, each of the other four configuration explorers exhibited some degree of pitch-up problem, ranging from moderate to severe. Of all five aircraft, the only one having generally pleasant flying characteristics was the D-558-2. The others exhibited the following behavior characteristics, which generally stemmed either from the peculiar configuration or the lack of a powerful enough engine:

Aircraft	**Behavior Problem**
Northrop X-4 #2	Poorly damped "hunting" motion about all three axes; "washboard road" motion.
Convair XF-92A	Sluggish and underpowered.
Bell X-5 #1	Dangerous stall approach and spin tendencies.
Douglas X-3	Sluggish and very underpowered.

After exploring the basic behavior of the aircraft and its characteristics, the NACA generally made aerodynamic modifications to the design to evaluate whether certain concepts such as wing leading edge extensions or wing fences would improve the behavior. If they did, the NACA concluded that these were generally applicable design features that could improve the behavior characteristics of that type of configuration. Such modifications were not attempted with the X-3 and X-5 because of concerns about cost and complexity. Modifications evaluated on the other three aircraft were:

Aircraft	**Modifications**
Douglas D-558-2 #3	Various wing-slat and wing-fence combinations, leading-edge extensions.
Northrop X-4 #2	Increasing the thickness of the trailing edge of the wing and elevon.
Convair XF-92A	Various combinations of wing fences.

Aside from a coupled-motions instability investigation on the X-3, the problem of greatest interest to the industry and military services was pitch-up, encountered in various forms by the D-558-2, the XF-92A, and the X-5.

The NACA's early X-series fleet (from viewer's left): Douglas D-558-2 Skyrocket, Douglas D-558-1 Skystreak, Bell X-5, Bell X-1, Convair XF-92A and Northrop X-4. (NASA photo EC-145)

Pitch-up was a problem inherent in any sweptwing or delta airplane. As a sweptwing airplane approaches a stalled flight condition—either at low speed by flying at an increasingly higher nose-up angle of attack and a lower and lower speed, or at high speed in an abrupt turning maneuver at a high g-loading—the natural tendency of the airflow around the wing to flow outward toward the wing tips (spanwise flow) is accentuated, promoting the development of so-called "separated" airflow, causing a loss of lift at the wing tips. As the stall condition progresses, the area of the stall moves progressively "up" the wing toward the wing root, followed by the center of lift of the wing. Put another way, the zone of wing lift becomes smaller and smaller and concentrated toward the fuselage, hence further "forward" along the plane's longitudinal axis. The change of lift vector to a point further forward along the length of the plane causes the plane to nose abruptly upward—that is, to pitch up.[72]

Pitch-up could be overcome by several "fixes": a "sawtooth" leading-edge extension to promote the formation of "active" airflow, defeating the tendency of the wing to exhibit spanwise flow; wing fences, literally small "fences" running in a chordwise (leading edge to trailing edge) direction to divert the spanwise flow into chordwise flow; and open wing slats (dating from the 1920s) to delay the onset of turbulent separated airflow over the wing at high angles of attack. All these were examined on the Skyrocket and XF-92A. The best solution was to place the horizontal tail low on the aft fuselage of an aircraft, where it would be below the wing wake and downwash of the wing. The Skyrocket and the X-5 both had highly placed horizontal

[72] This discussion of pitch-up is based on interpretation of flight test reports from the Skyrocket, X-5, and XF-92A programs, as well as from Robert L. Carroll, *The Aerodynamics of Powered Flight* (New York: John Wiley & Sons, Inc., 1960), p. 171, and Daniel O. Dommasch, Sydney S. Sherby, and Thomas F. Connolly, *Airplane Aerodynamics,* (New York: Pitman Publishing Corp., 4th ed., 1967), pp. 432-433, 467-476.

tails, giving them particularly objectionable pitch-up characteristics. The RAPP suggested adding a low horizontal tail to both for evaluation purposes, but the problems and cost outweighed the potential benefits. In any case, the obvious conclusion from NACA's testing regarding the desirability of the low horizontal tail surface led to that configuration's becoming standard on the first-generation supersonic sweptwing fighters such as the North American F-100 Super Sabre and the Vought F8U Crusader. Sweptwing supersonic aircraft lacking such a feature—such as the McDonnell F-101 Voodoo—proved to have dangerous and mission-limiting pitch–up characteristics. Tail changes, of course, could not be made with the triangular or delta wing configuration; rather, designers had to rely on various combinations of wing fences.

The several NACA programs on the Skyrocket, the X-4, XF-92A, X-5, and X-3 went relatively smoothly from a standpoint of data collection, analysis, and reporting, but maintenance often proved troublesome. Highly complex experimental aircraft, then and now, are notoriously difficult to keep up, even under the best of circumstances. Workload, program, and weather considerations also played roles, often forcing a stretch-out of planned flights. The Skyrocket, for example, took 27 months to complete 29 pitch-up research flights. In the following comparison, contractor and military test flights prior to the NACA's acquisition of the aircraft are excluded.

Aircraft	**Number of Flights**	**Duration of NACA Research**
Douglas D-558-2 #3	66	1950-1956
Northrop X-4 #2	82	1950-1953
Convair XF-92A	25	1953
Bell X-5 #1	133	1952-1955
Douglas X-3	20	1954-1956

The NACA research pilot staff at Edwards approached all of these aircraft with caution.[73] The Skyrocket had its quirks but was generally pleasant. The X-4 could be annoying. The XF-92A required good piloting skills. The X-3 and X-5 had truly vicious characteristics, particularly the latter's violent stall-spin instability, which eventually killed Air Force test pilot Ray Popson. Not unexpectedly, research aircraft often have characteristics that are demanding, but the X-5 was simply a flawed design, although this had nothing whatsoever to do with the feature the craft was developed to verify, the variable-sweep wing. Rather, it had to do with its aerodynamic layout, especially the poor position of the tail and vertical fin. An excerpt from a pilot report gives some idea of its qualities as the plane approached a stall:

> As the airplane pitches, it yaws to the right and causes the airplane to roll to the right. At this stage aileron reversal occurs; the stick jerks to the

[73] NACA pilots Champine and Griffith were joined by A. Scott Crossfield, Walter P. Jones, Joseph A. Walker, Stanley P. Butchart, John B. McKay, and Neil A. Armstrong.

One NACA attempt to remedy pitch-up: chord extensions on the NACA D-558-2, photographed in February 1953. (NASA photo E-927)

> right and kicks back and forth from neutral to full right deflection if not restrained. It seems that the airplane goes longitudinally, directionally, and laterally unstable in that order.[74]

During one flight, pilot Joe Walker lost 20,000 feet while recovering from a stall; fortunately, the stall had occurred at 40,000 feet.[75]

Despite its faults, the X-5 was an outstandingly productive airplane. Early testing of the craft had demonstrated that the variable-wing-sweep principle worked—that it endowed a plane with good low-speed performance when the wing was fully extended for takeoff and landing, and that it offered good high-speed performance as well when the wing was swept fully aft. The actual mechanism by which the X-5 "translated" its wing from fully extended to fully swept and back again was quite another matter, for it was complex and hindered the utility of the design. Indeed, variable-sweep aircraft did not become a practical reality until after the conceptualization of the outboard wing pivot by NACA engineers at Langley in the mid-1950s. The NACA was not too concerned over the variable-sweep aspect of the plane once it had been proved to work. Rather, the NACA and the RAPP viewed the unique advantage

[74] X-5 pilot report by Joe Walker, 31 Mar. 1952, NASA LaRC files.

[75] Recollection of Walter C. Williams in conversation with Hallion.

The NACA's Bell X-5 variable-sweep research aircraft with wings fully extended at minimum sweepback. (NASA Photo E95-43116-12)

Wings of the X-5 are fully swept to the maximum sweep position. (NASA photo E95-43116-13)

of the X-5 to be its ability to provide (in effect) a whole range of sweptwing research aircraft in one vehicle. Since the wing could be swept to many different positions, a variety of measurements was possible over a wide range of sweep angles up to 60 degrees, the same angle as the XF-92A delta.[76] The pitch-up investigation on the X-5 complemented the extensive work undertaken on the Skyrocket, especially since the craft could furnish aerodynamic information on how a variety of sweptwings reacted as a plane approached its pitch-up point. The NACA also used the X-5 as a chase plane for other research aircraft, because it could vary its flying characteristics to suit the airplane it was chasing. It was retired from service in late 1955.[77]

NACA's research program on the little X-4 was particularly fruitful for reasons not expected when the airplane was under development. The NACA had never been a strong supporter of the X-4, a sweptwing airplane designed for transonic flight minus a horizontal tail to damp out any pitching tendencies. Indeed, the craft came about as a result of two factors: John Northrop's own firm belief in the value of the tailless concept and German interest in the idea, which had spawned the wartime Messerschmitt Me 163, a plane with very poor high-speed behavior. In the postwar climate of military research, any idea the Nazi government had been working on often assumed an imagined worth all out of proportion to its true value. Then, the X-4 encountered development delays, and the first prototype proved, in Walt Williams's own words, a "lemon." But the second was quite reliable mechanically, and the NACA program proceeded smoothly, following the first NACA flight in November 1950.

At first the NACA program concentrated on the X-4's dynamic-stability problems. At about Mach 0.88, it began a longitudinal pitching motion of increasing severity; test pilots compared it to riding over a washboard road. But it also exhibited combined pitching, rolling, and yawing motions of increasing severity, a "hunting" about all three axes marked by inadequate motion damping as Mach number increased. The Edwards project team decided to thicken the trailing edge of the wing in an effort to cure the motions, not difficult since the X-4 had huge speedbrake surfaces above and below the wing that could be wedged open, the gap between their surfaces forming the thicker trailing edge. In 1952, the engineers went further and thickened the trailing edge of the elevons (the control surfaces the X-4 used for pitch and roll control) using balsa wood attachments. The thickening worked in part, increasing the craft's roll rate by 25 percent, and longitudinal control effectiveness was improved as well. But the persistent motions still appeared above Mach 0.9; and at Mach 0.94, they were so severe the plane porpoised along at vertical accelerations of ± 1 1/2 g. Clearly the

[76] In 1949, Hartley Soulé, with the assistance of NACA-USAF liaison officer William J. Underwood, emphatically stated this viewpoint in refuting allegations by the Air Force Aircraft Laboratory at Wright Field that the X-5 was an unnecessary duplication of the XF-92A delta program—a strange charge, and puzzling considering that the strongest X-5 supporters were Air Force partisans who felt its development should be accelerated to provide a lightweight export fighter. Military interest in the X-5 was, not surprisingly, always oriented toward applications, not research. See letter, Hartley A. Soulé to Hugh L. Dryden, 29 Nov. 1949, NASA LaRC files.

[77] Mulac X-5 chronology; summary of X-5 monthly progress reports, in the files of the DFRC.

The NACA's Northrop X-4 semitailless research aircraft, one of the smallest ever flown. (NASA photo E-17349)

semitailless configuration was unsuitable for transonic applications if one chose any shape resembling the Me 163, D.H. 108, or X-4.[78]

Although the X-4 configuration itself proved unsuitable, the amount of research data returned was substantial, particularly on the interactions of combined pitching, rolling, and yawing motions—an interaction soon to be of critical concern with high-performance military fighters. The blunt elevon research with the X-4 directly benefited the Bell X-2 then under development, which featured ailerons having a blunted trailing edge on the basis of models tested at Wallops. The NACA High-Speed Flight Research Station was able to verify the full-scale concept by demonstrating the pronounced benefits the blunted trailing edge gave the X-4 in rolling performance. Finally, Williams and his researchers recognized that the X-4's speedbrake enabled the plane to vary its lift-to-drag ratio to such a degree that it could simulate the approach of what are now termed "lifting reentry spacecraft." The X-4 had a minimal lift-to-drag ratio of less than 3, giving it X-15-like performance. And, indeed, it was with the upcoming generation of X-15-like craft in mind that the NACA undertook approach and landing studies of their predicted behavior using the X-4. It ended its days as a pilot trainer before being retired in 1954.[79]

[78] Summary of X-4 monthly progress reports in DFRC Historical Reference Collection.

[79] Interview of Williams and subsequent comments.

The Convair XF-92, a delta-wing aircraft. (NASA photo E95-43116-14)

By NACA standards, the HSFRS program on the XF-92A delta-wing research aircraft was brief, lasting only six months in 1953 with 25 flights. The XF-92A had an interesting past, for it was not originally conceived as a research craft at all, but rather as a testbed for a proposed interceptor that failed to materialize. Once the Air Force had abandoned the proposed interceptor, the service continued to support development of the XF-92A (only one of which was built) as a delta testbed. NACA interest in the plane was immediate, for the delta-wing planform offered exceptional wing area plus a thin-airfoil cross section and low aspect ratio, combined with low weight and high structural strength—all desirable attributes for a supersonic airplane. Even before the first flight of the XF-92A in 1948, the NACA had tested the plane in the full-size low-speed tunnel at Ames. The RAPP closely followed the actual Convair and Air Force program on the airplane, which the Air Force relinquished to the NACA in early 1953.

Besides validating the thin delta principle, the XF-92A played a major role in supporting the development of the Convair F-102A interceptor, the Air Force's first attempt at an all-weather supersonic interceptor. The XF-92A had surprisingly violent pitch-up characteristics during turns, often exceeding 6 g and once going above 8 g. NACA technicians at Edwards equipped the craft with various wing-fence combinations planned for the F-102, which had a similar wing planform, and the Air Force's Wright Air Development Center requested that the NACA send any data from its flight program that might prove beneficial to the F-102 program. This the NACA did, especially with regard to fence combinations to alleviate pitch-up. Eventually, however, the F-102 faced major redesign anyway, to take advantage of the Whitcomb area rule principle derived at Langley and the conical wing camber concept derived at Ames. Nevertheless, the contributions of the XF-92A to the F-102–and through the F-102 to the XF2Y-1 Sea Dart, the F-106 Delta Dart, and the B-58 Hustler–were substantial. (It is interesting to note that, like Convair with the XF-92A, F-102, F-106, and B-58, French aircraft manufacturer Marcel Dassault followed a similar development path, going from a small delta testbed, the Mirage I, to the Mirage III fighter family, and then to the B-58-class Mirage IV supersonic bomber.) The XF-92A was retired in October 1953, the progenitor of America's delta aircraft.[80]

Of NACA's configuration explorers, the only disappointment was the best-looking of the lot, the Douglas X-3. Conceived for supersonic research above Mach 2, the X-3 had been victimized by an experimental engine installation that failed to live up to its promise. Rather than two powerful turbojets, the X-3 had to be completed with puny (by comparison) Westinghouse J34s, which could not propel the airplane past Mach 1 in level flight. The X-3 proved frustrating for the NACA. It had perhaps the most highly refined supersonic airframe of its day as well as other important advances, including one of the first machined structures and the first use of titanium in major airframe components. It had a long fuselage, giving it a high fineness ratio, and a low aspect ratio (low ratio of span to chord) wing having a thickness/chord ratio of only 4 1/2 percent. Despite this potentially supersonic configuration, the maximum speed ever attained by the X-3 was Mach 1.21, during a dive. For a while, the RAPP thought about replacing the jet engines with two rocket engines and after fairing over the plane's air intakes, launching it from a modified jet bomber to reach Mach 3.5. But the X-3 was overtaken by events—namely, the development of the F-104, a genuine Mach 2 airplane to which it directly contributed.

The X-3 had made its first flight in October 1952. It was so badly underpowered that on the first flight its test pilot, Bill Bridgeman, complained into his mike, "This thing doesn't want to stay in the air," which might have been taken as an epitaph for the whole program. In July 1954, the Air Force completed its own brief evaluation of the craft, by now regarded as a glamorous "hangar queen," and turned it over to the High-Speed Flight Station, whose engineers judged the plane to have only "limited" research utility. It did have some contributions it could make. One—not to be minimized—was in tire studies: the plane routinely shed its small tires during high-speed landing and taxi runs, forcing revision of tire design criteria for high-performance aircraft.[81]

[80] Summary of XF-92A monthly progress reports in DFRC Historical Reference Collection; Mulac, XF-92A chronology.

[81] Summary of X-3 monthly progress reports in DFRC Historical Reference Collection; Mulac X-3 chronology; Williams's comments to Hallion.

The Douglas X-3, something of a hangar queen. (NASA photo E-1546)

The X-3 completed its first NACA flight in August 1954, and by late October, the HSFS X-3 project team expanded the planned program on the aircraft to include investigating its lateral and directional stability and control during abrupt rolls with the pilot holding the rudder "fixed" (centered). These studies had particular significance, for the X-3 closely approximated the then-current generation of military fighters entering testing or production. The fighters had a short wingspan and a long fuselage, with the aircraft "loaded" primarily along the fuselage rather than along the wing. This lack of spanwise loading greatly increased the plane's inertial characteristics in yaw and pitch. On 27 October 1954, NACA research pilot Joe Walker had the dubious honor of demonstrating just how dangerous flight-testing can unexpectedly be and how courage must be the constant attribute of the successful test pilot. As planned for this flight, Walker initiated an abrupt left roll at Mach 0.92 and an altitude of 30,000 feet. The plane rolled rapidly, but as it did so, the nose rose in pitch and simultaneously slewed in sideslip, reaching combined values of 20 degrees in pitch and 16 degrees in sideslip. After five wildly gyrating seconds, Walker regained control. He had every reason to call it a day and land, but that was not Joe Walker's style. His curiosity aroused, Walker accelerated in a shallow dive past Mach 1 and then executed an abrupt left roll. This time the reaction was more than violent; it was berserk, with the plane attaining a sideslip angle of 21 degrees, imposing a transverse load of 2 g. Simultaneously, the plane pitched violently downward, reaching -6.7 g, then violently pitched upward to +7 g before Walker could regain control. Fortunately, the rolling motions subsided; without further difficulty, Walker damped the yawing and pitching motions and landed immediately. Postflight analysis indicated that the fuselage had sustained (but fortunately had not exceeded) its maximum-limit load, while the high altitude prevented the wing from reaching its limit load. Joe Walker was a skillful—and lucky—man that day.[82]

The NACA wisely decided not to duplicate the flight conditions Walker had encountered that exciting day over the Mojave. The "inertial coupling" phenomenon that Walker encountered had first appeared in very mild form in the dynamic instability of the X-4 at transonic speeds. Concurrent with the X-3 experience, however, were a series of Air Force accidents occurring on the first-production F-100A Super Sabre jet fighters. Although attributable in part to a serious lack of directional stability, the progression of violent motions mimicked the X-3 experience closely. Inertial coupling, also called "roll coupling" or "roll divergence," had first been predicted by William H. Phillips of Langley Laboratory in a classic theoretical study.[83] One important cure was to increase the wing area and, especially, the tail surface area of the aircraft. Such a cure turned the F-100 from a killer into a reliable airplane. Walker's experience, like the early pitch-up encounters of Champine and Griffith, gave agency engineers their first "real-world" appreciation of how serious the inertial coupling problem could be. What pitch-up was to the early sweptwing jet aircraft, inertial coupling became to the first-generation supersonic airplanes. Generally speaking, 1970s to 1980s aircraft having twin vertical fins and generous wing areas plus other aerodynamic refinements were monuments to the lessons learned from the X-3 and its brethren.

In many ways, Walker's flight remained the apex of the X-3 program. Although it returned the X-3 to the air, the NACA was most reluctant to probe its lateral (roll) stability and control characteristics further, and finally retired the craft in 1956.[84]

One little-known configuration program run by the NACA High-Speed Flight Station involved a special investigation for the Navy and the Atomic Energy Commission (AEC) on the transonic drag characteristics of bomb and tank shapes—"external stores"—hung off the wing of an airplane. The bomb and tank shapes of the early 1950s did not differ appreciably from those of World War II, and aerodynamicists faced serious flow-interference problems generated by hanging these bulky shapes on otherwise streamlined airplanes. "Low drag" external-stores shapes compatible with the new generation of attack aircraft—those designed to carry nuclear weapons—were still largely a thing of the future. The military services and agencies such as the AEC that had to generate new weapons wondered how these new shapes would affect the transonic drag rise of high-speed attack and fighter aircraft. The dan-

[82] NACA Research Memo RM-H55A13, "Flight Experience with Two High-Speed Airplanes Having Violent Lateral-Longitudinal Coupling in Aileron Rolls," NASA HSFS, 4 Feb. 1955, passim; Walker's pilot report, in DFRC Historical Reference Collection; comment of Hubert Drake on the original chapter, 7 Feb. 2000.

[83] NACA TN-1627, "Effect of Steady Rolling on Longitudinal and Directional Stability" (1948).

[84] Walter C. Williams and Hubert M. Drake, "The Research Airplane: Past, Present, and Future," *Aeronautical Engineering Review* (Jan. 1958), pp. 38-39; HSFS annual report to the RAPP for 1956, p. 3; copy in NASA DFRC Historical Reference Collection.

ger, of course, was that the shapes would impose unacceptable penalties in range and maximum speed.

In 1951-1952, the AEC and Navy urged the NACA to study the problem. Walt Williams proposed a program to add stores pylons to the Skyrocket, the aircraft used for NACA's pitch-up research, and test bomb shapes and fuel-tank shapes produced by Douglas. Douglas was a natural choice and was in on the program from the start, for it had designed the Navy's string of first-line attack aircraft, the AD Skyraider, A3D Skywarrior, and A4D Skyhawk. The RAPP quickly assented to Williams's proposal, and D-558-2 #3 began its stores research program in the summer of 1954, continuing until December 1955, when NACA engineers concluded they had sufficient information. The data were delivered to the Navy and AEC for use in weapon design.[85] Nine months later, this Skyrocket was retired from service, the last of the Round One configuration explorers to fly.

Make-Work or Valuable Contributions?

The progression of aeronautical technology has been accompanied by radical changes in the shapes of aircraft. Certainly, designers in the mid-to-late 1940s and early 1950s faced conflicting choices of configurations for high-speed aircraft. There were some general trends, such as lengthening a plane's fuselage to increase its fineness ratio while reducing the wingspan to lower its aspect ratio, reducing the thickness of wings, and placing the horizontal tail clear of the wing wake. But a diversity of choices and decisions faced designers as well: should a plane have a moderately sweptback wing of, say 35 degrees, or a sharply sweptback wing of 45 degrees or more? How thin should a wing be? Should supersonic aircraft employ delta wings? Should the plane have a horizontal tail? What high-lift devices would work best on a sweptwing plane for low-speed flight? These and many other questions required answers, answers that the Round One configuration explorers provided.

The NACA always maintained that its work on the configuration explorers was of critical importance to postwar aircraft design. A few critics in industry (perhaps motivated, as historian Alex Roland has suggested, by a "not invented here" syndrome) believed that the postwar X-series program did not materially assist the design development of subsequent high-performance aircraft. Specifically, these critics of the program attacked it on three general grounds:

- The program was expensive, time-consuming, and distracted the industry and military services from developing practical, operational supersonic aircraft.

- The program failed to generate any improvements to turbojet engine propulsion systems.

[85] HSFS annual report to the RAPP for 1954, in NASA DFRC Historical Reference Collection; progress reports for the D-558-2 #3, May 1954-Sept. 1956; Mulac D-558-2 chronology; Williams's comments to Hallion.

• The stability-and-control information gathered was not applicable to advanced aircraft design because it was gathered from shapes not representative of what future high-performance aircraft would look like.[86]

The first charge is easy to refute. Industry and military researchers had no clear ideas of what a "practical," "operational" supersonic aircraft should look like. In retrospect, the designs they generated prior to access to X-series information were almost always wildly impractical.[87] The leading service aircraft of the 1950s and 1960s (especially the Air Force's "Century Series" fighters, the F-100, F-101, etc.) were all designed to incorporate the features recommended as the result of the X-series testing program.

The second charge is really a nonissue. The X-series program began as an aerodynamic research program concerned with transonic and supersonic flight conditions, including stability and control and flight loads. To acquire these data, rocket-propelled aircraft had to be designed, because conventional turbojets lacked the necessary power to propel craft past Mach 1. Had the services insisted upon jet propulsion for these aircraft, perhaps some acceleration of jet engine development would have taken place, but it is doubtful. Instead, it is likely that the acquisition of supersonic flight data merely would have been delayed. The first supersonic jet flew in 1953, and by that time the rocket-propelled advanced X-1 was pushing Mach 2.5. In any case, responsibility for advanced tubojet studies was not a concern of the X-series; it was a separate issue, involving industry, the military services, and, within the NACA, the specialists of the Lewis laboratory.

The third charge is simply false. Of all the Round One configuration explorers, only the X-4's weird semitailless shape did not appear on subsequent high-performance aircraft—and for good reason. Much of the stability and control information gathered from these aircraft warned designers what to adopt and, per-

[86] The most concise—and damning—example of this criticism of the NACA is in a letter from Lockheed designer Clarence "Kelly" Johnson to Milton B. Ames, 21 Oct. 1954. Johnson himself was a master of blending the research and operational aspects of a program into a single airframe (For example, the P-80, U-2, and later SR-71). However, his F-104 was heavily dependent upon the X-3 for its aerodynamic shape, performance analysis, and other matters, disproving his general contentions in the letter. A RAND Corporation study of Lockheed F-104 development concluded that:

> The F-104 history illustrates that research and development in one program can have a great carry-over value in another. Lockheed's success in building and flying a prototype less than a year after go-ahead would very probably not have been possible without the knowledge derived from the Douglas X-3 program. Although the value of this experimental effort in the F-104 effort could hardly have been anticipated when Air Force money was advanced to finance the program, nevertheless the value to the Air Force of the X-3 program extended far beyond the immediate results achieved with it.

Thomas A. Marschak, *The Role of Project Histories in the Study of R&D* (Santa Monica: The RAND Corporation, Jan. 1964), pp. 85-86, 90.

[87] For example, a proposed Convair interceptor powered by a combined rocket-ramjet engine and launched from a takeoff dolly.

haps more important, what to avoid: high horizontal tails, small and inadequate vertical fins, configurations prone to pitch-up or inertial coupling, etc. And this was the purpose of the program.

The cost criticism is not to be taken seriously either. The original XS-1s cost approximately $500,000 apiece, equivalent to the purchase of five production Lockheed P-80 subsonic jet fighters, a reasonable price for the information gained. What was often annoying about the X-series was how demanding their maintenance could be, but that has been and continues to be a facet of research aircraft operation. Even this could be misleading. Sometimes the rocket research aircraft were grounded for extended periods of time because of engine maintenance for their launch aircraft—the B-29 family always had a history of troublesome engine problems—or for water on the "dry" lakebed, which could be wet for considerable periods in the winter and spring. In sum, the X-series program, and especially the configuration explorers, did not constitute a drain or a waste of valuable research resources. In fact, quite the opposite is true.

Finally, one must remember that a most important NACA function was its communication of research results to industry and other government branches. The results of X-series research did not lie buried in the files of the High-Speed Flight Station but entered the technical literature through the standard NACA reporting format, chiefly the research memorandum (RM). Usually, slightly less than a year would pass between the gathering of results from a research flight and its publication in RM form. Informally, many NACA reports were circulated to the other NACA laboratories and to industry in advance of their actual publication date. Even the few critics of the X-series program admit that it was standard design practice for industry to rely on NACA reports for data and information. This same pattern was repeated with the reports generated by X-series testing, including tests of X-series aircraft in NACA wind tunnels.[88]

In conclusion, the X-series aircraft program and the extended NACA testing of these aircraft constituted an important and valuable aspect of post-1945 American aviation. The work that the NACA did on the early supersonic configuration testbeds gave the United States a commanding lead in the field of supersonic aircraft design so that by the end of the 1950s the military services were equipped in numbers with a wide range of combat aircraft capable of supersonic operation.

[88] Melvin B. Zisfein, a member of the F-104 design staff, clearly remembers the enthusiasm with which Lockheed designers welcomed the latest development reports on the X-3 from the NACA. Conversation with Hallion.

4

Through Mach 2 and 3
1951-1959

While exceeding the speed of sound had been a great unknown, there were other no less important unknowns involved at double or triple the velocity of sound. What particularly interested researchers were the potential problems of stability and control that might arise at Mach 2 or 3. They already recognized that above Mach 2, aerodynamic heating would become an increasingly serious problem to conventional aluminum aircraft structures and would require more exotic alloys and structural materials.

There was no organized program to develop specialized Mach 2 research vehicles, as there had been to develop the XS-1 and Skystreak, nor did there need to be, for the X-1 family proved perfectly amenable to the task, as did Skystreak's follow on, the D-558-2 Skyrocket. To see why this was so, consider the technical development of the X-1 and Skyrocket families.

When first designed by Bell, the original XS-1s were planned for a maximum speed potential of around Mach 2, thanks to a large fuel capacity that gave the craft about four minutes of powered flight time. But during development, troubles with a special kind of fuel pump (a turbine-driven device powered by steam generated by the decomposition of concentrated hydrogen peroxide passed over a catalyst) forced Bell to complete the first two airplanes with a fuel feed system incorporating high pressure nitrogen; this reduced the amount of fuel that could be carried and limited the design to a maximum speed of about Mach 1.45. Bell retained the third of the three planned XS-1s for later completion with a turbopump system, if it became available. Thus equipped, the X-1-3 (as it was known) would be capable of exceeding Mach 2, possibly reaching Mach 2.4. In 1948, the Air Force began development of the advanced X-1s, which incorporated turbopump fuel systems that were lengthened to give even greater fuel capacity for potential performance well in excess of Mach 2, possibly beyond Mach 2.5. The advanced X-1s were intended for Air Force military-related testing. With these aircraft under development, the Air Force lost all interest in the incomplete X-1-3, canceling its development. NACA interest in acquiring a Mach 2 X-1 to continue the work begun with its own X-1 caused the Air Force to reconsider its decision. The X-1-3 arrived at Edwards for contractor-testing in 1951, the same time that the first of the advanced X-1s (the X-1D) arrived at the dry lake.

The Skyrocket program had taken a different turn. Douglas had completed the first and second of the sweptwing planes with only jet engines, pending installation of a rocket engine in each when it became available. As events turned out, Douglas was not able to fly the plane with a rocket engine until 1949. After takeoff from the ground, even using both jet and rocket engines, the Skyrocket could reach only a disappointingly low maximum speed of Mach 1.08 in level flight at 40,000 feet. There were also safety problems. The heavily laden Skyrocket, brimming with fuel, required a 7,500 to 10,500 foot ground run for takeoff, imposing severe strain on its landing gear. Douglas sometimes fired the rocket engine to assist the takeoff, but this burned valuable fuel and limited the plane to about Mach 0.95 at altitude—a speed the jet only NACA airplane could already reach in a dive. NACA engineers recommended modifying the NACA Skyrocket to an all-rocket, air-launched research airplane. First, air-launching would improve safety. Second, the conserved rocket fuel would enable the plane to exceed Mach 1.5 far higher than it could attain from the ground. Third, an all-rocket version of the sweptwing Skyrocket could substitute in part for the lagging Bell X-2 program, already falling behind schedule (as will be discussed).[89] In 1949 Hugh Dryden proposed to the Navy that the NACA Skyrocket then being tested at Muroc be modified accordingly, giving it potential Mach 1.6+ performance. The Navy agreed to sponsor the project, amended the Skyrocket development contract, and in early 1950 the NACA Skyrocket left Muroc for the Douglas plant, returning as an all-rocket research airplane in November of that year.

The Year of Promise

The year 1951 offered the possibility that either the Navy-sponsored Skyrocket or the Air Force-sponsored advanced X-1 (the X-1D) would be the first aircraft to exceed Mach 2. The sense of scientific urgency attending the first Mach 1 flights did not exist for the Mach 2 mark, more of a psychological goal than a critical technological challenge. It was the NACA's nature to undertake a detailed flight-research program step by step, increasing Mach number slowly, until Mach 2 was attained. The NACA really did not have any control over the situation, except in the case of its own X-1-3, newly arrived at the dry lake and awaiting its first tests. Otherwise, exceeding Mach 2 was in the hands of the Navy and the Air Force. The NACA had always been torn between the public relations payoff of record-setting and the dangers it entailed for expensive research aircraft. With detailed research programs ready for the Skyrocket, the RAPP and Williams's station engineers could do little but watch and wait, concerned at the delay before they would get the plane and also concerned that this bit of rivalry between the Navy and the Air Force might lead to recklessness. Ironically, the goal others so eagerly sought was reached in 1953 by the NACA High Speed Flight Research Station.

The Douglas test team on the Skyrocket had the first shot at Mach 2. Throughout the spring and summer of 1951, Douglas test pilot Bill Bridgeman piloted the D-558-2 at increasingly higher speeds. On 7 August, he attained Mach 1.88

[89] It must be remembered throughout the story of the postwar high-speed aircraft research program that design of new fighter aircraft followed hard on the heels of flight-testing of research aircraft. Not until the large advances in research data provided by the X-15 in the 1960s did the data base outdistance the needs of the day. Throughout the 1950s, a gain of even a few months in the availability of high-speed research data could mean marked improvement in operational aircraft then in design.

The B-50 Superfortress launch vehicle hoisted above the Bell X-1-3 in preparation for mating. (NASA photo E-593)

(1,260 mph), well above the previous 957 mph attained by Chuck Yeager in the X-1, but still short of Mach 2. During its supersonic flights, the Skyrocket exhibited a highly objectionable and possibly extremely dangerous rolling motion (lateral instability) that was apparently aggravated by a basic flaw in the craft's dynamic-stability characteristics. The NACA studied its behavior in detail before attempting its own high Mach flights in the plane. Douglas wisely never attempted to go beyond Mach 1.88 delivering the plane to the NACA toward the end of the summer.[90]

All eyes turned to the Air Force and the X-1D. Though this plane had made but one contractor test flight—a glide flight at that—the Air Force was so eager to break Mach 2 that the test pilot, Frank "Pete" Everest, had been advised to "see what it could do wide open."[91] On 22 August 1951, the launch plane went up carrying the X-1D but had to cancel the planned launch because of mechanical problems. On the

[90] Various D-558-2 #2 progress reports, Dec. 1950-Aug. 1951 in NASA LaRC files; the Douglas program is discussed from Bridgeman's perspective in William Bridgeman and Jacqueline Hazard, *The Lonely Sky* (New York: Henry Holt and Co., 1955), passim; see also Hallion, *Supersonic Flight* , pp. 154-159.

[91] Frank E. Everest as told to John Guenther, *The Fastest Man Alive* (New York: Pyramid Books, 1959), p. 129.

way back to base, the X-1D exploded and caught fire, forcing the crew of the launch aircraft crew to jettison it hastily into the desert, fortunately without injury to anyone in the planes or on the ground. With the demise of the X-1D, Air Force hopes to break Mach 2 vanished before year's end. Then, on 9 November 1951, following a "captive" flight, the X-1-3 blew up under its own launch airplane, seriously injuring Bell test pilot Joe Cannon. Accident investigators blamed the loss of the X-1D on electrical ignition of fuel vapor and the loss of the X-1-3 on possible fracturing of a high pressure nitrogen gas storage system used to purge propellants from the rocket plane's tankage and propellant lines. While investigators may have been correct about the loss of the X-1-3 (independent testing by HSFRS confirmed the tendency of its nitrogen "bundle" to fracture when jolted, and scattered tubing was discovered as far as 250 feet from the accident site), such was not the case with the X-1D. The explosion of the X-1D was the first in a series of three accidents that finally would be attributed to explosive gasket material used in its liquid oxygen system.[92]

The year 1951 ended, then, with the loss of two valuable research planes and one launch aircraft and with the injury of a test pilot who fortunately recovered to fly again. Mach 2 remained unattainable for the near future, pending the arrival of the remaining advanced X-1s or the resumption of high Mach flights by the Skyrocket. In any case, the NACA was still several years away from acquiring a Mach 2 straight-wing research aircraft, a most frustrating and annoying situation.

Through Mach 2: the NACA in the Limelight

On 20 November 1953, NACA research pilot A. Scott Crossfield became the first human to exceed Mach 2. The NACA had seized the chance to surpass Mach 2 before the Air Force succeeded in doing so with the Bell X-1A. It came as a logical result of a two-year flight research program that had so thoroughly explored the Skyrocket's behavior above Mach 1 that no nasty surprises would await the Skyrocket team as the plane approached Mach 2.

The research had revealed that the Skyrocket's major difficulties above Mach 1 stemmed from dynamic instability when the plane flew at low angles of attack with low load factors—for example, pushing over into level flight from a climb while having less than a 1 g force on the airplane. Under these circumstances, the craft's lateral stability decreased markedly and it would manifest the dangerous rolling characteristics that Bridgeman had noticed during the Douglas program. In August 1953, Crossfield equaled Bridgeman's earlier Mach 1.88 mark. Then, the Navy entered the scene.

By the summer of 1953, the advanced X-1A had arrived at Edwards for testing, intensifying the racetrack atmosphere that permeated the base. Much like the Skyrocket

[92] The cause, discovered by HSFS in 1955, will be discussed subsequently in relation to NACA's X-1A program.

and X-1D rivalry of two years ealier, Skyrocket and X-1A seemed locked in a friendly but serious rivalry to be the first to exceed Mach 2. None of this would have meant much had the traditional NACA posture of leaving record-setting to others remained in effect. But now two factors changed this, one from the Navy, the other from inside the High Speed Flight Research Station itself.

The year 1953 held special significance for the American aviation community, for it was the 50th anniversary of the Wrights' first flight at Kitty Hawk. The Navy's Bureau of Aeronautics requested that Marine test pilot Marion Carl be allowed to make a series of high-altitude, high-Mach flights using the NACA's Skyrocket, support facilities, and launch team. Williams recognized that the flights were more for publicity than science. He had been petitioning NACA headquarters (to no avail) for permission to exceed Mach 2 for scientific purposes, and he was not enthusiastic about the Navy reentering the program with a new pilot. Williams sidestepped the Navy's first request. But the NACA was pursued by increasingly higher circles within the Navy. "School ties," Williams recollected later, "started flying all over the country." The Marine pilot's flights received NACA's go ahead.[93] There was a legitimate research objective: Carl would be testing an experimental pressure suit for high-altitude flight. Though he came close, Carl did not exceed Crossfield's speed mark even when he reached a new unofficial altitude record of 83,235 feet. By the end of August 1953, the Skyrocket once again seemed out of the running for Mach 2—unless the NACA tried its hand.

By then the record-setting bug clearly had bitten the NACA Skyrocket test team, especially its project pilot, Scott Crossfield. After Carl's flights, the Skyrocket team had added extensions to the nozzles on the plane's rocket engine, boosting its thrust by a small but important amount and preventing the exhaust flow from impinging upon the vertical stabilizer at supersonic speed, improving the plane's chances for Mach 2. On 14 October 1953, six years to the day since Yeager's historic flight, Crossfield touched Mach 1.96. The NACA pilot and plane were now the fastest in the world, but Hugh Dryden immediately clamped secrecy on the accomplishment and told Crossfield not to attempt Mach 2.[94] Dryden would have had to ground Crossfield and disband the rocket team to stop them, however. Mach 2 had become their Holy Grail.

Crossfield set out to work around Dryden's restriction. He approached an old friend who worked for the Navy's Bureau of Aeronautics. The friend spoke to former Navy test pilots at higher levels in the Pentagon. Within a week of Crossfield's entreaty, Hugh Dryden notified Williams that the Skyrocket was cleared to attempt a Mach 2

[93] Williams, conversation with Hallion.

[94] Crossfield's role in orchestrating the Mach 2 attempt is highlighted in A. Scott Crossfield with Clay Blair, Jr., *Always Another Dawn: The Story of a Rocket Test Pilot* (New York: The World Publishing Co., 1960), pp. 167-169. See also Mulac, D-558-2 flight chronology. As the negotiator of the interagency agreements on the research aircraft program, Hugh Dryden was well aware of the services' intention that records were to go to them, data to the NACA.

flight. Of course, Williams had been pressing for such a clearance for months as the next logical step in the ongoing high-speed investigations.[95]

The race was on. Over on the Air Force side of the base, a technical team readied the X-1A. The Air Force had definite plans to exceed Mach 2 before the anniversary of the Wrights' flight and, of course, before the Skyrocket as well, if possible. Yeager, the Air Force's best rocket-airplane pilot, had been instructed to make the attempt. But the X-1A was brand-new and required a lot of preparation. The Skyrocket team, on the other hand, was used to operating its aircraft and had learned its operating quirks and problems. The Skyrocket had the first crack at the Mach 2 mark, and the NACA team did not miss its shot. In preparation for the flight, project engineer Herman O. Ankenbruck computed an optimal flight path for the aircraft so that it would waste neither fuel nor energy. Technicians chilled the plane's alcohol fuel so that the craft could carry more of it and then insulated it by taping all panel cracks before covering the plane with a coat of wax.

At midmorning on 20 November 1953, the Skyrocket took off from Edwards under its Superfortress launch aircraft. The climb to launch altitude took over an hour, during which time Crossfield—sick with flu—entered the plane and readied it for flight. Finally came the launch, and the Skyrocket dropped away from the bomber, its sleek waxed shape glistening in the sun. Crossfield fired the engine and began its carefully programmed climb, neither too steep nor too shallow. At 72,000 feet, he began a pushover into level flight, continuing until the Skyrocket was in a shallow dive. The Machmeter edged toward 2. Everything worked: the nozzle extensions provided extra thrust, Ankenbruck's flight plan was the right one, the engine ran longer than normal because of the extra fuel, and Crossfield's piloting was excellent. At 62,000 feet, the Skyrocket nosed past Mach 2, reaching Mach 2.005. The engine continued to run for a few more seconds before starving itself. The deceleration jerked Crossfield forward in the straps, the plane having a lot of drag. He edged out of the shallow dive and set up a deadstick approach to the dry lake. While coasting down, he exuberantly victory-rolled the airplane before landing. The Skyrocket never again approached Mach 2; the NACA could not again justify the extensive preparations. In any case, the plane simply had no additional performance left in it. It soldiered on for a few more years in less exciting but still important research tasks until its retirement in mid-1957. The X-1 had reached Mach 1, but Mach 2 belonged to the Skyrocket.[96]

The Air Force was not about to let the NACA's record stand for any length of time. By early December, Yeager was fully checked out in the X-1A. The friendly rivalry between the Skyrocket and X-1A teams at Edwards in 1953 did not damage the close cooperation between the NACA and the Air Force on the actual flight-testing of the rocket airplanes. The X-1A depended for instrumentation support upon the

[95] Crossfield with Blair, pp. 168-169. Williams's conversation with Hallion.

[96] Crossfield with Blair, pp. 171-178; James A. Martin, "The Record Setting Research Airplanes," *Aerospace Engineering* (Dec. 1962), p. 51; for Ankenbruck's role, see Hallion's comments in *Toward Mach 2: The Douglas D-558 Program*, ed. J. D. Hunley (Washington, DC: NASA SP-4222, 1999), p. 34.

The NACA's D-558-2 #2 Skyrocket, piloted by Scott Crossfield, was the first aircraft to fly twice the speed of sound. (NASA photo E-1442)

NACA station, and though the NACA engineers were not able to instrument the aircraft as thoroughly as they would have had it been a NACA vehicle, they did install an airspeed altitude recorder and an accelerometer that later and unexpectedly proved quite valuable. They also provided radar tracking for the flight. On 12 December 1953, Yeager set out to break the Skyrocket's records.[97]

Were this simply a good story, it would play little part in the history of Dryden; but Yeager's flight was one of the most significant of the early rocket flights. It highlighted the serious stability difficulties that could be encountered at Mach 2 speeds, a subject of vital interest to NACA and particularly the Edwards station.

NACA Langley wind-tunnel studies, data taken from previous flights, and analog simulations using a Bell Corporation performance analyzer led program engineers to suspect that the X-1A and other advanced X-1 aircraft would have rapidly deteriorating directional stability above Mach 2.3. During Yeager's flight, the X-1A reached Mach 2.44 at an altitude of about 75,000 feet. At that altitude, despite the plane's speed, the

[97] Edwards AFB Historical Report, 1 July - 31 Dec. 1953, in the files of Air Force Materiel Command Historical Office, Wright-Patterson AFB, OH, and the Air Force Historical Research Agency, Maxwell AFB, AL; HSFRS annual report to the RAPP for 1953, p. 3, in the DFRC Historical Reference Collection.

The Bell X-1A glides to a landing at Edwards after a research flight, trailed by a North American F-86 Sabre flying chase. (NASA photo E-2490)

dynamic pressure was so low that the X-1A's controls were not completely effective in damping any sudden motions the craft might begin. The expected deterioration in directional stability simply reflected the need for much larger vertical fins for high-speed flight. Now, at Mach 2.44, the plane suddenly went out of control, beginning a slow roll to the left. As Yeager corrected for this, the roll reversed and the plane began a rapid roll to the right. He attempted to correct for the second departure from his intended flight path, but the X-1A violently snapped to the left and then tumbled completely out of control, throwing Yeager about the cockpit. In the process, he cracked the inside of the canopy with his helmet. In 51 seconds the aircraft fell about 50,000 feet, decelerated from 1,600 to 170 mph, and encountered accelerations of more than 8 g. As Yeager kept seeing the Sierra Nevadas flash by, he wondered where the plane would hit. It eventually wound up in an inverted spin. Thanks to his consummate piloting skills, Yeager was able to recover into a normal (upright) spin, and then into level flight at very low altitude. He glided back onto Rogers Dry Lake. The NACA accelerometer told an eloquent story of the forces the plane had encountered.

The High-Speed Flight Research Station issued a summary report by Hubert M. Drake and Wendell H. Stillwell, explaining as fully as possible the difficulties likely to be encountered by a straight-wing airplane with a relatively small vertical tail area during a flight in the tricky regions near Mach 2. The X-1A flew one more time above Mach 2.[98] Its sister ship, the X-1B, made one flight to Mach 2.3 a year later. Its wings rocked as much as 70 degrees before test pilot Pete Everest cautiously slowed the plane and regained stability. The advanced X-1s might be capable of reaching Mach 2 safely, but any edging beyond was risky at best.[99]

So 1953 ended with Mach 2 having been attained a mere six years after the X-1 had achieved Mach 1. As with the first Mach 1 flights, however, the attainment of Mach 2 still left a great deal of research to be done on particular flight conditions at this speed. Indeed, the detailed work still lay in the future, with other aircraft programs. The NACA looked forward, for example, to the X-1A, which it hoped to use in a detailed program of high-altitude Mach 2 research. The NACA also hoped to use the X-1B for a study of aerodynamic heating conditions near Mach 2. But events can have a funny way of working out. In this case, the NACA's plans would fall completely apart.

The Demise of the X-1A

In 1951, the X-1D and X-1-3 had blown themselves out of existence, and in May 1953 the second X-2 did likewise. The X-2's accident was truly tragic, for the explosion occurred as it was being carried in the bomb bay of its Superfortress mothership over Lake Ontario. The rocket plane vanished in a fiery red blast that killed its Bell test pilot and another Bell flight crewman. The launch plane returned to base, mangled by the blast. On 8 August 1955, it was NACA's turn, with the X-1A.

The NACA High-Speed Flight Station had made only one flight with the X-1A before the accident. That day, at 31,000 feet and less than one minute from launch, the X-1A's liquid oxygen tank burst from an internal low order explosion, expelling a shower of debris that fractured the canopy of one of the chase planes. NACA test pilot Joe Walker scrambled back into the B-29, and the Superfortress's crew began a steady descent, anxiously watching the steaming rocket plane. For a while, it appeared that they might be able to land; Dick Payne, the X-1A's crew chief, entered the X-1A's cockpit to jettison the remaining fuel. However, the blast had also caused the rocket plane's landing gear to extend, making a landing attempt questionable. For over half an hour, the Superfortress and its potentially deadly cargo cruised east of Rogers Dry Lake, as the NACA flight crew pondered what to do, with Joe Vensel and Scott Crossfield offering advice from the ground. But there was no real option. Resigned, Vensel radioed Stan Butchart, the B-29 pilot, "Butch, you might as well drop it. Pick a good place." They did, over the Edwards bombing

[98] NACA RM-H55G25, "Behavior of the Bell X-1A Research Airplane during Exploratory Flights at Mach Numbers near 2.0 and at Extreme Altitudes," NACA HSFS, 7 July 1955. Like RM-H55A13 (the X-3 and F-100 stability study), this RM was widely circulated throughout industry and was very influential.

[99] Details on Yeager's flight are from ltr., Yeager to author, 12 Oct. 1971; RM–H55G25, *passim*; Lundgren, *Across the High Frontier*, pp. 278-284; and the HSFRS annual report to the RAPP for 1953, pp. 3-4.

range.[100] The X-1A entered a flat spin and fell into the desert, exploding in an orange ball of flame and starting a small brush fire.[101]

And now the task of sorting out the cause began. Walt Williams was away fishing in the mountains, so De Beeler formed an accident board under his direction, consisting of representatives from the High-Speed Flight Station, the Air Force Flight Test Center, Bell Aircraft Corporation, the Air Force Office of the Inspector General, NACA's Langley laboratory, and the Air Force's power plant laboratory at Wright Patterson Air Force Base (AFB). Fortunately, the X-1B, sister of the X-1A, was available for examination, having just returned from Langley where it had been instrumented for aerodynamic heating studies. The X-1A's wreckage—what was left of it—was placed in the HSFS loads calibration hangar, and the X-1B was wheeled alongside it for comparison. Investigators quickly ruled out electrical detonation of fuel vapor (blamed previously for the loss of the X-1D), or fatigue-fracturing of the liquid oxygen tank. The tank pressure regulators were recovered in good condition, ruling out inadvertent overpressurization. The craft's nitrogen tanks had survived the ground impact, so they could not have triggered the explosion. Yet something had blown the oxygen tank apart.

HSFS staff member Donald Bellman discovered the vital clue. As members of the NACA board peered into the liquid oxygen tank of the X-1B, they noticed a slimy, oily residue coating the bottom of the tank. "What's that?" one of them asked. "Oh," a Bell representative replied, "We have that all the time. We just wipe it out." Suspicious, Bellman gathered up the sludge in small bottles and sent one sample to a highly touted laboratory in Los Angeles, another to the Air Force's chemical laboratory at Edwards. The Los Angeles laboratory returned a superficial report stating, in essence, that the residue was a hydrocarbon product that had no business being around liquid oxygen. However, the Air Force's chemists did a detailed analysis, identifying the substance as TCP (tricresyl phosphate), a substance used to impregnate leather. All of the rocket planes, those that had been destroyed, as well as those still flying, had gaskets made of Ulmer leather—leather impregnated with a 50-50 mix of TCP and carnauba wax.

Subsequent experiments showed that when compressed between flanges and allowed to stand overnight at room temperatures, the TCP would separate from the leather and wax, running and pooling as it had in the X-1B's lox tank. Commercial bottled gas experts informed Bellman that at high pressures and low temperatures, Ulmer leather could be extremely dangerous, exploding at a comparatively low impact. As early as 1950-1951, this information, on the basis of laboratory tests, had been known to commercial bottled gas companies. Bellman supervised construction of a

[100] Subsequently Butchart, Payne, and Walker received the NACA Exceptional Service Medal for their work in this connection. B-29 crewmen Charles Littleton and John Moise received the NACA Distinguished Service Medal, and crewmen Jack McKay, Rex Cook, Richard De More, and Merle Woods received letters of commendation. NACA also commended chase pilot Maj. Arthur "Kit" Murray (pilot of the damaged chase plane) in a letter from the Committee chairman to the secretary of the Air Force.

[101] Flight and accident details are from NACA HSFS, "Report of Investigation into the Loss of the X-1A Research Airplane on August 8, 1955," Nov. 1955; copy in DFRC Historical Reference Collection.

test apparatus to drop a five pound steel bar 10 feet onto lox-soaked samples of Ulmer leather and on frozen drops of TCP. The results of 30 tests were 30 explosions.[102]

The accident board theorized that when the gaskets compressed under pressure, the TCP exuded and ran into all available crevices. In the supercold environment of the lox tank, abrupt movements of the tank bulkhead or lox tubing could detonate this residue. Reexamination of the other rocket airplane explosions found a lot of supporting evidence for the Ulmer leather theory, especially in the location and sequence of the explosions. The board's final report blamed explosive gaskets for the loss of the X-1A and concluded that it could have caused the previous explosions as well. Thus was identified the probable cause of a series of accidents that had cost two lives and one serious injury, the destruction of four rocket research airplanes and two launch aircraft, and a two-year delay in the first Mach 2 flight.[103] Never again did any of the early rocket research aircraft suffer a catastrophic blast. The Air Force, anxious to begin flight-testing on the more powerful X-2, went ahead with renewed confidence on that behind-schedule program. The NACA, on the other hand, was still frustrated—it had lost yet another Mach 2 X-1. The program so carefully planned for the X-1A had to be abandoned, with some portions taken over by the X-1B and others by the X-1E, a Mach 2 "homebuilt" designed at the HSFS. (These two programs will be discussed subsequently). Meanwhile the Air Force, as the possessor of the fastest flight research aircraft at the dry lake, the X-2, set its sights on Mach 3.

The *Götterdämmerung* of the X-2

No program caused the NACA, especially the engineers of the High-Speed Flight Station, more frustration and disappointment than the X-2. It highlighted the terrible effects of underestimating the technical complexities involved in developing a radical new aircraft. It also highlighted the dangers of succumbing to the pressure to set records in the guise of research. The X-2 program was in many ways a failure, despite achieving both altitude and speed records. It failed to return as much high-speed aerodynamic heating information as had been anticipated from the program, but the X-2 program did yield some information in that area and on conditions of flight at high altitudes. Two X-2 aircraft were built; both were destroyed and caused three fatalities.

The X-2 was the most exotic and complex of the early rocket-propelled research aircraft. Designed for supersonic tests of the sweptwing shape, the plane had an estimated performance in excess of Mach 3. The first plane designed to withstand the rigors of aerodynamic heating, its structure was fabricated from stainless steel and a nickel alloy. To be air-launched and propelled by a two-chamber rocket engine, it would land on retractable landing skids. Bell had hoped to complete the first aircraft in 1948, but construction delays caused by the complex alloy structure and problems

[102] Ibid.; interview with Donald R. Bellman, 4 Mar. 1977, an unedited transcript of which is available in the DFRC Historical Reference Collection.

[103] Both sources listed previously (notes 101 and 102).

with its explosion prone Curtiss-Wright 15,000-pound rocket engine stretched the development program by years.

As time went on, NACA's interest in the airplane declined markedly. By 1953, much of the sweptwing information that the X-2 could have provided had already been derived from the Skyrocket. Initial glide trials with the first of two X-2s took place in 1952, demonstrating that the plane flew well at low speeds. (Its engine was still not ready for installation). The loss of the second X-2 in 1953 over Lake Ontario delayed the program even more. Problems with its planned electrical flight-control system forced a change to a conventional hydromechanical system patterned on that of the F-86 fighter. The sole surviving X-2 flew again on another series of glide trials in 1954—still lacking its rocket engine—which forced redesign of the landing skids and shock-absorbing strut system. At last, the Curtiss-Wright engine was ready for installation, and the X-2 arrived back at Edwards in the summer of 1955, ready for its powered flight trials. Then the loss of the X-1A and the subsequent accident investigation grounded the X-2 for replacement of its dangerous gaskets.

Management responsibilities for the X-2 lay between the Air Force and Bell. The NACA participated in some X-2 support research, primarily Langley wind-tunnel studies and Wallops rocket model tests, and the RAPP made many recommendations, suggesting unsuccessfully that its trouble-prone Curtiss-Wright engine be replaced. By October 1955, the Air Force had lost patience with the program and issued an ultimatum: if the X-2 did not complete a powered flight before the end of the year, the project would be terminated.[104] The NACA still retained a little enthusiasm for the plane, wanting it for aerodynamic and structural-heating studies. The X-1B was making similar studies, but the X-2 could go far beyond the X-1B's speed, up to Mach 3. Even though it recognized that the X-2 would soon be overshadowed by the X-15 then under development, the NACA still believed that the near-term availability of the plane would furnish much information unavailable from other flight-testing programs on the heating conditions encountered at Mach 3.

The X-2 completed its first powered flight on 18 November 1955. Piloted by Air Force test pilot Pete Everest, it featured a brief but not damaging fire in the engine bay. Nevertheless, the Air Force ruled the test a success, giving the program its reprieve. For various reasons, the plane did not fly again until March 1956. During these Air Force trials, the plane remained the property of the Bell Aircraft Corporation, which did not deliver it to the Air Force until 23 August 1956. Walt Williams and his engineering staff, watching patiently from the sidelines, were occasionally asked to furnish technical assistance.

On the advice of the NACA, the Air Force had bought a special computer, the Goodyear Electronic Differential Analyzer, which would predict aircraft behavior by

[104] Everest and Guenther, pp. 197-211; Crossfield and Blair, pp. 149-152, 213; *Ad Inexplorata: The Evolution of Flight Testing at Edwards Air Force Base* (Edwards AFB, CA: AFFTC History Office, 1996), p. 17. There is now also Henry Matthews, *The Saga of Bell X-2, First of the Spaceships: The Untold Story* (Beirut, Lebanon: HPM Publications, 1999), is a useful source of information about the program.

The ill-fated Bell X-2 rocket research aircraft after drop from the B-50 mothership. (NASA photo E-2820)

extrapolation of results from test flights. This would give engineers and pilots some indication of what to expect as they flew higher and faster. Richard Day, the HSFS program engineer for the X-2, helped with the new computer, providing equations and motions data. Day routinely briefed project pilot Pete Everest and, later in 1956, Iven Kincheloe and Mel Apt, Everest's replacements, all of whom "flew" the simulator.[105]

The simulations confirmed predictions from NACA wind-tunnel tests that the X-2 would have rapidly deteriorating directional and lateral (roll) stability near Mach 3. Aileron deflection (to roll the plane) could lead to an aerodynamic condition known as "adverse yaw," followed by increasingly rapid rolling until the rolling motions reached a "critical roll velocity," the point where the plane would roll into inertial

[105] Everest and Guenther, pp. 211-218; interview with Richard Day, 24 Feb. 1977; Richard Day and Donald Reisert, "Aerodynamics Section: History of Events Prior to the X-2 Flight of September 27, 1956," *X-2 Accident Investigation* (Edwards: HSFS, 17 Oct. 1956), pp. 1-2; interview of Williams and subsequent conversations; Richard E. Day, *Coupling Dynamics in Aircraft: A Historical Perspective* (Edwards, CA: NASA SP-532, 1997), pp. 8-11; personal accounts of Day and Donald Reisert, who worked with Day on the GEDA simulator, in Gene L. Waltman, *Black Magic and Gremlins: Analog Flight Simulations at NASA's Flight Research Center* (Washington, DC: NASA SP-2000-4520), pp. 137-143.

coupling and tumble. During 1956, as Pete Everest moved up in speed in the airplane, NACA's Dick Day and Hubert Drake anxiously watched the directional stability curves, compared them to flight data, and urged the Air Force to move in smaller increments, not in great leaps of half a Mach number.[106] In May 1956, Everest achieved Mach 2.53, making the X-2 the fastest aircraft in the world. By this time, the NACA's patience was running somewhat thin. In early June, at a meeting Of NACA, the Air Force and Bell at Edwards, the NACA representatives requested that the X-2 be delivered to NACA sometime between 15 September and 1 October so that the High-Speed Flight Station could complete a few flights before winter rains flooded the lakebed. The Air Force agreed, stating that their program would be "to expand the speed and altitude envelope to at least nominal values"—100,000 feet and Mach 3.[107] Everest came close to this on 23 July, when he reached Mach 2.87 (roughly 1,900 mph), his last flight before moving to a staff assignment in Norfolk, Virginia.

Following Everest's final flight, the Air Force momentarily lost interest in Mach 3 in favor of attaining the craft's maximum altitude. Test pilot Iven Kincheloe flew the plane to 126,200 feet, the first flight above 100,000 feet. At that altitude, aerodynamic controls were useless. The X-2's behavior in this region of low dynamic pressure ("low q" in engineer's shorthand) pointed to the need for reaction controls. Above 100,000 feet, still in a ballistic arc, the X-2 began a left bank which Kincheloe wisely did not attempt to correct, for fear of tumbling the airplane. He experienced less than 0.05 g for approximately 50 seconds, a foretaste of weightless spaceflight; popular science writers dubbed the pilot the "First of the Spacemen."[108] In late August, the Air Force had taken delivery of the X-2 and then extended its program for an additional month (before the plane would be turned over to the NACA), the purpose being "to obtain an incremental value of the high speed performance of the X-2 airplane."[109] Into the cockpit stepped a new Air Force pilot, Capt. Milburn G. Apt.

Though he had flown chase on many X-2 missions, Mel Apt had never flown a rocket-powered airplane. He was perhaps the most experienced pilot at Edwards on the phenomenon of inertial coupling, having flown many inertial coupling research flights in the F-100 fighter. Apt had received computer-based briefings on 29 July and 24 September, but the briefings had a flaw. The X-2 flights had accumulated useful data only up to Mach 2.4. Engineers extrapolated all data beyond that, and the predictions were dubious. One study, at a simulated Mach number of 3.2 at 70,000 feet, showed the aircraft "diverging" (going out of control) during lateral (rolling)

[106] Interview of Day; Day and Reisert, pp. 1-2.

[107] Ronald Stiffler, *The Bell X-2 Rocket Research Aircraft: The Flight Test Program* (Edwards: AFFTC, 12 Aug. 1957), pp. 61-62.

[108] I. C. Kincheloe, "Flight Research at High Altitude, Part II," *Proceedings* of the Seventh AGARD General Assembly, the Washington AGARD conference, 18-26 Nov. 1957; Stiffler, pp. 70-72; James J. Haggerty, Jr., *First of the Spacemen: Iven C. Kincheloe, Jr.* (New York: Duell, Sloan and Pearce, 1960), pp. 107-121.

[109] Memo, Richard E. Day to Walter C. Williams (WCW), 28 Sept. 1956.

maneuvers. Being extrapolations, none of these studies could be conclusive. Apt had also flown the simulator and practiced recovery methods applicable to inertial coupling, but the simulator was new, and he may have distrusted it. In any event, on 27 September 1956, Mel Apt dropped away from the Superfortress mothership in the X-2 at 8:49 a.m. His flight plan called for "the optimum maximum energy flight path," one certain, if successful, to exceed Mach 3. However, in a postflight question and answer session, a senior program official said, "Captain Apt was instructed to make no special effort to obtain maximum speed but rather to stay within previous limits and to concentrate on the best flying technique possible."[110] Clearly some confusion existed in the minds of mission planners. And, then, there was the matter of experience. Apt had not even had the benefit of a glide flight in the X-2. His sole time in the cockpit was spent in several ground engine runs and posing for publicity photographs with Kincheloe. He had been cautioned to decelerate rapidly if he encountered stability difficulties and not to make rapid control movements above Mach 2.7.

As Apt climbed away after launch, he followed a predetermined schedule matching the airplane's g loading versus altitude, based on code numbers radioed from ground radar tracking. He reached high altitude, nosed over and dived past Mach 3, reaching Mach 3.2 (over 2,000 mph) at about 65,500 feet. His rocket engine burned for another 10 seconds, longer than previously. The flight had been flawless, but victory suddenly turned to ashes. Apt began an abrupt turn back for the lakebed. Perhaps he believed the X-2 was traveling slower than it was. Like all early X-series aircraft, the X-2 had lagging instrumentation. The cockpit camera film showed the Machmeter indicating Mach 3 for over 10 seconds. As the X-2 turned, it started a series of rapid rolls and the "critical roll velocity," an engineering construct, now became a brutal reality. The X-2 coupled, tossing Apt violently about the cockpit. Apt tried to regain control, then jettisoned the craft's nose section in preparing to bail out. The drag chute opened, and Apt succeeded in releasing the Plexiglas canopy, but time had run out. Before he could step out of the capsule and activate his personal parachute, the X-2's nose plunged into the desert, killing him instantly. The rest of the X-2 spun into the desert five miles away. Barely three minutes after launch, Mel Apt had become the first pilot to reach Mach 3, and then died. Kincheloe's voice continued on the radio, "Mel, can you read me, Mel?"[111]

A valued pilot had died. A research airplane had crashed just as it might have begun fully justifying its development. A record had been set, but to little purpose. The accident illustrated the acute need for reliable cockpit instrumentation for high-speed flight research, and this eventually helped spawn the special gyro stabilized inertial guidance system used on the X-15. Some tried to point to "research

[110] Stiffler, pp. 81-82; Day's memo to WCW; Day and Reisert, *X-2 Accident Investigation*, pp. 1-2; Day, *Coupling Dynamics*, p. 11; Day's personal account in Waltman, *Black Magic*, pp. 138-140.

[111] Stiffler, pp. 81-82; Day and Reisert, pp. 1-2; Haggerty, pp. 122-133; e-mail correspondence, Richard E. Day to J. D. Hunley, 20 July 1997. A lot of technical data on the accident is now available in Day, *Coupling Dynamics*, pp. 8-15.

accomplishments" of the X-2, citing limited heating data acquired from seared samples of temperature-sensitive paint, data that could have been acquired more easily from rocket models. In reality, its research was limited. Groping for significance, the Edwards historian asked one program official, "I imagine the X-2 program contributed greatly to aeronautical knowledge, didn't it?" "More than ever before," answered the official, "we appreciate the requirement of providing the pilot with the information he needs to do his job."[112] Back in Washington, the NACA staff fired off a series of messages to Walt Williams, fearful lest the High-Speed Flight Station had condoned the flight. One, from Hugh Dryden's deputy, got right to the point:

WHAT DOES OPTIMUM MAX ENERGY FLT PATH MEAN
PD SGND CROWLEY

The Air Force Flight Test Center issued its accident report in November 1956, concluding that the fatal turn at peak velocity had led inevitably to coupled motion instability.[113]

The loss of the X-2 once again robbed the NACA of a research tool just at a time when it might have proved worthwhile. Previously, the NACA had lost its planned programs on the X-1-3 and X-1A because of the gasket explosions. The X-2 fiasco removed the last chance to get substantial Mach 3 heating data prior to the X-15. The NACA had to make do with the X-1B, capable only of approaching Mach 2, an unpleasant price to pay for a speed record. It was particularly galling because Apt's flight was to have been the last Air Force flight before the X-2 was turned over to the NACA. The only consolation was that the tragic accident, together with simulator studies, contributed to an understanding of coupling dynamics.[114]

Otherwise, the X-2 program was a disaster masquerading as a research program, and subsequent program reviewers could not ignore the facts. The Air Force's program historian argued that Mel Apt had certainly needed at least one low supersonic familiarization flight in the X-2, questioning why "a pilot with limited experience like Captain Apt [was] shoved into the cockpit of the X-2 on an optimum flight at the last minute."[115] The NACA would certainly have agreed with his overall conclusion:

> Only one conclusion can be reached and that is that the Air Force in its determination to attain a record speed and altitude with the X-2 which it did achieve assumed a calculated risk of losing the pilot and the aircraft in the process . . . . Fatigue, miscalculations, and poor judgment entered into the

[112] Stiffler, p. 87.

[113] Crypto message, Crowley to WCW, 2 Oct. 1956; crypto message, Williams to Dryden, 28 Sept. 1956; letter, Williams to Dryden, 3 Jan. 1957; memo, Day to Williams, 2 Nov. 1956; interview of Williams and subsequent conversations.

[114] See Day, *Coupling Dynamics*, esp. pp. 4, 8-15.

[115] Stiffler, p. 28.

program at a time when unhurried flights were in order and good judgment should have directed and supervised the program.[116]

Round One's Twilight Years

The NACA High-Speed Flight Station continued flying the two remaining rocket research airplanes, the X-1B and the X-1E, until mid-1958. Both assumed some of the tasks envisioned for the lost X-1A and X-2, but they also took on new ones as well. After the Air Force had delivered the X-1B to the High-Speed Flight Station in 1954, the NACA shipped the airplane to Langley for installation of 300 thermocouples and related instrumentation for measuring structural temperatures. The X-1B had arrived back at Edwards in time to assist in the X-1A accident investigation and did not fly until August 1956, embarking on its heating research program the next month. In January 1957, NACA test pilot Jack McKay extended the investigation to Mach 1.94, bringing the program to a conclusion. Project engineers believed the data to be representative of heating conditions that could be expected on future Mach 2 military aircraft. The maximum heating rate experienced was about 34 degrees Fahrenheit (F) per second, with a maximum skin temperature of 185 degrees F being recorded on the forward point of the nose. Internal heat "sinks" and sources appreciably affected skin temperatures. While the skin next to the liquid oxygen tank had a temperature of only 50 degrees F, that just ahead of the tank was 122 degrees F. The flight results generally agreed with estimated temperatures derived by calculation. This X-1B study was the first major aerodynamic heating flight research study undertaken in the United States, and, alas, was a good example of the kind of work the High-Speed Flight Station had expected from the X-2.[117]

Iven Kincheloe's high-altitude flight in the X-2 demonstrated the inadequacies of conventional aerodynamic controls for flight in regions of low dynamic pressure. One solution was the installation of small reaction control thruster jets for maintaining proper vehicle attitude in regions of low dynamic pressure, or "low q." In 1956 the High-Speed Flight Station began researching reaction controls in support of the X-15 program. Writing nearly a decade later, engineer Wendell Stillwell of the HSFS stated, "The transition from aerodynamic control to jet control loomed as the most difficult problem for this vast, unexplored flight regime."[118] The X-1B offered an ideal testbed for a trial reaction control installation. In preparation, HSFS technicians built an iron frame simulator, dubbed the "Iron Cross," which matched the dimensions and inertial characteristics of the X-1B, installing small reaction control thrusters on it and then

[116] Stiffler, p. 30.

[117] Richard D. Banner, NACA Research Memorandum RM-H57D18b, "Flight Measurements of Airplane Structural Temperatures at Supersonic Speeds," NACA HSFS, 7 June 1957; X-1B progress reports, Oct. 1955 to Jan. 1957; Mulac, X-1B flight chronology.

[118] Wendell H. Stillwell, *X-15 Research Results* (Washington, DC: NASA SP-60, 1965), p. 26.

mounting it on a universal joint so that a test pilot could maneuver it in pitch, roll, and yaw. NACA's test pilots "flew" this simulator extensively. In November 1957, the NACA finished installing reaction controls on the X-1B itself, and test pilot Neil A. Armstrong made three flights in the plane before it was grounded in the summer of 1958 because of fatigue cracks in its fuel tank. The NACA subsequently transferred the reaction control research program to a Lockheed F-104 Starfighter. This aircraft and its reaction control system played a major role in training pilots for the X-15.[119]

The X-1E was the last of the hardy X-1 breed to retire. An extensive modification of the NACA's original X-1 aircraft, the X-1E had been rebuilt with a low pressure fuel system and a special low aspect ratio wing having a thickness/chord ratio of only 4 percent. Much of the design work was undertaken by the High-Speed Flight Station staff, which saw the craft as an opportunity to get information at speeds above Mach 2 on this wing configuration—similar to that of the X-3—information that the X-3 had been unable to obtain because of its inadequate propulsion system. The wing had no less than 200 pressure-distribution-measurement orifices cut into it, as well as 343 strain gauges baked into the wing surface for structural load and heating research. At one point, after the loss of the X-2, NACA engineers Hubert Drake and Donald Bellman proposed boosting the X-1E's engine performance to enable the plane to reach Mach 3, but NACA opted to wait for the X-15 instead. The X-1E suffered hard luck in its flight research program, experiencing two landing accidents, one of them severely damaging the airplane. It did complement the heating research undertaken by the X-1B, but by the time of its flight trials, the Lockheed F-104 with a generally similar wing configuration was already flying and could more easily acquire data at Mach 2. The rocket aircraft required time-consuming preparations. As research engineer Gene Matranga recalled, "We could probably fly the X-1E two or three times a month, whereas Kelly [Johnson] was flying his F-104s two or three times a day into the same flight regimes, so it really didn't make sense for us to be applying those kind[s] of resources to [obtain] that kind of information."[120] The X-1E completed its last (26th) NACA flight in November 1958. It is now permanently exhibited in front of the Dryden center, perched at a jaunty angle.

Round One in Retrospect

The conclusion of Round One in 1958-1959 brought the era of the supersonic breakthrough to a close. Figure 1 compares rocket research aircraft, military fighter prototypes, and military fighters in service. As figure 1 indicates, the X-series never led the prototypes by less than 0.6 Mach; by 1956, this had increased by a whole

[119] Interview with Neil A. Armstrong, 26 Jan. 1972; X-1B progress reports, June 1957 to June 1958; Hubert M. Drake, "Flight Research at High Altitude Part I," *Proceedings* of the Seventh AGARD General Assembly, the Washington AGARD Conference, 18-26 Nov. 1957.

[120] Interview with Gene Matranga, 3 Dec. 1976; Williams and Drake, "The Research Airplane," pp. 36-41; X-1E progress reports, 1955-1958 inclusive.

Mach number. The differences between X-series performance and aircraft in service at the time is even more pronounced (for example, the X-1A vs. F-86F Sabre of 1953).

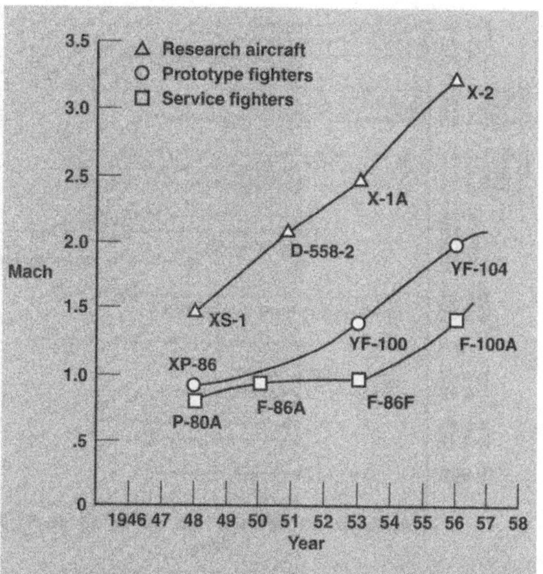

Figure 1. *Leader-follower relationship among research aircraft, military fighter prototypes, and military fighters in service.*

Figure 2 places the rocket research aircraft program within the context of speed trends throughout aviation history. The figure illustrates the interesting relationships among different growth curves. Notice that as piston engine technology approached its limits, a new technology revolutionized the field, the jet engine. The rocket research aircraft curve exhibits rapid growth over a short period of time. The jet fighter curve is a classic example of the "S," or biological curve: slow progress initially ("infant problems"), a period of very rapid growth (from Mach 1 to Mach 2), and then, beginning just beyond Mach 2, the rate of development slows because of a variety of factors, including propulsion efficiency, aerodynamic and heating constraints, cost of such complex systems, and questionable mission utility above Mach 2. Several leader-follower relationships are illustrated: piston military fighters led piston transports, jet fighters led jet transports (a continuance of the earlier trend), and the rocket research aircraft led the development of jet fighters.[121] Significantly, the growth curve for the rocket research aircraft is open-ended. Beyond

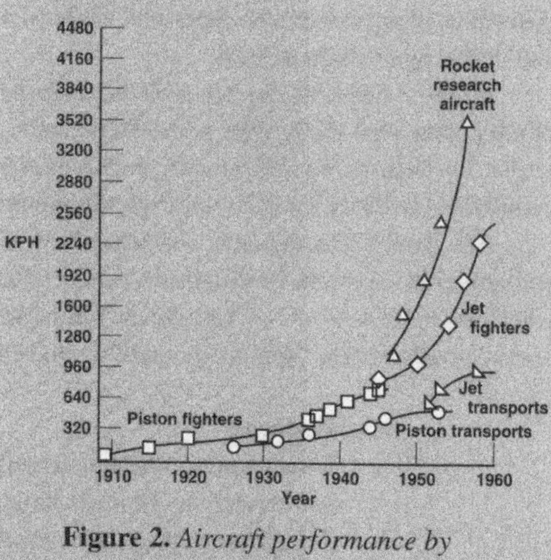

Figure 2. *Aircraft performance by aircraft category.*

[121] Applicability of growth and trend curves to the study of technological questions and as exploratory forecasting methods is examined by Joseph P. Martino, "Survey of Forecasting Methods, Part I," *World Future Society Bulletin* (Nov.-Dec. 1976), pp. 4-7.

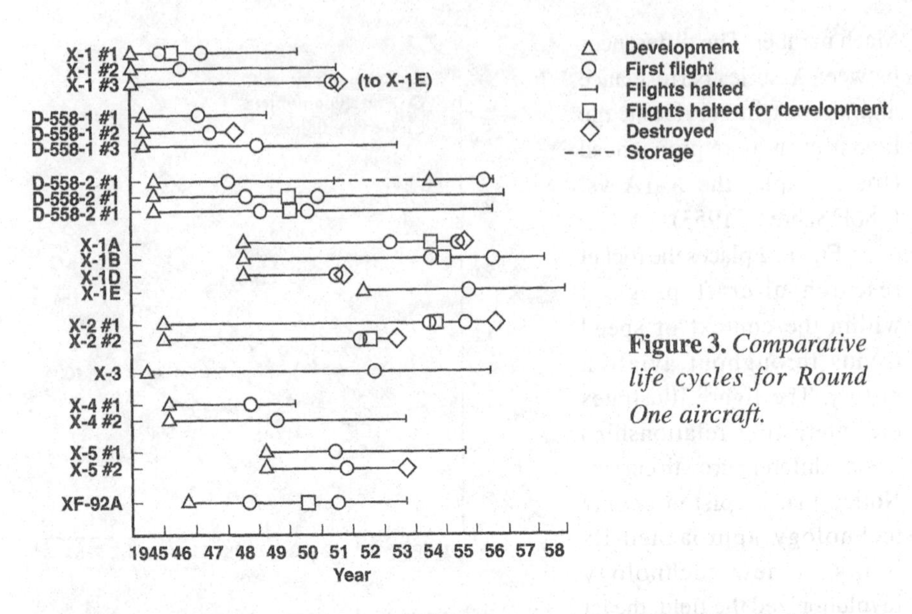

Figure 3. *Comparative life cycles for Round One aircraft.*

these Round One vehicles was the Mach-6 X-15; beyond it the logical successor system was some form of lifting-reentry spacecraft such as the present-day Space Shuttle. The Round One rocket aircraft, besides contributing markedly to the acquisition of information on supersonic flight, were pointing toward piloted suborbital and orbital spaceflight as well.

Figure 3 indicates the life cycle histories of the Round One aircraft, including development time along with active flight status. The relatively short development cycles for such successful aircraft as the original X-1 series forms an interesting comparison to those for such relatively disappointing programs as the X-2 and X-3.

The Round One aircraft, including both the rocket research aircraft and the configuration explorers, investigated a variety of topics and problems. Walter Williams and Hubert M. Drake of the High-Speed Flight Station tabulated them into four broad areas–aerodynamics, flight loads, stability and control, and operations:

Aerodynamics

Validation of transonic tunnel design
Interpretation of tunnel testing data
Aerodynamic heating at supersonic speeds
Lift and drag studies
Inlet and duct studies

Flight loads

Load distribution
Effect of wing sweep upon gust loads
Gustiness at high altitudes
Buffeting
Aeroelastic effects
Effect of stability reduction upon flight loads

Stability and control

Longitudinal control
Blunt trailing-edge control surfaces
Alleviation of pitch-up by wing devices
Effect of principal inertial axis upon lateral stability
Exhaust jet impingement effects upon stability
Inertial coupling
Directional instability
Reaction controls

Operations

High-speed flight exploration
Speed loss in maneuvers
High altitude problems
Pressure-suit research and use
Airspeed measurement
Variable-wing-sweep operation[122]

The diverse range of research areas offers yet another example of the serendipitous character of the Round One aircraft. Conceived largely for aerodynamic and loads research at the speed of sound, they contributed markedly in other areas as well, influencing subsequent aircraft design practice—use of vortex generators and all moving horizontal stabilizers, placing the horizontal tail low, increasing the size of vertical fin surfaces for high-speed flight, and alleviating pitch-up by a variety of wing leading-edge devices, to name just a few. Their development acted as a "forcing" function, encouraging the development of improved ground research methods, notably the transonic slotted-throat tunnel.[123] And, of course, there was the very real psychological benefit accruing from removing the "sound barrier" as a fixation from the minds of engineers.

In some respects, the Round One aircraft were disappointing. Maintenance demands limited most rocket research aircraft to an average of one or two flights per month. Then there is the sad chapter of the explosions, which robbed the NACA, at critical moments, of the X-1-3 and the X-1A. The X-2 story of delay and misuse, and the sad tale of the X-3 and its propulsion problems are classic examples of programs that got out of hand. They pointed to the need for greater coordination and cooperation among the NACA, industry, and branches of the military. The complexities of X-15 development provided an opportunity to exercise this tighter control, and the rising costs of aeronautical research and development implicitly dictated that the days where a program like the X-2 would be *allowed* to continue were at an end.

Round One was a flight research program; as such, it was almost exclusively the accomplishment of the NACA High-Speed Flight Station under the direction of

[122] Williams and Drake, "The Research Airplane," pp. 36-41.

[123] Becker, *The High-Speed Frontier*, p. 95.

These people and this equipment supported the flight of the NACA's D-558-2 Skyrocket at the High-Speed Flight Station at South Base, Edwards AFB. Note the two Sabre chase planes, the P2B-1S launch aircraft, and the profusion of ground support equipment, including communications, tracking, maintenance, and rescue vehicles. Research pilot A. Scott Crossfield stands in front of the Skyrocket. (NASA photo E-1152)

Walter Williams, with the support of Hartley Soulé as research airplane projects leader and Hugh Dryden as NACA director of research. The High-Speed Flight Station had used the Round One airplanes to undertake and consolidate the supersonic breakthrough. By 1959, as the early X-series passed into eclipse, all eyes at Edwards turned to the sleek black X-15. Round One had been a success, and the production aircraft then aloft, from passenger-carrying jet transports to Mach 2 military fighters, were the beneficiaries of its technical bounty.

5

Testing Service Aircraft 1953-1959

The NACA had a long history of assisting the military and other government agencies with research on new aircraft development projects. As early as the 1920s, it was not at all uncommon for the NACA to participate in flight tests of military aircraft. Indeed, on occasion, NACA pilots flew test flights for the contractor. One example is Langley's Bill McAvoy, who flew some of the hazardous spin and dive tests on the XF3F for Grumman and the Navy. During World War II, the NACA had joined in many flight research investigations related to improving the combat potential of American military aircraft such as the Republic P-47 fighter, the Curtiss SB2C dive bomber, and the North American P-51 Mustang. Of course, these aircraft, and many others as well, were also studied extensively in NACA wind tunnels, a traditional form of its support to the military and industry.

This cooperative role continued after World War II as well, encouraged by several factors including an official Air Force policy of "concurrency" testing whereby a large number of initial production-model aircraft were tested at laboratories and field sites around the country, including those of the NACA. Concurrency testing accelerated the testing process and reduced the chances of encountering problems that might arise should an experimental design be committed to production following only a few tests of prototypes. Further, with the new generation of transonic and supersonic fighter and bomber aircraft drawing on more new technology than ever before, there was greater need to deliver preproduction or early production models to the NACA for evaluation and for uncovering defects. After completing these projects for the military or manufacturers, the NACA either returned the aircraft or kept them for flying on a variety of "research opportunity" tasks, often for many years. Although military aircraft appeared at various NACA laboratories during the 1950s, the major focus of such research was, not unnaturally, the High-Speed Flight Station at Edwards AFB, since Edwards was the Air Force's center for flight-testing and research.

Early Work

As early as 1950, the NACA station at Edwards had participated in an Air Force development program, its engineers assisting the Air Force and the Republic Aircraft Company on the XF-91 experimental interceptor. Since the XF-91 was a radical departure from conventional aircraft design for the day, NACA research on the craft could be regarded as motivated as much by the desire to conduct pure research as by the need to assist the military. In 1952, however, the NACA station at Edwards made a major contribution to saving a military fighter program in serious trouble, the Northrop F-89 interceptor.

The F-89 was a high-priority air defense program. In the early months of 1952, six F-89s lost their wings in flight. With production beginning, the Air Force faced a serious crisis, for the F-89 was considered a major element in the North American air defense structure. At the request of both the contractor and the Air Force's Wright Air Development Center, the NACA station at Edwards entered the investigation. Since the aircraft had obviously suffered structural failures, Walt Williams loaned a NACA team to Northrop to determine the F-89's inflight loads. The NACA team installed strain gauges on an experimental F-89 and then studied the data acquired from test flights. As a result, Northrop discovered a serious weakness in the wing structure and redesigned it to impart greater strength. The F-89 subsequently went on to a long and useful service career, the NACA's assistance on the program enhancing the NACA's reputation with the military and industry flight-testing community.[124]

The NACA station at Edwards followed the F-89 experience with a major investigation of another Air Force aircraft, the B-47 jet bomber. Unlike the F-89, however, the B-47 was not in apparent difficulty. Rather, the NACA had asked for the loan of one of the planes to study aeroelastic wing-flexing. The B-47, a shoulder wing monoplane, had six podded jet engines and a very thin swept wing. An airplane with a large, thin, flexible wing could have peculiar aerodynamic and structural load responses as a result of interactions between wing and tail deflections and transonic airflow changes. The field of aeroelasticity, while not new, took on added importance with the large sweptwing aircraft then under development or in production, especially the B-52, an urgent defense program, and a Boeing tanker transport design that eventually spawned the KC-135 tanker and the 707 airliner. Two NACA laboratories had an interest in the B-47: Langley wished to study the impact of aeroelasticity upon structural loads, and Ames was interested in the impact of aeroelasticity upon dynamic stability. Operation of the aircraft from either center was dubious because of runway length. Accordingly, the RAPP sent it to the NACA station at Edwards, where it flew from May 1953 to 1957. NACA's B-47 testing revealed some serious design deficiencies: buffeting problems limited the plane to speeds no greater than Mach 0.8 and certain lift values. In late 1953, the NACA requested that the Air Force provide a B-52 as soon as possible so that the research gathered with the B-47 could be extended through Mach 0.9+ and up to about 50,000 feet in altitude. The NACA never got the B-52 for this research project but did secure permission to instrument a B-52 being flown by Boeing. A loads investigation sponsored by Boeing included the special maneuvers called for by the NACA, gathering much of the desired data. The B-47

[124] A comprehensive collection of HSFRS correspondence among NACA HQ, other NACA centers, the Air Force, and Northrop on the F-89 is in the records collection of the Ames Research Center, held by the Federal Archives and Records Center, San Bruno, CA.

testing resulted in reports prepared jointly by the High-Speed Flight Station and the Ames and Langley laboratories that gave engineers and design teams around the country access to reliable information on the dynamic behavior and response characteristics that could be expected of large, flexible sweptwing airplanes.[125]

The High-Speed Flight Station later continued its large jet aircraft studies using a Boeing KC-135 tanker, starting with one aircraft loaned by the Air Force in 1957. But flight tests were suspended after a near disastrous midair collision between the plane and a jet trainer from the Air Force Test Pilot School. The KC-135 staggered down to a safe landing on Rogers Dry Lake. The trainer, whose civilian pilot apparently never saw the transport, crashed, killing the student. The Air Force delivered a second KC-135 on ninety-day loan; it completed a number of flights before being returned in 1958. The KC-135 flights had been requested by the NACA Subcommittee on Flight Safety, in response to a plea from the Civil Aeronautics Administration (CAA). The CAA needed information that might be useful in writing regulations on cloud ceiling and minimum landing approach visibility for the new generation of jet transports then under development. NACA research on the plane evaluated high-altitude cruise performance, landing approaches including "instrument only" conditions, and the effect of the jet's wing spoilers on its glide path during landing approaches.[126]

The NACA and the "Century Series"

The High-Speed Flight Station's major service testing activities supported the Air Force's "Century Series" of fighter and interceptor aircraft. Table 2 lists the Century Series aircraft evaluated at the HSFS from 1954 onward. The F-100, F-102, and F-104 initially were sent to the HSFS in support of the military development of those aircraft. The F-101 and F-105 appeared at Rogers Dry Lake only briefly, so that pilots could familiarize themselves with the characteristics of those aircraft. The F-107 program was an abortive attempt by a contractor to develop a Mach 2 fighter bomber. The F-105 won the production order. The NACA acquired the F-107s to study some of their design features in support of the XB-70 and X-15 efforts.

During the early 1950s, the Air Force's procurement policy that stressed "concurrency" testing was formalized into the so-called Cook-Craigie plan, named after Gen. Orval R. Cook and Gen. Laurence C. Craigie, the deputy chiefs of staff for development and materiel. Cook-Craigie assumed that if a design appeared to warrant

[125] For example, see Henry A. Cole, et al., NACA Technical Report TR 1330, "Experimental and Predicted Longitudinal and Lateral-Directional Response Characteristics of a Large Flexible 35° Swept-Wing Airplane at an Altitude of 35,000 Feet" (Washington, DC: NACA, 1957). See also Williams, "A Brief History of the High-Speed Flight Station," p. 6; NACA HSFRS annual report to the RAPP for 1953 (pp. 14, 43) and 1954 (pp. 11-12, 42); Hartman, *Adventures in Research*, p. 261.

[126] Edwards AFB historical chronology, in the files of the AFFTC History Office; Mulac, KC-135 flight chronology; memos, Glenn H. Robinson to Walter C. Williams, 6 Dec. 1957, 27 Dec. 1957, and 5 Mar. 1958.

production, then a relatively large number of prototype aircraft should be built—say, 30 to 40—and tested extensively, the changes incorporated on newly emerging production aircraft. This approach avoided the time delays that might be expected if only a few prototypes were refined extensively, the design was committed to production. In actual practice, Cook-Craigie proved expensive, was prone to cause problems in "configuration control" of production models, left large numbers of early production aircraft having little relationship in systems or combat capabilities to later production models, and was as time-consuming as the older method of prototype evaluation followed by production. One of Cook-Craigie's strengths, however, was its endorsement of concurrency testing. Typically, various models of a new design were assigned to weapon-testing, engine-testing, systems-testing, flight-(aerodynamic) testing, and the like. As a result, the NACA received prototypes of new service aircraft for its own evaluations in support of Air Force development effort.

Table 2
Century Series Aircraft

Aircraft	Speed (Mach)	Period
North American F-100A	1.3	1954 — 1960
North American F-100C (#1)	1.4	1956 — 1957
North American F-100C (#2)	1.4	1957 — 1961
McDonnell F-101A	1.7	1956
Convair YF-102	0.98	1954 — 1958
Convair F-102A	1.2	1956 — 1959
Lockheed YF-104A[a]	2.2	1956 — 1975
Lockheed F-104A (#1)	2.2	1957 — 1961
Lockheed F-104A (#2)[b]	2.2	1959 — 1962
Lockheed F-104B[c]	2.2	1959 — 1978
Republic F-105B	2.0	1959
North American YF-107A (#1)	2.0	1957 — 1958
North American YF-107A (#2)[d]	2.0	1958 — 1959

[a]Completed 1,439 research missions in a 19-year career.
[b]Lost in accident; pilot safe.
[c]Transfer from NASA Ames.
[d]Lost in accident; pilot safe.

In 1954, the first two of the Century Series aircraft arrived at the High-Speed Flight Station, the F-100A and the YF-102. The F-100 was a supersonic aircraft having a low tail and a sharply swept wing mounted on a long, rakish fuselage. The YF-102 was basically an enlarged XF-92A delta. The F-100A was an early production airplane, and the YF-102 was a preproduction model of a proposed Air Force interceptor. The NACA, the Air Force, and the manufacturer already knew from wind-tunnel tests that the YF-102's configuration rendered it incapable of meeting the interceptor performance specification, calling for supersonic speed. Even as the High-Speed Flight Station acquired the YF-102, Convair was busily redesigning the airplane on the basis of the "area rule" principle developed by Langley's Richard Whitcomb to give it supersonic performance. While HSFS personnel were interested in using the YF-102 to extend the

The NACA's work horse, the North American F-100A Super Sabre, which flew in studies of inertial coupling. (NASA photo E-2087)

data on delta performance already derived by the XF-92A, they eagerly awaited the area-ruled version of the plane (the F-102A), which eventually arrived at the station in 1956. In any case, the YF-102 soon took a back seat to the F-100A at Edwards, as the F-100A program suddenly encountered serious difficulties.[127]

A series of mysterious crashes of F-100A fighters in 1954 claimed the lives of several airmen, including George Welch, North American's chief test pilot. His F-100A had suddenly yawed more than 15 degrees and broken up while making a rolling pullout from a dive at supersonic speeds. Evidently, the Air Force had placed the F-100A in production too quickly; Pete Everest, the Air Force's project pilot on the F-100, had recommended that it be modified to overcome supersonic directional-stability problems. He was overruled at Air Force headquarters, following a series of evaluation flights by fighter pilots of the Tactical Air Command. The fighter "jocks" were not trained test pilots and saw the F-100 only as a big improvement in performance over the older F-86s they had been flying. Now Everest's report came back to haunt those who had committed the new fighter to service.[128] The Air Force, with hundreds

[127] F-100 and YF-102 development is summarized by Marcelle Size Knaack in *Encyclopedia of the U.S. Air Force Aircraft and Missile Systems,* vol. I, *Post-World War II Fighters, 1945-1973* (Washington, DC: USAF, 1978), pp. 113-133, 158-173.

[128] Everest, *The Fastest Man Alive,* pp. 12-13, 19-20.

of the F-100As on order, had no choice but to ground the aircraft until investigators could find out what had happened and modify the design. Maj. Gen. Albert Boyd, commander of the Air Force Air Research and Development Command, detailed a senior officer on his staff to meet with Walt Williams of the HSFS to get the NACA's ideas on the crisis.[129]

Williams and several NACA engineers, including Joseph Weil and Gene Matranga, met with Air Force and North American representatives and mapped out a research program. Up to this time, the NACA had been primarily concerned with evaluating the F-100's general stability and control. Now, in light of the station's concurrent experience of violent inertial coupling with the X-3, the engineers decided to study not only the F-100A's directional-stability problems but also its roll-coupling tendencies, the latter having been identified as the cause of one of the crashes. North American already had an idea for a fix: enlarge the area of the craft's vertical fin and add more area to the plane's wing tips. Under NACA and Air Force pressure, North American cut its planned delivery schedule for the larger tail from over 90 days to just nine days, a measure of the urgently with which correcting the F-100's problems was viewed. Williams's engineers went to the Langley laboratory to run a computer simulation of the F-100's behavior—the first such simulation done by the HSFS—in conjunction with William Phillips, the NACA's acknowledged expert on coupled motion instability. The simulation confirmed the F-100's dangerous directional-stability and roll-coupling problems; one NACA engineer termed its directional-stability characteristics as "damn poor."

From October through December 1954, HSFS pilot Scott Crossfield flew the NACA's F-100A on a series of flights defining the coupling boundaries of the airplane. Williams reported to the RAPP on the results of one flight with the plane in its original (small fin) configuration: "a violent divergence in pitch and yaw occurred on the F-100A airplane during an abrupt aileron roll at a Mach number of 0.70 and an altitude of 30,000 feet in which a negative load factor of 4.4 g and a sideslip angle of 26 [degrees] were reached."[130] Had a sideslip of that magnitude occurred at supersonic speed, the negative load factor would have multiplied and the aircraft probably would have disintegrated.

This flight occurred on 2 November 1954. Many years later, Crossfield himself described it:

> We were to do exploratory small step inputs in roll at 0.8 Mach number at 30,000 feet to start building the airplane characteristics base. A chain on the stick was to limit the step input. The chain broke and I inadvertently put in full aileron. All hell broke loose. We yawed 15 degrees, hit three negative g's and performed an undefinable maneuver new to my experience. I was told it

[129] Interview of Williams and subsequent conversations.

[130] HSFS report to the RAPP for 1954, p. 37. Mulac F-100A flight chronology; Crossfield with Blair, *Always Another Dawn*, p. 198.

> bent the airplane. Fortunately we were high enough to not be able to break it. Now amusing is that I was totally baffled by the joy, the glee, the delight that Joe Weil and company expressed at my having scared myself to death. I do not take kindly to unanticipated maneuvers. The delight came from the fact that we had accidentally and unintentionally obtained the first recorded fully developed roll coupling data ever with a fully instrumented airplane.[131]

At the suggestion of both Williams and North American, the NACA added a larger vertical fin to the plane in December 1954, adding 10 percent more surface area initially and then 27 percent more surface area for flights in 1955. Eventually North American installed the fin having 27 percent greater area, as well as wingtip extensions, and the F-100 series went on to a long and distinguished service life. The F-100 data were incorporated in the same research memorandum (RM-H55A13, February 1955) that covered the X-3's experience, a warning not to underestimate the difficulties that could be expected with airplanes having insufficient tail area combined with long fuselages and narrow wing planforms. The NACA later used the F-100A for a variety of research projects. The HSFS evaluated the behavior of a pitching motion damper system on the first F-100C received in 1956. As expected, the damper further increased the plane's resistance to coupling. The other F-l00C arrived at the HSFS in 1957 and was used for general research support, including chase flights and pilot proficiency flights.[132]

In contrast to the F-100A experience, the High-Speed Flight Station's research on the YF-102 was more prosaic. Since Williams's engineers had more interest in the definitive F-102A just around the corner they used the YF-102 primarily to extend the data acquired on the basically similar XF-92A. Station engineers did a complete drag survey of the airplane, especially under various conditions of lift, this information greatly assisting researchers interested in correlating results taken from flight research with results from wind-tunnel tests of the configuration. The results constituted an effective measure of the accuracy of the wind-tunnel findings for aircraft design prediction. The YF-102 did experience some inertial coupling tendencies, the first encountered on a delta-wing airplane, but not as seriously as with the F-100. The NACA tested the production F-102A received in 1956 so that researchers could compare differences in drag between two generally similar configurations, one having

[131] Letter, Crossfield to Hunley, 18 Feb. 2000; Crossfield's pilot report, F-100A No. 778, 2 Nov. 1954, available in the DFRC Historical Reference Collection. The pilot report corresponds largely with Crossfield's recollections, but as Williams had reported to the RAPP, the Mach number was 0.7 rather than 0.8, although a previous right wind-up turn had been performed at 0.8 Mach number.

[132] Mulac, F-100A and F-100C flight chronology; RAPP meeting minutes, 4-5 Feb. 1957; Interview of Armstrong; Interview of Williams and subsequent conversations; HSFS Annual Report to the RAPP for 1954, p. 12; pilot reports on flights of the F-100A in Jan. 1955, available in the DFRC Historical Reference Collection. In his comments on the original chapter 18 Feb. 2000, Scott Crossfield stated,'"I believe that we (NACA) recommended the 27% vertical tail area increase when NAA [North American Aviation] was planning & manufacturing the 10% increase." He added in his cover letter, "I believe that it was Joe Weil who recommended the 27% additional vertical tail area while at NAA they were pushing the 10% tail."

area rule (the F-102A, serial number 54-1374) and the other lacking it. As the 1950s drew to a close, the NACA sought information on the low-speed approach and landing characteristics of unpowered delta-wing aircraft, information applicable to the design of future winged spacecraft such as the Round Three Dyna-Soar. Deltas have peculiar low-speed and approach characteristics, including high induced drag, and their combined ailerons and elevators (elevons) work under a disadvantage: deltas require so much elevon deflection during landing approaches that they have very little available elevon "travel" left for good lateral control. This limitation could seriously compromise safety during the landing approach of a delta-wing spacecraft, especially one having the inherent performance limitations imposed by reentry design constraints (i.e., sinking like a rock). Before NASA retired the F-102A in 1959, test pilots Jack McKay and Neil Armstrong flew a series of landing approaches under various lift-to-drag and power conditions, in preparation for the ill-fated Dyna-Soar program.[133]

The third Century Series program to get under way at the High-Speed Flight Station involved the Lockheed F-104, an airplane having a configuration roughly similar to the Douglas X-3. The program began in 1956, the start of an association with this hot fighter that continued nearly three decades later.[134] With an alluring Mach 2 design, high T-tail, long fuselage, narrow wingspan, and troublesome J79 jet engine, the F-104 posed numerous challenges. The long, pointy configuration—public relations agents dubbed the F-104 the "Missile with a Man"—promised to give roll-coupling problems, and the tail hinted ominously at pitch-up, though Lockheed designed a stick-shaker and "stick-kicker" into the controls to prevent an unwary pilot from getting into pitch-up difficulties. Both areas, of course, were ones in which the NACA had a vital interest. The Research Aircraft Projects Panel had sought an F-104 for the HSFS since 1954; in late summer of 1956, the station received a preproduction YF-104A (soon redesignated a JF-104A).[135]

The Air Force/Lockheed flight-test program on the F-104 did not go at all smoothly, largely because of powerplant problems and equipment failures. It eventually took twice as long as expected, with a number of accidents and incidents, some causing fatalities. At one point, Lockheed had lost all of its instrumented test airplanes and NACA's JF-104A was the only instrumented airplane left. The Air Force asked for its return, but the NACA countered with the proposal that it run the Lockheed test program on the YF airplane, using NACA pilots. Lockheed and the Air Force agreed, and the roll-coupling study began in May 1957. NACA engineer Thomas Finch was detailed to work with the Lockheed test team and aerodynamicists on analog studies of the F-104's expected rolling characteristics to predict what might happen in flight, while

[133] Williams, "A Brief History of the High-Speed Flight Station," p. 7; Mulac, YF-102 and F-102A flight chronologies; interview of Matranga; interview with Armstrong.

[134] The last of Dryden's F-104s had its final flight on 31 Jan. 1994. Over the years, 11 F-104s at Dryden flew a great variety of missions, including safety chase, basic research, airborne simulation, and service as aerodynamic testbeds.

[135] Mulac, F-104A flight chronology; interview of Finch.

The NACA's Lockheed Starfighter, the YF-104A, flies with an experimental test fixture. (NASA photo ECN-785)

NACA test pilot Joe Walker flew the trials. Over the next nine months, the station's JF-104A completed more than 60 roll investigations, showing the aircraft to be generally acceptable. Flight-test results and Finch's analog studies indicated that transonic and supersonic rolls near zero g "entry" conditions could lead to "autorotation," a tendency for the plane to continue rolling despite the pilot applying corrective aileron, with accompanying pitching and yawing motions. If this occurred, Finch recommended that the pilot use the stabilizer to damp out any tendency of the plane to couple. The NACA further recommended that Lockheed limit the aileron's "travel" (displacement) at transonic and supersonic speeds, only permitting "full" aileron "authority" with the plane in the low-speed, landing gear-and-flaps down configuration. This confirmed impressions at Lockheed, and the company built mechanical limits into the plane, added a yaw damper, and put cautioning notes in the plane's operational handbook.[136] Wary pilots still treat the F-104 with caution.[137]

The NACA and NASA later flew the JF-104A on a variety of research tasks. Equipped with reaction controls, it flew as a trainer for X-15 pilots. It performed other special aerodynamic investigations, such as a boundary layer noise research program for Ames Research Center, before being retired in 1975. While the NACA's program with the JF-104A constituted the High-Speed Flight Station's major early involvement with the F-104, the facility also flew three other F-104s acquired later—two

[136] Interview of Matranga; Interview with Finch; conversations with Michael Collins, Milton Thompson, and Thomas McMurtry, 5 Jan. 1978; Mulac, F-104A flight chronology; Thomas W. Finch, "Briefing on the NACA F-104A Roll-Coupling Program Presented to Wright Air Development Center and Lockheed Aircraft Corporation Personnel," 21 Oct. 1957.

[137] In a note on p. 95 of the original edition of this book, Hallion wrote: "There is a popular tale around Dryden of a local well known former naval aviator up in one of the two-seat F-104s who exuberantly initiated a rapid roll far above Mach 1, only to have the NASA pilot wrench the stick away with an oath, exclaiming, 'Not in *this* airplane you don't!'"

F-104As and a two-seater F-104B from Ames—on a variety of research tasks including tests of the Mercury spacecraft's drogue parachute and studies of boundary-layer-formation transition from laminar (smooth) to turbulent flow. The NASA Flight Research Center acquired a number of F-104s in the 1960s and 1970s for flight research and safety chase.[138]

In contrast to the F-100, F-102, and F-104 programs, the NACA's involvement with the F-107 did not involve support of a major defense production program. Rather, the F-107 was what Williams was fond of referring to as a "target of opportunity," an aircraft possessing some interesting features that the NACA wished to examine in detail.

The F-107 started out as a "growth" version of the F-100, with an estimated speed of Mach 2+ and some radical design elements. It featured a large inlet located above the fuselage, a very sophisticated stability augmentation system, and an all-moving vertical fin. The inlet and the fin designs were what interested the NACA. After the F-107 lost out to the F-105 for a major Air Force production contract, the NACA acquired the first and third YF-107s built. The first proved mechanically unreliable and completed only four flights before NASA grounded it. The third completed 40 flights during 1958 and 1959 before being damaged in a takeoff accident, fortunately without injury to the pilot, Scott Crossfield. During this time, the engineers at the High-Speed Flight Station modified it with a so-called side-stick flight-control system to gain experience using such a system, which was planned for the upcoming X-15 program. In this way, a number of designated X-15 research pilots gained experience with such a system before having to try it out in the actual X-15.[139] The side-stick program was the NACA's major accomplishment with the craft. The proposed inlet and fin studies went by the wayside after the retirement of the first F-107, in part because the complex inlet, with its movable inlet ramps and variable inlet control, caused so many problems that technicians were eventually forced to fix the inlet into a position that limited the plane to a maximum speed of Mach 1.2. The third F-107 was retired following the damage in the takeoff accident.[140]

An Assessment

The NACA's assistance to the Air Force and industry on the F-100 and F-104 contributed significantly to ensuring that both were safe and effective combat aircraft. The NACA's program on the YF-102 and F-102A was less important in this regard but offered an excellent opportunity to evaluate the accuracy of wind-tunnel predictions against full-scale data taken from flight research and also to study the direct benefits of Whitcomb

[138] See previous note.

[139] "In his review of the original chapter of this book, Scott Crossfield wrote: "The F-107 side-stick mechanization bore little or no resemblance to the X-15['s] other than it was a side-stick . . . ; [it] had no influence on the X-15 side-stick design. The X-15 side-stick was designed to a very rigid and unique criterion which was never duplicated to my knowledge in any other airplane."

[140] Mulac F-107A chronology; interview of Bellman; interview of Matranga; interview with William Dana, 3 Dec. 1976; F-107A flight chronology by Peter Merlin.

area ruling on an aircraft's configuration. The wind-tunnel predictions and expected benefits of area ruling were confirmed during tests. The F-107 helped X-15 development move smoothly along, giving confidence in the side-stick flight- control system.

For the most part, research on military aircraft took a definite second place at the High-Speed Flight Station to research on the X-series; the only exception came during the F-100 crisis. Also, toward the end of Round One, research on military aircraft picked up, in part because after 1955 the NACA station had more time to invest in them. Though this could smack of "make work," in fact many of these new aircraft had features of interest that did not appear on the X-series. They included innovations such as the all-moving vertical fin on the F-107 and complex stability augmentation systems. Generally, as soon as the High Speed Flight Station had finished with the military-related testing of the aircraft, the engineering staff would set to work on a program related more to NACA interests, such as reaction control studies with the JF-104A. Williams and his engineers, in pursuit of their larger mission of advancing supersonic and hypersonic research, consistently sought to place the testing of military aircraft not only within the framework of military and industry needs, but also within the NACA's interests in high speed research, such as the low-speed ground-approach studies of the F-102A in support of the Round Three Dyna-Soar. This often led to novel proposals and research trips far from the desert. Williams once advocated installing ramjets on the wingtips of a Lockheed F-104A to acquire information that could benefit the design of supersonic ramjets; despite interest from the Lewis engine laboratory and the Air Force's Air Research and Development Command, the project died for lack of other support. Because the X-15 had a so-called "rolling tail"—the tail surfaces functioned for pitch and roll control as well as for directional control—Williams and a test team journeyed to France to study a French airplane having such a feature, the Sud Ouest Trident experimental interceptor. Test pilots Joe Walker and Iven Kincheloe flew the French craft to become familiar with the performance and effectiveness of such a configuration. The experience gave the NACA added confidence in the capabilities of such a design feature for controlling the X-15.[141]

Flying these military aircraft was often as potentially hazardous as the regular X-series. For example, on the first flight of the NACA F-100A, Scott Crossfield had to make a "deadstick" (powerless) landing following a warning of engine fire, something North American's own test pilots doubted could be done, because the early F-100 lacked flaps and landed "hot as hell." Crossfield followed up the flawless approach and landing by coasting off the lakebed, up the ramp, and then through the front door of the NACA hangar, frantically trying to stop the plane, which had used up its emergency brake power. Crossfield missed the NACA X-fleet, but crunched the nose of the F-100A through the hangar's side wall. Chuck Yeager then proclaimed that while the sonic wall had been his, the hangar wall was Crossfield's.[142] Test pilot

[141] Interview of Williams and subsequent conversations. Walker and Kincheloe were designated pilots for the X-15. However, Kincheloe perished in an F-104 accident and never flew the X-15.

[142] HSFS annual report to the RAPP for 1954, p. 12; Crossfield with Blair, pp. 194-199.

Milt Thompson had a close call in one of the F-104As years later when one of its flap actuators failed, causing only one flap to lower. The F-104A began rolling crazily, but Thompson fortunately was at high altitude. He stayed with the plane through four rolls of increasing rapidity and coupling tendencies, then ejected. On the ground, observers heard Thompson radio, "It's going!" The tower at Edwards reported smoke in sight on the bombing range, but no parachute. "The gloom was so thick," one engineer recalled, "you could cut it with a knife." Meanwhile, Thompson landed in the desert, gathered up his chute, and flagged down a NASA vehicle for a ride back to the center before breakfast.[143]

After 1958, NASA's involvement with military-aircraft testing at Edwards was greatly reduced for several reasons. The NASA Flight Research Center, as the High-Speed Flight Station had been renamed in 1959, was heavily committed to the X-15 program, so the engineering staff lacked the manpower, resources, and time to become involved with other projects.[144] Changes in the military's procurement of aircraft also played a role. Military aircraft acquisition and development declined. By 1960, the aircraft that America would rely upon for its defense and with which it would go to war in Southeast Asia were in service or under development, so there was less for NASA to do in military-related testing, just as there was less for the Air Force as well. In effect, the military simply stopped building new airplanes for a while.

There were also changes in the procurement policy. The idea of building a number of prototypes and preproduction machines and testing them widely was replaced by heavy reliance upon paper studies and proposal analysis—"read before buy" rather than "fly before buy." This questionable practice also came to an end at the close of the 1960s, when the pace of acquisition stepped up. Then, the FRC again actively supported military aircraft projects with flight research.

Symbolically, the High-Speed Flight Station's activities on military testing completed the cycle of NACA involvement in the early era of supersonic flight. In the 1930s and early 1940s, engineers on the ground had generated the concept of a transonic and supersonic research aircraft program. In the late 1940s, a specialized NACA facility had been created and the research successfully undertaken. In the 1950s, the frontiers beyond Mach 1 were explored with a variety of instrumented, piloted research tools. And then, using much of the information derived by NACA ground and flight testing, manufacturers and the military created a new generation of turbojet driven combat aircraft and placed them in service, with the NACA station at Edwards (and NACA laboratories around the country) offering the military traditional NACA support. Supersonic flight had gone from the theoretical, to the experimental, to the practical. The next frontier was space.

[143] Mulac F-104A chronology; interview of Milton Thompson, 5 Jan. 1978; Milton O. Thompson, *At the Edge of Space: The X-15 Flight Program* (Washington, DC: Smithsonian Institution Press, 1992), pp. 119-122.

[144] The FRC's heavy emphasis on the X-15 was perceived in some quarters as evidence that the "single-mission" center could be closed down following the X-15 program. This perception actually led to a congressional proposal to close the FRC in 1965, as will be discussed.

II

Into Space and Back Again

6

The X-15 Era 1959-1968

On 1 October 1958, High-Speed Flight Station employees Doll Matay and John Hedgepeth put up a ladder in front of the Station's building at the foot of Lilly Avenue and took down the NACA emblem, a winged shield, from over the entrance door. NASA had arrived in the desert, bringing with it a new era of space-consciousness, soaring budgets, and publicity.

The changes had been long in coming, and the post-Sputnik furor only accelerated the process. For the past five years, advanced planners at the High-Speed Flight Station (HSFS) had devoted increasingly greater amounts of time to studying the possibility of hypersonic (Mach 5+) aircraft and winged spacecraft. Within the station's research division, winged spacecraft problems and conceptions clearly dominated the staff's thinking, not unexpected in light of the increasingly heavy commitment to the upcoming X-15. The orientation at the HSFS dovetailed nicely with the new emphasis on unpiloted and piloted spaceflight implicit in the charter of the National Aeronautics and Space Administration (NASA).

The change in the station's research emphasis during the 1950s can be seen by comparing the activities of the three branches of the station's research division—the stability and control branch, aero-structures branch, and airplane performance branch—in the three years between 1955 and 1958.[145] The change is even more evident in records of how the professional research staff spent its time in 1955, 1957, and 1959.[146]

Branch	Research Emphasis (1955)	Research Emphasis (1958)
Stability & control	Inertial coupling	Hypersonic boost-gliders
	Roll-rate requirements for Mach 2 fighters	Reaction control studies
		Winged spacecraft & satellites
Aero-structures	Transonic airload distribution	Structural loads of hypersonic boost-gliders
Airplane performance	Transonic drag rise	Boost-gliders

[145] NACA HSFS, *Review of Aeronautical Research* (HSFS, n.d.), originally found in the files of the office of the deputy director, DFRC.

[146] Ibid.

Research Area	**1955**	**1957**	**1959**
Satellite studies	5%	11%	16%
Ballistic missile research	1	1	3
Boost-glide aircraft	15	18	35
Anti-ICBM studies	1	1	2
Surface-to-air missiles	3	4	4
Advanced fighter aircraft	33	32	16
Supersonic bombers & transports	23	19	18
Subsonic bombers & transports	18	12	3
Special projects (VTOLs, etc.)	1	1	3

Figures might not add up to 100 percent because of rounding.

This indication of professional interests mirrored trends within the NACA-NASA as a whole. By the late 1950s, the Ames and Langley laboratories were devoting more effort to studying the problems of hypersonic flight and reentry from space than they were on aeronautics per se. By 1965, over 80 percent of NASA's research went to space-related projects.[147]

The Old Order Changeth

On 1 October 1958, the day the National Aeronautics and Space Administration officially came into being, the High-Speed Flight Station had a personnel complement of 292. The new agency employed some 8,000 civil servants, 3,368 of whom were at Langley. In contrast to the tiny NASA station at Edwards AFB, the Air Force contingent there numbered over 8,000. But like the rest of the newly created NASA, the High-Speed Flight Station was on the verge of rapid growth. The station's increasingly heavy emphasis upon the problems of winged spaceflight was an unusual, but certainly understandable, legacy from its pioneering days of trying to "break the sound barrier." HSFS Director Walter Williams identified a dozen problems affecting design and piloting of future hypersonic craft, problems that logically grew out of the Round One experience and would be encountered on both Round Two (the X-15) and any Round Three orbital vehicle:

Design Problems

Aerodynamic heating and heat transfer
Aerodynamic interference

[147] See NASA Office of Aeronautics and Space Technology, *OAST Research Centers: Charter, Contributions, Resources—A Foundation for Institutional Planning* (Washington, DC: NASA, Sept. 1973), pp. 3-1 to 3-3.

Aerodynamic efficiency
Structural design
Crew survival

Piloting Problems
Poor landing configuration
Large accelerations
Reaction control operation
Large changes in control effectiveness
Large changes in stability
Inertial coupling
Presentation of piloting information

These were all problem areas that the station could be expected to work on in the years ahead.[148]

Williams and the professional staff at the HSFS had recognized for several years that their activities had broadened considerably beyond those envisioned for the Muroc Flight Test Unit back in the 1940s. The station had a major new role to play with the X-15, which—together with the upcoming Project Mercury program—represented essentially a two-pronged approach to studying the problems of piloted spaceflight. Williams had always sought laboratory status for the station, making it equal organizationally with the other NACA laboratories. Of course the scope of its work and the size of the station were smaller than those of Langley or even Ames. Nevertheless, after independence had been achieved from Langley in 1954, laboratory status had been the next logical step. To Williams, it was important for reasons of morale, making the station's employees feel equal in prestige and value with the laboratories, even though it would not actually affect administration. When NASA came into existence, the traditional laboratories, Langley, Ames, and Lewis, were redesignated as research centers, to reflect their primary role in NASA's coming activities. Williams's continued pressing for a redesignation of the High-Speed Flight Station now paid off, for the scope of the X-15 program and NASA's heavy priority on it argued for a name change. On 27 September 1959, NASA headquarters redesignated the High-Speed Flight Station as the NASA Flight Research Center (FRC). That name continued until it was renamed the Dryden Flight Research Center in 1976 in honor of Hugh L. Dryden.[149]

By the time the station became a center, Walter Williams was gone. At the behest of NACA's Director Hugh Dryden, in September Williams had joined Project Mercury, America's first man-in-space venture, as its operations director. His appointment was

[148] Williams and Drake, "The Research Airplane: Past, Present, and Future," p. 40.

[149] Interview of Williams and subsequent conversations; Van Nimmen and Bruno, *NASA Historical Data Book*, Vol. I, p. 271; Robert L. Rosholt, *An Administrative History of NASA, 1958-1963* (Washington, DC: NASA SP-4101, 1966), p. 79.

Paul F. Bikle, Center director 15 September 1959-27 April 1971, with his familiar cigar in hand. (NASA photo E95-43116-19)

indicative of NASA's emphasis upon placing individuals with flight-test experience in positions of managerial and administrative responsibility for America's growing piloted spacecraft program. Williams would be missed, and not simply because he had been an excellent station director. He had influenced the local community as well; he had worked for high-quality elementary and secondary education in the Antelope Valley school systems and had encouraged station employees to take an active part in civic affairs.

In Williams's place came Paul F. Bikle, a Pennsylvanian with long experience in flight-test projects. Bikle had been Williams's choice for the job, the two men somewhat akin in temperament and outlook. After graduation from the University of Detroit with a B.A. in aeronautical engineering in 1940, Bikle had joined the staff at Wright Field as a civilian flight-test engineer. Well known in military flight-testing circles, he was serving as technical director of the Air Force Flight Test Center (AFFTC) when he joined NASA. Like Williams, Bikle had little use for unnecessary paperwork; he often remarked that he would stay with NASA as long as the paperwork level remained below what he had experienced in the Air Force.

Bikle replaced Williams on 15 September 1959, oversaw the HSFS in its transition to the NASA Flight Research Center (FRC), and remained for the next dozen years. The center was fortunate in having two such excellent administrators sequentially presiding over its activities. Bikle was a short, stocky individual who loved poker and cigars. He had a natural affinity for flying and flight-testing. A sailplane pilot of unusual ability, in February 1961 he set a world's altitude record for sailplanes by soaring to 46,269 feet, a record still standing more than two decades later.[150] Bikle believed in doing things quietly and with a minimum of fuss and outside attention. "Under Paul Bikle," one FRC engineer recalled, "we were well aware that headquarters was 3,000 miles away." He was at home with the engineers, the test pilots, the crew chiefs, and the mechanics. Every day he would walk through the building and hangars, asking questions, expecting answers, and constantly checking. The careless and unprepared could wind up in the "Bikle barrel" very quickly. Like Williams before him, Bikle impressed those around him with his bluntness, drive, and canny engineering sense. "He'd sit in a meeting, listen to us, and say 'Do this,'" one FRC veteran remembered. "We'd all think 'Why the hell didn't I think of that?'" Genuinely liked around the center, Bikle was known (but not to his face) as "the ole Man," and his retirement party at the Antelope Valley Country Club long triggered warm memories in the minds of Dryden staffers.

Bikle's immediate challenge involved shifting the center from planning for the X-15 program to operating it. He needed people and began wiping out manpower-consuming projects to get the force necessary to run the new program efficiently. As one of his first moves, Bikle asked NASA's Ira Abbott for 80 new positions and added them to the rapidly growing X-15 team. In accordance with NASA's center management policy, he elevated De E. Beeler from chief of research (and de facto deputy) to deputy director of the FRC, a position Beeler held until his retirement. The FRC's budget, personnel, and facilities expanded throughout the 1960s, as did NASA's as a whole, and these expanded resources added to Bikle's administrative tasks. The center's budget went from $3.28 million in 1959 to $20.85 million in 1963 and to $32.97 million in 1968. Staff went from 292 to a peak of 669 in 1965. Its facilities expanded as well. The AFFTC in conjunction with the FRC had built a special high-speed flight-test corridor for the X-15 (discussed later). The FRC had added a communications building in 1963, a runway noise measurement system in 1964, and a high-temperature loads calibration laboratory in 1966, which proved very useful during the YF-12 Blackbird program.[151]

The center's organization remained largely unchanged from that of the 1950s. There were four main divisions, later designated as directorates: administration, research, data systems, and flight operations. In November 1965, Bikle added a biomedical program office of equal stature with the divisions,[152] and in 1969 he added a safety director. Bikle also added a projects and program management office that evolved, after he left the FRC, into a directorate of its own. The four main directorates—

[150] On 17 Feb. 1986, Robert Harris set a new absolute-altitude record in a Class D glider of 49,009 feet. As of 1992, Bikle still held the record for altitude gained of 42,303 feet, however. National Aeronautic Association, *World and United States Aviation and Space Records (as of December 31, 1992)* (Arlington, VA: NAA, 1993), p. 160.

[151] Data are from the section on the FRC in Van Nimmen and Bruno, NASA Historical Data Book, Vol. I.

[152] After Bikle's departure, biomedical dropped from directorate level, becoming a branch of the center director's office. See Appendix A for FRC organization during the Bikle and post-Bikle eras.

research, data systems, flight operations, and administration—continued to predominate under center director Lee Scherer until 1976, when his successor David Scott added a directorate for shuttle operations and shifted the projects office into the directorate for aeronautical projects. In 1978, center director Isaac Gillam combined projects and research into a combined directorate for aeronautics.

Until 1963, the various NASA centers reported to *the* NASA associate administrator, Robert C. Seamans, Jr., and had a great deal of leeway in choosing projects within the areas of their expertise. In 1963, NASA authorized the Office of Advanced Research and Technology (OART) to supervise the five original laboratories and stations of the old NACA—Langley, Ames, Lewis, the FRC, and Wallops Island—and to act as their managerial liaison with NASA headquarters. The FRC thus now reported to the NASA associate administrator in charge of OART. During the 1960s, OART itself was locked into competition for resources and support with the Office of Manned Space Flight (OMSF) and the Office of Space Science and Applications (OSSA), often causing OART's engineers to mutter among colleagues that senior management had to remember NASA's "first A stood for Aeronautics." OART itself was expected to act within NASA much as the old NACA had acted for the military services and industry. OART "would have to anticipate problems, do preliminary studies, and carry its investigations to the point where the research could be usefully applied—in this case by NASA itself."[153] On this model, the FRC's relationship with OART was much like the earlier NACA/HSFS relationship, so that major upheavals in NASA itself only rocked the FRC when they affected OART. OART was always heavily oriented toward winged vehicles. In 1972, NASA changed the name of OART to the Office of Aeronautics and Space Technology (OAST) "to give adequate recognition to NASA's responsibilities in aeronautics."[154]

Above all, the decade of the 1960s at the FRC was the decade of Round Two (the X-15) and, to a certain extent, Round Three (the Dyna-Soar and its follow-ons) as well. When Bikle assumed leadership of the NASA station, Round One was at an end. All eyes and NASA's attention shifted to the rakish black rocketplane called the X-15, which would take research pilots from the center, the AFFTC, and the Navy to the fringes of space.

The Beginning of Round Two

The X-15's origins were complex, for its development was stimulated by both foreign and domestic research. A major initial influence was the prewar and wartime work of German scientists Eugen Sänger and Irene Bredt (later Irene Sänger-Bredt), who in 1944 had set forth a concept of a hypersonic rocket-powered aircraft that could be

[153] Arnold S. Levine, "An Administrative History of NASA, 1963-1969" (draft ms.), NASA History Office, 23 Aug. 1977, p. 96.

[154] *OAST Research Centers*, pp. 3-1 to 3-3.

boosted into orbit and then glide back to earth (hence the term boost glider). NACA's John Becker later wrote:

> Professor Sänger's pioneering studies of long-range rocket-propelled aircraft had a strong influence on the thinking which led to initiation on the X-15 program. Until the Sänger and Bredt paper became available to us after the war we had thought of hypersonic flight only as a domain for missiles. . . . From this stimulus there appeared shortly in the United States a number of studies of rocket aircraft investigating various extensions and modifications of the Sänger and Bredt concept. These studies provided the background from which the X-15 proposal emerged.[155]

The Sänger/Bredt study directly influenced the birth of the X-15. It also generated a climate from which sprang Round Three, the abortive Dyna-Soar effort. This occurred because Walter Dornberger, the wartime director of Germany's Peenemünde proving grounds, had joined the Bell Aircraft Corporation after World War II and used his position to propose various types of Sänger/Bredt-inspired boost gliders for military missions, including orbital strike and reconnaissance. The Air Force, generally receptive, sponsored a number of studies that coalesced in 1957 as the Dyna-Soar program, later designated X-20A, the Round Three of Hartley Soulè's research aircraft classification scheme. All parties recognized the advisability of first acquiring basic hypersonic flight data, especially on hypersonic aerodynamics and heating, from a special high-speed research airplane—and thus was born the X-15.

Within NACA, the first call for such a vehicle came from Robert Woods of Bell Aircraft, a member of the prestigious NACA Committee on Aerodynamics and the man most responsible for getting Bell involved with the X-1 program nearly a decade before. In two committee meetings in October 1951 and January 1952, Woods urged that the NACA study requirements for piloted Mach 5+ research aircraft. NACA took no action on this proposal at the time, but individual engineers at the Ames, Langley, and Edwards facilities undertook their own studies of suitable configurations. At Edwards, two of Williams's advanced planners, Hubert Drake and L. Robert Carman, began a series of configuration studies. Langley engineers proposed salvaging the X-2 for a hypersonic test program, using two jettisonable rockets for additional boost and adding reaction controls. NACA headquarters moved slowly and deliberately. In mid-1952, the Aerodynamics Committee endorsed a proposal for NACA to "devote a modest effort" to hypersonic studies, but Hugh Dryden, recognizing that a "modest effort" would stand little chance of accomplishing much and that NACA was already overcommitted to various projects, reduced it to a study for identifying the problems of hypersonic flight, rather than doing research on the problems themselves. In August

[155] John V. Becker, "The X-15 Program in Retrospect," *Raumfahrtforschung* (March-April 1969), p. 45. The study in question was E. Sänger and I. Bredt, "*Über einen Raketenantrieb für Fernbomber,* Deutsche Luftfahrtforschung UM 3538 (Ainring: Deutsche Forschungsanstalt für Segelflug, 1944).

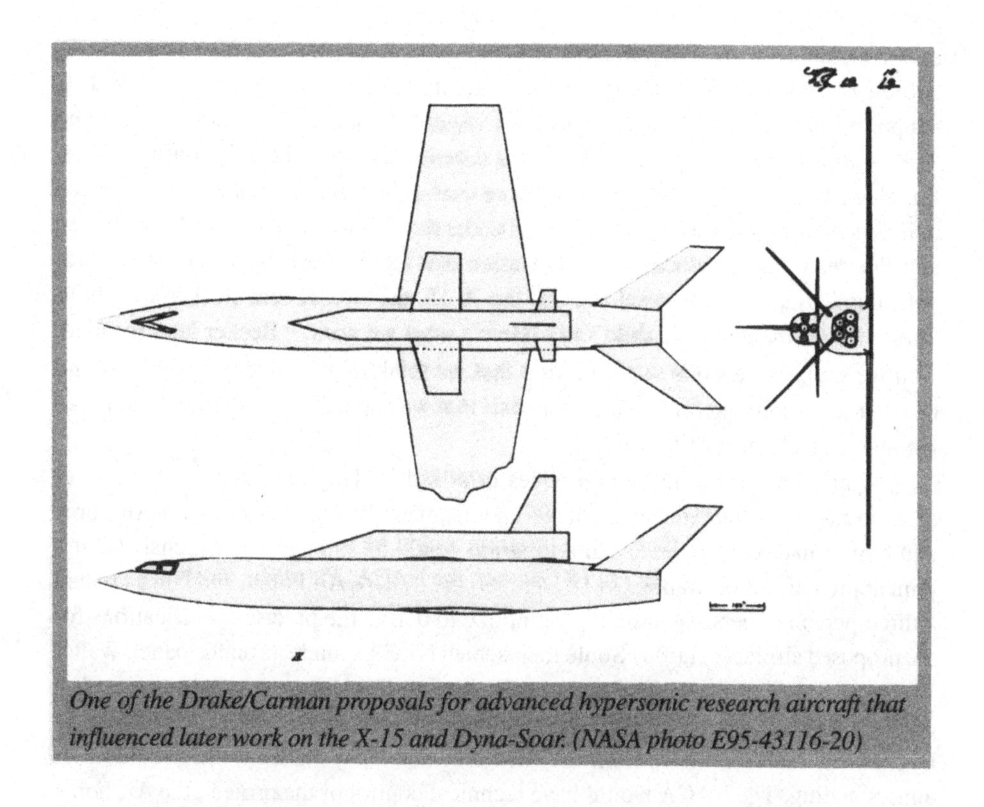

One of the Drake/Carman proposals for advanced hypersonic research aircraft that influenced later work on the X-15 and Dyna-Soar. (NASA photo E95-43116-20)

1953, Drake and Carman submitted a proposal from Edwards to headquarters for a five-phase hypersonic research program leading to an orbital winged vehicle. Dryden and his associate director for research, Gus Crowley, shelved the proposal as too futuristic, which indeed it was. Nevertheless, in its bold advocacy of a "piggy back" two-stage-to-orbit research craft, the Drake/Carman study constituted one of the earliest predecessors of the Shuttle. By the end of 1953, the notion of a hypersonic research aircraft had spawned two military study efforts, one by the Air Force Scientific Advisory Board, the other by the Office of Naval Research. The next year was the critical year of decision for the future X-15.[156]

At its annual meeting for 1954, the Research Airplane Projects Panel (RAPP) concluded that the NACA should procure a new hypersonic research aircraft. Just

[156] NACA Aerodynamics Committee meeting minutes, 4 Oct. 1951, 30 Jan. 1952, and 24 June 1952; NASA LaRC staff report, "Conception and Research Background of the X-15 Project" (Hampton, VA: LaRC, June 1962), pp. 2-4; Hubert Drake and L. Robert Carman, "A Suggestion of Means for Flight Research at Hypersonic Velocities and High Altitudes" (Edwards, CA: HSFS, n.d.); D. G. Stone to F. L. Thompson, 21 May 1952; NASA LaRC staff report (draft), "History of NACA-Proposed High Mach Number, High-Altitude Research Airplane" (Hampton, VA: LaRC, n.d.), p. 2, in files of NASA history office; Hubert M. Drake and L. Robert Carman, "Suggested Program for High-Speed High-Altitude Flight Research" (HSFS, Aug. 1953); Douglas Aircraft Co., Summary Report for Contract No. 1266(00), "High Altitude and High Speed Study" (El Segundo, CA: Douglas Aircraft Co., 28 May 1954); letter., Edward H. Heinemann to Hallion, 10 Feb. 1972; Thomas A. Sturm, *The USAF Scientific Advisory Board: Its First Twenty Years, 1944-1964* (Washington, DC: USAF, 1 Feb. 1967), p. 59; John V. Becker, "The X-15 Project," *Astronautics & Aeronautics* (Feb. 1964), pp. 52-61.

over a month later, on 9 March 1954, NACA headquarters directed the laboratories to submit their views to Washington for evaluation. Ames, Langley, and Edwards supported the concept. Lewis favored an unpiloted rocket that could be launched from Wallops. Only Langley and Edwards submitted proposed configurations. Since Langley's was in greater detail, it was more useful for planning. Langley had created a five-memeber configuration study panel under the direction of engineer John Becker, and this team had produced a configuration that closely resembled the later X-15. When soliciting bids for what became the X-15, the NACA sent Becker's study to interested companies. "We didn't say 'Here's what we want,'" Becker later recalled, "but we said, 'Here's one configuration that we think might solve the problems and be what we're looking for.'. . . the proposals that we got back looked pretty much like the one we had put in."[157]

A briefing of the military services arranged by Hugh Dryden on the Becker study in July 1954 garnered enthusiastic endorsement. By October 1954, the Air Force and NACA had realized that such a program would be so large and expensive that a joint approach was desirable. On 18 October, the NACA, Air Force, and Navy created a three-person hypersonic aircraft committee to derive the precise specifications for the proposed airplane; Hartley Soulè represented NACA's interests on the panel. Walter Williams's staff at the High-Speed Flight Station furnished the committee with a detailed study of the instrumentation requirements. On 23 December 1954 Hugh L. Dryden and representatives from the Air Force and Navy signed a memorandum of understanding. The NACA would have technical control of the project, the Air Force and Navy would fund the design and construction phases, and the Air Force would administer those phases. Upon completion of contractor testing, the aircraft would be turned over to the NACA (NASA, as it turned out), which would conduct the flight-testing and report results. The memo concluded that "accomplishment of this project is a matter of national urgency."[158]

The three parties created a Research Airplane Committee, an interagency body of senior-level executives—Hugh Dryden represented the NACA on the body—to supervise the project. Program participants recall that the committee, popularly known as the X-15 Committee, did not exert much influence or control. It served primarily a psychological and political function and was largely honorary. The committee did not dabble in the design of the airplane; this was left up to the laboratories—especially Langley and the High-Speed Flight Station—the contractor, and the earlier RAPP headed by Soulè. Instead, it offered high-level sanction of lower-level initiatives. As

[157] Interview of Becker.

[158] LaRC, "Conception . . .," pp. 5-12; Robert S. Houston, *Development of the X-15 Research Aircraft, 1954-1959*, v. III of *History of Wright Air Development Center, 1958* (Wright-Patterson AFB, Dayton, OH: WADC, June 1959), pp. 3-9; Aerodynamics committee minutes, 5 Oct. 1954; Becker, 47; interview of Benjamin S. Kelsey, 15 Mar. 1978; "Principles for the Conduct by the NACA, Navy, and Air Force of a Joint Project for a New High Speed Research Airplane," memorandum of understanding signed 23 Dec. 1954, copy in NASA history office files.

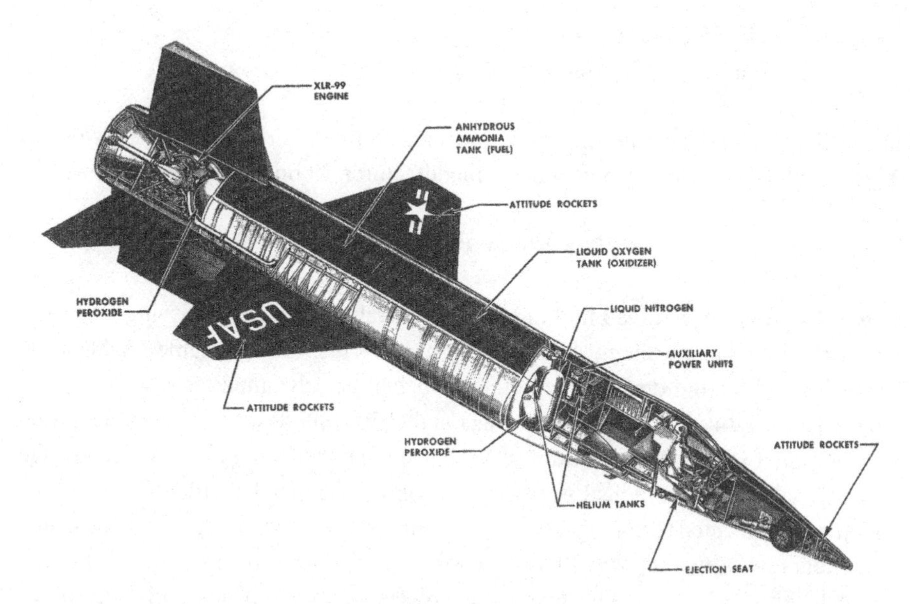

A cutaway drawing showing the internal details of the North American X-15 rocket-propelled research aircraft. (NASA photo E-4987)

one senior engineer recalled, it "met once in a while, but usually provided only a rubber stamp. And it was useful [to get a budget approved] to say 'And here's what the X-15 Committee wants to do.'"[159] The committee continued to exist until 26 October 1967, when OART closed it down. Its last significant action had been more than three years earlier on 18 February 1964, when committee members approved the Langley-developed Hypersonic Ramjet Experiment (HRE) for the X-15A-2.

The NACA/Air Force/Navy specification panel had stipulated by mid-December 1954 that the craft should be capable of attaining an altitude of 250,000 feet and an airspeed of 6,600 feet per second (Mach 6+). On 30 December 1954, invitations to bid on the contract were sent to 12 prospective contractors. Only four eventually submitted competitive designs: Bell, Douglas, North American, and Republic. For various technical reasons, Bell and Republic were quickly eliminated from serious consideration, and the competition became a neck-and-neck race between North American Aviation and Douglas. Douglas proposed a magnesium structure for the craft, but North American preferred Inconel, a nickel alloy, and this coincided with the dominant view at Langley. A final NASA/Air Force/Navy listing ranked the proposals in order:

1. North American (81.5 points out of a possible 100)
2. Douglas (80.1 points)

[159] Interview of Becker.

3. Bell (75.5 points)
4. Republic (72.2 points)

On 30 September 1955, the Air Force informed North American that it had won the X-15 competition. The X-15 now had a manufacturer. Round Two was under way.[160]

Necessary Preparations

The X-15 program involved building three research airplanes; modifying two B-52 bombers to air-launch them;[161] developing a powerful, fully reusable "man-rated" rocket engine for the craft; constructing a special aerodynamic test range running from Utah to Edwards, across the Nevada and California deserts; devising a special full-pressure flight suit; and building a special simulator within the Navy's centrifuge at the Naval Aviation Medical Acceleration Laboratory in Johnsville, Pennsylvania, which was connected to analog computing equipment. Eventually X-15 pilots spent 8 to 10 hours in a separate, fixed-base iron-bird simulator at Edwards practicing each 10- to 12-minute flight. All developments proceeded relatively smoothly, except for the development of the craft's throttleable rocket engine, rated at up to 250,000-newton (57,000-pound thrust), the Thiokol XLR-99.[162] Various delays and difficulties forced North American to substitute two of the older XLR-11 engines first used in the X-1 series, until the larger powerplant was ready for flight in late 1960. The X-15 airplane itself was ready in mid-1959.

One of the most important aspects of the X-15 effort, and one that the High-Speed Flight Station was intimately involved with, was creation of the X-15's tracking range, the so-called "High Range," short for High Altitude Continuous Tracking Radar Range. The NACA and Air Force cooperated in planning the range, with the High-Speed Flight Station's instrumentation staff under Gerald Truszynski determining its layout. Truszynski's staff informed the RAPP in November 1955 that the range should be at least 400 miles long, with three radar tracking stations able to furnish precise data on aircraft position, reentry prediction, geometric altitude, and ground speed. It required an air-launch site located over an emergency dry-lake landing area, intermediate dry-lake landing sites, intermediate launch (drop) sites, nearby airfields

[160] Houston, pp. 9-20; interview of Becker; review of Bell, Douglas, Republic, and North American proposal; LaRC, "Conception . . . ," pp. 25-54.

[161] The original launch vehicle was to have been a B-36. Hartley Soulè realized that it would be obsolete and impossible to maintain by the time the X-15 program got underway. Comments by Johnny Armstrong and Robert G. Hoey on the original chapter, 14 Feb. 2000. See also Robert S. Houston,"Transition from Air to Space: The North American X-15," *The Hypersonic Revolution: Case Studies in the History of Hypersonic Technology, Vol. I: From Max Valier to Project Prime (1924-1967)*, ed. Richard P. Hallion (Bolling AFB, Washington, DC: AF History and Museum Program, 1998), p. 123.

[162] There are reports that the XLR-99 engine could actually produce up to 60,000 pounds of thrust. See, for example, Milton O. Thompson, *At the Edge of Space: The X-15 Flight Program* (Washington, DC: Smithsonian Institution Press, 1992), p. 46.

that could be used for radar site support, and a "reasonably straight course." Truszynski and his staff concluded that the best course lay on a line from Wendover, Utah, to Edwards, with tracking stations at Ely and Beatty, Nevada, and at Edwards. The range would take the X-15 over some of the most beautiful, rugged, and desolate terrain in the Western Hemisphere. The X-15 would fly high over Death Valley before swooping down over the Searles basin to land at Rogers Dry Lake. In 1956, construction started on the High Range, and it was ready for operation in July 1958. It measured some 500 miles long, with a corridor width of about 50 miles. The Ely, Beatty, and Edwards tracking stations had radar and telemetry tracking with oscillograph recording, magnetic tape data collection, and console monitoring services. Each maintained a "local plot" of the X-15 as it passed, much as national Air Traffic Control centers process airliners on transcontinental flights. Edwards also had a master plot, and in the jargon of electronics engineers, the three sites had "interstation communication" via radio and telephone. Real-time data passed from one to the other as the X-15 sped along. On every flight, 87 channels, sampled 10 times per second, relayed information from the plane to the ground. The range would also prove beneficial to later NASA research involving vastly different aircraft. The three tracking stations did not come cheap: the Edwards station cost NASA $4,244,000; the other two combined cost about the same. The Air Force spent another $3.3 million on High Range construction.[163]

Aside from its involvement in the High Range, the High-Speed Flight Station in the years prior to the arrival of the X-15 supported the design and development stages of the program with such activities as the reaction control studies on the X-1B and later YF-104A, and the side-stick evaluation on the F-107. Station representatives reviewed progress on the development of the aircraft, attended meetings with the contractor, participated in mock-up inspection, and generally supported the NACA's—and later NASA's—involvement in the program with informed criticism and suggestions. When the first X-15 arrived at the High-Speed Flight Station, the station's technical staff was more than ready to begin work on it.

The first of the three X-15s arrived at the High-Speed Flight Station in mid-October 1958, trucked over the hills from the plant in Los Angeles. It was joined by the second airplane later that month. In contrast to the relative secrecy that had attended flight tests with the X-1 a decade before, the X-15 was pure theater.

The program inspired a great deal of public attention, coming, as it did, after Sputnik and during the race between the US and the USSR to orbit a man, a race won by the Soviet Union. North American erected a huge neon sign over its plant reading "Home of the X-15." Journalists flocked to Edwards for the first contractor test flights; international rivalry with the Soviet Union received less attention from the press than did the idea of the X-15's being a tool in America's "War against Space," as one

[163] Interview of Truszynski; interview of K. C. Sanderson, 3 Mar. 1977; "NASA Aerodynamic Test Range" (Edwards, CA: DFRC, n.d.), pp. iii-iv, 1-1; Houston, *Development of the X-15*, pp. 173-174; Stillwell, *X-15 Research Results*, pp. 41-44; Van Nimmen and Bruno, p. 273; USAF Systems Command, *System Package Program for X-15A Research Aircraft System 653A*, 18 May 1964, p. 3-5; Hoey comments.

The X-15 dropped from its B-52 mothership during one of the early contractor demonstration flights. (NASA photo E-4942)

journalist tagged it. Implicit in this were literary "How do they do it?" looks at the test pilots, writers waxing eloquent over the airmen going out and confronting the X-15 *mano a mano*. The project even gave rise to a ghastly Hollywood film, incorporating all the hackneyed stereotypes of celluloid test-flying. As Project Mercury moved from drawing board to launch pad, the camera crews and journalists left Edwards for Cape Canaveral; stayed there through Gemini, Apollo, and Skylab; and ventured back to the hinterlands of the high desert only when the Shuttle arrived on the scene.

The X-15's contractor program lasted two years, from mid-1959 through mid-1961. North American had to demonstrate the craft's general airworthiness during flights above Mach 2 and successful operation of its new XLR-99 engine before delivering the craft to NASA for research by the NASA/Air Force/Navy team. Anything beyond Mach 3 was considered a part of the government's research obligation. The task of flying the X-15 during the contractor program rested in the capable hands of Scott Crossfield, who had left NACA to join North American and help shepherd the craft through its long development. Crossfield completed the first captive flight on 10 March 1959 and first glide flight on 8 June. Just prior to landing, the plane began a series of increasingly wild pitching motions due to pilot-induced oscillations resulting from rate limiting of the horizontal-stabilizer actuators. The X-15 team subsequently increased the actuator rates from 15 to 25 degrees per second on all three X-15 aircraft to prevent a recurrence. The X-15 never again experienced the porpoising motions that had threatened it on its first flight. On 17 September, the X-15 completed its first powered flight as Crossfield flew the second airplane to Mach 2.11.[164]

A series of ground and in-flight accidents marred the X-15's contractor program, fortunately without injuries or even great delays for the program. On 5 November 1959, a small engine fire—always extremely hazardous in a volatile rocket airplane—forced Crossfield to make an emergency landing on Rosamond Dry Lake. The X-15 landed with a heavy load of propellants and broke its back, grounding the number two X-15 for three months. During a ground engine test with the third X-15 (the first one equipped with the large Thiokol engine), a stuck pressure regulator caused the craft to explode, requiring virtual rebuilding. The second X-15 was actually the first of the series to test-fly the large XLR-99 engine, and after adding the engine to the other two craft, North American delivered the last of the X-15s to NASA in June 1961. By that time, NASA, Air Force, and Navy test pilots had been operating the X-15 on government research flights for over a year.[165]

Researching the Fringes of Space

The government phase of the X-15's research program involved four broad objectives: verifing predicted hypersonic aerodynamic behavior and hypersonic heating rates, studying the X-15's structural characteristics in an environment of high heating and high flight loads, investigating hypersonic stability and control problems during atmospheric exit and reentry, and investigating piloting tasks and pilot performance. By late 1961, these four areas had been generally examined, though detailed research continued to about 1964 on the first and third aircraft, and to 1967 with the second (the X-15A-2). Before the end of 1961, the X-15 had attained its Mach 6 design goal[166] and had flown well above 200,000 feet (37 miles); by the end of the next year, the X-15 was routinely flying above 300,000 feet (55 miles). Within a single year, the X-15 had extended the range of winged aircraft flight speeds from Mach 3.2 to Mach 6.04, the latter achieved by Air Force test pilot Bob White on 9 November 1961.

The use of fixed-base simulators to support flight research of rocket-powered airplanes came into its own during this early phase of the X-15 program. Very short flight times (on the order of 10 minutes) meant that detailed and well-rehearsed flight

[164] Mulac, X-15 flight chronology; Crossfield and Blair, pp. 307-366; comments of Armstrong and Hoey; X-15 Program Summary, n.d., Milt Thompson Collection, Dryden Historical Reference Collection; on the pilot-induced oscillation, see Thomas W. Finch and Gene J. Matranga, "Launch, Low-Speed, and Landing Characteristics Determined from the First Flight of the North American X-15 Research Airplane" (Washington, DC: NASA TM-195, 1959), esp. Fig. 13 on p. 26; records of the X-15 kept by Roy Bryant.

[165] Mulac, X-15 flight chronology; Crossfield and Blair, pp. 366-405; X-15 series flight log assembled by Betty J. Love, NASA DFRC.

[166] The original design goals were a speed of Mach 6.6 and an altitude of 250,000 feet. Later the design speed was reduced to Mach 6.0 See Thompson, *Edge of Space*, p. 33.

The X-15 beginning its climb after launch, as shown by its contrail. (NASA photo EC65-884)

plans were essential to maximize the data returned from each flight. The X-15 team also used the simulator to develop and practice emergency procedures for various missed predictions about stability or system malfunctions.[167]

The intensive flight program on the X-15 revealed a number of interesting things. Physiological researchers discovered that the heart rates of X-15 pilots varied between 145 and 180 beats per minute on a flight, as compared to a normal of 70 to 80 beats per minute for test missions in other aircraft. Aeromedical researchers eventually concluded that prelaunch anticipatory stress, rather than actual postlaunch physical stress, influenced the heart rate. They believed, correctly, that these rates could be considered as probable baselines for predicting the physiological behavior of future pilot-astronauts. Aerodynamic researchers found remarkable agreement between the tunnel tests of exceedingly small X-15 models and actual results, with the exception of drag measurements. Drag produced by the blunt aft end of the aircraft proved 15 percent higher on the actual aircraft than wind-tunnel tests had predicted.[168] At Mach

[167] From comment by Hoey.

[168] Correlating full-scale flight-test measurements of base drag with predicted drag values from tunnel tests continues to pose serious challenges for engineers. Many aircraft continue to exhibit much higher base drag in actual flight than has been indicated by tunnels.

6, the X-15 absorbed eight times the heating load it experienced at Mach 3, with the highest heating rates occurring in the frontal and lower surfaces of the aircraft, which received the brunt of airflow impact. During the first Mach 5+ excursion, four expansion slots in the leading edge of the wing generated turbulent vortices that increased heating rates to the point that the external skin behind the joints buckled. As a solution, NASA technicians added small Inconel alloy strips over the slots, and the X-15 flew without further evidence of buckling. It was "a classical example of the interaction among aerodynamic flow, thermodynamic properties of air, and elastic characteristics of structure."[169]

Heating and turbulent flow generated by the protruding cockpit posed other serious problems; on two occasions, the outer panels of the X-15's heavy glass cockpit windshields fractured because heating loads in the frame overstressed the soda-lime glass. The X-15 team solved the difficulty by changing the cockpit frame from Inconel to titanium, modifying its configuration, and replacing the outer glass panels with high-temperature alumina-silica glass. Another problem concerned an old aerodynamics and structures bugaboo, panel flutter. Panels along the flanks of the X-15 fluttered at airspeeds above Mach 2.4, forcing engineers to add longitudinal metal stiffeners to the panels. These difficulties warned aerospace designers to proceed cautiously. John Becker, writing in 1968, noted in regards to the X-15 experience,

> The really important lesson here is that what are minor and unimportant features of a subsonic or supersonic aircraft must be dealt with as prime design problems in a hypersonic airplane. This lesson was applied effectively in the precise design of a host of important details on the manned space vehicles.[170]

A serious roll instability predicted for the airplane under certain reentry conditions posed a serious challenge to flight researchers. To successfully complete the reentry into the atmosphere from the design altitude (250,000 feet) without exceeding the aircraft's structural limits, the X-15 had to fly at angles of attack of at least 15 degrees. Yet the cruciform "wedge" tail,[171] so necessary for stability and control in other portions of the plane's flight regime, actually prevented it from being flown safely at angles of attack greater than 20 degrees in normal flight because of potential rolling problems.[172] By this time, FRC researchers had gained enough experience with the XLR-99 engine

[169] Stillwell, p. 65. Quotation from this source. Stillwell's book, published in 1965, is useful mostly for the information on the program prior to that date. Interview of Dale Reed by Hallion, 30 Nov. 1976.

[170] Becker, "The X-15 Program in Retrospect." See also Stillwell, p. 66; James E. Love, "X-15: Past and Future," a paper presented to the Fort Wayne Section, Society of Automotive Engineers, 9 Dec. 1964.

[171] Actually, only the lower ventral fin, not the wedge shape.

[172] A single failure in the roll damper would decrease the maximum angle of attack that could be flown safely to eight degrees.

The X-15 landing from a research mission on Rogers Dry Lake, followed by a Lockheed F-104A flying chase. (NASA photo E95-43116-22, copied from an official USAF photo)

to realize that fears of thrust misalignment—a major reason for the large vertical fin—were unwarranted. The obvious solution was simply to remove the lower half of the ventral fin, a portion of the fin that X-15 pilots had to jettison prior to landing anyway so that the craft could touch down on its landing skids. Removing the ventral produced an acceptable tradeoff. While it reduced stability by about 50 percent at high angles of attack, it greatly improved the pilot's ability to control the airplane, even with a failed roll damper. With the ventral off, the X-15 could now fly into the previously "uncontrollable" region above 20 degrees angle of attack with complete safety. Eventually the X-15 went on to reentry trajectories of up to 26 degrees, often with flight path angles of -38 degrees at speeds up to Mach 6, a much more demanding piloting task than the shallow entries flown by piloted vehicles returning from orbital or lunar missions. Its reentry characteristics were quite different from those of the later NASA Space Shuttle Orbiter, but the X-15 did demonstrate the basic idea of a controlled lifting entry.[173]

When Project Mercury took to the air, it rapidly eclipsed the X-15 in glamour. FRC's researchers and NASA Headquarters viewed the two programs as complementary, however. Mercury dominated some of the research areas that had first interested X-15 planners, such as "zero g" weightlessness studies. Reaction controls used to maintain a vehicle's attitude in space proved academic after Mercury flew, but the X-15 had already proved them in a near-space environment and would also furnish valuable design information on the use of blending reaction controls with

[173] Stillwell, pp. 51-52, 75-78; for a general technical review of the X-15 effort, see Joseph Weil, *Review of the X-15 Program* (Washington, DC: NASA TN D-1278, June 1962); Becker, "The X-15 Program in Retrospect"; comments of Hoey.

conventional aerodynamic controls during an exit and reentry, a matter of concern to subsequent Shuttle development. The X-15 experience clearly demonstrated the ability of pilots to fly rocket-propelled aircraft out of the atmosphere and back in to reasonably precise landings. FRC Director Paul Bikle saw the X-15 and Mercury as a

> parallel, two-pronged approach to solving some of the problems of manned space flight. While Mercury was demonstrating man's capability to function effectively in space, the X-15 was demonstrating man's ability to control a high-performance vehicle in a near-space environment. . . . considerable new knowledge was obtained on the techniques and problems associated with lifting reentry.[174]

Operationally, the X-15 gave the Flight Research Center staff a number of headaches. Because of the complexity of its systems, the plane experienced a number of operational glitches that delayed flights, aborted them before launch, or forced abandonment of a mission after launch. Early in the program, the X-15's stability augmentation and inertial guidance systems were two major problem areas. NASA eventually replaced the Sperry inertial unit with a Honeywell unit first designed for the Dyna-Soar. The plane's propellant system also had its own weaknesses. Pneumatic vent and relief valves and pressure regulators gave the greatest difficulties, followed by spring pressure switches in the auxiliary power units, the turbopump, and the gas generation system. NASA's mechanics routinely had to reject 24 to 30 percent of spare parts as unusable, a clear indication of the difficulties of devising industrial manufacturing and acceptance-test procedures when building for use in an environment at the frontier of science.[175] Weather posed another critical factor. Many times Edwards enjoyed fine weather with the lakebed bone-dry, while upcountry the High Range was covered with clouds, alternate landing sites were flooded, or some other meteorological condition postponed a mission. In one case, weather and minor maintenance kept one X-15 grounded from mid-October 1961 to early January 1962. When it finally flew, the pilot had to make an emergency landing up range. Weather and maintenance then grounded the plane until mid-April.[176]

The X-15 had its share of accidents, one of which killed an Air Force test pilot; another seriously injured a NASA research pilot. As previously mentioned, Scott Crossfield once made an emergency landing on Rosamond Dry Lake with an X-15 damaged by an engine fire; the plane broke its back on landing, necessitating lengthy repairs. The third X-15 blew up during ground testing of its XLR-99 engine, but it too was rebuilt. In November 1962, an engine failure forced Jack McKay, a NASA veteran

[174] Stillwell, p. iv; see also Walter C. Williams, "The Role of the Pilot in the Mercury and X-15 Flights," *Proceedings* of the 14th AGARD General Assembly, 16-17 Sept. 1965, Portugal.

[175] Weil, *Review of the X-15 Program*; James E. Love and W. R. Young, "Operational Experience from the X-15 Program," n.d., DFRC Historical Reference Collection.

[176] Betty J. Love, X-15 flight chronology, DFRC Historical Reference Collection.

of Round One, to make an emergency landing at Mud Lake, Nevada, in the second X-15; its landing gear collapsed and the X-15 flipped over on its back. McKay was promptly rescued by an Air Force medical team standing by near the launch site and eventually recovered to fly the X-15 again. But his injuries, more serious than at first thought, eventually forced his retirement from NASA. In November 1967, Maj. Mike Adams was killed in a strange accident in the third X-15 that will be discussed later in great detail. One of the most remarkable close calls in the X-15 program involved Air Force test pilot Maj. William J. "Pete" Knight. In June 1967, he experienced a complete electrical and hydraulic failure while climbing through 107,000 feet at Mach 4+. With no computed information and guidance, Knight continued to climb, suddenly reduced to "seat of the pants" flying technique. During reentry he managed to restart one of the auxiliary power units, restoring some instruments and hydraulic pressure for the aerodynamic flight-control system. He then made an emergency landing at Mud Lake, for which he received the Distinguished Flying Cross. This was an excellent demonstration of the value of simulator training for the X-15 pilot. Knight knew from practice on the simulator that he could complete the reentry with only one hydraulic system operational.[177]

The X- 15 Follow-on Program, 1963-1967

Within NACA and later NASA, developing the X-15 had been left largely in the hands of Langley, the center most closely involved in determining its mission and configuration, with important input from the other centers and especially from the High-Speed Flight Station. The flight research program was the province of the HSFS, which became the Flight Research Center during the X-15 era, with liaison and support from the Air Force Flight Test Center at Edwards. In the summer of 1961, as the X-15 approached its maximum performance during test flights, a new initiative began, one that sprang jointly from the Air Force's aeronautical systems division at Wright-Patterson AFB and from NASA headquarters: using the X-15 as a "testbed," or carrier aircraft, for a wide range of scientific experiments unforeseen in its original conception.

Pressures had existed, even before the X-15 first flew, to extend the scope of the program beyond aerodynamics and structural research. Researchers at the Flight Research Center had proposed using the airplane to carry to high altitude some experiments related to the proposed Orbiting Astronomical Observatory; others suggested modifying one of the planes to carry a Mach 5+ ramjet for advanced air-breathing propulsion studies. Over 40 experiments were suggested by the scientific community as suitable candidates for the X-15 to carry. In August 1961—after consulting with Bikle at the FRC, with NASA headquarters, and the Air Force's aeronautical systems division—NASA, Air Force, and Navy representatives formed an X-15 Joint Program Coordinating Committee to prepare a plan for a follow-on experiments program. Most of the suggested experiments were in space science, such

[177] Comments of Hoey and Armstrong.

as ultraviolet stellar photography. Others supported the Apollo program and hypersonic ramjet studies. A series of meetings held at NASA headquarters over the fall of 1961 that included the joint committee, Hartley Soulè, and John Stack (then NASA's director of aeronautical research) culminated in approval of the proposed follow-on research program and the classification of two groups of experiments. Category A experiments consisted of well-advanced and funded experiments having great importance; category B included worthwhile projects of less urgency or importance.[178]

In March 1962, the X-15 committee approved the "X-15 Follow-on Program," which NASA announced 13 April in a Headquarters news conference presided over by Stack and FRC planner Hubert Drake. Drake announced that the first task would be to fly an ultraviolet stellar photography experiment from the University of Wisconsin's Washburn Observatory. NASA had investigated the possibility of the X-15 carrying a Scout booster that could fire small satellites into orbit, the entire B-52/X-15/Scout becoming in effect a multistage satellite booster, but NASA eventually rejected the idea for reasons of safety, utility, and economy. The X-15's space science program eventually included 46 experiments ranging from astronomy to collecting micrometeorites, using wingtip pods that opened at 150,000 feet, and conducting high-altitude mapping, although not all of the experiments could be completed. Two of the follow-on programs, a horizon definition experiment from the Massachusetts Institute of Technology and tests of proposed insulation for the Saturn launch vehicle, directly affected navigation equipment and the thermal protection used on Apollo-Saturn. The FRC quickly implemented the follow-on program. In 1964, fully 65 percent of all data returned from the three X-15 aircraft involved follow-on projects; this percentage increased yearly through the conclusion of the program.[179]

NASA's major X-15 follow-on project involved a Langley-developed Hypersonic Ramjet Experiment (HRE). FRC advanced planners had long wanted to extend the X-15's speed capabilities, perhaps even to Mach 8, by adding extra fuel in jettisonable drop tanks and some sort of thermal protection system. Langley researchers had developed a design configuration for a proposed hypersonic ramjet engine. The two groups came together to advocate modifying one of the X-15's as a Mach 8 research craft that could be tested with a ramjet fueled by liquid hydrogen. The proposal became more attractive when the landing accident to the second X-15 in November 1962

[178] James E. Love, "X-15: Past and Future"; Becker, "The X-15 Program in Retrospect"; memo on X-15 follow-on program, Homer Newell to Hugh Dryden, 18 Dec. 1961; undated memo, Bikle to Soulè (believed Nov. 1961); memo, Stack to Dryden, 3 Jan. 1962; letter, Dryden to Maj. Gen. Marvin C. Demler, USAF, 12 July 1962; NASA news release 61-261; memorandum of understanding, X-15 Flight Research Program, 15 Aug. 1961, signed by Maj. Gen. Marvin C. Demler, Hugh L. Dryden, and John T. Hayward, for the Air Force, NASA, and the Navy, respectively, copy in the Dryden Historical Reference Collection.

[179] Air Force Systems Command, X-15 System Package Program, 6-37-48; James E. Love, "X-15: Past and Future"; letter, Dryden to Demler, 23 Mar. 1962; NASA news release 62-98; X-15 news release 62-91; letter, Dryden to Lt. Gen. James Ferguson, 15 July 1963; NASA news release 64-42; Carlton R. Gray, *MIT/IL X-15 Horizon Definition Experiment Final Report* (Cambridge, MA: MIT Instrumentation Laboratory, Oct. 1969), passim; NASA George C. Marshall Space Flight Center news release 68-69; Ronald G. Boston, "The X-15's Role in Aerospace Progress,"Hallion, ed., *Hypersonic Revolution*, Vol. I, pp. 173-177.

forced the rebuilding of the aircraft. The opportunity to make the modifications was too good to pass up. In March 1963, the Air Force and NASA authorized North American to rebuild the airplane with a longer fuselage. Changes were to be made in the propellant system; two huge drop tanks (3 feet, 2 inches by 23 feet, 6 inches) and a small tank for liquid hydrogen within the plane were to be added. Forty weeks and $9 million later, North American delivered the modified plane, designated the X-15A-2, to NASA in February 1964.[180]

The X-15A-2 first flew in June 1964, piloted by Air Force test pilot Maj. Bob Rushworth. Early proving flights demonstrated that the plane retained satisfactory flying qualities at Mach 5+ speeds, although on one flight thermal stresses caused the nose landing gear to extend at Mach 4.3, generating "an awful bang and a yaw," but Rushworth landed safely despite the blowout of heat-weakened tires on touchdown. In November 1966, Air Force Maj. Pete Knight set an unofficial world's airspeed record of Mach 6.33 in the plane. NASA then grounded it for application of an ablative coating to enable it to exceed Mach 7.[181]

The Flight Research Center's technical staff had evaluated several possible coatings that could be applied over the X-15's Inconel structure to enable it to withstand the added thermal loads experienced above Mach 6. NASA hoped that such coatings might point the way toward materials that could be readily and cheaply applied to reusable spacecraft, minimizing refurbishment costs and turnaround time between flights. Such a coating would have to be relatively light; have good insulating properties; be easy to apply, cure, and then remove; and be easy to reapply before another flight. On the FRC's advice, a joint NASA/Air Force committee selected an ablator developed by the Martin Company, MA-25S, in connection with some corporate studies on reusable spacecraft concepts. Consisting of a resin base, a catalyst, and a glass bead powder, it would protect the X-15's structure from the expected 2,000 degrees Fahrenheit (F). heating as the craft sped through the upper atmosphere. Martin estimated that the coating, ranging from half an inch thick on the canopy, wings, vertical and horizontal tail down to less than 2/10 of an inch on the trailing edges of the wings and tail, would keep the skin temperature down to a comfortable 600 degrees F. The first unpleasant surprise came, however, with applying the coating to the X-15A-2: it took six weeks. Because the ablator would char and emit a residue in flight, North American had installed an "eyelid" over the left cockpit window that would remain closed until just before approach and landing. During launch and climb-out, the pilot would use the right window, but residue from the ablator would render it opaque above Mach 6.[182]

[180] AFSC, X-15 System Package Program, 1-2; Robert A. Hoover and Robert A. Rushworth, "X-15A-2 Advanced Capability," paper presented at the annual symposium of The Society of Experimental Test Pilots (SETP), Beverly Hills, CA, 25-26 Sept. 1964.

[181] Betty J. Love, X-15 flight chronology; cockpit voice transcription for Rushworth flight, 14 Aug. 1964; X-15 Operations Flight Report, 19 Aug. 1964; Rushworth flight comments, n.d.

[182] Hoover and Rushworth, "X-15A-2 Advanced Capability," NASA news release 66-11; C. M. Plattner, "Insulated X-15A-2 Ready for Speed Tests," *Aviation Week & Space Technology* (24 July 1967), pp. 75-81.

X-15A-2 with ablative coating and external tanks in the configuration that carried Air Force Maj. Pete Knight to the program's top speed of Mach 6.7 (4,520 mph). (NASA photo EC68-1889)

Late in the summer of 1967, the X-15A-2 was ready for flight with the ablative coating. It had already flown with a dummy ramjet affixed to its stub ventral fin; the ramjet, while providing a pronounced nose-down trim change, actually added to the plane's directional stability. The weight of the ablative coating—125 pounds more than planned—together with expected increased drag reduced the theoretical maximum performance of the airplane to Mach 7.4, still a significant advance over the Mach 6.3 previously attained with the plane. The appearance of the X-15A-2 was striking, an overall flat off-white finish, the huge external tanks a mix of silver and orange-red with broad striping. NASA hoped that early Mach 7+ trials would lead to tests with an actual "hot" ramjet rather than the dummy now attached to the plane. On 21 August 1967, Knight completed the first flight in the ablative-coated plane, reaching Mach 4.94 and familiarizing himself with its handling qualities. His next flight, on 3 October 1967, was destined to be the X-15's fastest flight and the most surprising as well.[183]

That day, high over Nevada, Knight dropped away from the B-52, the heavy X-15A-2 brimming with fuel. Knight climbed under the full thrust of the rocket engine. When the external tanks were emptied, he jettisoned them and continued on the craft's

[183] William J. Knight, "Increased Piloting Tasks and Performance of X-15A-2 in Hypersonic Flight," paper presented at the annual symposium of the SETP, Beverly Hills, CA, 28-30 Sept. 1967; Betty J. Love, X-15 flight chronology.

internal supply, leveling off at slightly over 100,000 feet. It was a flight in the grand Edwards tradition of Yeager and Crossfield. The X-15A-2's engine burned more than 141 seconds and reached Mach 6.72, or 4,520 miles per hour—a mark that would stand as a record for winged vehicles until the return of the Space Shuttle *Columbia* from orbit in 1981. Unknown to Knight, however, all was not well with the plane. Preflight studies did not adequately predict the complex local heating conditions the aircraft would experience. Temperatures later determined to have been above 1,650 degrees Centigrade (3,000 degrees F.) damaged the structure around the dummy ramjet so that it fell off its pylon, searing a hole measuring seven by three inches into the ventral fin's leading edge. An airscoop effect channeled hot air into the lower fuselage and damaged the propellant jettison system. Knight eventually had to land the plane 1,500 pounds heavier than planned because he could not jettison residual fuel. If the heat had damaged the craft's hydraulics, Knight might have had to abandon the plane. Fortunately, that did not happen. Knight landed at Edwards, the plane resembling burnt firewood. It had been an eventful flight. The engineers then sat down and took a long look at what it all meant.[184]

What it really meant was the end of the refurbishable spray-on ablator concept. It was the closest any X-15 came to structural failure induced by heating. The ablative material was charred on the airplane's leading edges and nosecap. The ablator had actually prevented cooling of some hot spots by keeping the heat away from the craft's metal heatsink structure. On earlier flights without the ablator, some of those areas had remained relatively cool because of heat transfer through the heavy Inconel structure. Some heating effects, such as at the tail and body juncture and where shockwaves intersected the structure, had been the subject of theoretical studies but had never before been seen on an actual aircraft in flight. To John Becker at Langley, the flight underscored "the need for maximum attention to aerothermodynamic detail in design and preflight testing."[185] To Jack Kolf, an X–15 project engineer at the FRC, the X–15A–2's condition "was a surprise to all of us. If there had been any question that the airplane was going to come back in that shape, we never would have flown it."[186] The ablator had done its job, but refurbishing for another flight near Mach 7 would have taken five weeks. Technicians would have had great difficulty in ensuring adequate depth of the ablator over the structure. Obviously, a much larger orbital vehicle would have had even greater problems. The spray-on ablator concept thus died a natural death. The unexpected airflow problems with the ramjet ended any idea of using that configuration on the X-15. After the flight, NASA sent the X-15A-2 to its manufacturer for general maintenance and repair. Although the plane returned to Edwards in June 1968, it never flew again.

[184] Johnny G. Armstrong, *Flight Planning and Conduct of the X-15A Envelope Expansion Program*, AFFTC-TD-69-4; Betty J. Love, X-15 flight chronology.

[185] Becker, "The X-15 Program in Retrospect"; comments of Armstrong.

[186] Interview of Jack Kolf, 28 Feb. 1977.

The End of an Era

The third X-15 featured specialized flight instrumentation and displays that rendered it particularly suitable for high-altitude flight research. A key element of its control system was a so-called "adaptive" flight-control system developed by Honeywell. Automatically compensating for the airplane's behavior in various flight regimes, it combined the aerodynamic control surfaces and the reaction controls into a single control "package." This offered much potential for future high-performance aircraft such as the Dyna-Soar and supersonic transports.

By the end of 1963, the X-15-3 had flown above 50 miles, the altitude that the Air Force recognized as the minimum boundary of spaceflight. FRC pilot Joe Walker set an X-15 record for winged spaceflight by reaching 354,200 feet, a record that stood until the orbital flight of *Columbia* nearly a decade later. These flights, and others later, acquired reentry data considered applicable to the design of future "lifting reentry" spacecraft such as the present-day Space Shuttle. By October 1967, the X-15-3 had completed 64 research flights, 20 at altitudes above 200,000 feet. It became the prime testbed for carrying experiments to high altitude, especially micrometeorite collection and solar-spectrum analysis experiments, although the number one X-15 also flew some of the micrometeorite-collection flights.[187]

As had happened in some other research aircraft programs, a fatal accident signaled the end of the X-15 program. On 15 November 1967 at 10:30 a.m., the X-15-3 dropped away from its B-52 mothership at 45,000 feet near Delamar Dry Lake. At the controls was veteran Air Force test pilot Maj. Michael J. Adams. Starting his climb under full power, he was soon passing through 90,000 feet. Then an electrical disturbance distracted him and slightly degraded the control of the aircraft. Having adequate backup controls, Adams continued on. At 10:33 he reached a peak altitude of 266,000 feet. In the FRC flight-control room, fellow pilot and mission controller Pete Knight monitored the mission with a team of engineers. Something was amiss. As the X-15 climbed, Adams started a planned wing-rocking maneuver so an onboard camera could scan the horizon. The wing rocking quickly became excessive, by a factor of two or three. When he concluded the wing-rocking portion of the climb, the X-15 began a slow, gradual drift in heading; 40 seconds later, when the craft reached its maximum altitude, it was off heading by 15 degrees. As the plane came over the top, the drift briefly halted, with the plane yawed 15 degrees to the right. Then the drift began again; within 30 seconds, the plane was descending at right angles to the flight path. At 230,000 feet, encountering rapidly increasing dynamic pressures, the X-15 entered a Mach 5 spin.[188]

In the flight-control room, there was no way to monitor heading, so nobody suspected the true situation that Adams now faced. The controllers did not know that

[187] See Appendix 1 of Thompson, *Edge of Space*.

[188] Donald R. Bellman et al., *Investigation of the Crash of the X-15-3 Aircraft on November 15, 1967* (Edwards, CA: NASA FRC, Jan. 1968), pp. 8-15.

the plane was yawing, eventually turning completely around. In fact, control advised the pilot that he was "a little bit high," but in "real good shape." Just 15 seconds later, Adams radioed that the plane "seems squirrelly." At 10:34 came a shattering call: "I'm in a spin, Pete." A mission monitor called out that Adams had, indeed, lost control of the plane. A NASA test pilot said quietly, "That boy's in trouble." Plagued by lack of heading information, the control room staff saw only large and very slow pitching and rolling motions. One reaction was "disbelief; the feeling that possibly he was overstating the case." But Adams again called out, "I'm in a spin." As best they could, the ground controllers sought to get the X-15 straightened out. They knew they had only seconds left. There was no recommended spin recovery technique for the plane, and engineers knew nothing about the X-15's supersonic spin tendencies. The chase pilots, realizing that the X-15 would never make Rogers Dry Lake, went into afterburner and raced for the emergency lakes, for Ballarat, for Cuddeback. Adams held the X-15's controls against the spin, using both the aerodynamic control surfaces and the reaction controls. Through some combination of pilot technique and basic aerodynamic stability, the plane recovered from the spin at 120,000 feet and went into a Mach 4.7 dive, inverted, at a dive angle between 40 and 45 degrees.[189]

Adams was in a relatively high altitude dive and had a good chance of rolling upright, pulling out, and setting up a landing. But now came a technical problem that spelled the end. The Honeywell adaptive flight-control system began a limit-cycle oscillation just as the plane came out of the spin, preventing the system's gain changer from reducing pitch as dynamic pressure increased. The X-15 began a rapid pitching motion of increasing severity. All the while, the plane shot downward at 160,000 feet per minute, dynamic pressure increasing intolerably. High over the desert, it passed abeam of Searles Lake, over the Pinnacles, arrowing on toward Johannesburg. As the X-15 neared 65,000 feet, it was speeding downward at Mach 3.93 and experiencing over 15 g vertically, both positive and negative, and 8 g laterally. It broke up into many pieces amid loud sonic rumblings, striking northeast of Johannesburg. Two hunters heard the noise and saw the forward fuselage, the largest section, tumbling in the sky. On the ground, NASA control lost all telemetry at the moment of breakup, but still called to Adams. A chase pilot spotted dust on Cuddeback, but it was not the X-15. Then an Air Force pilot, who had been up on a delayed chase mission and had tagged along on the X-15 flight to see if he could fill in for an errant chase plane, spotted the main wreckage northwest of Cuddeback. Mike Adams was dead, the X-15 destroyed. NASA and the Air Force convened an accident board.[190]

Chaired by NASA's Donald R. Bellman, the board took two months to prepare and write its report. Ground parties scoured the countryside looking for wreckage, any bits that might furnish clues. Critical to the investigation were the cockpit camera and its film. The weekend after the accident, a voluntary and unofficial FRC search

[189] Ibid.; confidential interviews.

[190] X-15 accident report, passim.

party found the camera; disappointingly, the film cartridge was nowhere in sight. Engineers theorized that the film cassette, being lighter than the camera, might be further away, to the north, blown there by winds at altitude. FRC engineer Victor Horton organized a search, and on 29 November, during the first pass over the area, W. E. Dives found the cassette. It was in good condition. Investigators meanwhile concentrated on analyzing all telemetered data, interviewing participants and witnesses, and studying the aircraft systems. Most puzzling was Adams' complete lack of awareness of major heading deviations in spite of accurately functioning cockpit instrumentation. The accident board concluded that he had allowed the aircraft to deviate as the result of a combination of being distracted, misinterpreting his instrumentation display and, possibly, vertigo.[191] The electrical disturbance early in the flight degraded the overall effectiveness of the aircraft's control system and further added to pilot workload. The X-15's adaptive control system then broke up the airplane on reentry. The board made two major recommendations: install a telemetered heading indicator in the control room, visible to the flight controller, and medically screen X-15 pilot candidates for labyrinth (vertigo) sensitivity. As a result of the X-15 crash, FRC added a ground-based "8 ball" attitude indicator, displayed on a TV monitor in the control room, which furnished mission controllers with "real time" pitch, roll, heading, angle-of-attack, and sideslip information available to the pilot, using it for the remainder of the X-15 program.[192]

So passed the third X-15. The program itself did not long survive. NASA had grounded the X-15A-2 for major repairs by North American Rockwell, a grounding that became permanent. Only the first X-15 remained, and it soldiered on. Opinion within NASA had long been split as to whether the X-15 program should continue. The ramjet and a proposed X-15 delta conversion offered hope to zealots that the program might last until 1972 or 1973, but the loss of two of the three aircraft ended that hope. As early as March 1964, after consultation with NASA, Brig. Gen. James T. Stewart, director of science and technology for the Air Force, had determined that the program would end in December 1968.[193] The X-15-1 had just about exhausted its research ability, and each flight cost roughly $600,000. Even FRC Director Paul Bikle believed that the program had continued beyond its point of useful return. "X-15" and "FRC" had become such synonymous terms that, according to uninformed speculation, when the X-15 stopped flying, the FRC would cease to exist. In fact, many other FRC programs could benefit from the resources needed to fly the X-15—programs such as the lifting bodies and the YF-12A advanced supersonic Mach 3 airplanes. NASA's

[191] During testing for the Manned Orbiting Laboratory (MOL) program, Adams had shown an unusual susceptibility to vertigo and had experienced vertigo throughout boost to reentry on earlier X-15 flights. Other X-15 pilots often experienced vertiginous tendencies during boost.

[192] X-15 accident report; R. Dale Reed, "RPRV's: The First and Future Flights," *Astronautics & Aeronautics* (Apr. 1974), pp. 26-42.

[193] USAF HQ. Development Directive No. 32, 5 Mar. 1964, repr. in X-15 System Package Program, p. 13-7.

OART recognized this, so support for continued X-15 operations was not strong. NASA did not request funding for X-15 operations after December 1968.[194]

During 1968, Bill Dana of NASA and Pete Knight of the Air Force took turns flying the first X-15. A variety of weather, maintenance, and operational problems caused rescheduling and cancellation of a number of flights. On 24 October 1968, Bill Dana completed the first X-15's 81st flight, the 199th flight of the series. The plane attained Mach 5.38 at 255,000 feet, carrying a variety of follow-on experiments. Two months remained before funding would end, and FRC engineers hoped to get the 200th flight in before the program closed down. In spite of every effort, however, maintenance and weather problems intervened. After several abortive attempts and repeated changes in the flight plan, the FRC at last had the X-15 and B-52 ready for flight on 20 December—and it began to snow. The support helicopters didn't have the visibility to get airborne and go up range. Technicians demated the pair for the last time, then left with the rest of the personnel for a wake at Juanita's saloon in Rosamond. Betty Love, assembling the log of the X-15 trio for the FRC aerodynamics structures office, closed with a final notation: "This ends an era in flight research history. " And indeed it did.

The X-15 in Retrospect

It was unfortunate that Hugh Latimer Dryden did not witness the conclusion of the X-15 program at the center soon to bear his name. Dryden had seen much of aviation history, from the early days of transonic research in the 1920s when he studied airflow around moving propeller tips, through the heyday of the X-1, X-15, and into Apollo. His voice had been an important one in design of several major systems in the X-15. "It is fair to state," Jerome Hunsaker and Robert Seamans have written, "that Dryden's 1920 work on supersonic aerodynamics led consistently to operational supersonic airplanes, the famous rocket-propelled X-15, and successful manned space flight."[195] But Dryden was dead. Exploratory surgery in 1961 had revealed a serious malignancy. Dryden continued working almost to the end, living to see the X-15 hailed as the most successful research airplane of all time. His death on 2 December 1965, at the age of 67, was a great loss to the nation and to NASA. He left a rich legacy and an outstanding reputation. Nothing could have satisfied him more than the three X-15s flying in desert skies.

Tabulating the X-15's statistics is easy. Assessing its significance to postwar aerospace research and development is more difficult. In 199 flights, the X-15 spent more than eighteen hours above Mach 1, some twelve hours above Mach 2, nearly nine hours above Mach 3, nearly six hours above Mach 4, more than one hour above

[194] James E. Love and William R. Young, "Survey of Operation and Cost Experience of the X-15 Airplane as a Reusable Space Vehicle" (Washington, DC: NASA TN D-3732, Nov. 1966), p. 7; interviews of Bikle and Finch.

[195] J. C. Hunsaker and R. C. Seamans, "Hugh Latimer Dryden: 1898-1965" in *Biographical Memoirs of the National Academy of Sciences* (1969), p. 50.

Mach 5, and scarcely more than one minute above Mach 6. It flew to a speed of Mach 6.72 and reached an altitude of over 67 miles. Twelve pilots flew it. Starting as a hypersonic aerodynamics research tool, the X-15 became much more than that. What, then, did it accomplish?

In October 1968, John Becker enumerated 22 accomplishments from the research and development work that produced the X-15, 28 accomplishments from its actual flight research, and 16 from testbed investigations. As of May 1968, the X-15 had generated 766 technical reports on research stimulated by its development, flight-testing, and test results, equivalent to the output of a typical 4,000-person federal research center working for two years. As the X-1 had provided a focus and stimulus for supersonic research, the X-15 furnished a focus and stimulus for hypersonic studies. A sampling of its accomplishments indicates their scope:

> • Development of the first large restartable "man-rated" throttleable rocket engine, the XLR-99.
> • First application of hypersonic theory and wind-tunnel work to an actual flight vehicle.
> • Development of the wedge tail as a solution to hypersonic directional stability problems.
> • First use of reaction controls for attitude control in space.[196]
> • First reusable superalloy structure capable of withstanding the temperatures and thermal gradients of hypersonic reentry.
> • Development of new techniques for the machining, forming, welding, and heat-treating of Inconel X and titanium.
> • Development of improved high-temperature seals and lubricants.
> • Development of the NACA [servo-actuated] flow-direction sensor for operation over an extreme range of dynamic pressures and a stagnation air temperature of 1,900 degrees C. [3,452 degrees F.].
> • Development of the first practical full-pressure suit for pilot protection in space.
> • Development of nitrogen cabin conditioning.
> • Development of inertial flight data systems capable of functioning in a high-dynamic pressure and space environment.
> • Discovery that hypersonic boundary layer flow is turbulent and not laminar.
> • Discovery that turbulent heating rates are significantly lower than had been predicted by theory.

[196] This claim depends for its validity on the definition of space that one applies. If the NASA and international definitions of 62 miles constitute the criterion, the Mercury capsule performed limited attitude control with reaction controls before the X-15. If the definition is a low dynamic pressure comparable to that in space, the claim is valid. In any event, as already pointed out, the X-15 was the first vehicle to blend reaction controls with normal atmospheric flight-control surfaces.

- First direct measurement of hypersonic skin friction, and discovery that skin friction is lower than had been predicted.
- Discovery of "hot spots" generated by surface irregularities.
- Discovery of methods to correlate base drag measurements with tunnel test results so as to correct wind-tunnel data.
- Development of practical boost guidance pilot displays.
- Demonstration of a pilot's ability to control a rocket-boosted aerospace vehicle through atmospheric exit.
- Development of large supersonic drop tanks.
- Successful transition from aerodynamic controls to reaction controls, and back again.
- Demonstration of a pilot's ability to function in a weightless environment.
- First demonstration of piloted, lifting atmospheric reentry.
- First application of energy-management techniques.
- Studies of hypersonic acoustic measurements used to define insulation and structural design requirements for the Mercury spacecraft.
- Use of the three X-15 aircraft as testbeds carrying a wide variety of experimental packages.[197]

The X-15 also made its mark in many other ways. When the NACA began its development, the science of hypersonic aerodynamics was in its infancy; the few existing hypersonic tunnels were used largely for studies in fluid mechanics. Aerodynamicists feared that there might be a hypersonic "facility barrier," much like the earlier transonic tunnel trouble that led to the Bell X-1 and Douglas D-558, so that hypersonic tunnel tests might prove of little value in predicting actual flight conditions. The X-15 disproved this. Predicted wind-tunnel data and data from flight-testing of the airplane generally showed remarkable agreement. Proving that hypersonic laminar-flow conditions did not develop led to the disappearance of this "technical superstition" as well as to the recognition that the small surface irregularities that prevent laminar flow at low speed also prevent its formation at hypersonic speeds. Like the earlier X-1, the X-15 encouraged a great deal of ground research and simulation techniques. So successful were these methods and so great was the engineers' confidence in these methods and the X-15's flight results that the program wound up actually decreasing the likelihood of NASA's developing any future hypersonic research aircraft with the prime justification being the generation of unique and otherwise unobtainable data. Any future research aircraft would be built more for "proof of concept" purposes than for acquiring information unobtainable by other means. At the conclusion of the X-15 program, the German Society of Aeronautics

[197] John V. Becker, "Principal Technology Contributions of X-15 Program" (Hampton, VA: NASA LaRC, 8 Oct. 1968), copy on file with the NASA history office.

and Astronautics presented the NASA X-15 team with the Eugen Sänger Medal, a fitting and appropriate honor. In his acceptance address on behalf of the team, John Becker stated that "no new exploratory research airplane can ever again be successfully promoted primarily on the grounds that it will produce unique flight data without which a successful technology cannot be achieved."[198]

The X-15 story had another side: its effect upon the people on the team. Their intense and devoted work was recognized in numerous honors that, at one time or another, went to the X-15 team or its members: the Sänger Medal, the Collier Trophy, the Harmon Trophy, the Octave Chanute Award, the NASA Medal for Exceptional Bravery, the Thomas D. White Space Trophy, the NASA Exceptional Service Medal, the NASA Distinguished Service Medal, the NASA Medal for Outstanding Leadership, the Iven C. Kincheloe Memorial Award, the FAI Gold Air Medal, the Lawrence B. Sperry Award, the Sylvanus Albert Reed Award, the Haley Astronautics Award, the Flight Achievement Award, the David C. Schilling Trophy, the NASA Group Achievement Award.

The public had little understanding of the X-15 and, after the early fanfare, saw only the occasional items in newspaper back pages on new speed and altitude marks—as if that was all the X-15 did. The uninformed public could not understand what went into a flight: the mission planning; the hours of simulator time; the flight practice; the endless maintenance; the annoying delays for weather; the excitement as the B–52 took off; the long wait to drop or, disappointingly, to an abort; the moment of launch, with ignition and boost, or an abort and emergency landing; the tension in the control room; the hypersonic glide back; the chase and X-15 coming in like a flock of ducks; the resounding smack as the X-15's skids thumped on the dry lakebed; and, the maintenance, debriefing, data analysis, and planning for the next mission. They could not know the strong bonds the program forged, nor the collective worry produced by an errant flight or an emergency condition, nor the heartache generated by the death of Mike Adams. They could not fathom the emotional and psychological release of the parties at Juanita's. For almost a decade, the Flight Research Center sustained this effort, and its personnel found new kinship and dedication. When the X-15s left Rogers Dry Lake for the last time, a little bit of the center and its personnel went with them. But there were other programs, other vehicles to follow. The frontier of flight research had not yet reached its end.

[198] Becker, "The X-15 in Retrospect."

7

Serving Gemini and Apollo 1962-1967

NASA's major priority in the 1960s was, of course, space. The agency's activities were related to three major piloted spacecraft projects, Mercury, Gemini, and Apollo. Beyond Apollo, the agency had at best vague plans for some sort of semi-permanent orbital space station supplied by an earth-to-station "shuttle." In truth, however, NASA had not formulated long-range plans beyond the lunar landing. Because of the intensity of the space program, particularly during the early 1960s, NASA channeled the activities of all field centers and stations toward some aspect of it. The Flight Research Center during this time concentrated its efforts on various means of returning men from space—such as the Paresev and lifting body, discussed later—and analyzing how to land on the moon. This work was directly related to Gemini and Apollo and to the later Space Shuttle as well. However, the Flight Research Center labored under one serious handicap during the 1960s, a handicap that almost cost the center its existence. In an agency dominated by space flight, the FRC appeared to be anachronistically obsessed with aeronautics.

Whither the FRC?

In point of fact, the FRC's research during the 1960s was oriented primarily toward spaceflight, though with a heavy aeronautical flavor: hypersonic flight within the upper atmosphere and into and back from space, the low-speed handling qualities of spacecraft, lifting reentry schemes, and support of space research at other NASA centers, such as high-altitude drop tests of Mercury spacecraft's drogue parachute. Since the FRC relied heavily upon research aircraft—vehicles having wings—the center seemed to be concentrating its activity on the airplane in the era of the spacecraft. But these aircraft were actually being used as tools for studying problems that were basically space technology. Research on the X-15, for example, clearly benefited spaceflight studies more than, say, supersonic aerodynamic research. This point tended to be missed by individuals not familiar with the true scope of the center's research. The FRC suffered simply because no spacecraft were being managed from the center, no boosters were being developed by its engineers, and no rockets were being launched from Edwards.

As early as 1957, the percentage of the FRC's research staff involved with space studies had begun to grow. This fact, too, was missed, possibly because most of these internal studies went no further than the FRC's front office, in part because many of them were speculative and not directly involved with NASA's mainstream budget

Table 3

Distribution of FRC Permanent Personnel by Program, 1960-1968
(Number Assigned and Percentage of Total)

Program	1960[a]	1962	1964	1966	1968
Manned Space Flight	0 0.0%	6 1.2%	50 8.3%	34 5.6%	0 0.0%
Space Applications	0 0.0%	0 0.0%	0 0.0%	0 0.0%	0 0.0%
Unmanned Space Investigations	0 0.0%	1 0.2%	3 0.5%	1 0.2%	0 0.0%
Space Research & Technology	0 0.0%	45 8.6%	51 8.4%	104 17.2%	92 16.2%
Aircraft Technology	-- 90%	443 84.5%	344 56.8%	308 51.1%	325 57.4%
Supporting Activities[b]	-- 10%	29 5.5%	157 26%	156 25.9%	149 26.3%

Notes:

[a] Actual positions unavailable for FY 1959-60; percentages from NASA Office of Programming, Budget Operations Div., History of Budget Plans, Actual Obligations, and Actual Expenditures for FY 1959 through 1963 (NASA, 1965), sect. 8.

[b] Includes tracking and data acquisition, data analysis, and technology utilization staff.

Source: Nimmen, Bruno, and Rosholt, *NASA Historical Data Book, 1958-1968*, vol. 1, *NASA Resources* (Washington, DC: NASA SP-4012, 1976), Table 6-31, p. 304.

items. Even in the 1960s, the actual percentage of FRC personnel involved in space-related research appears from available internal evidence to have been higher than shown in published NASA statements. Table 3 shows the distribution of permanent personnel at the FRC by fiscal year and budget activity, as set forth in NASA's budget estimates, at two-year intervals from 1960 through 1968. A closer examination of these data, however, raises serious questions about their accuracy, possibly indicating a source of misinformation that might well have convinced many within and outside NASA—including Congress—that the FRC was far less in step with the times than it actually was. For example, the figures in Table 3 for Manned Space Flight reflect only the personnel working on the Lunar Landing Research Vehicle (discussed later) and not the much larger numbers working on the X-15. Yet, as indicated inChapter 6, the X-15 flew to the edge of space and provided a great deal of data about such disciplines as high-speed aerodynamics and aerodynamic heating that were of great value to the space program.

The NASA budget estimate for 1960 stated that approximately 90 percent of the FRC's staff was engaged in "aircraft technology." However, by 1959, the personnel breakdown in the internal planning documentation for the HSFS indicates that no more than 40 percent were working on aircraft studies. Fully 35 percent were studying boost-glide (that is, orbital) aircraft, another 16 percent were examining satellites (both occupied and unoccupied), 5 percent were engaged in ICBM and anti-ICBM research, and 4 percent were studying antiaircraft missiles.[199] These figures certainly could not have changed in favor of aeronautics in one year. Clearly budget-request statistics from 1960 to 1968 are misleading because they lump together such major activities as the X-15 and lifting body programs as "aircraft technology" when, in fact, these programs were space-related. In 1962, fully 84.5 percent of the FRC's staff were officially listed under "aircraft technology," but it is doubtful if more than 20 percent were engaged on purely aeronautical projects (i.e., flight entirely within the atmosphere) at that time, and no more than 40 percent in 1968. Since so many of the X-15 and the lifting-body programs were related to piloted spaceflight, it is inconceivable that the FRC in 1964 had only 8.3 percent of its staff investigating piloted spaceflight activities. This figure should have been in the author's calculations, about 40 percent as well, based on the flight research activities surrounding these projects, the number of employees engaged in them, and the amount of paperwork (an indication of administrative prioritizing of projects) generated by them.

These statistics from NASA's budget requests are misleading in another way: they ignore the trait of "ad hocracy" a phrase coined by Alvin Toffler that has always characterized the FRC's administrative style to some degree. Its small staff has never been entirely divided by rigid administrative lines and networks separating programs, authorities, and administrative units. Instead, specialized small work forces have been formed to accomplish certain projects or goals, such as the Paresev, lifting-body, and lunar-landing simulator.[200] Workers ostensibly assigned to aeronautics projects might suddenly be called upon to participate in a space-related project. They might still show up on organizational charts as "aeronautics" personnel, when, in fact, they often flitted back and forth from "aero" to "space" as the research need arose.

Doubtless, responsibility for the failure to portray adequately the wide-ranging air and space interests of the FRC divided equally between the FRC and NASA headquarters. The FRC's casual though highly effective administration showed little inclination to set up a sharply structured bureaucracy that would clearly divide the center's activities between aeronautics and astronautics, or spaceflight. Because of the small size of the center and the need to shift people to meet constantly changing

[199] NACA HSFS, *Review of Aeronautical Research* (HSFS, n.d.) originally seen in the files of the office of the deputy director, DFRC; comments of Wayne Ottinger and Gene Matranga on the original chapter, ca. 1 Feb. 2000.

[200] See, for example, the discussion of "ad hocracy" in Burt Nanus, "Profiles of the Future: The Future-Oriented Corporation," *Business Horizons* (Feb. 1975), passim. Nanus's essay could equally apply to the FRC's history in the 1960s and 1970s. As the organizational chart for 1948 in Appendix A of this book would suggest, the Muroc Flight Test Unit was more organized by project than became the norm in the 1950s, but even then, some individuals such as the "computers" worked on more than one project. By sometime in the 1960s, despite the reemergence of project offices, the FRC had shifted increasingly to what is called a matrix organization in which individuals worked increasingly on more than one project at a time, but the shift was never abrupt. From the beginning, elements of the later matrix organizational style had existed to some degree. Conversations of J. D. Hunley with William H. Dana, Roy G. Bryant, Robert R. Meyer, Jr., Marta Bohn-Meyer, and Betty J. Love, Mar. 2000.

project structures, such a bureaucracy would have made little sense anyway. Unfortunately, the persons in NASA headquarters charged with preparing and submitting budget requests to Congress typecast the FRC as an "aeronautics" center. Communication between the FRC and headquarters was inadequate, although Paul Bikle recognized the danger of being perceived as an aeronautics-only facility and worked hard to move FRC into the mainstream of space-related programs. Paresev, lifting bodies, and the Lunar Landing Research Vehicle were Bikle's initiatives. The FRC's staff may well have failed to grasp just how single-minded the non-NASA governmental community, especially Congress, was when it came to emphasizing aeronautics or astronautics in the early 1960s.

All of this would constitute little more than a curious footnote to the FRC's managerial style and visibility during the 1960s had it not been for a critical event: an attempt by some in Congress to close down the center at the conclusion of the X-15 program.

As mentioned earlier, the X-15 dominated the FRC's activities to such an extent in the early 1960s that some saw the FRC and X-15 as so intertwined that the end of the latter would spell the demise of the former. This perception reached Congress. In the summer of 1963, during consideration of NASA's 1964 budget, the influential House Committee on Science and Astronautics (later the House Committee on Science and Technology) recommended closing the Flight Research Center since, in members' judgment, "no known future aircraft projects will specifically require the continued existence of the Flight Research Center beyond the date when the X-15 project will be completed."[201] Dr. Raymond Bisplinghoff, OART director, worked hard over the next few weeks to save the FRC, pointing out that NASA envisioned its participation in a range of programs in both aeronautics and space activities. Fortunately, the Senate Committee on Aeronautical and Space Sciences restored funding for the FRC, on grounds that it would be vital to the upcoming American supersonic transport (SST) testing program. By the end of the summer, the FRC was safe, having survived a serious attempt to legislate its demise.[202]

In hindsight, it is ironic that the FRC was saved at a critical juncture of its existence by its anticipated need to support the American SST, since the SST program fellbeneath the congressional axe in 1971. By that date the FRC was again well established with a variety of research projects, primarily in aeronautics, that necessitated its continued existence. Further, its major role in flight-testing the upcoming Space Shuttle was already mapped out.

[201] U.S. Congress, House Committee on Science and Astronautics, *Authorizing Appropriations to the National Aeronautics and Space Administration*, Report No. 591 (Washington, DC: GPO, 1963), p. 176. For a history of the committee, see Ken Hechler, *Toward the Endless Frontier: History of the Committee on Science and Technology, 1959-1979* (Washington, DC: GPO, 1980).

[202] U.S. Congress, Senate Committee on Aeronautical and Space Sciences, *NASA Authorization for FY 1964*, Report No. 385 (Washington, DC: GPO, 1963), p. 187. See also "Outlook Better for Flight Research Center," *Missiles and Rockets* (8 July 1963), p.16.

That the FRC was so well established again by 1971 stemmed from a variety of factors but chiefly from the aggressive policies and initiatives of its director, Paul Bikle. In 1963, at Bikle's urging, De Beeler and senior FRC staffers prepared a comprehensive five-year plan for the future direction of the center.[203] This document served as a general guide for center activities through the end of the decade. The plan (Table 4) emphasized continuing four on-going air and space activities while developing six new initiatives. Some of these, such as lifting bodies, continued into the 1970s.

By the mid-1960s, then, the Flight Research Center clearly knew where it was going in the future, even if others elsewhere were not so certain.

Early Space Research at the FRC

The FRC's research in support of NASA's space program began in 1959 when, at the request of the Space Task Group, the center flew a series of F-104 flights to drop-test versions of the Project Mercury spacecraft's drogue parachute from altitudes above 50,000 feet. As a result of these tests, critical design problems were discovered and corrected before the spacecraft first flew.[204] The center's greatest early space effort, however, was on the planned Dyna-Soar program, the X-20A.

Dyna-Soar, the Round Three that followed the X-15, was a Sänger-like boost-glider designed to be lofted into orbit by a Titan III booster. Dyna-Soar had three major objectives: to demonstrate controlled lifting reentry from space and acquire data useful for the development of other lifting reentry spacecraft, to investigate a pilot's ability to perform useful tasks in space, and to explore piloted maneuverable reentry, including landing at conventional airfields.[205] Its general configuration was as a hypersonic slender delta, a flat-bottom glider using radiative cooling. Under development for the Air Force by Boeing, Dyna-Soar was pushing technology in many areas, including high-speed aerodynamics, high-temperature structural materials, and reentry protection concepts. Eventually, questions about its utility, research potential, and safety forced cancellation of the craft in December 1963. Nevertheless, Dyna-Soar was a generally useful design exercise, much of the research encouraged by this program significantly influencing subsequent Shuttle studies. Like others of the X-series before it, the X-20A acted as an important research focal point.[206]

[203] NASA FRC-*Five-Year Plan* (July 1963), DFRC Historical Reference Collection.

[204] See, for example, Loyd S. Swenson, Jr., James M. Grimwood, and Charles C. Alexander, *This New Ocean: A History of Project Mercury* (Washington, DC: NASA SP-4201, 1966), p. 198.

[205] "X-20 Will Probe Piloted Lifting Re-entry," *Aviation Week & Space Technology* (22 July 1963), pp. 230-240.

[206] Interview of Becker. The development of the X-20 in the context of lifting reentry research from 1952-1980 is examined in R.P. Hallion, "The Antecedents of the Space Shuttle," *AIAA Student Journal* (Spring 1980), pp. 26-35.

Table 4
FRC Five-Year Research Plan, 1963

Continuing Activities

Aeronautics Technology
- Studies of SST operational problems using modified service aircraft.

Space Technology
- X-15 flight operations (X-15 follow-on program)
- Paresev studies.[a]
- Active support and research on Dyna-Soar.[b]

New Initiatives

Aeronautics Technology
- Renewed military service testing, starting with F-111.
- Development of a multipurpose airborne simulator using a modified Lockheed JetStar transport to simulate a wide range of aircraft, from hypersonic reentry vehicles to SST
- Investigation of the handling qualities of light airplanes to improve general aviation safety.

Space Technology
- Flight testing of M2-F1 lifting body and development of supersonic lifting bodies to assess the low-speed handling qualities and approach and landing characteristics of lifting-body spacecraft.
- Development of a lunar lander simulator to serve as a training device for the Apollo program.
- Studies of an advanced hypersonic research vehicle successor to the X-15.

Notes:
[a] A kite-like landing system for spacecraft, to be discussed later.
[b] Dyna-Soar was terminated at the end of 1963.

The Dyna-Soar project office, in conjunction with NASA, had selected an FRC pilot, Milt Thompson, as the only NASA pilot to fly the craft. Further, the FRC had complete responsibility for stipulating the X-20A's instrumentation requirements. Center engineers had already prepared papers on Dyna-Soar's expected operational problems and the possibility of air-launching it from B-52

Neil A. Armstrong prepares to fly a Dyna-Soar abort simulation in one of the Flight Research Center's Douglas F-5D-1 Skylancer aircraft. (NASA photo EC62-128)

and B-70 motherships.[207] In early 1961, the FRC had received two "castaways," prototypes of the Douglas F5D-1 Skylancer, an experimental Navy fighter that had not been placed into production. Since the F5D-1 had a wing planform very similar to that projected for Dyna-Soar, FRC pilot Neil A. Armstrong recognized that the Skylancer could be used to study Dyna-Soar abort procedures. How to save the pilot and spacecraft in the event of a launch-pad booster explosion was a problem of great concern to the Dyna-Soar team. The X-20A Dyna-Soar had a small escape rocket to kick it away from its booster, but no one really knew what kind of separation flight path and landing approach would best bring Dyna-Soar safely to earth. Armstrong developed a suitable maneuver using the F5D-l. It consisted of climbing vertically to 7,000 feet, pulling on the control column until the X-20A was on its back, rolling the craft upright, then setting up a low-lift-to-drag-ratio approach, and touching down on a part of Rogers Dry Lake that was marked like the 10,000-foot landing strip at Cape Canaveral. Following Dyna-Soar's cancellation, the FRC continued to fly the F5D in support of lifting-body and SST studies, before retiring the aircraft in 1970.[208]

[207] Hubert M. Drake, Donald R. Bellman, and Joseph A. Walker, NACA RM H58D21, "Operational Problems of Manned Orbital Vehicles," NACA HSFS, 12 Apr. 1958; Donald R. Bellman and Harold P. Washington, NASA TM X-636, "Preliminary Performance Analysis of Air Launching Manned Orbital Vehicles," NASA FRC, 9 Oct. 1961.

[208] Interview of Dana; Larry Grooms, "No Gold Watch for Old Faithful NASA Retiree," Antelope Valley (CA) *Ledger-Gazette*, 28 Apr. 1970. See also Gene J. Matranga, William H. Dana, and Neil A. Armstrong, NASA TM X-637, "Flight-Simulated Off-the-Pad Escape and Landing Maneuver for a Vertically Launched Hypersonic Glider, " NASA FRC, Mar. 1962.

Paresev: A Space-Age Kite

The Flight Research Center's major support activities in space research concerned the Paresev and the Lunar Landing Research Vehicle (LLRV), developed and flown at the FRC in support of the Gemini and Apollo programs. Paresev was an indirect outgrowth of the kite-parachute studies by NACA Langley engineer Francis M. Rogallo. The "Rogallo wing" had a diamond profile with a flexible covering attached to a V-shaped leading edge (with the point foremost) and a longitudinal keel. As with a parachute, the air filled out and shaped the sail-type surface. In the early 1960s, this shape seemed an excellent means of returning a spacecraft to earth. A spacecraft could streak in through the atmosphere and then, at much lower altitudes and subsonic speeds, deploy a stowed Rogallo wing, enabling the astronauts on board to fly it down to an airplane-like landing, obviating the need for a water landing and recovery flotilla. NASA engineers had begun studying how the agency could apply the Rogallo wing to current spacecraft projects, especially one tentatively designated Mercury Mark II.[209]

In January 1962, Mercury Mark II became the Gemini program, America's second major man-in-space venture, involving a two-man crew and encompassing extravehicular spacewalks, rendezvous, and docking. In May 1961, while Mercury Mark II was slowly evolving, Robert R. Gilruth, director of NASA's Space Task Group, requested studies of an inflatable Rogallo-type "parawing" for spacecraft. Several companies responded. North American Aviation produced the most acceptable concept, and development was contracted to that company. At a 28-29 November 1961 meeting, NASA headquarters launched a paraglider development program, with Langley doing wind-tunnel studies and the Flight Research Center supporting the North American test program. NASA grafted the parawing scheme onto the Mercury Mark II program.[210]

Paraglider development involved solving major design difficulties in stowing and deploying the wing, ensuring that the crew would have adequate control over the parawing-equipped craft, and providing satisfactory stability, control, and handling qualities. The Flight Research Center's technical staff was never convinced that the scheme was workable. Eventually, because of poor test results and rising costs and time delays, the parawing was dropped from Gemini in mid-1964. FRC engineers and pilots had believed that any vehicle so equipped might present a pilot with a greater flying challenge than contemporary advanced airplanes. They thought that NASA should acquire some sort of baseline experience before attempting development and flight of a parawing on a returning spacecraft. After returning to Edwards, they continued their discussions among themselves.

The best way to acquire such experience, of course, was by building and flying a parawing. FRC research pilots Neil Armstrong and Milt Thompsonactively favored such an approach. They approached Paul Bikle, who liked the idea but recognized

[209] D. S. Halacy, *The Complete Book of Hang Gliding* (New York: Hawthorn Books, 1975), pp. 24-27.

[210] James M. Grimwood, Barton C. Hacker, and Peter J. Vorzimmer, *Project Gemini: Technology and Operations—A Chronology* (Washington, DC: NASA SP-4002, 1969), pp. 8, 17.

that both pilots had heavy Dyna-Soar commitments. The FRC could not spare their services elsewhere, even to a project as interesting as the proposed parawing. Instead, Bikle called in a group of center engineers under the direction of Charles Richard, a team composed of Richard Klein, Vic Horton, Gary Layton, and Joe Wilson. Bikle's instructions were characteristically short and to the point: build a single-seat paraglider and "do it quick and cheap." All of this took place just before Christmas 1961. The team, now totaling nine engineers and technicians, set to work on this "Paraglider Research Vehicle," conveniently abbreviated Paresev. Seven weeks later, after expending $4,280 on construction and materials, the team rolled out the Paresev I. It resembled a grown-up tricycle, with a rudimentary seat, an angled tripod mast, and, perched on top of the mast, a 150-square-foot Rogallo-type parawing. The vehicle weighed about 600 pounds, had a height of over 11 feet, and a length of about 15 feet from front to back of the center wing spar (the fuselage being much shorter). The pilot sat out in the open, strapped in the seat, with no enclosure of any kind. He controlled the descent rate by tilting the wing fore and aft and turned by tilting the wing from side to side using an overhead control stick. NASA registered the Paresev, the first NASA research airplane to be constructed totally in-house, with the Federal Aviation Administration on 12 February 1962. Flight-testing started immediately.[211]

At first, engineers cautiously tested the Paresev by towing it behind a utility vehicle. Technicians drove a tow vehicle up to 50 knots on the lakebed. The Paresev lifted into the air at about 40 knots, followed by a dusty gaggle of "chase" cars and motorcycles. Milt Thompson, one of the two project pilots (the other being NASA's Bruce Peterson), would let the plane float along a few feet off the ground as he gained familiarity with the vehicle. The original configuration had several faults. For one, the control system had built-in lag. Pilots used to the sensitivity of modern jet aircraft found that the Paresev flew as if "controlled by a wet noodle."[212] Because cloth-covered airplanes often used Irish linen, the Paresev design team decided to use it for the wing surface. Dick Klein and Gary Layton visited a sailmaker in Newport Beach, who cast a quizzical eye at the material and suggested Dacron instead. The team stuck with linen, found it did indeed have a number of problems including flutter at the trailing edge, and changed to Dacron at a later date.[213] The Paresev was difficult to fly—Thompson considered it more demanding than the later lifting bodies. He made a great many ground tows and 60 air tows, recollecting later that "it was a lot of fun."[214] But flying the Paresev had its moments of danger, too. During one ground tow on 14

[211] FRC, *Five Year Plan*; Hubert M. Drake, "Aerodynamic Testing Using Special Aircraft," (Edwards: NASA FRC, n.d.), pp. 8-9; NASA FRC *X-Press*, 26 Jan. 1968; interview of Richard Klein, 1 Mar. 1977; interview of Joseph Wilson, 23 Feb. 1977; interview of Milton Thompson, 29 Mar. 1976.

[212] Interview of Klein.

[213] Ibid. Interview of Wilson.

[214] Interview of Thompson; see also Milton O. Thompson, "I Fly Without Wings," *Air Progress* (Dec. 1966), pp. 10-13, 80-82.

The Paresev I-A (Paraglider Research Vehicle) and one of its tow airplanes, a rented Stearman sport biplane, on Rogers Dry Lake, Edwards, California. (NASA photo E-8712)

March 1962, Bruce Peterson got out of phase with the lagging control system and the Paresev developed a rocking motion that got worse and worse. Just as the tow truck started to slow, the Paresev did a wingover into the lakebed, virtually demolishing the Paresev and injuring Peterson, though not seriously.

The accident ended the days of the Paresev I. FRC technicians salvaged only the tripod from the wreck. They totally rebuilt the vehicle, this time with a much more sophisticated control system, using a stick and pulley arrangement with cables to move the wing. They took the sailmaker's advice and used a Dacron wing. The rebuilt vehicle became the Paresev I-A. Ground tows quickly indicated this paraglider handled better than its predecessor, and NASA moved on to flight tests. To tow the I-A, the FRC rented a Stearman biplane from a Tehachapi sailplane operator. Later, a Cessna L-19 Bird Dog was acquired on loan from the U.S. Army Reserve, and a Super Cub and Boeing helicopter were also used as tow planes. During Paresev I-A tests, tow planes dragged the vehicle to 10,000 feet before release. For a research pilot used to the confined but comforting environment of a supersonic jet, it was an eerie sensation to sit out in the open, like a pre-World War I aeronaut, strapped in the seat. Besides, Milt Thompson and Bruce Peterson, others flew the little aircraft, including Neil Armstrong, Emil "Jack" Kluever (an Army pilot detailed to the FRC), Charles Hetzel (a North American Aviation test pilot), astronaut Gus Grissom, and Langley research pilot Bob Champine, a Muroc old-timer. The vehicle underwent two further modifications. As the Paresev I-B, it featured a smaller Dacron membrane (100 square feet); as the I-C, it incorporated a half-scale version of an inflatable Gemini-type wing. The I-C was especially difficult to fly with extremely high stick forces, even after a rudder pedal was added to boost roll control. NASA ended flight research on the Paresev in 1964, having completed over 341 flights of which Bruce Peterson flew 228 (ground and air tows).

The Paresev program is a good example of the Bikle low-cost do-it-quick approach. Originally scheduled as a two-month flight research project, the program became interesting enough to warrant running for two years. Eventually engineers evolved a useful vehicle having moderately acceptable handling characteristics, although it was always difficult to fly and required a great deal of pilot attention. Nevertheless, it was a big step from this simple technology demonstrator with a rigid and fixed wing framework to a stowable, inflatable parawing on an actual spacecraft that could be relied on to return a crew safely to earth. At the same time that NASA's Paresev was concluding, North American's complex Gemini Paraglider program had already forced a test pilot to abandon one of the vehicles in flight—hardly encouraging. The long process involved in making the relatively unsophisticated Paresev an acceptable craft indicated the magnitude of the task awaiting those developing such devices for spacecraft.[215]

NASA's Flying Bedsteads

NASA's major undertaking in the 1960s was the Apollo program, an ambitious and breakneck-paced effort to place astronauts on the moon by the end of the decade. It is difficult now to relive those hectic days, to imagine the level of activity at centers around the country, the frantic pace of meetings, the sense of mission that pervaded NASA and its workers. Virtually every worker felt privileged to work for NASA, and even wing-oriented NACA old-timers did their best to contribute to the national space effort.

One of the many critical issues in the Apollo program was the descent to the lunar surface. The descent vehicle would be operating in a gravity only one-sixth of earth's, but the airless moon dictated a strictly propulsion-borne descent, not an aerodynamic descent. Grumman was the subcontractor for the landing vehicle, the LEM, later shortened to LM.[216] Nobody wanted the first lunar landing, with its attendant high pilot workload and psychological stress, to be also the first time an astronaut team flew a lunar landing descent profile. Some sort of exotic simulator was needed to give the crew useful experience before they tackled the task of setting down on the moon. There were several possible ways of doing this. One would be an electronic

[215] Memo, Bruce Peterson to FRC Director, 16 Mar. 1962; "Paresev," described in the NASA Dryden Photo Gallery Index, Sept. 1998, on the Web at URL: http://www.dfrc.nasa.gov/gallery/photo/Paresev/HTML/index.html; interviews of Klein and Wilson; FRC, *Five-Year Plan;* NASA FRC *X-Press*, 26 Jan. 1968; comments by Bruce Peterson on the original chapter, 25 Jan. 2000 and subsequent conversations; Milton O. Thompson and Curtis Peebles, *Flying Without Wings: NASA Lifting Bodies and the Birth of the Space Shuttle* (Washington, DC: Smithsonian Institution Press, 1999), pp. 32-56.

[216] Lunar Excursion Module, later shortened to Lunar Module.

simulator. Another would be a free-flight test vehicle. Yet a third would be a tethered device, suspended beneath some sort of framework. NASA decided to be conservative and followed all three routes. The most ambitious of the three was the free-flight vehicle. As might be expected, this was the Flight Research Center's contribution to Apollo.

The FRC staff conceived the idea for a free-flight lunar landing simulator. In early 1961, Hubert Drake had convened a group of FRC engineers to investigate simulating a lunar landing. Drake contacted Walt Williams, then associate director of the Manned Spacecraft Center. Williams offered his enthusiastic support, recommending that the FRC propose such a vehicle to NASA headquarters. At the same time, unknown to the FRC group, Bell Aerosystems Company (heir of Bell Aircraft Corporation, which had built so many of the early X-series aircraft) was also examining ways of building a free-flight simulator. When Drake and FRC engineers Gene Matranga and Donald Bellman learned from NASA headquarters that Bell was interested, they invited company representatives to the FRC for consultation; this culminated in a $50,000 study contract to Bell, which the FRC awarded in December 1961. At the time, FRC was thinking of the vehicle primarily for research, rather than as a training aid.

At the same time, the Langley Research Center was supporting a much less ambitious concept involving a tethered rig. When constructed, the large gantry (about 400 feet long, 240 feet high) supported five-sixths of the test vehicle's weight. Rockets supported the remaining one-sixth. The Langley Lunar Landing Research Facility cost $3.5 million and started operations in June 1965. By that time, the FRC had already amassed considerable flight experience with its own (free-flight) lunar landing simulator, the remarkable Lunar Landing Research Vehicle, the LLRV.[217]

A jet engine supported five-sixths of the LLRV's weight. Rockets lifted the remainder, simulating the descent propulsion system of an actual lunar lander. Attitude control thrusters allowed the pilot to control the vehicle–aerodynamic controls played no part. It was not a new idea, but an old idea serving a new purpose. Aircraft companies had built and flown similar vehicles, dubbed "flying bedsteads," to acquire information needed for designing vertical-takeoff-and-landing (VTOL) aircraft. Dr. A. A. Griffith, a pioneer in British VTOL technology, had built the first such rig, powering it with a pair of Rolls Royce Nene turbojets. Such rigs invariably had an open framework supporting the pilot, his instrumentation, the fuel system, the engines, and a variety of "puff pipes" running hither and yon to control the attitude of the vehicle. Griffith's "Flying Bedstead" first flew in August 1954, gaining a great deal of attention in the aviation and popular press. The FRC's engineers naturally considered this vehicle when conceiving the LLRV.[218]

[217] Michael D. Keller, Langley chronology, pp. 84, 89; interview of Bellman; FRC *Five-Year Plan*; NASA *FRC X-Press,* 2 Mar. 1962, 1 Feb. 1963, and 19 Feb. 1963.

[218] Oliver Stewart, *Aviation: The Creative Ideas* (New York: Praeger, 1966), pp. 144-146; William Green and Roy Cross, *The Jet Aircraft of the World* (Garden City, NY: Hanover House, 1957), p. 172.

Bell was nearly the only firm in the United States that had a great deal of experience in the design and construction of VTOL aircraft using jet lift for takeoff and landing. The FRC's engineers consulted with Bell personnel before drawing up the specifications for their vehicle. In early 1962, following award of the Bell study contract, Donald Bellman (head of the project), Gene Matranga, and Lloyd Walsh (the FRC's contracting officer) went to Bell to interest the company in fabricating such a vehicle for NASA. While at Buffalo, they rode company helicopters on simulated lunar descents. Stopwatch and notepads in hand, they quickly learned that a helicopter could not match the expected descent rates and paths of a jet-lift lander. The tests quickly silenced those who thought NASA could simulate the lunar landing mission aerodynamically by using helicopters. Following the Buffalo visit, Bellman passed along their tentative findings to Walt Williams at Houston. Williams endorsed the concept.

Out of this came support from the Manned Spacecraft Center and NASA headquarters. On 1 February 1963, NASA awarded Bell a $3,610,632 contract for the design and fabrication of two lunar landing research vehicles capable of taking off and landing under their own power, attaining an altitude of 4,000 feet, hovering, and flying horizontally. Bell had 14 months to build and deliver the first vehicle, the second to follow two months later. NASA intended using them for studies of piloting and operational problems during the final phase of a lunar landing and the initial phase of a lunar takeoff. The tests would permit study of controls, pilot displays, visibility, propulsion control, and flight dynamics. Each LLRV would carry about 150 pounds of research equipment.[219]

Bell unveiled the first of the two LLRVs during ceremonies at its plant in Wheatfield, New York, on 8 April 1964. Kenneth L. Levin of Bell oversaw the development. C. Wayne Ottinger of the FRC served as NASA resident representative. The completed LLRV weighed about 3,700 pounds, stood slightly more than 10 feet high, and had four aluminum truss legs spread about 13 feet. A General Electric CF-700-2V turbofan engine provided 4,200 pounds of thrust, enough to boost the LLRV to altitude. Then the engine would automatically adjust to support five-sixths of the vehicle's weight, and the pilot would use two lift rockets capable of modulation from 440 to 2,200 newtons (100 to 500 pounds) of thrust for controlling the "lunar descent." The lift rockets burned hydrogen peroxide. Sixteen smaller rockets, arranged in eight pairs, controlled pitch, yaw, and roll. To permit the turbofan engine to maintain vertical thrust when the vehicle assumed other than a horizontal attitude, Bell gimbaled the engine below the apex of the vehicle's legs. The LLRV had six backup rockets capable of 2,200 newtons (500 pounds) of thrust for emergency use if the turbofan engine quit. The pilot sat out in the open, behind a Plexiglas shield, on an emergency "zero-zero" ejection seat—a wise precaution, as things turned out. The LLRV could remain aloft 14 minutes at full thrust, though safety considerations dictated a more

[219] Interview of Bellman; NASA FRC *X-Press*, 1 Feb. 1963; Bell Aerosystems Co., *Uprated LLRV Pre-Design Configuration Studies for Research and Training* (July 1964), p. 3.

Lunar Landing Research Vehicle No. 1 lifting off the ramp at South Base for a research flight in 1964. (NASA photo ECN-506)

prudent limitation of 10 minutes. It used an electronic analog fly-by-wire (FBW) control system connected to a conventional aircraft-type center stick for pitch and roll control and rudder pedals for yaw control. There were no aerodynamic control surfaces. The system provided direct electronic control—with no mechanical linkages, even as a backup safety system—of the attitude rockets. An electronic fly-by-wire backup system did provide attitude control in the event the primary system failed, however. The LLRV also realistically simulated the actual vehicle motions and control system response that an astronaut could expect to encounter while piloting a descending lunar module.

After unveiling the craft to the press, Bell sent both LLRVs to Edwards in partially disassembled and incomplete condition to save money and expedite NASA's installation of instruments and performance of other tasks. The two LLRVs arrived at the Flight Research Center in the spring of 1964, and center personnel immediately set to work preparing the first vehicle for flight.[220]

By August 1964, the FRC had LLRV No. 1 ready for its first trials, mounted on a fixed tilt table constructed by the center's aircraft modification and repair group. Joe Walker first tested out the craft in this manner. The vehicle itself had complete freedom of movement, being restricted only from flight. The jet engine, however, was mechanically constrained and could not rotate from the vertical. The tilt-table tests proceeded more or less smoothly; by the fall of 1964, the LLRV research team was ready for free-flight trials. Test operations were set up at Edwards' South Base, scene of the old High-Speed Flight Station. On 30 October 1964, center research pilot Joe Walker took the craft on its first flight, making three separate liftoffs and landings, reaching a peak altitude of ten feet and remaining aloft for a total free-flight time of just under one minute. The craft took off, as Walker subsequently described it, "just like going up in an elevator." At liftoff, with the CF-700 wailing, Walker maintained proper attitude by firing short bursts of reaction controls; they hissed loudly, swathing the craft in peroxide steam, enhancing the Rube Goldberg appearance. By the end of the year, Donald Mallick, a new FRC pilot who had transferred from Langley, joined Walker. Mallick completed his checkout on 9 December. Over the next year, LLRV No. 1 continued its flight program. By the end of August 1966, it had completed 187 flights, flown by Walker, Mallick, and the Army's Jack Kluever.

In preparation for an LLRV training program for the Apollo astronauts at Houston, Manned Spacecraft Center research pilots Joseph Algranti and H. E. "Bud" Ream checked out in the strange vehicle. On 11 March 1966, piloted by Don Mallick, LLRV No. 1 flew with a three-axis sidearm controller, making it comparable to the actual Grumman LM control system. NASA also moved the LLRV's control panel from the center of the cockpit to the right side, again matching the LM configuration, and planned to reduce the amount of pilot visibility to give the craft the same visual characteristics as the lunar lander. In January 1967, Jack Kluever completed the FRC's first flight in LLRV No. 2, which had an enclosed cockpit like the LM. LLRV No. 2 completed five more flights and No. 1 ran its total up to 198 before the FRC concluded its program on the two vehicles in the winter of 1966. By this time, the LLRVs had flown as long as 9.5 minutes and attained altitudes nearing 800 feet.[221]

The FRC shipped LLRV No. 1 to Houston on 12 December 1966 and followed with No. 2 on 17 January 1967. Kluever flew LLRV No. 1 at Houston's Ellington

[220] Francis J. O'Connell, "Bell Unveils Lunar Training Craft," Buffalo, NY, *Courier Express*, 9 Apr. 1964; "Lunar Landing Research Vehicle," *Space World*, D-5-41 (May 1967), pp. 11-14; Bell, *Uprated LLRV* Pre-Design..., pp. 4-7; DFRC vehicle summary; comments of Cal Jarvis, Gene Matranga, and Wayne Ottinger on the original chapter.

[221] LLRV Flight Log, n.d.; FRC NASA *X-Press*, 25 Sept. 1964, 6 Nov. 1964, 11 Dec. 1964, 19 Feb. 1965, 25 Mar. 1966, 1 Jul. 1966, 26 Aug. 1966, 13 Jan. 1967, 27 Jan 1967; Walker quote from Harold R. Williams, "Training for a Lunar Touchdown," *Rendezvous*, VII, No. 1 (1968), pp. 5-7; "Lunar Landing Research Vehicle," p. 11; Jarvis comments.

AFB in March 1967. Afterward, Joe Algranti and Bud Ream, who would act as instructor pilots for the astronauts, also flew the craft. A month later, Robert R. Gilruth, director of the Manned Spacecraft Center, in an official commendation of the FRC's LLRV project team, said the flights at Edwards had "yielded important information on vehicle handling qualities and piloting techniques and procedures necessary for a successful lunar landing. . . . The LLRV program has and will continue to contribute much to the United States' efforts for a manned lunar landing."[222]

Gilruth's words "will continue to contribute" referred to an extension of the two-vehicle LLRV program. In mid-1966, the Manned Spacecraft Center had ordered three more lunar landing simulators from Bell, these designated as LLTVs, Lunar Landing Training Vehicles. Each cost about $2.5 million. Incorporating modifications that resulted from experience with the LLRVs, the LLTV weighed about 5,000 pounds and could attain an altitude of 400 feet. With the cockpit display and control system modeled on the lunar module, the pilot's visibility was restricted to match what the LM would offer. The first LLTV arrived at Houston in October 1967 and first flew in October 1968. The Manned Spacecraft Center modified the two original LLRVs to more closely resemble LLTV aircraft as well. The new vehicles ordered straight from Bell became LLTVs Nos. 1, 2, and 3. Houston's pilots made the initial LLTV flights at Houston and acted as instructor pilots to the astronauts. The Manned Spacecraft Center quickly evolved an astronaut training program. Potential LM crewmen first went to helicopter school for three weeks, then to Langley's Lunar Landing Facility, then on to 15 hours in a ground simulator, and finally to the LLTVs, which they flew from nearby Ellington AFB.[223]

The LLTVs proved extremely useful. Indeed, as astronaut chief Donald "Deke" Slayton noted, there was "no other way to simulate moon landings except by flying the LLTV."[224] All prime and backup commanders of lunar landing missions practiced on the LLRV and LLTV vehicles. Gene Cernan completed the last LLTV flight on 13 November 1972. Commenting to newsmen following an LLTV training flight on 16 June 1969, a month before liftoff of Apollo 11, mission commander Neil Armstrong remarked, "We are very pleased with the way it flies. . . . I think it does an excellent job of actually capturing the handling characteristics of the lunar module in the landing maneuver. . . . we're getting a very high level of confidence in the overall landing maneuver."[225]

Houston's LLTV operations, however, were not without difficulties. In fact, three of the five vehicles crashed. On 6 May 1968, Neil Armstrong took off in LLRV No. 1.

[222] Letter, Gilruth to Bikle, 28 Feb. 1967; NASA *FRCX-Press*, 10 Mar. 1967.

[223] NASA MSC release 72-230; "Training for a Lunar Touchdown," pp. 5-7; Zack Strickland, "Series of Lunar Landings Simulated," *Aviation Week & Space Technology* (30 June 1969), p. 55; minutes of Flight Readiness Review Board, Lunar Landing Training Vehicles, 12 Jan. 1970.

[224] Williams, "Training for a Lunar Touchdown," p. 7.

[225] Strickland, "Series of Lunar Landings Simulated," p. 55.

While hovering about 30 feet above the ground, the vehicle suffered a loss of helium pressure in the propellant tanks, causing shutdown of its attitude-control rockets. It started nosing up and rolling over, and Armstrong immediately ejected. His zero-zero seat kicked him away from the stricken craft, which tumbled into the ground and exploded as the astronaut safely descended by parachute. It was a sad fate for the LLRV No. 1, a pioneering flight craft. On 8 December 1968, gusty winds forced LLTV No. 1 out of control, and MSC pilot Joe Algranti safely ejected just one second before the wobbling simulator crashed. Finally, on 29 January 1971, LLTV No. 2 suffered an electrical system failure that caused loss of attitude control. MSC pilot Stu Present abandoned this sick bird safely.[226]

The LLRV/LLTV program is a remarkable example of how the Flight Research Center's bias toward free-flight testing helped NASA achieve a spectacular success: the first piloted lunar landing. Naturally, when discussions turned toward putting astronauts on the moon, this bias had triggered a desire on the part of the Flight Research Center engineers to build a specialized flight research testbed. Other centers, dominated by ground-based laboratory thinking, had favored less radical, more traditional, and less satisfactory methods, such as fixed simulators and semimobile rigs. The combination of these methods produced the successful lunar landings, which went off flawlessly. Two of these craft still exist: the LLRV No. 2 and the LLTV No. 3. It is difficult now to conceive of such strange and grotesque hardware making a worthwhile contribution to any development effort, but that they did contribute, and handsomely, is beyond dispute.

By the time Neil Armstrong set foot on the moon, however, the Paresev and LLRV programs were rapidly fading memories at Edwards. The FRC was busy on other space-related projects in an area of traditional FRC interest: hypersonic lifting reentry from space. At the heart of this effort was a strange group of test vehicles, the lifting bodies. They come as a postscript to the early days of Round Two and Round Three and as a prelude to the Space Shuttle.

[226] Ibid.; NASA MSC release 72-230; NASA MSC release 68-182.

8

Return from Space: The Lifting Bodies 1962-1975

During the late 1950s and early 1960s, two camps emerged among those studying reentry from space. One group favored so-called "ballistic" reentry, literally dropping out of orbit and transiting the atmosphere much like a plunging stone. The other camp favored "lifting" reentry, a longer passage from space to earth that would enable a crew to fly a spacecraft to a conventional landing at an airfield. A lifting reentry spacecraft was a far more demanding—but potentially far more useful—technology than a ballistic capsule. Designers would have to develop a configuration with adequate structural strength to withstand the rigors of a missile-like launch, with a reusable or refurbishable thermal protection system for reentry, and with adequate hypersonic, supersonic, transonic, and subsonic flying qualities—no mean feat. The X-20A Dyna-Soar project was a premature attempt to develop such a craft.

Dyna-Soar was not the only lifting reentry approach to orbital flight. There were also weird wingless shapes known, for want of a better name, as "lifting bodies." The lifting-body concept dated back to the blunt-body studies of H. Julian "Harvey" Allen, an imaginative engineer at Ames Aeronautical Laboratory. Allen conceived the blunt body theory in 1951. Together with Alfred Eggers, Allen concluded that a ballistic missile warhead having a blunt, rounded nose (as opposed to a pointed shape) would better survive the intense heat generated as it entered the atmosphere from space at near-orbital velocities. The blunt shape produced a strong, detached bow shock wave that, in effect, gave the following warhead excellent thermal protection. Allen's work remained highly classified, but the fruits of it appeared on the Atlas missile's deadly nose.

Necessarily, the blunt body had a very low lift-to-drag ratio, far less than 1. It flew a ballistic descent path having a minimal "cross-range footprint."[227] Allen and Eggers, together with Clarence Syvertson, George Edwards, and George Kenyon, recognized that designers might be able to combine the blunt body with a piloted orbital vehicle in such a way that it had an acceptable lift-to-drag ratio, on the order of 1.5. This could reduce reentry g loadings from the 8 g experienced by a blunt body to 1 g and give a cross-range footprint in excess of 1,500 miles from the initial point of atmospheric entry. Eggers deduced that one desirable shape for such a vehicle would

[227] "Cross range" refers to the distance a reentry vehicle could travel laterally from its initial entry path. This was important for landing on a runway, as was foreseen in the case of the lifting bodies. The footprint was simply the area encompassing the lateral and longitudinal landing capabilities of a given vehicle—that is, the region within which it could land.

be a modified half-cone (flat on top) with a rounded nose to reduce heating. Working at Ames, Eggers, Syvertson, Edwards, and Kenyon refined the concept in 1958, deriving the M2 configuration, a 13-degree half-cone with a rounded nose having a lift-to-drag ratio of 1.4 at hypersonic speeds. At subsonic speeds, however, its woefully inadequate stability characteristics made it prone to tumbling end over end. Eventually the Ames engineers "boat-tailed" the top and bottom of the shape, giving it an airfoil cross-section and curing most of the stability difficulties. This final M2 version had a protruding canopy and twin vertical fins—the fins earning it the nickname "M2 Cadillac." By 1960, the lifting-body work at Ames was far from fruition, but engineers had chosen a basic shape.

Ames was not the only NASA center engaged in lifting-body studies. The High-Speed Flight Station did not have the hypervelocity tunnels, guns, and shock tubes needed for such research, but the staff kept in touch with colleagues at the larger centers and were aware of what was going on. The HSFS engineers would make their own contributions soon enough, originating the flight-research programs for the lifting bodies. At Langley, engineers favored a more traditional approach over sawing a cone in half. They opted for modified delta configurations. Eventually, as a result of the work of Eugene S. Love, Langley devolved the shape for the HL-10—*HL* standing for horizontal lander. It first appeared on Langley drawing boards in 1962 as a piloted lifting reentry vehicle. Though still working on Dyna-Soar, the Air Force considered other lifting reentry schemes and in the early 1960s, commissioned a series of studies that eventually spawned the Martin SV-5D shape, a configuration between the cone-like M2 and the modified delta HL-10. The Ames M2, Langley HL-10, and Air Force/Martin SV-5D shapes were outgrowths of the same climate of research that had created the Dyna-Soar program; their roots in Round Three thinking.

The First Lifting Body

R. Dale Reed, an FRC research division engineer, was fond of building flying models. While recognizing that models are limited in the range of information they can return, he knew they could validate the basic stability and control characteristics of a new configuration. Reed had followed with interest the Ames work on the M2, noting that while it had potentially excellent hypersonic characteristics, doubts existed that the M2 could successfully fly to a landing because of difficulties in handling at transonic and subsonic speeds. Other NACA engineers had suggested in the 1958 HSFS research assessment that NACA develop low-speed testbeds of proposed hypersonic shapes to determine their landing behavior. In February 1962, Reed built a two-foot model of the M2, which he launched from a larger radio-controlled "mothership" having a 10-foot wingspread—a typical FRC approach scaled down in size. Reed's wife filmed some of the flights to show to Paul Bikle, De Beeler, and Alfred Eggers. Reed also flew small lifting-body models down the corridors at FRC, causing raised eyebrows among skeptics. But Eggers promised the use of wind tunnels at Ames, and Bikle

In the foreground, lifting-body configurations that were launched from "Mother," the large, twin-engine, radio-controlled model airplane in the background. (NASA photo ECN-1880)

authorized a six-month feasibility study of a cheap, piloted, lightweight M2 glider, the "next step" suggested by Reed, who also flew sailplanes as a hobby.[228]

In September 1962, Bikle authorized the design and construction of a piloted M2 glider. Under Dale Reed's direction, Victor Horton headed the effort, assisted by Dick Eldredge and Dick Klein. FRC engineers built the tubular steel structure, and Gus Briegleb of the Sailplane Corporation of America built the plywood outer shell. At first, Reed, Horton, Eldredge, and Klein wished to test various lifting-body shapes, including M1, M2, and a lenticular "flying saucer" concept. The M2 seemed the most practicable, however, and was the only one the FRC proceeded with. Technicians set aside floor space in a hangar, walled it off with canvas, and put up a sign reading "Wright's Bicycle Shop." The project team drew on many other FRC staffers for assistance, especially the large local NASA community of aircraft homebuilders, mostly members of the Experimental Aircraft Association. Bikle ran the project out of local funds on a nickel-and-dime basis, fearing he could not secure headquarters support rapidly enough to permit a quick development program. Bikle's concern over complicating the project by working through the system was well founded: one major

[228] Interview of Reed; NASA FRC *X-Press*, 10 Mar. 1967; interviews of Klein and Williams; Dale Reed's comments on original chapter, 1 Mar. 2000.

aircraft company informed the FRC M2 team that it would have cost $150,000 for the firm to build such a vehicle. By using in-house funding and exacting cost control, FRC engineers kept expenditures on the design and fabrication of the M2 glider (not including an ejection seat and solid-propellant rockets to provide extra lift over drag for landing, but including other support) under $30,000. Briegleb's own construction team, consisting of three mechanics and a draftsman working at El Mirage Dry Lake, built the mahogany plywood body shell (less than an inch thick) in 120 days. The FRC design team had stipulated that the body shell weigh less than 300 pounds; Briegleb's team managed to complete it at 273 pounds.[229]

The FRC/Briegleb team finished the M2 glider, which the FRC designated the M2-F1, early in 1963. A tubby vehicle, it was about twenty-two feet long, nine and a half feet high, and fourteen feet wide. It had two vertical fins, just like the earlier Ames "M2 Cadillac" study, with stubby elevons mounted on the fins. The body had trailing-edge flaps for trimming purposes and landing gear wheels from a Cessna 150 airplane. With its ejection seat and instrumentation, the M2-F1 weighed 1,250 pounds. The pilot sat under a large bubble cockpit. Though at first the craft had no provisions for emergency ejection, the FRC added a lightweight Weber rocket-propelled zero-zero seat. Later the craft also had a 1,070-newton (24-pound-thrust) solid-propellant rocket developed by the Naval Ordnance Test Station at nearby Inyokern, California, to assist in the prelanding "flare" maneuver, if this became necessary. The craft was trucked to Ames for low-speed testing in the 40-by-80-foot wind tunnel. The wind-tunnel tests, completed in March 1963, were very encouraging. NASA project pilot Milt Thompson often sat in the cockpit of the M2-F1 during the studies, "flying" the rigidly mounted craft in the cavernous maw of the full-scale wind tunnel. Satisfied, Ames gave the shape its blessing, and the FRC took it back to Edwards in preparation for its first flights, a series of Paresev-like ground tows.

Strange enough already, the M2-F1 program now took a real turn toward the bizarre. Obviously, the shape had a lot of drag, requiring a tow vehicle with great power and speed. NASA's general-purpose trucks and vans just could not do the job; a specialized high-performance tow car was needed. The solution did not take long. Out in the desert lived a number of racing aficionados, many of whom worked at the FRC. After consulting with the FRC M2 team, one of them, veteran dirt-biker Walter "Whitey" Whiteside, bought a stripped-down Pontiac convertible with the largest engine available, a four-barrel carb, and a four-speed stick shift, capable of towing the M2 to 110 miles per hour in 30 seconds. Then the convertible was turned it over to Bill Straup's famous hot-rod shop near Long Beach, California, where technicians fine-tuned the engine, added rollbars, put on dual exhausts, and tested it on a dynamometer. Mickey Thompson's garage supplied the Wheels and tires, a set of drag-racing slicks. Technicians at the FRC installed radio equipment, turned around

[229] Interview of Gus Briegleb, 18 May 1977; interviews of Reed, Klein, and Bikle; NASA FRC *X-Press*, 10 Mar. 1967.

The plywood-shelled M2-F1 lifting body in towed flight over Rogers Dry Lake. The tow-line is barely visible next to the nose gear. (NASA photo ECN-225)

the right passenger bucket seat to face aft, and removed the rear seats, installing another bucket seat for a second observer facing sideways.

Lest anyone hastily conclude that the souped-up Pontiac was somebody's private toy paid for with government funds, the team painted "National Aeronautics and Space Administration" on the sides, then sprayed the hood and trunk high-visibility yellow, like any other flight-line vehicle.[230] The NASA engineers added a tow rig and some airspeed measuring equipment. Whitey then then took it to the Nevada desert, with its (then) anything-goes speed limits, to calibrate the speedometer—just like any research airplane. Team members fondly recall the strange head-shaking stares of California and Nevada highway patrolmen as the exotic auto rumbled along, driven by Walter Whiteside, engine exhausts roaring. Its gasoline mileage wasn't good—just four miles per gallon. Finally, by the spring of 1963, all was ready. Milt Thompson ventured out on the dry lake, the M2-F1 rigged behind the Pontiac on a tow line for its first excursion into the air.[231]

[230] In about 1968, NASA shipped the Pontiac to Langley for tests at Wallops Island. "No longer," mourned a commenter in the FRC *X-Press* with only modest exaggeration, "can we drive along the lakebed and pass the airplanes in flight."

[231] Interview of Walter W. "Whitey" Whiteside by Betty J. Love and Robert Hoey, 22 July 1994, Lancaster, CA, pp. 8-9. See also R. Dale Reed with Darlene Lister, *Wingless Flight: The Lifting Body Story* (Washington, DC: NASA SP-4220, 1997), pp. 34-35. Interview of Klein; NASA FRC *X-Press*, 18 May 1968; comments of Dick Klein and Bertha Ryan, late Feb. 2000.

The M2-F1 completed its first airborne ground tows on 5 April 1963 and ended up making 45 others by the month's end. In all until the first air tows, the little lifting body made almost 100 tows, an accumulated air time of a little over an hour. Generally speaking, the M2-F1 had acceptable flying qualities, warranting its being air-towed to altitude and released, but Thomas Toll, the FRC's chief of research and one of the men responsible for the X-15 concept, had serious misgivings. He had become especially concerned after Thompson's first airborne ground tow, when the pilot had encountered a dangerous lateral oscillation. The lifting-body team tested the M2-F1 in the Ames 40-by-80-foot wind tunnel before resuming ground tows. They had also looked at movies of the airborne ground tow. As a result, they changed the roll-control procedures. Afterward, Bikle approved air tows. The FRC had a Douglas C-47 "Gooney Bird" assigned for general duties. The C-47, the military version of the legendary DC-3, had been an excellent glider tug during World War II in such campaigns as those in Sicily and Normandy. Vic Horton of the FRC's M2 team scrounged up a C-47 tow mechanism from a junkyard. It was installedon the plane, and on 16 August 1963 Milt Thompson was piloting the little lifting body as the C-47 towed it off the dry lakebed. On this and other flights, the C-47 generally climbed at about 120 miles per hour to over 10,000 feet, the M2-F1 trailing on a 1,000-foot towline. The towplane would release the glider above its intended landing spot on Rogers Dry Lake, and Thompson would guide the rapidly sinking craft to a touchdown about two minutes after release, landing at about 85-90 miles per hour. On 3 September, the FRC unveiled the craft to aviation news reporters. The lifting-body concept at once became a hot item in the media, acquiring the nickname, "Flying Bathtub."[232]

The first flights of the M2-F1 had proved that the lifting-body shape could fly. As early as mid-April 1963, Bikle was convinced enough to bring NASA headquarters into his confidence. He told Milton Ames, NASA's director of space vehicles: "The lifting-body concept looks even better to us as we get more into it. We also recognize a rising level of interest in the concept at Ames and at Langley."[233] There was a rising level of interest on Capitol Hill as well, as word got back to Washington. By mid-April 1963, many congressmen were quizzing NASA headquarters officials on the M2 flight program, causing consternation among some Department of Defense officials who apparently had no idea that the M2 was flying. Some congressmen feared the low-budget M2 might soar overnight into a major multi-billion-dollar post-Apollo development program. Others later suspected that the program was a way for NASA to circumvent the decision to cancel Dyna-Soar. Hugh Dryden

[232] M2-F1 interflight worksheets, in DRFC pilot office files; interviews of Reed and Klein; "M-2 Research Craft Landed Seven Times," *Aviation Week* (9 Sept. 1963), p. 34; Russell Hawkes, "M-2 Flight Successes Spur Interest in Lift Re-entry," *Missiles and Rockets* (9 Sept. 1963), pp. 14-15; Thompson, "I Fly Without Wings," pp. 12-13; NASA *X-Press*, 10 Mar. 1967; Thompson and Peebles, *Flying Without Wings*, pp. 72-88; Robert G. Hoey, *Testing Lifting Bodies at Edwards*, published as part of the Air Force/NASA Lifting Body Legacy History Project, Sept. 1994, p. 164, and available on the Internet at http://www.dfrc.nasa.gov/History/Publications/LiftingBodies/contents.html; Reed with Lister, *Wingless Flight*, pp. 36-37.

[233] Letter, Bikle to Milton Ames, 24 Apr. 1963.

and OART's Raymond L. Bisplinghoff defended the FRC effort, and the M2 program continued.[234]

At Edwards, nine other pilots checked out in the M2-F1: NASA test pilots Bruce A. Peterson, Donald L. Mallick, Fred W. Haise, Jr., and William H. "Bill" Dana; as well as the Air Force's Col. Charles E. "Chuck" Yeager, Capt. Jerauld R. Gentry, Maj. James W. Wood, Capt. Joseph H. Engle, and Maj. Donald M. Sorlie. Col. Yeager clambered out of the craft after his first flight exclaiming, "She handles great!" He hoped to use similar vehicles, powered by small jet engines, as lifting-body simulation trainers at the Aerospace Research Pilots School, which he commanded.

Eventually the little M2-F1 completed 77 flights and approximately 400 ground tows before being retired to the Smithsonian's National Air and Space Museum. (It is now held in storage at Dryden for the museum). The FRC did have to make some modifications to the craft. On one flight, NASA's Bruce Peterson landed with sufficient force to shear off the landing wheels, and the M2-F1 sustained minor damage; during the tow to altitude, the automobile-type shock absorbers had become chilled, and the cold hydraulic fluid simply failed to function properly on touchdown. NASA replaced the Cessna 150 landing gear with more rugged gear from a Cessna 180. On two other flights, Jerauld Gentry became involved in some extremely hazardous rolling maneuvers. On one occasion, Vic Horton glanced out of the C-47 in time to see Gentry and the M2-F1 rolling inverted on the towline; for several seconds, the launch crew in the C-47 did not know if the errant lifting body had ploughed in. When they next saw it, however, it rested safely on the dry lakebed: Gentry had cast off, stabilized the M2-F1, flared, and landed—just another close call. Not wishing to take further chances, Bikle shut down the M2-F1 program. It had served its purpose: it proved that the lifting-body shape could fly and encouraged further research with supersonic rocket-powered lifting bodies, to determine if the shapes so desirable for hypersonic flight could safely fly from supersonic speeds down to landing, through the still tricky area of transonic trim changes. When the tubby M2-F1 completed its last air-tows on 16 August 1966, work was already well along on two "heavyweight" aluminum follow-ons.[235]

Establishing a Joint Lifting-Body Program

Encouraged by the M2-F1, the FRC pressed forward in its lifting-body studies, which eventually led to the Northrop M2-F2 (and later the M2-F3) and the Northrop HL-10. Air Force interest resulted in formation of a joint NASA/Air Force lifting-body

[234] U.S. Senate, Committee on Aeronautical and Space Sciences, *Hearings: NASA Authorization for Fiscal Year 1964*, Pt. 1: *Scientific and Technical Program*, 88th Congress, 1st Session, Apr. 1963, pp. 40, 608-634; Pt. 2: *Program Detail*, 88th Congress, 1st session, June 1963, pp. 919-920.

[235] NASA news release 32-63; "Pilot Declares 'Flying Bathtub' Makes Fine Re-Entry Vehicle," *The Oregonian*, 6 Dec. 1963; interview of Klein; DFRC monthly flight activities files for 1963 and 1964, DFRC pilot office files; interview of Jerauld Gentry, 11 Sept. 1975; Reed with Lister, *Wingless Flight*, pp. xviii, 193-94.

The "heavyweight" Northrop M2-F2 lifting body on the Dryden ramp. Note the water on the normally dry lakebed. (NASA photo ECN-1088)

program. The Air Force Flight Test Center and the NASA Flight Research Center issued a memorandum of understanding on the program in April 1965.

Early in 1963, as the M2-F1 was taking shape at Edwards and El Mirage, Dale Reed's M2 team had preliminary studies under way on an air-launched, mission-weight, rocket-propelled, Mach 2 lifting body using off-the-shelf systems and equipment. This research vehicle, informally dubbed "Configuration II," could return useful information on the supersonic and transonic behavior of such craft, piloting problems and workloads, and approach and landing characteristics of a mission weight lifting body. The earlier lightweight M2-F1 had a wing loading only one-fifth of that expected with a fully developed and operational space-rated lifting body. Oddly, NASA recognized from the outset that the lightweight lifting body would be considerably more difficult to land than the heavyweights. Even though both had the same lift-to-drag ratio, the lightweight M2 had an inherently shorter time between the pilot's landing flare and touchdown than the heavyweight would have. This increase in time available before touchdown was desirable from a piloting standpoint, but the heavyweight vehicles also landed much faster. The FRC M2 team had decided to proceed with the lightweight M2 tests, even though the vehicle would be difficult to fly, because its low touchdown speed (around 85 miles per hour) reduced the risk of pilot injury.[236]

[236] Drake, "Aerodynamic Testing Using Special Aircraft," pp. 8-10; NASA FRC, *Five-Year Plan.*

Paul Bikle's almost-covert M2 operation at Edwards proved a big success in boosting the lifting-body concept. The flight-research results encouraged greater participation by other NASA centers and headquarters through the Office of Advanced Research and Technology under NASA Associate Administrator Raymond Bisplinghoff. On 15 and 16 September 1964, Bisplinghoff and some of his staff met with Bikle and the M2 team at Edwards. What came out of this meeting was a directive to the NASA center directors asking that they document "existing research effort on entry vehicles of the lifting-body class," with a view toward possible construction of a hypersonic lifting body. OART now strongly supported the lifting-body research program at Edwards. Bisplinghoff wrote, "I believe it is essential that we have a strong in-house research effort covering all the technical problem areas of importance to lifting-body vehicle design and operation."[237]

By this time the "heavyweight" program was under way. In February 1964, the FRC solicited proposals from 26 firms for two heavyweight, lowspeed lifting-body gliders. NASA would test them in the full-scale Ames wind tunnel and also air-launch them from a B-52 flying at 45,000 feet. The firms had five weeks to submit proposals. OART would supervise the program, with Ames, Langley, and the FRC participating. One glider would be an M2, and the other would be Langley's own proposed HL-10 modified delta shape. Only five companies submitted proposals. The FRC selected the Norair division of the Northrop Corporation to build the vehicles. On 2 June 1964, the center awarded a fixed-price contract to Northrop for the fabrication of the M2 and HL-10 heavyweight gliders at $1.2 million apiece. Northrop would deliver the M2-F2 in the late spring of 1965, with the HL-10 following six months later.[238] The lifting-body program had moved into its next phase. The FRC and Headquarters still favored going beyond gliders to powered supersonic lifting-body trials. In early August 1964, Bikle, Bisplinghoff, and Bisplinghoff's deputy Alfred Eggers agreed on incorporating provision for XLR-11 rocket engines in the two new gliders.[239]

What most influenced Bikle and the FRC project team in their selection of Northrop were the elements of simplicity and costs. Northrop, a company in the midst of a highly successful private fighter venture (the F-5 program), assured the FRC that it could build the two gliders cheaply. Richard Horner, who had worked with Bikle first at Edwards, then from NASA headquarters, was now executive vice president of Northrop. The two men dispensed with red tape and unnecessary paperwork. As a result, the vehicles, which one industry spokesman had predicted could cost $15 million apiece, wound up costing just $1.2 million apiece, unheard of for complex research airplanes. Bikle assigned FRC engineer John McTigue as NASA program manager, while Northrop assigned Ralph Hakes as Norair's program manager. The two men

[237] Letter, Bisplinghoff to NASA center directors, 28 Oct. 1964.

[238] NASA news releases 64-41 and 64-93, and NASA FRC release 14-64.

[239] Letter, Eggers to Bikle, 18 Aug. 1964; Bisplinghoff memo for NASA associate administrator, 20 Aug. 1964.

devised a joint action management plan to minimize paperwork and the number of employees working on the project, to make decisions by individuals and not by committees, to locate the project in one area where all necessary resources could be easily and quickly directed to it, and to fabricate the vehicle using a conservative design approach. As Hakes recalled,

> We never had more than a handful of engineers. . .They were all twenty-year men who had worked to government specifications all their lives and knew which ones to design to and which to skip. McTigue's people and ours would talk things over and decide jointly what was reasonable compliance with the specifications. Decisions were made on the spot. It didn't require proposals and counter-proposals.[240]

Because of his long Air Force association, Paul Bikle worked closely and effectively with his Air Force Flight Test Center counterparts, much as Walt Williams had before him. He recognized that, like the X-15, the lifting-body program required some sort of joint-operations agreement because the program was getting too large for NASA to manage and operate alone. He knew that the NASA/Air Force/contractor flight-testing relationship was a close one. As with the NACA in the late 1940s at Muroc, there were few if any disagreements among the working-level personnel. Such disagreements as existed were imposed from above. Bikle saw that the Air Force and NASA had similar interests in the lifting-body concept. Over the early spring of 1965, he met with Maj. Gen. Irving Branch, commander of the Air Force Flight Test Center at Edwards. Out of these meetings came a memorandum of understanding on 19 April 1965. The memo drew on previous X-15 program experience, alluding to the similarities between the programs and the excellent working relationships that had existed between Air Force and NASA personnel assigned to the X-15 program. The memo created the Joint FRC/AFFTC Lifting-Body Flight-Test Committee composed of 10 members: the director of the FRC (chairman), the commander of the AFFTC (vice-chairman), NASA and Air Force pilots, NASA and Air Force engineers, NASA and Air Force project officers, a NASA instrumentation representative, and a medical officer from the Air Force.

The joint flight-test committee had overall responsibility for the research program and assumed responsibility for all outside relations and contacts. The FRC had responsibility for maintenance, instrumentation, and ground support of the craft, while the AFFTC assumed responsibility for the launch aircraft, support aircraft, medical support, the rocket power plant, and the pilots' personal equipment. The AFFTC and the FRC assumed joint responsibility for planning research flights, analyzing flight data, test-piloting, range support, and overall flight operations.[241] Bikle and Branch

[240] Northrop, "Lifting Bodies: Coming In on a Vanishing Wing," p. 7; interview of Bikle.

[241] AFFTC-FRC memo of understanding, 19 Apr. 1965, repr. as Appendix V in FRC-AFFTC, *Lifting Body Joint Operations Plan* (Edwards, CA: 1 Sept. 1969), pp. 8-9 to 8-9b.

issued the memo two months before Northrop rolled out the M2-F2. But the M2-F2 and the HL-10 were no longer the only "heavyweights" under construction. A year and a half later, on 11 October 1966, the AFFTC and the FRC amended the memo to cover NASA participation in an Air Force-sponsored lifting-body program, the Martin SV-5P.[242]

The Martin SV-5P had a complex origin. In 1960, the Air Force had begun examining piloted, maneuverable, lifting-body spacecraft as alternatives to the ballistic-type orbital reentry concepts then in favor. This investigation became Project START (*S*pacecraft *T*echnology and *A*dvanced *R*eentry *T*ests), although this name emerged only much later. START involved a three-phase program, with ASSET (*A*erothermodynamic/Elastic *S*tructural *S*ystems *E*nvironmental *T*ests), PRIME (*P*recision *R*ecovery *I*ncluding *M*aneuvering *E*ntry), and PILOT (*Pi*loted *Lo*wspeed *T*ests) as its eventual constituents.

In May 1961, the Air Force Flight Dynamics Laboratory awarded the McDonnell Aircraft Corporation a contract for a suborbital lifting-body reentry vehicle called ASSET. The craft measured over five feet long and generally resembled the canceled X-20A. McDonnell built six of these vehicles, launching them down the Eastern Test Range from Thor-Delta and Thor boosters between September 1963 and March 1965. These shapes reached speeds between 8,864 and 13,295 miles per hour while making lifting reentries from 200,000 feet over the South Atlantic. All of the vehicles survived reentry, though some were lost at sea before recovery crews could pick them up.

The next step, PRIME, began in November 1964 when the space systems division of the Air Force Systems Command gave the Martin Company a contract to design, fabricate, and test a maneuvering reentry vehicle to demonstrate whether a lifting body could, in fact, be guided from a straight course and returned back to that course. Martin already had been studying lifting-reentry vehicles for some time—the company had, after all, been in the Dyna-Soar competition—and had put more than two million man-hours into lifting-entry studies. The outcome had been the SV-5 body shape, resembling a finned potato. Company engineers built a five-foot radio-controlled model and flew it at the Martin plant in Middle River, Maryland. They raised it to altitude under two balloons, then dropped it and guided it to a landing. These quick-and-dirty trials proved the shape could fly. Eventually Martin refined the design into the SV-5D, an 895-pound aluminum vehicle with an ablative heat shield. The Air Force ordered four of the SV-5D PRIME vehicles, designating them X-23A and launching three of them between December 1966 and mid-April 1967 over the Western Test Range using Atlas boosters that blasted them at some 15,000 miles per hour toward Kwajalein. The three vehicles performed so well that the Air Force canceled the last launch to save money. The PRIME project demonstrated that a maneuvering lifting body could indeed successfully alter its flight path upon reentry.

The Air Force and Martin had further expanded upon the company's PRIME work and had derived PILOT—a proposed "low-speed" (Mach 2) research vehicle

[242] Addendum to memo of understanding, 11 Oct. 1966, repr. as part of Appendix V in ibid., p. 8-9c.

that the Air Force could test to determine its supersonic, transonic, and subsonic-to-landing behavior. Martin designated this vehicle as the SV-5P.[243]

Martin also proposed a low-speed lifting-body trainer, the SV-5J, to be powered by a small turbojet, for use at the Air Force test pilot school. Nothing came of this, although the company built the shells of two such vehicles and tried to entice a NASA pilot—one of the FRC's best—to fly it if and when it was completed.[244] On the other hand, the SV-5P development program went smoothly. The Air Force awarded Martin a contract for one SV-5P vehicle in May 1966, and the company began development under the direction of engineers Buz Hello and Lyman Josephs. Martin completed it a little over a year later, rolling it out of its Baltimore plant on 11 July 1967. The Air Force designated the craft as the X-24A. It soon journeyed to Ames for comprehensive wind-tunnel testing, and from there to Edwards, where the other lifting bodies, the M2-F2 and HL-10, had already flown.[245]

The "Heavyweights" Fly

Without a doubt, the lifting bodies were the ugliest of the postwar research aircraft. Only two were passingly handsome: the HL-10 was pleasingly plump and the X-24B, with its laundry-iron shape, had rakish lines that hid the tubby bulge of its X-24A ancestry. Despite their lines, they generally flew satisfactorily. "Lifting bodies," one test pilot remarked at the unveiling of the X-24A, "fly a lot better than they look." For convenience, the aircraft will be discussed in the following order: the M2-F2 and M2-F3, the HL-10, and the X-24A and X-24B.

The M2-F2 rolled out of Northrop's plant in Hawthorne, California, on 15 June 1965 and was trucked from the Los Angeles basin over the San Gabriel Mountains to

[243] U.S. House, Committee on Science and Astronautics, *1967 NASA Authorization*, Pt. 2, 89th Congress, 2nd session, Feb.-Mar. 1966, pp. 1073-1077; ibid., *1968 NASA Authorization*, Pt. 2, 90th Congress, 1st session, Apr. 1967, pp. 1011-1012; "McDonnell Corp. Making Space Research Craft," St. Louis *Post-Dispatch*, 6 Jan. 1963; Frank G. McGuire, "First ASSET Launches Due in Summer," *Missiles and Rockets* (14 Jan. 1963), p. 18; "Space Glider Lost at Sea After Suborbital Flight," *Washington Post*, 24 Feb. 1965; USAF AFSC news release 31.65; "START and SV-5 Shuttle Gets Go-Ahead," *Space Daily*, 4 Mar. 1965; "A Decision on AF Manned Space Shuttle Nears," *Space Daily*, 20 Sept. 1965; William J. Normyle, "Manned Flight Tests to Seek Lifting-Body Technology," *Aviation Week & Space Technology* (16 May, 1966), pp. 64-75; "PRIME SV-5D-III Maneuvers and Recovered," *Space Daily* , 21 Apr. 1967; James J. Haggerty. "USAF Finishes PRIME Project," *Journal of the Armed Forces* (3 June 1967), p. 9.

[244] The SV-5J would be very underpowered—perhaps too underpowered to gain enough altitude to extend the landing gear before landing if the drag of retracting the landing gear after takeoff precluded gaining enough altitude for a full flight. During consultations, the pilot (Milt Thompson) waggishly proposed putting a two-by-four board across the Edwards runway. The SV-5J would hit the board, bounce, and the pilot would raise the gear if the vehicle began flying or land if it did not. Fortunately, for the sake of Martin's reputation, it allowed the SV-5J program to die. See Thompson and Peebles, *Flying Without Wings*, pp. 158-159.

[245] Evert Clark, "Rocket Plane May Let Astronauts Land at Airfields," *The New York Times*, 12 July 1967; USAF AFSC new releases 85-66 and 59-67.

Edwards the next day. The M2-F2 resembled the earlier M2-F1. At its unveiling, it still lacked the planned XLR-11 rocket engine. NASA would fly it first as a glider and then modify it for powered flight. Fabricated from aluminum, the M2-F2 weighed 4,630 pounds and measured 22 feet in length, with a span of 9.4 feet. Like the earlier M2-F1, it had two vertical fins, but lacked the earlier craft's horizontal control surfaces. Unlike the M2-F1, it had a retractable landing gear, assembled from off-the-shelf components, including the main landing gear of a Northrop T-38 trainer and the nose gear of a North American T-39 Sabreliner. High-pressure nitrogen would blow down the gear just prior to touchdown. It had a complex series of body flaps: a full-span ventral flap controlled pitch, while split dorsal flaps controlled roll (lateral) motion through differential operation and pitch and trim through symmetrical operation. Half-rudders mounted on the outboard side of the twin vertical fins provided directional (yaw) control and also acted as speed brakes. The M2-F2 had a stability augmentation system to assist the boosted control system in damping out undesirable vehicle motions. The pilot could use four throttleable hydrogen peroxide rockets rated at 400 pounds apiece for instant lift during the prelanding flare. If the craft proved unmanageable or some other calamity struck, the M2-F2 had a modified zero-zero ejection seat from an F-106 Delta Dart.

At the FRC, technicians checked out the aircraft, added research instrumentation, and then trucked it to Ames for two weeks of tests in the full-scale wind tunnel. Ames completed 100 hours of testing in August 1965. Apart from a correctable high-frequency oscillation of the upper surface flaps, the M2-F2 received a clean bill of health. It returned to Edwards for its initial flight trials. Northrop furnished a special 22-foot adapter so that the M2-F2 could launch from the B-52 mothership's existing X-15 launch pylon. On 23 March 1966, the M2-F2 completed its maiden captive flight. Following a series of similar checkouts, NASA readied the craft for free flight.[246]

Fixed-base simulators had predicted a potential lateral-control problem for the M2-F2. Consequently, before the vehicle's first flights, the FRC launched a cooperative pilot training and aircraft simulation program with the Cornell Aeronautical Laboratory of Buffalo, New York. Earlier, the FRC had flown Cornell's highly modified variable-stability Lockheed NT-33A jet trainer to simulate the low lift-to-drag reentry characteristics of the X-15. Now, in the spring and summer of 1965, the FRC again flew Cornell's NT-33A, this time in lifting-body studies, using the M2-F2 as the reference type. The variable-stability NT-33A—in its own right, one of America's most successful postwar research aircraft—had "drag petals" installed on its wingtip tanks. These petal-shaped surfaces, extended in flight, varied the lift-to-drag ratio of the aircraft from the NT-33A's normal 12-14 to as low as 2, the approximate ratio of an M2 lifting body. Typical lifting-body approaches were executed by Cornell test pilot Robert Harper and by FRC pilots Milt Thompson, Bruce Peterson, Bill Dana, and Fred Haise. The NT-33A tests confirmed that the M2-F2 aircraft would have

[246] NASA FRC *X-Press*, 23 Apr. and 18 June 1965; comments of Robert G. Hoey on the original chapter, 1 Feb. 2000.

undesirable lateral-control characteristics under certain conditions—a fact that later assumed critical importance. In addition, NASA's pilots simulated lifting-body approaches and landings using the center's F-104s and the amenable Douglas F5D.[247]

The M2-F2 completed its maiden flight on 12 July 1966. NASA pilot Milt Thompson dropped away from the B-52 mothership at 45,000 feet, flying at 450 miles per hour. During the brief flight—not quite four minutes—Thompson made a 90-degree turning descent, performed a practice landing-flare maneuver at 22,000 feet, made another 90-degree turn onto final approach, increased his gliding speed to 333 miles per hour, initiated the landing flare at 1,200 feet, reducing his rate of descent from 245 feet/second to 10 feet/second, lowered the landing gear, and touched down exactly at the planned aiming point on Rogers Dry Lake at 195 miles per hour, coasting 1.5 miles across the lakebed. The M2-F2's first flight was an unqualified success. By mid-November 1966, the craft had completed an additional 12 flights, piloted by Thompson, Bruce Peterson, Jerauld Gentry, and Donald Sorlie. Following flight 14 on 21 November, NASA grounded the M2-F2 for installation of its XLR-11 rocket engine. On 2 May 1967, the M2-F2 made its first flight carrying, but not using, the rocket engine, another glide flight piloted by Gentry. Along with all of the other pilots who had flown the craft, Gentry did not like the M2-F2's poor lateral-directional stability characteristics. At low angles of attack and high speeds, it often developed a rolling motion that increased in severity. If the pilot increased the angle of attack, this motion damped out. On the very next flight, this behavior contributed to a major accident that set back the entire lifting-body program and seriously injured Peterson.[248]

On 10 May 1967, Peterson launched away from the B-52 at 44,000 feet, heading to the north and flying east of Rogers Dry Lake. All went well as the M2-F2 sank like a stone, until the wingless craft reached 7,000 feet. Then, flying with a "very low" angle of attack, the M2-F2 began a "Dutch roll" motion, rolling from side to side at over 200 degrees per second. Peterson, who earlier had turned a nearly uncontrollable first flight in the HL-10 into a brilliantly successful landing, was an excellent pilot. He quickly and instinctively raised the nose, damping out the lateral motions. However, the recovery had carried the craft away from its intended flight path. Peterson realized he was too low to reach the planned landing site near lakebed Runway 18 and was rapidly sinking toward a section of lakebed that lacked the visual runway reference markings needed to estimate height above the lakebed with accuracy.

At this point, a rescue helicopter appeared in front of the M2-F2. Overburdened and disoriented from the rolling motions, Peterson now had an additional worry. He

[247] NASA FRC *X-Press*, 7 May 1965; NASA FRC news release 12-65; G. Warren Hall, "Research and Development History of USAF Stability T-33," *Journal of the American Aviation Historical Society* (Winter 1974); Hoey comments.

[248] NASA FRC *X-Press*, 15 July 1966; chronology of lifting-body flights prepared by Nancy Brun, NASA history office; interview of Gentry: NASA release 66-329; NASA FRC *X-Press*, 5 May 1967; Cf. Reed with Lister, *Wingless Flight*, pp. 87-92, 104-106, 195.

called, "Get that chopper out of the way," followed seconds later with "That chopper's going to get me, I'm afraid." FRC chase pilot John Manke, flying an F5D, assured Peterson the helicopter was clear, and it did chug off, out of Peterson's path. Realizing he was very low, Peterson fired the landing rockets, and the M2-F2 flared nicely. He lowered the landing gear, which needed only 1.5 seconds to deploy from up-and-locked to down-and-locked. But time had run out. Before the gear locked, the M2-F2 hit the lakebed, shearing off its telemetry antennas. In the control room, engineers saw the needles on their instrumentation meters flick to their null points. Startled, they looked up to the video monitor—just in time to see the M2-F2, as if in a horrible nightmare, rolling over and over across the lakebed at more than 250 miles per hour. It turned over six times before coming to rest on its flat back, minus its canopy, main gear, and right vertical fin. Peterson, who by all expectations should have died in the accident, was badly injured. Rescue crews pulled him from the wreckage, rushed him to the Edwards hospital for emergency surgery, then to the hospital at March Air Force Base, and several days later to UCLA's University Hospital. Peterson pulled through, but he lost the sight of one eye. He remained at the FRC as the center's director of safety, continuing to fly as a Marine reservist.[249]

Instead of simply trucking the wrecked M2-F2 to a scrapyard, NASA returned it to Northrop's Hawthorne plant. Technicians placed it in a jig to check alignment, removing the external skin and portions of the secondary structure. The inspection took 60 days. In March 1968, NASA's Office of Advanced Research and Technology authorized Northrop to restore the primary structure and return the vehicle to the FRC. There it sat, while lifting-body advocates from Ames and the FRC determined its future. In light of its poor handling characteristics, the M2-F2 obviously needed to be modified. By this time the rival HL-10 was already demonstrating superior handling qualities. Nevertheless, the M2 shape still appeared worth studying. On 28 January 1969, NASA headquarters announced that the M2-F2 would be repaired, modified, and returned to service as the M2-F3.[250]

The rebuilt aircraft, returned to Edwards and was first flown in 1970, looked much like its predecessor, except for a short stubby vertical fin located midway between the two large vertical fins. This center fin acted as a large "flow fence" to improve lateral control. The craft had a new jet-reaction roll-control system, which NASA hoped might be used on future lifting-body spacecraft so that the pilot could rely on a single control system all the way from orbit to landing, rather than the multiplicity of systems used on such craft as the X-15. NASA planned to use the M2-F3 as a testbed for research on the lateral control problems encountered by lifting-body vehicles.

[249] A comprehensive collection of official NASA accident reports and pilot interviews was published as "Pilot Work Load Cited in M2-F2 Crash," *Aviation Week & Space Technology* (1 Oct. 1967); interview of Bruce Peterson, 26 Mar. 1976; interview of Wen Painter, 8 Aug. 1977; NASA FRC *X-Press*, 19 May 1967. The accident eventually inspired a popular novel and TV series.

[250] NASA FRC news release 10-68; NASA news release 69-15; NASA *Ames Astrogram*, 30 Jan. 1969.

On 2 June 1970, Bill Dana completed the M2-F3's first flight, a glide flight to evaluate how the modifications changed the plane's performance. A planned powered flight on 25 November went awry when the engine shut down prematurely. Air Force test pilot Jerauld Gentry–the only pilot at Edwards to fly the M2-F2, HL-10, and M2-F3 (not to mention the M2-F1 and X-24A, which he also flew)–flew the plane on 9 February 1971 and said it flew as well as the HL-10. This was high praise, for the HL-10 flew much better than had the unmodified M2-F2. NASA and the Air Force then embarked on a joint program of incrementally increasing its speed and altitude performance, with the last two flights of the M2-F3 setting the fastest and highest marks. On 25 August 1971, Bill Dana had made the craft's first supersonic flight, attaining Mach 1.1. Over a year later, on 13 December 1972, Dana attained Mach 1.613 (1,064 miles per hour), the fastest M2-F3 flight. On the last flight of the craft on 20 December, FRC research pilot John Manke reached an altitude of 71,500 miles, an M2 record. On only one occasion did trouble occur. On its tenth flight, 24 September 1971, the M2-F3 experienced an engine ignition malfunction; Dana shut down the XLR-11 engine, but a small amount of propellant flared briefly in the engine bay before extinguishing itself. Dana made "a hard but otherwise uneventful landing" on Rosamond Dry Lake, the alternate emergency landing site. Toward the end of the craft's flying career, FRC technicians installed and evaluated a rate command augmentation control system, a kind of fly-by-wire system that used an analog computer and a sidearm control stick in addition to the regular control stick. Altogether, the M2 completed 43 flights, 16 as the F2 and 27 as the F3. Retired at the end of December 1972, the plane subsequently was sent to the Smithsonian Institution.[251]

NASA complemented the M2-F2 and M2-F3 trials with an extensive evaluation of the Northrop HL-10. In contrast to the accident-marred M2 flight-test program, HL-10 research moved along quite smoothly—once the aircraft had been modified after a very frightening first flight. The HL-10, product of Eugene Love's work at Langley, was one of the most successful lifting bodies. Indeed, when the Space Shuttle began to take shape, the consensus among NASA engineers at the Flight Research Center was that it should look like the HL-10. Unlike the M2, which had a cone-shaped underside, the HL-10 had a flat bottom and a rounded top; it was, in effect, an inverted airfoil in cross-section, with a delta planform. It had three vertical fins, two of them angling outwards from the body, and a tall center fin. The flush canopy did not protrude above the body lines of the vehicle. Like the M2-F2, it measured 22 feet in length, but it was wider (15 feet) and higher (11.5 feet). It used many off-the-shelf components from the T-38, T-39, and F-106, among others. The control system consisted of upper body surface and outer fin flaps for transonic and supersonic trim, blunt trailing-edge elevons, and a split rudder on the center vertical fin. It had a stability augmentation system in all three axes, landing rockets, and provisions for an XLR-11

[251] Brun chronology; M2-F3 progress and flight reports; OART M2-F3 flight reports.

engine, although the engine was not installed at rollout from Northrop's Hawthorne plant on 18 January 1966.[252]

Northrop shipped the vehicle to Ames for testing in the 40-by-80-foot full-scale wind tunnel. The tests proceeded uneventfully, although some tests hinted at flow separation over the outer vertical fins, a condition engineers did not consider serious. At the Flight Research Center on 22 December 1966, NASA pilot Bruce Peterson completed the craft's first glide flight. It was anything but routine. During the three-minute descent to landing, Peterson discovered that he had minimal lateral control over the HL-10. Flow separation was much worse than anticipated. Peterson managed to set the HL-10 down safely on Rogers Dry Lake, no small tribute to his piloting skills. NASA immediately grounded the HL-10 for study, taking the opportunity to install its rocket engine. The first flight, in the words of Langley engineers, "once again demonstrated the value of flight tests as proof-of-concept."[253] Langley undertook a series of wind tunnel tests. As a fix, NASA engineers modified the leading edge of the outer vertical fins so as to direct more air over the control surfaces. Technicians added the new leading edges, constructed of fiberglass, late in 1967, smoothing over the installation with epoxy paint. The HL-10 experience reemphasized to engineers that aerodynamically shaping lifting-body designs for good subsonic performance could lead to potentially disastrous flow-separation problems in the absence of thorough design analysis. "This experience," Langley engineers concluded, ". . . pointed up the significance of seemingly minor shape changes. . . ."[254]

When the HL-10 took to the air again on 23 October 1968, it handled very nicely. What was to have been the first HL-10 powered flight had to be aborted after launch when only one of the XLR-11's chambers fired, Jerauld Gentry making an emergency landing on Rosamond Dry Lake. On 13 November, however, everything clicked. NASA pilot John Manke reached Mach 0.84 (524 miles per hour) using two of the engine's four thrust chambers. NASA now began incrementally working toward the craft's maximum performance. The HL-10 went supersonic for the first time on 9 May 1969 with John Manke as the pilot. This was the first supersonic flight of any piloted lifting body and a major milestone in the entire lifting-body program. The craft exhibited acceptable transonic and supersonic handling characteristics. On 18 February 1970, Air Force test pilot Maj. Peter C. Hoag reached Mach 1.86 (1,228 miles per hour), the fastest lifting-body flight ever made. Nine days later, on 27 February 1970, Bill Dana reached an altitude of 90,303 feet, another record for the lifting-body program. The HL-10 thus became the fastest and highest-flying piloted lifting body ever built.[255]

[252] NASA FRC *X-Press*, 14 and 28 Jan. 1966; "HL-10 Delivered Today," *Space Daily*, 18 Jan. 1966.

[253] Robert W. Rainey and Charles L. Ladson, "HL-10 Historical Review" (Hampton, VA: LaRC, July 1969).

[254] Ibid. See also Reed with Lister, *Wingless Flight*, pp. 95-102.

[255] Brun chronology; HL-10 progress and flight reports; OART HL-10 flight reports.

Toward the end of the HL-10 flight research program, NASA embarked on a series of powered landing trials. By 1970, the Space Shuttle was being discussed. One critical question was whether it should make unpowered landing approaches or, like a conventional transport aircraft, fly a powered approach and landing. Engineers had several schemes for the powered landing, the most popular being pop-out retractable turbojet landing engines that the Shuttle crew could deploy at subsonic speeds while approaching earth. Advocates thought the landing engines would give the Shuttle a shallower descent angle, reducing pilot workload and enhancing overall mission safety. While popular with many industry and government engineers who had little background in the Round One, Round Two, and lifting-body programs, this scheme was not at all popular at Edwards. Research pilots and engineers there recognized the complexity that landing engines would add to any Shuttle design, as well as the danger to a Shuttle crew if one of the engines failed during the final and most critical portions of flight. Because of the popularity of this idea elsewhere, however, FRC engineers embarked on a powered-landing program using the HL-10.

In February 1970, following the record altitude and speed flights, NASA grounded the HL-10 and replaced its XLR-11 rocket engine with three 2,200-newton (500-pound-thrust) Bell Aerosystems hydrogen-peroxide rocket engines. NASA planned to launch the HL-10 from the B-52 in the vicinity of Palmdale, California. The pilot of the HL-10 would ignite the rocket engines as the lifting body passed through an altitude of 6,500 feet. The rockets would reduce the approach angle of the aircraft from its customary 18 degrees to 6 degrees and give the HL-10 an airspeed in excess of 350 miles per hour. At 200 feet above the lakebed, the pilot would shut down the rockets and extend the landing gear, executing a routine landing. The HL-10 completed two of these flights piloted by Pete Hoag on 11 June and 17 July 1970, the latter flight being the craft's final mission. The flights gave much more encouragement to the Edwards viewpoint than to those in favor of landing engines. The shallow descent angle actually had increased pilot workload and degraded mission safety. Hoag found he had more trouble in determining the landing aiming point, and the higher approach speed aggravated control-sensitivity problems.

The HL-10 tests carried the day for advocates of a dead-stick Shuttle reentry, approach, and landing. As Milton O. Thompson, a test pilot with experience in numerous low lift-to-drag research aircraft, subsequently stated,

> The shuttle, whether it has landing engines or not, must be maneuvered, unpowered, to a point near the destination because the engines cannot be started until the vehicle is subsonic and only limited fuel will be available. To us it seems ridiculous to maneuver to a position where power must be relied upon to reach the runway.

The HL-10, in large measure, contributed to the decision to design the Space Shuttle without landing engines.[256]

During its brief flying career, the HL-10 completed 37 flights. It is now on display on Lilly Drive just outside the Dryden Flight Research Center's main gate. It was a fine flying vehicle, and its flight research program encouraged Eugene Love of Langley to advocate the HL-10 design concept for any future NASA Shuttle. For a variety of reasons, this did not come to pass. Pilots who flew the craft uniformly praised its handling characteristics, reserving criticism only for its bubble Plexiglas nose. The lenticular-shaped nose acted as a giant "demagnifying" lens at low altitude, causing severe visual distortion and misleading pilots into thinking that they were much higher over the lakebed than they really were. Consequently, they sometimes waited too long before extending the landing gear. With experience, however, they learned to compensate for this distortion, and the problem disappeared.

At first, NASA had no role to play in the Air Force's X-24A program. In mid-1965, before the Air Force had issued Martin a development contract for the vehicle (then designated SV-5P), NASA Associate Administrator Ray Bisplinghoff and the Air Force's Alexander Flax had agreed in principle that the SV-5P should be added to the Air Force/NASA joint M2-F2 and HL-10 programs. OART was receptive to testing the SV-5P, but the then-uncertain state of the program prevented the SV-5P from being included in the joint program until October 1967, when NASA and the Air Force concluded a memo of understanding on use of the vehicle, now designated the X-24A. The memo also confirmed the earlier joint lifting-body program agreements established by Paul Bikle and Maj. Gen. Irving Branch. Branch subsequently died in the crash of a T-38 trainer, but his successors at the AFFTC had also approved participation in NASA's M2 and HL-10 programs.[257]

Martin had completed the X-24A at its plant in Middle River, Maryland, during the summer of 1967. The craft had little aesthetic appeal; indeed, it could lay claim to being the most unattractive of the lifting bodies. Its ultimate maturation into the sleek X-24B shape suggests the story of the ugly duckling that eventually turned into a swan. The body shape differed greatly from the M2 and HL-10. While the M2 was basically a modified boat-tailed half cone and the HL-10 was basically a delta derivative with negative camber (i.e., an inverted airfoil) and boat-tailing, the plump X-24A had positive camber. It had a landing weight of 6,270 pounds, a span of about 13.5 feet, and a length of roughly 24.5 feet. After rollout on 11 July 1967, Martin shipped the craft to Ames for full-scale-tunnel testing. That completed, NASA shipped the craft to Edwards in early 1969 for flight trials. Jerauld Gentry completed the maiden glide

[256] NASA FRC news release 8-70; NASA news release 70-71; Brun chronology; HL-10 flight reports. See also Thompson and Peebles, *Flying without Wings*, pp. 210-214, for further details.

[257] Memo, Bisplinghoff to NASA associate administrator, 30 June 1965; NASA Management Instruction 1052.96, "NASA-DoD (USAF) Memorandum of Understanding: Use of X-24A Research Vehicle in Joint Lifting Body Flight Research Program," 7 Nov. 1967.

Three lifting bodies lined up on Rogers Dry Lake (viewer's left to right): the Martin X-24A, the Northrop M2-F3 (rebuilt from the M2-F2 with a central fin), and the Northrop HL-10. The M2 and HL-10 were NACA or NASA concepts, while the X-24 came from the Air Force. (NASA photo E-21093)

flight on 17 April, the craft making nine more such flights before its first powered mission. Gentry flew the X-24A's first powered flight on 19 March 1970, reaching Mach 0.87, well into the transonic region. Following this flight, Gentry, NASA pilot John Manke, and Air Force test pilot Maj. Cecil Powell steadily opened the X-24A's performance envelope. On 14 October 1970, 23 years to the day since Chuck Yeager's first supersonic flight, Manke piloted the X-24A on its own initial excursion past Mach 1, reaching Mach 1.19 (784 miles per hour) at 67,900 feet. Not quite two weeks later, Manke flew the X-24A to 71,400 feet, simulating a Space Shuttle approach and landing from that altitude. On 29 March 1971, Manke reached Mach 1.60 (1,036 miles per hour), the X-24A's fastest research flight. On 4 June 1971, the 28th and final research mission was a disappointment because only two of the XLR-11 engine's four chambers ignited, limiting the craft to subsonic speeds.[258]

The little X-24A had no vices, although it once gave researchers a bad moment. The rocket engine shut down prematurely and a small fire erupted in the engine bay, but Gentry made an emergency landing. Damage to the four maneuvering flaps, wiring, and flap instrumentation kept the ugly duckling grounded for nearly two months. The X-24A did have one bothersome quirk: during boost, it exhibited a pronounced nose-up trim change that prohibited low angles of attack during powered flight. FRC engineers concluded that the aerodynamic effects of the rocket exhaust plume impinging on the craft caused the nose-up condition. They warned the designers of the Shuttle to beware of similar problems. Although such trim changes sound

[258] Brun chronology; X-24A flight reports.

innocuous, they could impose unacceptable aerodynamic loads on the Shuttle during its boost to orbit. Aside from this quirk, the X-24A flew very well and the pilots liked it. Like the M2-F3 and HL-10, the X-24A demonstrated that Shuttle-type hypersonic vehicles could make precise landings without power. The X-24A pilots found they could land the vehicle on lakebed Runway 18 with an average 250-foot longitudinal "miss" distance from the intended touchdown spot. Indeed, NASA lifting-body team members had no qualms about landing the X-24A on a confined concrete runway, such as the 15,000-foot runway at Edwards. Such landings had not been attempted with earlier lifting bodies only because they lacked nosewheel steering. All of the lifting-body trials gave great confidence to advocates of landing an unpowered Space Shuttle on a conventional runway after its return from space, the plan ultimately followed for the Space Shuttle and demonstrated at Edwards with the Orbiter prototype *Enterprise* in 1977.[259] Had this been all that the X-24A contributed, the program would have been satisfactory. Instead the ugly duckling X-24A turned into a swan, the sleek and significant X-24B.

The End of Another Era

The Martin X-24B was America's last postwar rocket research aircraft in the traditional sense.[260] Its story began in the late 1960s when engineers at the Air Force's Flight Dynamics Laboratory evolved a family of reentry shapes–the FDL-5, -6, and -7–having a reasonable lift-to-drag ratio (approximately 2.5) at hypersonic speeds and large internal volume. These configurations were suited to hypersonic flight from Mach 4 to orbital velocities, but they were tailored primarily for aircraft in the Mach 8-12 performance regime. The Air Force hoped that these shapes could be used for two applications: sustained hypersonic-cruise aircraft powered by advanced air-breathing engines, and unpowered orbital reentry vehicles capable of landing at virtually any convenient airfield. At first, of course, the Flight Dynamics Laboratory wished to verify the performance of the shapes on low-speed lifting-body vehicles.

In a bid to reduce costs, Air Force engineers thought of modifying one of the abortive Martin SV-5J shells into an FDL-7 body shape by gloving the FDL-7 around

[259] Interview of Kolf.

[260] The validity of this statement, of course, depends on the definition of "rocket research aircraft." At least some researchers at Dryden would consider the Space Shuttle as not simply an operational launch vehicle but also a research airplane. Its launch using solid rocket boosters is analogous to the launch of the X-15s or lifting bodies by a B-52. Like them, its main source of power is liquid-rocket engines. It reenters the atmosphere at a much higher speed but in a way similar to that of the X-15s, and it lands unpowered as did the X-15s and lifting bodies. In the process, it provides a great deal of data to researchers. An example of the treatment of the Shuttle as a research aircraft is Kenneth W. Iliff and Mary F. Shafer, "A Comparison of Hypersonic Flight and Prediction Results," AIAA paper 93-0311, presented at the 31st Aerospace Sciences Meeting & Exhibit, Jan. 11-14, 1993, in Reno, NV. This paper treats the X-15, Reentry F, the Sandia Winged Energized Reentry Vehicle Experiment (SWERVE—arguably another postwar rocket research aircraft), and the Shuttle, comparing actual flight data with predictions about them based on ground tests.

the SV-5J, retaining the three vertical fins, and designating this composite shape as the FDL-8. In January 1969, the Flight Dynamics Laboratory issued a proposed development plan for the project, the jet-powered craft to be air-launched from a B-52 mothership. As studies matured, however, the advantages of rocket propulsion became obvious. Consequently, the Air Force scrapped the SV-5J plan and, instead, built the FDL-7 shape around the X-24A then flying at Edwards. Because of the joint lifting-body agreements, Air Force engineers had consulted their NASA counterparts, including Paul Bikle at the FRC and Fred J. DeMeritte, NASA OART's chief of the lifting-body program, to secure tentative NASA support. In August 1970, the laboratory sent a memorandum describing the proposed program to all interested parties. By the end of the month, the directors of the AFFTC and the FRC had concurred, but Air Force Systems Command (AFSC) delayed approval pending arrangements for joint NASA/Air Force funding. On 11 March 1971, NASA transferred $550,000 to the Air Force to initiate acquisition of the aircraft. The Air Force pledged a similar amount, and on 21 April 1971 the AFSC's director of laboratories gave the program its go-ahead, five months later than supporters had desired. On 4 June 1971, the X-24A completed its last flight. On 1 January 1972, the Air Force awarded the modification contract to the Martin Marietta Corporation. The X-24B program was now officially under way. Modifying the existing craft for $1.1 million secured a research vehicle that could have cost $5 million if built from scratch. Hypersonic tests at the Air Force's Arnold Engineering Development Center near Tullahoma, Tennessee, indicated that the FDL-8 shape performed well at those speeds. However, as always, the big question was what happened when the vehicle decelerated to much lower velocities. As Fred DeMeritte stated at the beginning of the program, "We are looking for surprises as we go through [the] transonic [speed range]."[261]

Martin Marietta Corporation's Denver plant delivered the X-24B in the fall of 1972. It had grown about 5.6 feet in span and 14.1 feet in length and weighed about 13,800 pounds at launch. It had a 78-degree double-delta planform for good center-of-gravity control, a boat tail for favorable subsonic lift-to-drag characteristics, a flat bottom, and a sloping three-degree nose ramp for hypersonic trim. Like the earlier lifting bodies, the X-24B used several off-the-shelf components used earlier on the X-24A that were reinstalled after the structural modification. Portions of its landing gear, control system, and ejection system came from the Northrop T-38, Lockheed F-104, Martin B-57, Grumman F11F, Convair F-106, and North American X-15. It had an XLR-11 rocket engine and Bell Aerosystem landing rockets. Once the aircraft was back at Edwards, technicians reinstalled a research instrumentation package.

[261] Everly Driscoll, "The Shape of Things to Come?" *Science News* (15 Sept. 1973), pp. 171-172; Michael L. Yaffee, "X-24B Lifting Body Nearing Completion," *Aviation Week & Space Technology* (4 Sept. 1972), pp. 77-79; John A. Manke and M. V. Love, "X-24B Flight Test Program," 1975 Report to the Aerospace Profession, Society of Experimental Test Pilots, 26 Sept. 1975; Johnny G. Armstrong, *Flight Planning and Conduct of the X-24B Research Aircraft Flight Test Program*, AFFTC-TC-76-11 (Edwards AFB: AFFTC, 1977), pp. 12-14 (hereafter cited as Armstrong, X-24B report).

The Martin X-24B, its laundry-iron planform fitted around the existing Martin X-24A. (NASA photo ECN-3764)

Program managers Johnny Armstrong (for the Air Force) and Jack Kolf (for the FRC) supervised preparations for the first flight.[262]

John Manke completed the X-24B's first glide flight on 1 August 1973, launching from the B-52 carrier aircraft at 40,000 feet, coasting earthward at 460 miles per hour, and performing a series of handling-qualities maneuvers and a practice landing approach before making a 200-mile-per-hour landing on the lakebed. The flight initiated the usual sort of programs and investigations that accompany all new research aircraft. On succeeding missions, Manke and the Air Force project pilot, Maj. Michael V. Love, checked the vehicle's behavior in a variety of configurations. Following this series of glide flights, the X-24B made its first powered flight on 15 November 1973,

[262] Manke and Love, "X-24B Flight Test Program"; NASA Management Instruction 1052.96B, "Provision for the Use of the X-24B Research Vehicle in a Jointly Sponsored NASA-DoD (USAF) Lifting Body Flight Research Program," 4 Feb. 1972; Armstrong, X-24B report, pp. 16-26; NASA news release 71-139; NASA FRC news release 30-71; notes of telecon, C. Karegeannes with Ralph Jackson, 13 Dec. 1972, in NASA History Office X-24B project files; Hoey comments. As comes out in part in Reed with Lister, *Wingless Flight*, pp. 20-21 and elsewhere, one of the secrets of the FRC's success was its extraordinarily capable technicians, a fact not sufficiently stressed in most histories.

piloted by John Manke. As always, the pilots had practiced for their brief seven-minute sojourns in the X-24B with numerous simulations of lifting-body approaches in T-38 and F- 104 aircraft. By the end of the X-24B program, pilots had flown more than 8,000 of these simulated approaches in support of the entire lifting-body program. On the X-24B's sixteenth flight, on 25 October 1974, Mike Love reached Mach 1.75 (1,164 miles per hour), the craft's fastest flight. Manke followed this on 22 May 1975 by making the craft's highest approach and landing, coming down to the lakebed from a height of 74,100 feet. Both Love and Manke were pleasantly surprised by the handling qualities at all speed ranges, with and without engaging the control dampers in the stability augmentation system. Even in turbulence the aircraft flew surprisingly well, its handling qualities, including the landing approach, reminding pilots of the F-104. Its subsonic handling qualities in general earned the X-24B a rating of 2.5 on the NACA-developed Cooper-Harper pilot rating scale, a very high mark. In short, it was a fine airplane.[263]

By mid-1975, the Space Shuttle was well into its design phase. Mission planners were still interested in whether such unpowered low lift-to-drag reentry shapes could demonstrate successful landings on the relatively confined geographical and heading constraints of a fixed runway. John Manke was convinced that the X-24B could simulate such an approach and landing. He recommended that the lifting body—which, in contrast to its fellows, did have nosewheel steering—make a series of landings on the main 15,000-foot concrete runway at Edwards, Runway 04/22. Manke, Love, and others considered the demonstration important for developing confidence before attempting similar landings in the Space Shuttle itself. In January 1974, the X-24B Research Subcommittee approved the proposal. Manke and Love began a three-week familiarization program, flying F-104 and T-38 approaches that simulated the X-24B's characteristics. Manke alone shot over 100 such approaches. The payoff came on 5 August 1975, when Manke launched from the mothership B-52, ignited the XLR-11 engine, climbed to 60,000 feet, and began his descent. Seven minutes after launch, Manke touched down precisely at the planned target mark 5,000 feet along the Edwards runway. Afterward he said, "We now know that concrete runway landings are operationally feasible and that touchdown accuracies of ±500 feet can be expected. We learned that the concrete runway, with its distance markers and unique geographical features, provides additional 'how goes it' information not available on our current lakebed runways." Two weeks after Manke's first runway landing, Mike Love duplicated the feat. The runway landing program, a major accomplishment, brought the X-24B research program to a conclusion. The Air Force and NASA embarked on a series of pilot checkout flights.[264]

On 9 September 1975, Bill Dana completed the X-24B's last powered flight, a flight that also brought to an end the postwar American rocket research program. No

[263] NASA X-24B flight reports; Brun chronology; Manke and Love, "X-24B Flight Test Program"; Armstrong, X-24B report, pp. 89-97.

[264] Manke and Love, "X-24B Flight Test Program"; Brun chronology; interview of John Manke, 1 Dec. 1976; Armstrong, X-24B report, pp. 78-82.

more would the rumble of an igniting rocket engine echo along the lakebed. No more would the XLR-11 power some exotic airframe. Old-timers who had worked in the early days with Chuck Yeager and Walt Williams on the XS-1 recognized that a unique period had at last come to a close. Following Dana's flight, as the X-24B sat inert on the ground, the four chase planes, two T-38s and two F-104s, closed up in a tight diamond formation and dipped low in a noisy salute over the Flight Research Center. That night, center personnel reminisced until the wee hours at an "End of an Era" party at the Longhorn, outside Lancaster. Following Dana's flight, the X-24B completed a series of six pilot familiarization glide flight by Air Force Capt. Francis R. Scobee and NASA's Einar Enevoldson and Tom McMurtry. On 26 November 1975, the X-24B dropped from the sky for the last time, piloted on its 36th flight by McMurtry. The NASA flight report concluded laconically that "all objectives for this flight were attained." Up to the dedication ceremony the following spring renaming the FRC as the Hugh L. Dryden Flight Research Center, the X-24B remained at Edwards, resplendent in blue and white. Then it departed for the Air Force Museum, where it is currently exhibited. The lifting-body flight-test program gave way to the next phase: the Space Shuttle's approach and landing tests.

Beyond the X-24B?

Among the Flight Research Center's space-related activities in the 1960s, some of the most important and influential were the lifting-body studies. Evidence exists that the Soviet Union followed a similar course, air-launching a lifting-body shape reminiscent of the X-20 Dyna-Soar from a Tupolev Tu-95 mothership. The Flight Research Center's work on the other space-related projects—such as the Project Mercury drogue chute, the Paresev, and the LLRV/LLTV—was important, but the lifting bodies received the center's greatest attention. The fact that the lifting body per se did not dictate the Space Shuttle shape is no reflection on NASA's work with these shapes. Indeed, the FDL-8 shape used on the X-24B is considered ideal for a hypersonic sustained-cruise aircraft. Other considerations dictated the Shuttle's shape. These, together with new thermal protection systems, lessened the once-urgent need for pure blunt-body lifting reentry vehicles. Writing in 1968, lifting-body advocate Clarence Syvertson stated prophetically:

> A technology so new and challenging cannot be rushed. . . . But I believe that later in this century we will come to regard today's purely ballistic piloted capsules, splashing down in an ocean, as a relatively crude and inefficient way of returning from a space mission. The lifting body offers an alternative that is already proved in principle.[265]

[265] Syvertson, "Aircraft Without Wings," p. 50.

NASA's lifting body program led to two abortive research efforts, a mini-Shuttle and an air-breathing hypersonic follow-on to the X-24B. In the former case, center engineers proposed construction of piloted, 36-foot versions of the Space Shuttle for studying the most critical area of its flight, the deceleration from Mach 5 through the landing. Mach 1, 2, and 3 models were to be powered respectively by one, two, and three XLR-11 engines, or a Mach 5 model could be powered by an XLR-99. Such research aircraft, air-launched from a B-52, could fly in direct support of Space Shuttle development, especially by validating wind-tunnel predictions of stability, controllability, and performance at hypersonic, supersonic, transonic, and subsonic velocities. They could be used for astronaut training and for investigating launch-abort maneuvers. As with the earlier lifting bodies, FRC advocates of the subscale shuttle planned on using components from a variety of existing aircraft, including the M2-F3, F-4, YF-12, F-15, and X-15, as well as some Apollo hardware. It was hoped that, using this approach, costs could be kept down. An XLR-99-equipped Mach 5 subscale shuttle was estimated to cost $19.7 million. If NASA's Office of Aeronautics and Space Technology—the successor to OART—and the Office of Manned Space Flight had authorized immediate go-ahead, the mini-shuttle could have been flying by the end of 1975.[266]

This was a typical Flight Research Center proposal: do something that no other center could do, and do it in support of a broader research program. Unfortunately, the proposal came to grief. The major push for a subscale shuttle came in August 1972 with the preparation of a well defined and detailed proposal. Afterwards, Milton Thompson, Joe Weil, and other mini-Shuttle proponents traveled to the Manned Spacecraft Center and NASA headquarters to make presentations for the vehicle. It had some high-level support—Robert Gilruth of MSC was a strong advocate—but critics argued that the projected costs were far too low, that a realistic cost estimate would be more like $150 million. FRC supporters pointed to the costs in the earlier FRC-managed lifting-body program. They conceded that if the program went through conventional management procedures at headquarters, its costs would indeed rise. Other critics believed the FRC could not go it alone on the project and that it would ultimately involve people who were at work on the Shuttle. But the overriding difficulty seems to have been a matter of pride: the FRC justified the subscale shuttle on the basis of its use in validating and verifying the results of wind-tunnel testing—an old sore spot to tunnel devotees who passionately believed in their facilities. Despite strong industry support from Northrop and Martin (both with lifting-body experience) and Rockwell (the Shuttle contractor) the subscale shuttle succumbed to the cost argument. The actual Shuttle's hypersonic, supersonic, and transonic performance remained unchecked by actual results until the first all-out Mach 25 reentry from space.[267] No one seems to have proposed an unpiloted subscale, shuttle to be flown

[266] NASA FRC, "Subscale Shuttle" (17 Aug. 1972); interview of Milton Thompson, 9 Aug. 1977; interview of Joe Weil, 18 Aug. 1977.

[267] For the record, its behavior was highly satisfactory.

like the earlier ASSET and PRIME shapes. There is also no reason to believe that such a proposal would have won acceptance.[268]

The other proposal derived from the lifting-body effort, as well as from desires for an X-15 follow-on, was the X-24C, a strange aircraft subsequently redesignated as the National Hypersonic Flight Research Facility, or NHFRF, pronounced "Nerf." The Flight Research Center had high hopes for the development of this vehicle, a B-52 air-launched Mach 8 research aircraft equipped with rocket boost and designed for 40 seconds of sustained Mach 6+ cruise. The FRC, in conjunction with Langley's hypersonic ramjet research program, could use the aircraft to test scramjet (supersonic combustion ramjet) air-breathing engines. As early as the mid-1960s, De Beeler of the Flight Research Center had pressed hard for development of such a craft. With the conclusion of the X-15 program in 1968, calls from enthusiasts for an advanced hypersonic air-breathing research aircraft became clamorous. Langley launched two programs: HYFAC, the Hypersonic Research Facilities study, a Mach 12 design; and the less ambitious HSRA, a Mach 8 High-Speed Research Aircraft. The Air Force originated two proposals, one for a Mach 3-5 test vehicle, and the other for the Incremental Growth Vehicle, a test airplane initially designed for Mach 4.5, but which could be modified for flight at Mach 6, and later for Mach 9. Starting in July 1974, after recognizing the probable high costs of the program, NASA and the Air Force jointly conducted a series of design studies for an air-breathing hypersonic vehicle. The Flight Dynamics Laboratory FDL-8 body shape appeared ideal. Studies pursued this approach, encouraged by Air Force research on two proposed follow-on X-24 configurations, one with "cheek" air inlets, and the other with an XLR-99 rocket engine. In December 1975, NASA headquarters and the Air Force established an X-24C Joint Steering Committee, composed of the commanders of the Air Force's Flight Dynamics Laboratory and the AFFTC, and the directors of NASA Langley and the Flight Research Center. In July 1976, out of this joint committee came the NHFRF.[269]

The NHFRF came close to winning approval. It was strongly supported at Dryden at all levels. Langley's hypersonic aerodynamicists and propulsion team saw in it the fruition of all their work. They also saw it as a good opportunity to "cover the whole hypersonics waterfront and do it before we've lost all the hypersonic talent we developed from the X-15 program." There were certainly psychological overtones as well, primarily a desire to reassert and revitalize within NASA the role of aeronautics. NASA forecast a $200-million program involving construction of two aircraft, with 200 flights over a 10-year period. NASA and the Air Force would begin funding the

[268] See note 40.

[269] Donald P. Hearth and Albert E. Preyss, "Hypersonic Technology: Approach to an Expanded Program," *Astonautics & Aeronautics* (Dec. 1976), pp. 20-37; "Hypersonic Aircraft by 2000 Pushed" and "USAF to Begin Hypersonic Testing" in *Aviation Week & Space Technology* (17 Sept. 1973), pp. 52-57, 83-90; Marvin Miles, "Wingless Rocket Plane May be Converted for Faster Test," *Los Angeles Times*, 24 Dec. 1973; "Study Eyes Altering X-24 Lifting Vehicle, " Air Force Systems Command *Newsreview*, Jan. 1974; AFSC news release OIP 158.74; "Toward Hypersonics," *Flight* (30 Oct. 1975), pp. 657-658; F. S. Kirkham, L. Robert Jackson, and John P. Weidner, "Study of a High-Speed Research Airplane," AIAA *Journal of Aircraft* (Nov. 1975), pp. 857-863.

program in 1980, with the first airplane flying in 1983. To Dryden management, uneasily eyeing the future of the center after the Space Shuttle left the lakebed for the last time, the NHFRF seemed especially important for the 1980s. It would be the logical conclusion of two decades of X- 15/X-20A/X-24B work.[270]

What happened was a sad anticlimax. Discussions between the Air Force and NASA continued into 1977. As plans grew, so did the expected cost of the vehicles. A 40-second cruise requirement added complexity that translated directly into higher costs. Finally, despite the wishes of NHFRF supporters at Edwards and Langley, NASA headquarters canceled the program in September 1977. James J. Kramer, NASA's acting associate administrator for aeronautics and space technology, stated that "the combination of a tight budget and the inability to identify a pressing near-term need for the flight facility had led to a decision by NASA not to proceed to a flight test vehicle at this time."[271] The Air Force was in no financial or political position to go it alone on such an ambitious venture. The result hit Dryden hard. Center morale dropped precipitously. Some blamed over-management. Some blamed the cruise requirement. Others felt the FRC should have pressed harder for a no-frills off-the-shelf shape. It was all to no avail: NHFRF was gone.

It is ironic that the center's work with lifting bodies for reentry from space eventually spawned an abortive interest in hypersonic atmospheric flight. It was a joining together of two streams of research: the stream running from the X-15 through the X-20 and on to the HSRA; and the stream running from the Allen blunt body to the Eggers M2, the Love HL-10, the Martin SV-5, the FDL-8/X-24B, and the NASA HYFAC. Both streams pooled together in the NHFRF. The subscale shuttle was certainly spawned by the lifting-body program and constitutes a little puddle of its own to the side. It did not influence the work on what became NHFRF, although some of its technology was very close. Following the cancellation of the NHFRF, there was a general feeling among subscale-shuttle proponents that it might have evolved into a research tool, like the NHFRF, had NASA proceeded with development. That is indeed likely.

With cancellation of the NHFRF, the national program on transonic, supersonic, and hypersonic flight research using specialized rocket-propelled research vehicles was over. The actual Space Shuttle, of course, does not fit neatly into any of these research areas, although it has contributed a great deal of hypersonic data. As its enthusiasts claim, however, the Space Shuttle is primarily a space-age DC-3, a vehicle revolutionizing piloted and unpiloted orbital spaceflight. The cancellation of NHFRF came in the midst of the FRC's program on the Space Shuttle. The Shuttle was, for a

[270] "Hypersonic Propulsion Milestone Passed," *Aviation Week & Space Technology* (9 May 1977), pp. 49-51; Hearth and Preyss, "Hypersonic Technology"; letter of agreement among AFFTC, AFFDL, NASA LaRC, and NASA FRC, 15 Mar. 1977; AFFDL, "Technology Program Plan: NHFRF (AFSC, Jan. 1977), passim.

[271] Quoted in "NASA to End Manned Hypersonic Effort," *Aviation Week & Space Technology* (26 Sept. 1977), p. 24; confidential sources.

very brief time, a major center program. While Johnson Space Center (formerly MSC) had overall control of the program, Dryden furnished the technical expertise on flight-testing to validate the craft's approach and landing characteristics. The Shuttle program ultimately involved a great number of center personnel, plus others from Johnson, and brought Dryden its greatest public exposure. The odd sight of a 747 carrying and then launching a delta monstrosity the size of a DC-9 airliner could not help but draw attention. The Shuttle program involved a lot of preparation, including a special "mate-demate" facility, a microwave landing system, and work on the 747 mothership. Yet this transitory program was over almost as soon as begun and did not reappear until four years later, when the first Space Shuttle dropped out of the Mojave skies to land on the dry lakebed at Edwards AFB.

The center's involvement with space came as a prelude to the Space Shuttle. But the same years that witnessed the X-15, the Paresev, the LLRV, and the lifting bodies also saw a return to conventional aeronautics: flight at transonic and supersonic speeds. Although the Flight Research Center did not run an extensive number of military test programs in the 1960s and early 1970s, several aeronautical research projects were under way. Some of these, such as the Blackbird, XB-70A, Supercritical Wing, and the TACT program, became quite visible and were very important, both in terms of their technology contributions and in how they promoted the reputation of the center. Though they played second fiddle during the heyday of space, it has been these programs—and others like them—that subsequently emerged as Dryden's lifeblood.[272]

[272] Later, hypersonics research resumed in the ultimately abortive National AeroSpace Plane and then the Hyper-X programs, the latter discussed in Chapter 13 of this book. The lifting-body program eventually led to the X-33 and X-38 projects, see also Chapter 13 of this volume and Reed with Lister, *Wingless Flight*, pp. 179-192. As explained in Chapter 13, NASA cancelled the X-33 project in May 2001 before the vehicle flew, but data from the effort may still inform a future reusable-launch-vehicle program, perhaps with a lifting-body shape.

9

Aeronautics, the First "A" in NASA 1966-1976

As long as NASA had not yet fulfilled its mandate of landing men on the moon before the end of the 1960s, aeronautics had to take a second place to space within the agency. Yet even before Neil Armstrong's "one small step" at Tranquillity Base, a ground swell of renewed interest and support for aeronautics was building. Within the agency, engineers—especially those at the old NACA centers, Langley, Ames, Lewis, and the Flight Research Center—decried the imbalance. In May 1966, a congressional report bluntly stated that "any new or expanded aeronautical activity within NASA immediately has to compete for attention, money, resources, and manpower with an urgent, presidentially declared, national space goal. Under these circumstances it is perhaps surprising that NASA's aeronautical efforts have not suffered any more than they have."[273] And the Flight Research Center, so recently out of favor with some congressional staffers, now came back into the good graces of the legislative branch. Indeed, the same report credited the FRC with "a spectacular series of technological 'firsts.'"

Several major developments suggested the need for greater aeronautical research and development. First, a protracted war in Southeast Asia was revealing surprising problems with American aircraft and airpower doctrine. In one measure, the overwhelmingly favorable air-combat victory-loss ratios of earlier wars—8 to 1 against the *Luftwaffe,* and about 12 to 1 against North Korea—were missing. Indeed, at times the victory-loss ratio slightly favored the North Vietnamese. Advancing aircraft technology offered the hope that clear military air supremacy might be regained. Foreign military aircraft technology was moving rapidly, especially in the Soviet Union. NASA's OART, in an in-house 1971 study, concluded that "the U.S. traditional preeminence in military airpower has been lost in recent years. While progress in foreign airpower during the last decade has been rapid, few truly advanced aircraft have been developed in this country."[274] Second, new generations of jet transports—particularly supersonic jet transports—were being developed in the U.S., Europe, and the Soviet Union. NASA's aeronautical partisans and congressional supporters recognized a need to strengthen America's traditional position of leadership in civil air transportation. For the 1960s, this meant supporting the national supersonic transport

[273] Library of Congress, Legislative Reference Service, *Policy Planning for Aeronautical Research and Development,* U.S. Senate Committee on Aeronautical and Space Sciences, 16 May 1966, p. 9.

[274] OART, *NASA Research and Technology Objectives: Aeronautics in the '70's,* preliminary copy (Washington, DC: NASA OART, 16 Aug. 1971), p. 1.

(SST) effort. Further, sustained NASA research would be needed for vertical-takeoff-and-landing (VTOL) and short-take-off-and-landing (STOL) aircraft, new advanced wing designs for more efficient transonic and supersonic flight, and new concepts for flight-control systems.

The Langley Research Center remained, as it had always been, NASA's principal aeronautical research resource. Langley was NASA's "team leader" for advanced supersonic aircraft design, particularly in relation to the national SST program, then a joint effort between NASA, the Federal Aviation Administration (FAA), and American industry. When NASA began devoting more time and resources to aeronautics from the mid-1960s onward, Langley intensified its own research. The Flight Research Center followed these activities as much as possible, offering opinions and judgments as to what technical policies and programs NASA should support. The FRC remained heavily committed to the space-related efforts of the X-15, LLRV, and lifting bodies, so the amount of engineering talent available to work on non-hypersonic non-space-oriented programs was limited. Nevertheless, the FRC applied some resources to comprehensive supersonic research in support of the national SST effort and, later, to research on improving the efficiency and performance of transonic aircraft. This involved the FRC in four major aeronautical research programs: the XB-70A, YF-12, F-8 Supercritical Wing, and F-111 Transonic Aircraft Technology efforts. The first two programs involved research in sustained Mach 2.5-3+ flight. The latter two were concerned with transonic aircraft design. Figure 4 places these four programs within the context of selected FRC activities from 1959 through 1980. The gradual loss of emphasis on space research in favor of more traditional aeronautical research is obvious, and the FRC's experience mirrored that of Langley, Ames, and Lewis over the same period.[275]

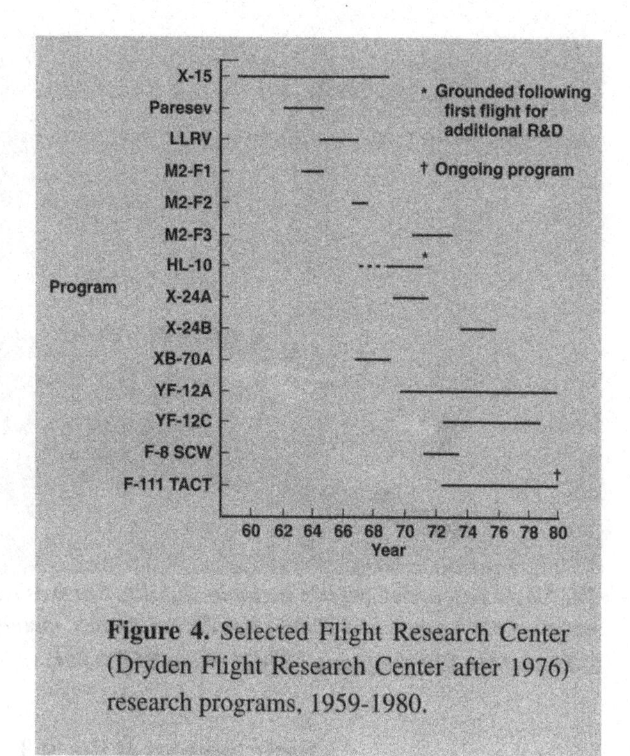

Figure 4. Selected Flight Research Center (Dryden Flight Research Center after 1976) research programs, 1959-1980.

[275] OAST, *OAST Research Centers*, pp. 3-4 to 3-5, 4-1 to 4-11.

The Flight Research Center's variable-stability North American F-100C Super Sabre was used for a range of airborne simulation studies, including some in support of the X-15 and the SST programs. (NASA photo EC62-144)

Early Support of the SST

The FRC's supersonic research during the 1960s emphasized support of the national SST program, a logical outgrowth of the center's research on Round One and the Century Series aircraft in the 1950s. The American SST program had begun in 1963 but dated to a Kennedy Administration initiative in 1961 that called for development of a Mach 3 supersonic transport. With hindsight, it is obvious that the goal was ill-chosen. The complexity of such a craft and its enormous costs made it at best a luxury and at worst a severe burden on the airline community expected to buy it. Proponents argued for an SST largely from a "pure" technology standpoint, with overtones of nationalism. The requisite technology base exists for such a craft, the argument went, as a result of the nation's supersonic research program in the 1950s and the development of such supersonic bombers as the B-58 and the XB-70; therefore, the country should do it. Often added to this line of argument was the thought that if the United States did not develop an SST, Europe or the Soviet Union would sweep past American technology with its own SSTs. Thoughtful arguments questioning a Mach 3 SST's cost, utility, and desirability were ignored, especially after 1963, when the federal government had committed itself to supporting development of such a craft as a major American aeronautical research and development initiative.

In 1963, the Flight Research Center was flying three military aircraft on SST studies. Because the Douglas F5D-1 Skylancer had a modified delta-wing planform similar to wing configurations suggested for a Mach 3 SST, center pilots flew the F5D-1 on SST landing studies, accumulating data on sink rates and approach

NASA's North American A-5A Vigilante was used for simulation studies of the supersonic transport in 1963. (NASA photo ECN-231)

characteristics. A North American F-100C Super Sabre, modified to have the variable-stability characteristics that would simulate the handling qualities of an SST, was acquired from Ames and flown to generate information on predicted SST handling qualities. The FRC also acquired a North American A-5A Vigilante attack bomber from the Naval Air Test Center at Patuxent River, Maryland, and flew it to determine the let-down and approach conditions of an SST flying into a dense air traffic network. During 1963, center pilots Milt Thompson and Bill Dana flew the Vigilante over remote areas around Edwards on expected supersonic transport flight profiles and even flew supersonic approaches into the terminal approach control zone for the Los Angeles International Airport.[276] The A-5A was returned to the Navy at the end of the year.

The FRC went beyond these efforts. The research staff planned to provide funds—and eventually did contribute approximately $2,000,000—to instrument the experimental North American XB-70A Valkyrie Mach 3+ bomber so that it could return supersonic-cruise research data. In February 1963, the center purchased a Lockheed JetStar four-engine business jet. The aircraft was extensively modified with flight-control servos in all three flight axes. A hybrid analog-digital computer provided the intelligence for a state-of-the-art flight-simulation system. It could simulate the handling characteristics of a wide range of aircraft, including SSTs. The FRC purchased the JetStar for $1,325,000 and sent it to the Cornell Aeronautical Laboratory at Buffalo, New York, for installation of the simulation equipment at a cost of an additional

[276] FRC, *Five-Year Plan;* interviews of Beeler, Thompson, and Dana; DFRC vehicle summary.

$1.3 million. Back at Edwards for test duty in November 1965, it was known as the general-purpose airborne simulator (GPAS). Many engineers believed that any SST would require a movable "droop" nose (such as later employed on the Concorde) for adequate pilot visibility in the high angle of attack assumed by such an aircraft on takeoff and landing. Others believed visibility could be provided by an extendable periscope-like binocular system. FRC engineers installed binocular optics in the center's two-seat F-104B, and center pilot Bill Dana evaluated it in flight. The press of concurrent X-15 work terminated the program. Eventually, the advocates of the "droop nose" carried the day.[277] By the mid-1960s, then, the FRC was definitely SST-minded in its aeronautical research. Its principal involvement with the SST program came with the XB-70A test program.

The XB-70A Accident

North American's XB-70A Valkyrie was a six-engine experimental bomber designed for Mach 3+ speeds. Generally, the two prototypes of the XB-70A closely resembled the aerodynamic configuration that could be expected of a large supersonic jet transport. At the time of its maiden flight on 21 September 1964, the Valkyrie was the world's largest experimental airplane, with a length of about 186 feet, a wingspan of 105 feet, and a height of 30 feet. It had two large vertical fins, a canard ("tail-first") horizontal control surface mounted behind the cockpit and in front of the wings on the fuselage, and a sharply swept delta wing, the tips of which could be lowered to furnish greater supersonic lateral (roll) and directional (yaw) stability. Constructed of titanium and brazed stainless steel "honeycomb" materials, it could withstand sustained temperatures on the order of 630 degrees F. as it cruised at high altitude and Mach 3. It was designed as an intercontinental bomber, but production in quantity was canceled before its first flight because of changes in Defense Department offensive doctrine. Instead, the government decided to complete the two prototypes and use them for Mach 3 research in support of the SST program. The first XB-70A had an FRC-funded package of test instrumentation capable of telemetering 36 separate measurements of aircraft performance and condition to ground stations. A further 900 measurements were recorded by digital pulse-code-modulation and analog frequency-modulation recording systems on magnetic tape at the rate of 20,000 samples per second—a far cry from the scratchy oscillograph film used on the old X-1.

During the first phase of its flight-test program, the XB-70A and its later sister ship were flown by North American and Air Force test pilots. The planes were routinely flying above Mach 3 by early June 1966. Turns required flight corridors hundreds of kilometers wide. Obviously an SST could not use conventional airway routes, a vital discovery. The first airplane proved to have poor stability characteristics above

[277] FRC, *Five-Year Plan*; interview of Dana; NASA FRC *X-Press*, 1 Feb. 1963, 25 Sept. 1964, 19 Nov. 1965; comments by Donald L. Mallick on the original chapter, 1 Feb. 2000.

North American's remarkable XB-70A Valkyrie takes off on a research mission. NASA flew this large supersonic aircraft in support of the national SST program. (NASA photo ECN-792)

Mach 2.5. On the basis of wind-tunnel studies at Ames, North American had added five degrees of dihedral to the wing of the second XB-70A. This airplane had much better stability characteristics above Mach 2.5, so researchers designated it the prime Mach 3 research airplane. The complex systems of the airplanes posed maintenance headaches. Also, poor bonding of the stainless steel skin on the wing sometimes allowed whole sections of the skin to peel off in flight. Landing gear retraction problems plagued the craft. In one case, because of partial gear failure, the plane veered almost a kilometer off a lakebed runway's centerline, causing the test pilot to scribble in his report, "This landing could not have been accomplished on any runway in this country. Thank God for Rogers Dry Lake. . . ." Despite all of these difficulties, the XB-70As were returning a great deal of useful information for SST designers—on noise, operational problems, control system requirements, validation of tunnel-test techniques by comparison with actual flight-test data, and the presence of high-altitude clear-air turbulence.[278]

[278] Betty J. Love, XB-70A flight log, in DFRC Historical Reference Collection; NASA *X-Press*, 12 Mar. 1965; interview with Donald Mallick, 10 Dec. 1976; Robert C. Seamans, Memo for Record; XB-70 Flight Research Program Procurement, 16 Feb. 1967; XB-70A #1 pilot report, flight 1-37, 7 Mar. 1966; Thomas G. Foxworth, "North American XB-70 Valkyrie," *Historical Aviation Album*, no. 7 & 8 (1969-1970), pp. 76-87, 164-175; Eldon E. Kordes and Betty J. Love, "Preliminary Evaluations of XB-70 Airplane Encounters with High-Altitude Turbulence" (Washington, DC: NASA TN D-4209, 1967), p. 2.

NASA's OART had already allocated $10 million for support of the XB-70A program, primarily for flight-test instrumentation on the first and second aircraft. Then in the spring of 1966, the Air Force and NASA announced a joint $50-million program to be run by the FRC and the Air Force Aeronautical Systems Division (part of Air Force Systems Command). To begin in mid-June 1966, at the conclusion of the North American airworthiness demonstration program, the joint NASA/Air Force program would study the problems of sonic booms and evaluate the aircraft during typical SST flight profiles. The FRC's Joe Walker was designated project pilot for the civilian agency.

On 8 June 1966, the second XB-70A took off from Edwards, piloted by North American test pilot Al White and a new copilot, Maj. Carl Cross, making his first flight in the plane. The XB-70A was to make a series of tower passes at various airspeeds to calibrate its onboard airspeed system, then make a single pass at Mach 1.4 and 31,000 feet to acquire sonic boom information during an overflight of a specially instrumented test range. There was another item on the flight plan, not a critical one: the Air Force had approved a request by General Electric for the XB-70A to lead a formation of aircraft equipped with General Electric engines. The XB-70A used GE J93s. Participating would be a Navy F-4B Phantom, an Air Force T-38A Talon, an Air Force YF-5A, and a NASA F-104N Starfighter piloted by Joe Walker. A Learjet would photograph the formation for publicity purposes. At the preflight briefing the day before, John Fritz, a GE test pilot who would be flying the YF-5A, advised the other pilots to fly a loose formation, with about one wingspan clearance between airplanes.[279]

The XB-70A took off from Edwards at 7:15 a.m. on 8 June, followed by a T-38A piloted by Pete Hoag and Joe Cotton. White and Cross made three tower flybys, aborted a fourth because they were not properly aligned with the course, and canceled the remaining eight because of low altitude turbulence. At 7:59 a.m. White and Cross climbed for altitude while Hoag and Cotton landed and refueled their T-38. White and Cross completed the sonic boom pass by 8:30 a.m. and headed for the formation flight rendezvous point, Lake Isabella. By 8:43 a.m., the F-4B from Point Mugu, the YF-5A, the NASA F-104N, the photo Learjet, and the now-refueled T-38 had joined up with the big white delta. The plan for the formation flight called for the XB-70A to lead the other aircraft on a racetrack pattern between Mojave and Mt. Whitney at 20,000 feet. White and Cross soon discovered that clouds precluded this original plan and changed to a racetrack pattern northeast of Rogers Dry Lake. The new track was much shorter: the formation covered the straight portion of the track in a little over a minute and then made a three-minute turn through 180 degrees. The Air Force T-38 and Navy Phantom rode off the XB-70A's left wing, with Walker's Starfighter and the YF-5A on the right. A two-seat Air Force F-104D returning from a test mission briefly joined the group while the rear-seat cameraman took high-speed motion pictures, using up his film. The visiting pilot noticed that the two right-hand aircraft, the NASA

[279] Donald R. Bellman, "Briefing for NASA Headquarters on XB-70/F-104 Collision," 7 July 1966 (hereafter referred to as Bellman, XB-70A accident report).

F-104N and the Air Force YF-5A, were flying a much tighter formation than the Navy F-4B and Air Force T-38. He suggested the T-38 tighten its position relative to the F-4B to improve the looks of the formation.[280]

The F-104D left for Edwards while the XB-70A flew along, followed by its flock. Joe Walker in the F-104N edged closer and closer to the mammoth research airplane. A B-58 on a test flight passed high overhead. Then Walker's plane closed with the XB-70A, its horizontal stabilizer touching the downturned tip of the Valkyrie's wing.

Why? There were a lot of possibilities but no certainties. For one thing, Walker was 40 feet ahead of the tail of the XB-70, a plane with an unusual protruding tail configuration, a T-tail that had its maximum width high above Walker's cockpit. Then there was the long, sharply swept leading edge of the XB-70A's wing: deltas are notoriously difficult to maintain formation on, and the chances for misjudging distance are high. Checking wing and tail clearances would have required Walker to resort to extreme neck craning. There was the possibility of pilot distraction: the group had held formation for 43 minutes, not unsafe or unusual for a loose formation, but dangerous for a tight one. Then there was the darting Learjet and the oncoming B-58. That initial F-104N motion was slight. Even the YF-5A pilot, off Walker's right wing, failed to detect a significant change in the Starfighter's position.

In any case, the top of the F-104N's left horizontal stabilizer (near the tip) contacted the underside of the right wing tip of the Valkyrie. Later salvage of the crash pieces revealed the imprint of the XB-70's wing tip running light on the upper horizontal stabilizer of the F-104N. This contact and the vortex off the XB-70's wing caused the F-104N to pitch up violently and roll over the top of the XB-70A, hooking its left wing tank on the Valkyrie's wing. The tip tank broke up, initiating a built-in sequence so that the F-104N's right tank immediately jettisoned. The Starfighter, still rolling over the XB-70A, smashed into the right and left vertical fins, exploded in flames, and impacted the top of the XB-70A's left wing. Walker was killed instantly. The F-104N fell away in bits of wreckage and flame; the XB-70A continued on, minus its vertical fins and with major damage to both wings—a doomed aircraft.[281]

The other aircraft reacted immediately. Hoag and Cotton radioed "Mid-air, mid-air!" followed by "You got the verticals, this is Cotton, you got the verticals—came off left and right. We're stayin' with ya, no sweat, now you're holdin' good, Al. . . ." The F-4B and YF-5A broke formation. The Learjet stayed away. The XB-70A continued to fly straight and level for 16 seconds. In the cockpit, White and Cross were unaware that they had been involved in a collision; White thought it might be two of the chase planes, and he missed the "s" on Cotton's "vertical*s*." Then the XB-70A abruptly yawed right and rolled right, tumbling over and over so violently that White thought the plane's nose would break off. Hoag and Cotton called "Bailout, bailout, bailout"

[280] Bellman, XB-70 accident report.

[281] Ibid.; Mallick comments.

over and over. Finally a parachute appeared; one pilot was out. In fact, the chute belonged to White, who had just waged a successful struggle to stay alive. After he initiated the escape sequence, the capsule's closing clamshell doors trapped his right elbow. He worked it free, but then the doors would not close, and he ejected in that condition. The doors had inflicted painful shoulder injuries, but White faced a more serious problem: the open doors prevented the capsule's built-in "shock attenuation bag" from deploying. The capsule struck the ground with a 45 g force, causing White severe internal injuries. Carl Cross died in the wreckage of the XB-70A. The copilot's ejection capsule never left the airplane.[282]

One minute and 11 seconds after the collision, the XB-70A spun into the ground and exploded, about twelve miles north of downtown Barstow. Walker's F-104, in several pieces, was already burning in the desert, two miles away to the northwest. White's capsule floated downward. Hoag and Cotton circled around and around, looking for another chute. Back at Edwards, ground monitors received the first word of the accident. It spread like wildfire through the FRC, a numbing shock. A casual observer of flight-testing might wonder why the participants are not hardened to death, but they are not. One aeronautical research technician reflected, "You just feel so defeated. You know what I mean? The life you can't replace. The loss of the aircraft was secondary. You can get another airplane, but you can't get another pilot like that." The word first came from Operations: an accident had occurred, an accident involving the XB-70A. Little knots of people came by. Operations knew only that an F-104 had hit the XB-70A. Then came confirmation. NASA F-104N 813 had collided with the XB-70A. Both aircraft were down. Joe Walker was presumed dead, as was one of the XB-70A crewmen. Then came the final word: Walker and Cross dead, White badly injured, two airplanes destroyed.[283]

Of course, there was an accident investigation. The Air Force Directorate of Air Safety established a team of more than 60 people and a smaller accident board as well. The board was under Air Force control, and NASA's official representative, FRC engineer Donald Bellman, was a nonvoting member. Wreckage analysis clearly indicated what had happened, and the XB-70A's telemetry system had transmitted data all the way down to impact. The XB-70A program had great national visibility, and the deaths of Walker and Cross called forth tributes from many quarters. NASA Deputy Administrator Robert Seamans cited Joe Walker for his many contributions to flight research, and President Lyndon Johnson issued a statement of tribute from the White House.[284] Charges over the wisdom of risking the XB-70A and the lives of test pilots merely to provide corporate publicity photographs flew back and forth. But none of this could change the unhappy situation: two test pilots had died and two

[282] Bellman, XB-70A accident report.

[283] Interview of Armistead; comments of Peter Merlin, who has visited both crash sites.

[284] "Coast Air Collision Kills X-15 Test Pilot," *New York Times*, 9 June 1966; see also Bellman, XB-70A accident report.

aircraft had been lost. NASA's Flight Research Center had lost a valued colleague, and the XB-70A program had received a serious setback.

The FRC's XB-70A Program

The destruction of the second XB-70A drastically altered plans for NASA's joint SST research program with the Air Force. After the first numbing shock, the FRC went back to work, assessing where the program now stood. The first XB-70A was down for maintenance, including modifications to its landing gear, instrumentation, and inlet system. It did not resume flying until November 1966. Meanwhile, the Air Force reassessed its own plans for the aircraft. The second XB-70A had been the better suited for the Phase Two flight tests planned by the Air Force Systems Command (AFSC) and NASA. It had a better wing configuration, better inlet ramp control system, and much better instrumentation. It was gone. AFSC doubted that the first XB-70A could meet the same goals and, indeed, when testing resumed, it never ventured beyond Mach 2.57. On 3 November 1966, Joe Cotton and NASA pilot Fitz Fulton[285] took the remaining XB-70A over an instrumented test range for boom assessment at Mach 2.1. The plane made 10 more flights by the end of January 1967.[286]

That same month the Air Force, after comparing cost with research utility, decided to transfer total program and funding responsibility for the XB-70A to NASA "as soon as possible." Following its last Air Force flight, the big aircraft remained down for maintenance for two and a half months. During that time, Air Force and NASA officials worked out the details of the transfer. On 15 March, NASA and Air Force representatives signed an agreement under which the Air Force would continue to run some XB-70A research projects and provide aircraft support and pilot participation. A week later, FRC Director Paul Bikle and AFFTC commandant Maj. Gen. Hugh Manson created a joint FRC/AFFTC XB-70A operating committee patterned on the very successful X-15 and lifting-body agreements. Expenditures up to this point had amounted to approximately $2 million per month. To stay within its available 1967 and 1968 spending rates, NASA limited its planned XB-70A monthly program expenses to $800,000 per month, which automatically cut back the planned flight program. NASA had requested $10 million in fiscal-year (FY) 1968 funding, sufficient to continue the program through 1968. Also, the FRC awarded an $8.9 million contract to North American for maintenance and support of the XB-70A while it was flown by NASA and a $1.9-million contract to General Electric for engine maintenance.[287]

[285] Fulton, a former senior Air Force test pilot on the XB-70A, had retired from the Air Force. He had launched most of the early "Round One" rocket airplanes from B-29 and B-50 motherships while flying for the Air Force. Fulton was a welcome addition to the FRC's pilots office, for he was the world's finest test pilot of large multiengine supersonic airplanes.

[286] Betty J. Love, XB-70A chronology.

[287] Seamans, memo for record, 16 Feb. 1967; Bikle and Manson, XB-70A working agreement, 22 Mar. 1967; NASA news releases 67-59 and 67-75; NASA FRC *X-Press*, 24 Mar. 1967 and 7 Apr. 1967.

During its 11 flights from November 1966 through January 1967, the XB-70A supported the National Sonic Boom Program. This program, begun in June 1966, had involved a number of military aircraft inflights over selected American cities. The XB-70A made these flights at different weights, altitudes, and Mach numbers over a test range at Edwards instrumented to record the "boom carpet" of the aircraft and its "overpressure" (pressure rise) on two specially constructed test houses.

Such studies were critical. While the boom of a supersonic fighter might do little more than annoy citizens, the possibility existed that a large heavy SST would lay down a boom of such magnitude that it might do serious damage. During the XB-70A's tests, the craft made one overflight at Mach 1.22, about 420,000 pounds weight, and an altitude of roughly 27,000 feet, generating an overpressure of 3.15 pounds/square foot lb/sq ft. Higher, at 70,000 feet, the XB-70A once generated a boom having 2.33 lb/sq ft overpressure directly underneath the aircraft, and an overpressure of 1.71 lb/sq ft up to eight miles to one side of the plane. An overpressure of 7.5 lb/sq ft is sufficient to damage some structures. During turns, the XB-70A's shock waves converged, often doubling the overpressure felt on the ground. The tests clearly indicated that much work remained on tailoring aircraft design to minimize shock-wave magnitude. Even though the booms were not materially damaging, they were annoying. Indeed, the XB-70A tests went far toward providing quantitative evidence that overland commercial SST operations at supersonic speeds would generate boom phenomena that simply would not be tolerated.[288]

When the remaining XB-70A returned to the air in April 1967 on its first NASA flights, the agency had mapped out another program for the airplane: acquiring flight data that could be used to correlate and validate the data from two SST simulators, a ground-based simulator at the Ames Research Center and the FRC's Lockheed JetStar general-purpose airborne simulator (GPAS). NASA also had the XB-70A aircraft instrumented to record information on aeroelastic response of the structure to gusts, handling qualities (especially during landing approach), and boundary-layer noise. NASA engineers believed that the combination of XB-70A tests and tests of the GPAS aircraft could benefit the development of Boeing's proposed SST in four key areas, including control in the event of engine failure at supersonic speeds, development of an SST stability augmentation system, derivation of longitudinal stability requirements, and the influence of "ground effect" upon the landing characteristics of an SST. Later FRC added other programs to investigate inlet performance and structural dynamics, including fuselage bending and canard flight loads.[289]

Fitz Fulton and Joe Cotton completed the XB-70A's first NASA flight on 25 April 1967. By the end of March 1968, the plane had completed a further 12 flights by

[288] XB-70A chronology; Foxworth, pp. 168-169.

[289] NASA FRC *X-Press*, 7 Apr. 1967; memo, Donald T. Berry to chief, FRC research division, 20 July 1967; XB-70A chronology.

Fulton, Cotton, Van Shepard, Lt. Col. Emil "Ted" Sturmthal, and NASA pilot Donald Mallick. Following the 73rd flight on 21 March, NASA grounded the airplane for installation of a structural dynamics research package dubbed ILAF for "identically located acceleration and force." Two small, thin exciter vanes extending about two feet outward from just in front of the crew compartment could rotate 12 degrees at a frequency up to 8 cycles per second. The vanes induced structural vibrations having a known frequency and amplitude. Accelerometers sensed the disturbances and signaled the aircraft's stability augmentation system to move the aircraft's controls and suppress the disturbance. NASA hoped the ILAF program would serve as a prototype for advanced systems that could be installed on SSTs, enabling them to fly with increased smoothness, reducing the fatigue experienced by both passengers and airframe. Previously XB-70A crews had frequently experienced annoying trim changes and buffeting from clear air turbulence and rapidly fluctuating atmospheric temperature. Test results indicated that the ILAF system reduced the buffeting associated with such conditions. The XB-70A made its first ILAF-equipped flight on 11 June 1968. From then until the end of the program in 1969, the aircraft acquired a great deal of information applicable to the design of future SST or large supersonic military aircraft.[290]

By the end of 1968, operating expenses and maintenance problems had caught up with the XB-70A. The research data gained from the plane no longer justified the resources needed to maintain and operate it. The Flight Research Center could look forward to operating another Mach 3+ airplane, the Lockheed YF-12A Blackbird, which represented a more advanced technology than that of the already dated XB-70A. On 13 January 1969, NASA headquarters announced the termination of the joint NASA/DoD XB-70A flight research program. The announcement rightly hailed the XB-70A as "a productive flight research vehicle for studying sonic boom, flight dynamics, and handling problems associated with the development of advanced supersonic aircraft."[291] On 4 February 1969, the Valkyrie made its last flight, arriving at Wright-Patterson AFB, Ohio, where it is now on exhibit at the Air Force Museum.[292] In total, two XB-70A aircraft had completed 129 flights. The first XB-70A had completed 83 of these. The total flying time for both airplanes had been 252 hours, 38 minutes. Of this total, 22 hours, 39 minutes were spent above Mach 2.5.[293] Today, visitors at the Air Force Museum can compare the XB-70A to other dinosaurs of flight. The Valkyrie is still an impressive sight.

Thus ended the XB-70A program. Without a doubt, the loss of the second aircraft hurt whatever results NASA and the Air Force could have expected to reap from this

[290] NASA FRC *X-Press*, 17 May 1968 and 12 Jul. 1968; XB-70A chronology.

[291] NASA FRC news release 1-69.

[292] William T. Gunston, *Bombers of the West* (London: Ian Allen, 1973), p. 260.

[293] XB-70A chronology.

extensive time- and budget-consuming project. Critics of the aircraft often fail to realize, however, just how ambitious the XB-70A was. It was the world's first large transport-size aircraft capable of sustained long-range supersonic flight. So intoxicating were its performance figures that in 1959, the FAA administrator, Gen. Elwood Quesada, recommended to President Eisenhower that the United States develop a commercial version of the aircraft. While this proposal went nowhere, North American naturally drew quite heavily on its XB-70A work when developing its own abortive SST plans. Critics also fail to recognize that the aircraft did return a great deal of information on sustained supersonic cruise. The data predicted SST behavior, which could be incorporated in simulators, and the structural and control requirements of such airplanes. The flight requirements for a Mach 3 SST are far more complicated than the requirements for a Mach 2 SST. The magnitude of the problems is easily determined by noting that the Anglo-French Concorde, a modest Mach 2 airplane, is the product of one of the greatest international cooperative industrial efforts conceived to this time. The problems of Mach 3 present an even greater engineering challenge. Designers of Mach 3 aircraft cannot use a conventional aluminum airframe. Rather, because of aerodynamic heating, they must use sophisticated and challenging material, such as titanium. Then there are the problems that go with controlling an aircraft moving at Mach 3 and integrating it into an air transport network with aircraft moving much slower than it does. It is remarkable that the XB-70A achieved the performance it did since it was the first U.S. venture into large supersonic aircraft design.

NASA's Flight Research Center engineers had always hoped that the center could play some role in the development and testing of Boeing's SST, seeing such activity as the logical conclusion of the FRC's work with the XB-70A and GPAS programs. In September 1967, FRC engineers prepared a rough proposal for the FAA and NASA headquarters enumerating a variety of areas where the FRC could assist the FAA and Boeing on development of the airplane. In some of the areas—such as studies on pressure drag, skin friction, surface roughness, shock-wave-boundary layer interaction, and boundary layer noise—Boeing and the FAA had no research efforts under way, while the FRC's experience and background were unique.[294]

The American SST fell further and further behind its European competitors as cost and complexity rose. Even before the XB-70A concluded its flying program, the first supersonic transport, the Soviet Tu-144, had completed its maiden flight. One month after the XB-70A retired, the Anglo-French Concorde took to the air. In contrast, the Boeing design was in serious difficulty, including numerous major design changes, such as going from a variable-sweep wing to a fixed modified delta, a bad sign. Although the American SST had the full support of three successive presidents—Kennedy, Johnson, and Nixon—it had numerous criticsas well, ranging from thoughtful spokesmen who questioned its economic utility to those simply biased against technology. To save the foundering program, the FAA created, at the behest of President

[294] FRC, "Proposal for NASA SST Flight Research Program," n.d., passim.

Nixon, an Office of the Supersonic Transport. This office, directed by William M. Magruder, a distinguished test pilot and, ironically, former technical director of Lockheed's SST design, did its best to keep the Boeing SST alive, but to no avail. On 24 March 1971, the Senate declined to appropriate $289 million for prototype fabrication, abandoning the field to the Concorde and the Tu-144.

NASA continued some supersonic research from the mid-1970s into the mid-1980s, but it was only in 1990 that a full-scale successor to the earlier SST research came along. Then, NASA initiated a High-Speed Research program, part of which was a concept for a High Speed Civil Transport (HSCT)—a supersonic passenger jet carrying 300 passengers at more than twice the speed of sound. It was intended to cross the Pacific or Atlantic oceans in less than half the time employed by modern subsonic jets, but with ticket prices less than 20 percent higher. In December 1995, the program selected a single aircraft concept to focus the technological development planned for the following three years. This concept evolved from separate McDonnell Douglas and Boeing designs. NASA centers and industrial partners did considerable research to reduce noise and emissions of undesirable oxides of nitrogen without sacrificing performance. Flight research was done with a variety of aircraft, including a Tu-144LL in partnership with Russia. (The Tu-144LL was a retired Russian supersonic commercial transport that was upgraded to a high-speed research testbed.) However, after Boeing absorbed McDonnell Douglas, it opted out of the partnership with NASA on the HSCT in late 1998. According to Robert Cuthbertson, manager of the HSCT for Boeing, "Economically and technically we felt the hurdles were too high to build a commercially viable supersonic aircraft. . . . Until we make more progress in the noise, environmental, and manufacturing areas, it's not clear anybody will build a replacement for the Concorde." As a result of Boeing's decision, NASA in 1999 announced that it was canceling its program as well. Nevertheless, the program had developed a lot of technologies that might find application in the subsonic, military, and space markets, including advanced engine designs and improved composite materials.[295]

Meanwhile, when the XB-70A program concluded, hopes were still high that the United States might produce an SST for the 1970s, and the program's end did not end the FRC's work on advanced supersonic cruise aircraft. Indeed, the center terminated the XB-70A to make way for an even more advanced vehicle: the Lockheed YF-12A Blackbird. The first NASA FRC research flight of the YF-12A took place in 1969. However, by that time, center engineers had already been supporting the Air Force on the Blackbird program for two years. That program quickly took up where the XB-70A program had left off.

[295] NASA fact sheet FS-1998-09-15-LaRC, "NASA's High-Speed Research Program," Sept. 1998, available on line at http://hsr.larc.nasa.gov/FactSheets/HSR-Overview.html; NASA fact sheet FS-1999-03-62-DFRC, "The Tu-144LL: A Supersonic Flying Laboratory," Aug. 1999, available on line at http://www.dfrc.nasa.gov/PAO/PAIS/HTML/FS-062-DFRC.html; quotation as well as information from James Schultz, "HSR Leaves Legacy of Spinoffs," article excerpted from *Aerospace America*, Sept. 1999, in the [Langley] *News Researcher*, 8 Oct. 1999, pp. 1, 5.

A low sun angle highlights the blended-wing-body configuration of NASA's Mach 3+ YF-12C (actually, an SR-71) Blackbird. (NASA photo ECN-3516)

NASA and the Blackbirds

Lockheed conceived the Blackbird series of aircraft to fulfill a requirement for a Mach 3+ strategic reconnaissance aircraft. The program spawned several similar configurations, including the YF-12A, an abortive interceptor, and the SR-71A, a long-range reconnaissance aircraft. Both were Mach 3+ vehicles capable of flying at over 85,000 feet.[296]

The Blackbirds came out of the Lockheed Advanced Development Projects Group, the famed "Skunk Works" headed by Clarence "Kelly" Johnson. Given the scope of the technical challenges, the Blackbirds offered unparalleled design difficulties overcome by Johnson and his team of fewer than 200 engineers. Because of the sustained high temperatures that the planes would encounter during Mach 3 cruise, Johnson chose a largely titanium airframe. All supporting systems and fluids, including lubricants and fuels, had to be developed from scratch. During Mach 3+ cruise, the afterburning turbojet engines functioned more as ramjets than as gas turbines. The

[296] A useful source published since the original edition of this book is Dennis R. Jenkins, *Lockheed SR-71/YF 12 Blackbirds* (North Branch, MN: Specialty Press, 1997).

first Blackbird flew at a remote airstrip in 1962, and flight tests generally went smoothly. Although flown in single- and two-place versions, Lockheed standardized on a two-place configuration, with a pilot and navigator-systems operator. The plane featured a distinctive blended wing-body shape, with long chines running along the fuselage sides from the wing roots. Each engine was located at mid-span, and each nacelle was surmounted by a large, inwardly canted, vertical fin. For additional directional stability, the YF-12A had a folding ventral fin and two smaller fixed ventral fins as well. In February 1964, President Lyndon Johnson announced the existence of the plane. The first of the definitive reconnaissance variants, the SR-71A, flew later that same year.[297]

The Flight Research Center's involvement with the Blackbird program began in 1967. The Ames Research Center had opened negotiations with the Air Force for access to the early YF-12 wind-tunnel data that had been generated at Ames under extreme secrecy. The Air Force agreed, in return for NASA assistance on the flight-test program then under way at Edwards. This arrangement closely dovetailed with the plans of OART, which saw the Blackbird as a means to advance high-speed technology, especially that necessary to build SSTs. In the summer of 1967, the Air Force and NASA agreed to Flight Research Center participation. Paul Bikle and FRC research chief Joseph Weil asked engineer Gene Matranga to represent NASA on the Blackbird test force. Matranga, then busily involved in general aviation studies, thought about it over a weekend and agreed to go. Bikle, Weil, and Matranga assumed the center would work with the Air Force on the project for about six months. The exposure would give FRC engineers data to compare with the flight results coming from the XB-70A program. Matranga began working on Blackbird stability and control and soon brought a small team of experienced FRC engineers to labor along with him. Much good will among the Air Force, Lockheed, and NASA test-force team members ensued.

The Air Force team needed assistance in several technical areas. The Air Force wanted to get the SR-71A fully operational with the Strategic Air Command as quickly as possible. NASA wanted an instrumented SR-71A to use for its own research. If that were not possible, NASA was willing to install an instrument package on the Air Force SR-71A stability-and-control test aircraft. The Air Force declined, but offered NASA use of two YF-12A aircraft then in storage at Edwards. NASA quickly accepted, even taking the unusual step of paying the operational expenses of the airplanes, using funds made available by termination of the X-15 and XB-70 programs. The Air Force would also furnish a test team from the Air Defense Command for maintenance and logistics support. A memorandum of understanding was signed 5 June 1969, followed by public announcement on 18 July. Matranga and the FRC team immediately

[297] Clarence L. Johnson, "Some Development Aspects of the YF-12A Interceptor Aircraft," AIAA *Journal of Aircraft* (July-Aug. 1970), pp. 355-359; Richmond L. Miller, "Flight Testing the F-12 Series Aircraft," AIAA *Journal of Aircraft* (Sept. 1975), pp. 695-698; Marcelle Size Knaack, *Post-World War Two Fighters,* Vol. I of *Encyclopedia of U.S. Air Force Aircraft and Missile Systems* (Washington, DC: USAF, 1978), pp. 333-334.

set to work instrumenting the two YF-12A aircraft and mapping out a joint program with the Air Force.[298]

At the FRC and Ames, interest was high in Kelly Johnson's Blackbird. Most engineers expected to see its airframe, propulsion system, and related equipment on future Mach 3 airplanes. It was an ideal vehicle for assessing the state of the art of wind-tunnel prediction, aerodynamics, propulsion, and structural design. The plane also could carry experimental research packages, which the FRC considered a secondary objective, at least at first. Langley engineers were interested in running fundamental aerodynamics experiments and tests of advanced structures. Lewis was interested in propulsion research. Ames, a vital partner to the FRC, was interested in inlet internal aerodynamics and the correlation of wind-tunnel and flight data. The Flight Research Center had the challenging task of organizing all of these interests into a single unified research program. At first, it concentrated on aerodynamic loads and structural effects because instrumentation was available for those investigations. Much time-consuming work remained to be done before one of the Blackbirds could be instrumented for propulsion tests. So when the Air Force brought the two YF-12As out of storage, FRC technicians installed strain gauges and thermocouples. They instrumented the wing and fuselage for aerodynamic loads and the left side of the aircraft for temperature measurements to better define the craft's thermal environment.[299]

NASA and Air Force technicians spent three months readying the first Blackbird for flight. On 10 December 1969, the joint flight-research program got under way with a successful maiden flight. The first YF-12A that was ready quickly became the program's workhorse, while technicians readied its stablemate.[300] With the first flight out of the way, the NASA/Air Force team got down to the serious business of acquiring data points. While the Air Force concentrated on military applications, such as studying bomber penetration tactics against an interceptor having YF-12A capabilities, NASA pursued a loads-research program. FRC and Langley engineers were interested in measuring the flight loads, which depended on both the actual load conditions and the effects of structural heating. At some future date, FRC engineers planned to move the airplane into the FRC's high temperature loads laboratory, heat it, and determine how much of the load stemmed from thermal heating of the structure. This is not as innocuous as it sounds. When an airplane's structure is heated, the induced thermal stresses change the shape of the structure even without loads being applied. The

[298] Interviews of Matranga and Klein: YF-12 test program memoir, originally seen in the files of the external affairs office, DFRC.

[299] See note above.

[300] These two aircraft were the second and third YF-12As actually built; the second, serial number 60-6935, became NASA's long-lived YF-12A. The third, 60-6936, crashed and was replaced by the "YF-12C," 60-6937, which actually was SR-71A 61-7951 but was given the serial number of a Lockheed A-12 as its NASA tail number to disguise the fact that it was an SR-71. This number was selected because it followed the sequence of tail numbers assigned to the YF-12A aircraft.

changed airframe shape then has a much different load distribution pattern. When actual flight loads are added, the importance of knowing how the structure reacts to temperature and load is self-evident. To predict loads and structural response, NASA had developed two computer modeling programs using a technique known as finite element analysis. Both programs, FLEXSTAB and NASTRAN, were applied to the YF-12A. One of the major objectives of the flight tests on the Blackbirds was to compare the actual flight-test results with the predicted data. Technicians also installed a Hasselblad camera within the fuselage of the YF-12A to photograph the structure during high g maneuvers, recording the deformation of the aircraft. Under certain conditions, the camera revealed that the plane experienced as much as six inches of deflection at the aft end of the fuselage.[301]

While the program on aircraft 935 went smoothly, the program on 936, the other YF-12A, ended badly. The aircraft had completed 61 successful flights, mostly to gather data for Air Force operational requirements, when it crashed. During a flight on 24 June 1971 to acquire operationally useful information, this Blackbird experienced fatigue failure of a fuel line and a fire in the right engine. Lt. Col. Ronald J. Layton and systems operator Maj. Billy A. Curtis debated whether they could land the burning Blackbird. They wisely elected to eject, and the YF-12A smoked down to an explosive finale.[302] The loss of the YF-12A did not seriously affect the NASA structures program, which was almost finished; however, it did delay plans for the propulsion research program. NASA had wanted to add a third aircraft to the YF-12A joint test program, solely for propulsion tests. Even before the loss of the YF-12A, the Air Force had made 937 available. This aircraft, which was designated a YF-12C, was in fact an SR-71A. Because the SR-71 program was shrouded in the highest security classification, the Air Force restricted NASA to using the aircraft solely for propulsion testing. The YF-12C, which looked like the SR-71A, was thus an oddball. For the NASA programs on both the YF-12A and YF-12C, the Flight Research Center had designated pilots Fitz Fulton and Don Mallick and flight-test engineers Vic Horton and Ray Young. As the program developed, generally Fulton and Horton flew together as one team, Mallick and Young as the other. At Beale AFB, the pilots received familiarization flights in a humpback SR-71B having a second pilot cockpit in place of the navigator-systems operator's cubicle.[303]

On 24 May 1972, Fulton and Horton crewed the YF-12C on its first NASA flight. By this time, NASA had already accumulated 53 flights in the YF-12A and had grounded the airplane for testing in the FRC's high temperature loads laboratory. It

[301] See note 26 above; also see historical chronology of Edwards AFB, in the files of the History Office, AFFTC; "The Lockheed YF-12," Fact Sheet-1999-11-047-DFRC, available on the Internet at http://www.dfrc.nasa.gov/PAO/PAIS/HTML/FS-047-DFRC.html.

[302] YF-12 program memoir; interview of Victor Horton, 26 Feb. 1977; comments on original chapter by Peter Merlin of the Dryden history office, 9 Feb. 2000.

[303] Interviews of Matranga, Klein and Horton.

remained in the lab for over a year, not flying again until July 1973. As a result of the correlation between flight tests and tests in the heat laboratory, FRC engineers were confident that they had developed instrumentation and test procedures that would allow the aircraft industry to proceed with assurance on the development of other high-temperature aircraft.

NASA engineers approached the propulsion program on the YF-12C with a similar purpose in mind: "provide a baseline of information that can be used in future times as well as the present time to assess the validity of current prediction and wind tunnel test techniques."[304] Together with Pratt & Whitney (the engine manufacturers) and Lockheed, FRC engineers assembled a computer model of the engine and inlet system. In conjunction with the Ames, Langley, and Lewis research centers, the FRC compared flight data of the aircraft with data taken from tests of scale-model inlets. Furthermore, a full-scale inlet was tested in the Lewis 10-by-10-foot tunnel in early 1972. One surprise was the discovery that a strong vortex, coming from the fuselage chines, streamed into the middle of the inlet. These studies were very detailed, examining such questions as what percentage of airflow through the inlet left through bypass doors in the inlet and what percentage actually passed through the engine. The FRC team also examined inlet "unstart"—if the airflow was not properly matched to the engine, internal pressure would force the standing normal shock wave from inside to outside the inlet. This action lost the thrust provided by inlet pressure recovery. The thrust imbalance generated a large yawing motion, as well as residual pitching and rolling tendencies. The first time one NASA crewman encountered unstart, the aircraft motions and accelerations were so violent that he expected the YF-12 might break up. Obviously, this condition could not be tolerated on an SST aircraft. NASA devoted a great deal of attention to unstart in an attempt to learn how to control it, deliberately inducing unstarts on test flights. Automatic inlet sensing and control was one method of combating it. The production SR-71A's system worked so well that the Air Force had to induce the phenomenon to familiarize pilots with it during training. NASA installed a special manual inlet control override system that allowed the research pilot to select and control inlet by-pass doors and spike positions precisely. This permitted a multitude of test points, including deliberate unstarts. NASA's YF-12 crews became so familiar with unstarts that they could sense when one was imminent even before the instrumentation showed it.[305]

The YF-12 engineers pioneered another important propulsion advance. After successful flight research on the F-111A Integrated Propulsion Control System (IPCS), during the late 1970s Dryden technicians replaced the mechanical control apparatus aboard the NASA YF-12 with a cooperative digital system. A single computer governed inlet control, autopilot, autothrottle, air-data, and navigation functions. Subsequent test flights found the new system capable of speedier and more precise response

[304] YF-12 program memoir.

[305] Ibid.; interview of Horton; Mallick comments.

computations than the one it supplanted, resulting in a five-percent increase in the aircraft's range. Eventually, the entire SR-71 fleet incorporated this improvement, which in the operational context actually extended the range by *seven* percent and, at the same time, largely resolved the dangerous unstart phenomenon.

The FRC's YF-12 program was ambitious. The aircraft flew an average of once a week unless down for extended maintenance or modification. Program expenses averaged $3.1 million per year just to run the flight tests, and Ames, Lewis, and Langley were heavily involved in the program as well. The YF-12A program dominated the annual FRC *Basic Research Review* reports that the center prepared for OAST's research council during the 1970s.[306] The scope of what was involved in a YF-12 flight was enormous. Technical preparation and briefings aside, the flights required coordination of the highest order between NASA, the FAA, and the Air Force. The crew would suit up an hour and a half before takeoff. Using a special Air Force aeromedical van, the crew would drive out to the flightline and enter the aircraft. For what seemed an interminable time, they would run up the engines and check out systems. The Blackbirds—sometimes both would fly together—would sit on the ramp, engines oddly muted, exhaust waves shimmering over the lakebed. Other FRC personnel would ready an F-104, and maybe a slower T-38 as well, to follow the craft on takeoff and acceleration to Mach 2. Further north, at Beale AFB, the Air Force would send aloft a KC-135Q tanker with a load of the Blackbird's special JP-7 fuel. Finally, all would be ready. One after another, the aircraft would taxi from the Flight Research Center to the 15,000-foot runway. After final safety checks, the Blackbirds would scoot down the runway and rumble into the air with a shattering roar. The chase planes would follow. The YF-12A would accelerate to about Mach 0.9, enter a shallow dive (the most efficient way to exceed Mach 1), nose upward, and accelerate to the maximum speed selected for the flight, outrunning and outranging the chase. After one gigantic circuit over the western U.S. (with the Air Force and the FAA keeping watch to make certain that the craft did not wander around other SR-71As or U-2s tooling about in the sky), the Blackbirds would decelerate and descend, take on a load of fuel from the KC-135Q, again go supersonic, make another circuit, then return and land. This allowed the Blackbird crews to gather over two hours of flight-research data points—the maximum capacity of the precision onboard data recorders.

NASA's Blackbird program had its exciting moments, routine unstarts aside. On one YF-12C flight, Don Mallick and Ray Young experienced a stuck inlet spike, which caused the airplane to burn prodigious amounts of fuel, necessitating an emergency landing at Fallon Naval Air Station, Nevada. Another time, during a stability test at Mach 0.9 with the craft's roll and yaw stability augmentation system deliberately off, they lost the folding ventral fin from NASA's YF-12A. Fortunately, this fin is needed only at high supersonic speeds. At Mach 3 the effect would have been much

[306] William S. Aiken, FRC YF-12 Research Aircraft RTOP, 766-72-01, 9 Mar. 1972, p. 2; see, for example, *Basic Research Reviews* [*BRR*] for NASA OAST Research Council, 20 Aug. 1974, 20 Oct. 1975, and Dec. 1976; Frank W. Burcham, Ronald R. Ray, Timothy R. Conners, and Kevin R. Walsh, NASA/TP-1998-206554, "Propulsion Flight Research at NASA Dryden from 1967 to 1997," DFRC, July 1998, pp. 7-8.

NASA's YF-12A Blackbird cruises over the desert, carrying a Coldwall heat-transfer experiment under the fuselage. (NASA photo ECN-4728)

more serious, probably loss of the airplane. Mallick and Young skillfully brought the ailing airplane back to Edwards. The departing fin had damaged the wing, aft fuselage, and stability augmentation system; it also ruptured a fuel tank, causing it to dump its contents overboard in a long silver trail.

Tests of a proposed Coldwall experiment package gave bad moments as well. The Coldwall, a Langley-supported heat-transfer experiment, consisted of a stainless steel tube equipped with thermocouples and pressure-sensing equipment. A special insulation coating covered the tube, which was chilled with liquid nitrogen. At Mach 3, so planners hoped, the insulation could be pyrotechnically blown away from the tube, instantly exposing it to the thermal environment. Its data could be compared with results taken from testing a similar tube using ground-based wind-tunnel facilities and would validate ground research methods. Eventually researchers did get a successful test, but the experiment caused numerous inflight difficulties. On one of the Coldwall flights, for example, the YF-12A experienced a simultaneous unstart on both engines, followed by rough engine operation after firing the Coldwall. As it descended, anxiously followed by the YF-12C photo chase plane, the latter aircraft also experienced multiple unstarts. For a brief while, test monitors at Dryden worried for the safety of both crews. Both aircraft limped back to Edwards at reduced power. NASA grounded them for extended inspection.[307]

[307] YF-12A and YF-12C flight chronology, DFRC YF-12 project office files; information from Ray Young; interview of Horton; Report of YF-12 Ventral Incident Investigation Board, 25 Apr. 1975; interview of Klein; Mallick and Merlin comments.

Flight research with the YF-12 aircraft furnished some interesting data. For example, at Mach 3 fully 50 percent of the aircraft's total drag came from simply venting air overboard through the inlet bypass doors. Also, a gray area was discovered between stability and control, on the one hand, and propulsion, on the other. Inlet components were almost as effective as ailerons and rudders in influencing aircraft motion at high speeds. Inlet spike motion and bypass door operation could alter the aircraft's flight path under some conditions. The airflow dumped overboard through the inlet louvers entered a "stagnation area" just ahead of the louvers and actually flowed *forward* along the outside of the nacelle for a brief distance before mixing with the Mach 3 airstream and moving aft—a weird effect. Most serious, however, was a problem that had earlier cropped up on the XB-70A: unwanted altitude changes while cruising at high altitude and high speed.

In fact, the main stability-and-control area of interest to NASA researchers was the ability to hold a desired cruise altitude. At high speeds and altitudes, without stability augmentation, the plane could change attitude slightly. Since the plane was moving at Mach 3, any nose-up or nose-down change immediately produced major changes in altitude. The plane entered porpoising motions for up to three minutes, during which altitudes changed by as much as plus or minus 3,000 feet. Such operation would certainly be prohibitive from an air-traffic-control standpoint with a commercial SST aircraft. At the altitudes the YF-12s and SR-71As operated, there was no other traffic aside from an occasional U-2 or fellow SR-71A, but that situation could change with time. The thought of fleets of SST aircraft wobbling about their flight paths is not comforting. The YF-12's very ability to attain high speeds and altitudes contributed to the problem. At Mach 3, it covered distance quickly, passing through local pressure and temperature changes that would affect a slower aircraft much more gradually. Since Mach number is a function of pressure and temperature, the rapid variations caused velocity changes. Correcting for these changes by adjusting inlet controls or aerodynamic controls produced large altitude deviations. In future supersonic transports, such a situation would pose problems for air traffic controllers and in some circumstances could cause the aircraft to exceed its operating limits.[308]

As one potential solution to the altitude-holding problem, FRC engineers developed a new autopilot and flight-tested it on the YF-12s. Traditional autopilots moved aerodynamic control surfaces to maintain speed or altitude. The experimental YF-12 system compensated for various pressure-sensitive instrumentation that influenced altitude deviations. After further modifications, it linked the aircraft's central air-data computer to the autopilot, the inlet control system, and the engine throttle system. The combination of aerodynamic surface controls and throttle control, together with more advanced data sensing equipment, worked well on actual flight tests, even during extended high-Mach cruise.[309] Such integrated systems would almost certainly be used on future SSTs.

[308] YF-12 test program memoir; NASA DFRC *X-Press*, 3 June 1977.

[309] NASA DFRC *X-Press*, 3 June 1977; Glenn B. Gilyard, "Supersonic Cruise Autopilot Operation," *BRR*, 20 Aug. 1974, pp. 23-24; Paul J. Reukauf, "Integrated Propulsion/Flight Control," *BRR*, Dec. 1976, pp. 23-25.

NASA performed a variety of research on the Blackbirds. For example, technicians installed a special computerized checkout system in the aircraft, the central airborne performance analyzer (CAPA). CAPA monitored a number of parameters dealing with aircraft maintenance, including the craft's electrical system, inlet control system, and hydraulic system. Although just a research project itself, CAPA offered great promise for such future projects as the Space Shuttle and commercial and military aircraft. During flight, the system could actually diagnose a problem, informing the pilot whether he should abort. At the end of the flight, technicians could check the CAPA readout to determine the maintenance required before the next flight.

Another program investigated the temperature, pressure, and other physical characteristics of the upper atmosphere, since they would have great impact on the performance and operation of future aircraft. The FRC examined high-altitude turbulence, which the YF-12s encountered at virtually all altitudes, and researchers supported the work with statistical studies at the National Climate Center and the University of California at Los Angeles. Biomedical researchers took physiological and biomedical measurements of the flight crews on most YF-12 flights to derive a better understanding of physiological stress. Researchers used the airplane as a sort of "flying wind tunnel" carrying experiments and instrumentation for studying boundary-layer flow and noise, heat transfer, skin friction, and base pressure measurements. Under Langley Research Center's supersonic cruise aircraft research program, the FRC evaluated a number of advanced structural techniques on the YF-12. Engineers replaced a panel on the airplane with a series of Langley-designed experimental panels of advanced design. The flight research complemented laboratory work on small test specimens. Technicians chose a test panel, 28 by 16 inches in size on the inboard upper surface of the wings between the nacelle and the fuselage. Between 1974 and 1976, they evaluated three lightweight structures there: a weld-brazed titanium-skin stringer panel, a titanium honeycomb-core sandwich panel, and a sandwich panel faced with boron-aluminum and having a titanium honeycomb core. All exceeded required strengths. In all of these ways and more, the Blackbirds contributed to flight technology. Because of the tight security restrictions on the program, engineers could get information only on a need-to-know basis. Nevertheless, in June 1974, the Flight Research Center hosted a major conference attended by 150 representatives from government and the aviation industry to report on the YF-12 loads research. Although it had been over a decade since the first flight of the Blackbirds, they still represented advanced state-of-the-art technology.[310]

Blackbirds, Bye-Bye

By the beginning of 1977, the YF-12 aircraft had completed over 175 flights, much of the time above Mach 3. Although still the pride of Dryden's hangars, the two Blackbirds were becoming increasingly expensive to maintain and more difficult to

[310] DFRC YF-12 fact sheet, n.d.; YF-12 test program memoir; *BRR* for 1974, 1975, and 1976.

justify. Other programs—notably the center's McDonnell F-15 Eagle research aircraft—could lay greater claim to funding. The axe fell during an OART center directors' management council meeting in the spring of 1977. Residual funding enabled the YF-12C to fly through October 1978, continuing tests of an integrated aerodynamic and propulsion control system. The oldest YF-12 still flying, 935, would end its research program a year later.[311] Dryden's most visible program thus ended far sooner than most YF-12 partisans had assumed. Previously, the center had planned to operate the Blackbirds into the 1980s.[312]

The decision was not popular with the YF-12A team. Team members saw themselves as linked to the center's elite program and tended to view the cancellation as more evidence that Dryden's golden age was past. Many decried what they felt to be a growing tendency within NASA to homogenize the centers, reducing center control over research. The fiercely independent NACA veterans shook their heads in frustration. Partisans grumbled that the decision was simply the latest in a long line afflicting a technological hiatus on American aviation. They pointed to the SST cancellation as evidence. When the B-1 and the National Hypersonic Flight Research Facility (NHFRF) joined the ranks of cancelled programs, the YF-12 became just one more name on a growing list. For the time being, Dryden was through with Mach 3.

NASA's Blackbird program proved one of the most useful programs ever flown at the FRC. It was the major air-breathing propulsion work done at Dryden and helped change the center's image away from a rocket bias toward a more balanced mix of research. The Blackbird program was certainly much more productive than the XB-70A. The two YF-12s proved surprisingly free of chronic maintenance problems, aside from some difficulties with fuel tank leaks. The program generated a great deal of information that will prove useful to future Mach 3 sustained-cruise designs. To those unacquainted with the flight research process, it often seems odd that so much effort should be spent testing modified versions of production designs. Surely, the criticism goes, the contractor and the user have already obtained all the information of value from the aircraft by the time another agency or group acquires it. The YF-12 program is a good example of how incorrect this supposition is. The contractor and the user were naturally much more interested in assuring that the aircraft was safe to operate and met its performance specifications. When the SR-71A entered full service with the Strategic Air Command, Air Force interest in the craft's research utility quickly cooled. A contractor is never in a financial position to run an extended flight-research program, no matter how beneficial it might be later on. And so it fell to NASA to use the Blackbirds as research instruments. The Blackbird teams derived an important database for subsequent aircraft design. Interest in supersonic flight was already ebbing at Dryden by 1979, but the engineering staff was busily working on a variety of other problems from transonic research to remotely-piloted research vehicles, maintaining the center's reputation for diversity.

[311] Appendix P contains a chronology of YF-12 flights.

[312] Conversation with Gene Matranga and Ming Tang, 10 Apr. 1978.

III

Pragmatic Pursuits

10

A Mansion of Many Rooms 1970-1976

During the 1960s and 1970s, the Flight Research Center initiated aeronautical studies in a wide range of research areas. This work continued the FRC's tradition of simultaneously running diverse research programs, supporting civil and military aircraft testing and development, and supporting research in progress at other NASA centers, the latter usually by "proof of concept" flight-testing. During these years, engineers and technicians evaluated the Whitcomb supercritical wing and winglet; developed a new flight research tool, the remotely-piloted research vehicle; developed and evaluated a radical new method of flight control using electronics; studied wake-vortex formation and clear-air turbulence, two areas of importance to aviation safety; supported development of new military aircraft systems; and entered a number of other fields as well, including design configurations for long-distance trailer trucks and flight-testing of advanced rotorcraft. Although much of the center's research was applied, more of it was basic, exploring and deriving new data on the often mysterious and perplexing conditions and phenomena that influence how flight vehicles perform.[313] The full spectrum of the FRC's research in aeronautics is shown in Figure 5.

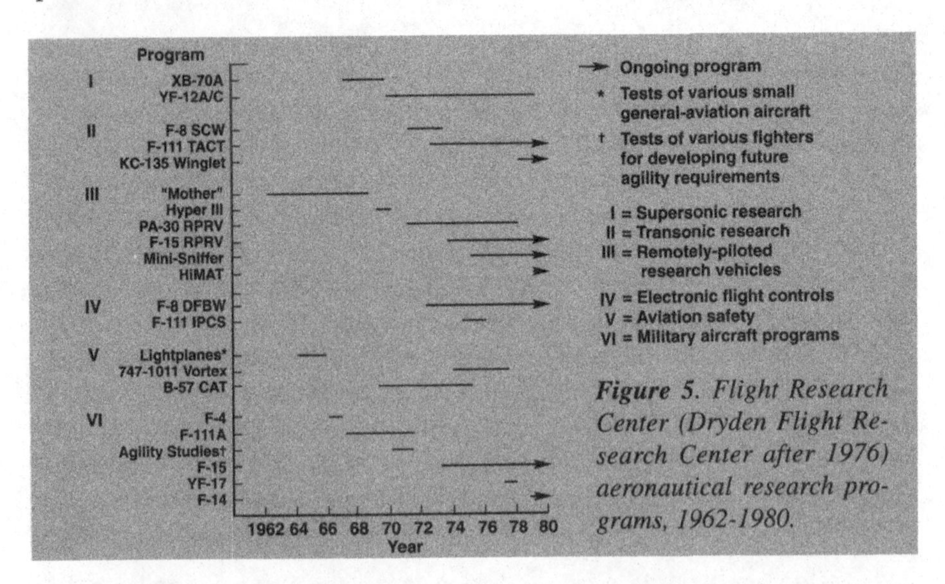

Figure 5. Flight Research Center (Dryden Flight Research Center after 1976) aeronautical research programs, 1962-1980.

[313] Interview of Krier by Hallion; interview of Edwin Saltzman, 6 Dec. 1976; FRC *Basic Research Reviews,* passim; Fitzhugh L. Fulton, NASA TM X-56021, "Pilot's Flight Evaluation of Concorde Airplane No. 002" (FRC: Jan. 1974); Fitzhugh L. Fulton, "Flight Evaluation: Ford Tri-Motor, Model 4AT-B" (Nov. 1970); E. J. Saltzman and Robert R. Meyer, NASA TM X-56023, "Drag Reduction Obtained by Rounding Vertical Corners on a Box-Shaped Ground Vehicle," (FRC: Mar. 1974).

The FRC and the Supercritical Wing

In 1978, over three decades after Chuck Yeager exceeded Mach 1, John Anderson, a noted aerospace engineering educator and historian, wrote:

> The analysis of transonic flows had been one of the major challenges in modern aerodynamics. Only in recent years, since about 1970, have computer solutions for transonic flows over airfoils come into practical use; these numerical solutions are still in a state of development and improvement. Transonic flow has been a "hard nut to crack."[314]

Although the transonic regime had long disappeared as a barrier in the minds of engineers, it continued to fascinate aerodynamicists. In the transonic regime, an airplane experiences mixed subsonic and supersonic flow patterns. At some point, which varies with the design of the plane, the flow over the wings goes supersonic. A little faster and standing shock waves dance across the wing. Then the drag of the plane rises sharply with concomitant losses in efficiency. It is also in the transonic regime that most commercial jet airliners fly, so the intricacies of transonic aerodynamics are part of the real world for aircraft designers.

One individual who devoted the major portion of his NACA-NASA career to transonic research was Langley's Richard T. Whitcomb, an engineer fond of remarking, "We've done all the easy things—let's do the *hard* ones." In the 1950s, Whitcomb had derived the concept of transonic area rule, which gave an entire generation of aircraft a "wasp waist," or pinched look. An engineer equally at home with a slide rule at his desk or shaping a wind-tunnel model for testing, Whitcomb demonstrated an uncanny ability for visualizing configuration changes to enable airplanes to fly more efficiently at transonic speeds. He ultimately conceived two other means of improving that efficiency: the supercritical wing (SCW) and the wingtip "winglet." All three advances went to the High-Speed Flight Station (later called the Flight Research Center, and later still, Dryden) for "proof of concept" flight-testing.

During the early 1960s, Whitcomb investigated a technique for tailoring airfoil designs to raise the drag-divergence Mach number as close to the speed of sound as possible. Such airfoils would have a "supercritical" Mach number, the point at which the airflow over the airfoil exceeds the speed of sound. They would have less drag, because the design would discourage shock-wave formation. In other words, if two transports of similar design cruised at the same speed, differing only in that one had a conventional airfoil and the other a supercritical airfoil, the transport with the supercritical airfoil should have less drag and, consequently, should use less fuel. It should also have higher speed potential and, because of its fuel efficiency, greater

[314] John D. Anderson, Jr., *Introduction to Flight: Its Engineering and History* (New York: McGraw-Hill Book Co., 1978), pp. 171-172.

range. Whitcomb estimated that such airfoils could raise the cruising speed of long-range jetliners by as much as 100 miles per hour. He embarked on a four-year wind-tunnel study program at Langley. The shape he finally selected had a flattened top surface, with a downward curve at the trailing edge; it looked somewhat like a tadpole. The flattened top reduced any tendency of the wing to generate shock waves, and the downward curve at the trailing edge restored the lift lost by flattening the top. Whitcomb spent many hours in the tunnel, hunched over development models, refining his concept. By 1967, he was convinced that he had a major breakthrough. Wind-tunnel tests indicated that the new shape would greatly improve the transonic performance of transport aircraft. Would the wing perform in flight as advantageously as those tests indicated? Flight validation was obviously required. Whitcomb and other Langley researchers started looking for a suitable aircraft to serve as a testbed for a supercritical wing.

The airplane chosen was the Vought F-8A Crusader, a single-seat, single-engine, obsolescent Navy jet fighter. The Crusader had been an excellent aircraft. Capable of Mach 1.7 speed and equipped with both cannon and missiles, it had formed the backbone of naval aviation during the late 1950s and early 1960s. Indeed, at the very time that NASA contemplated modifying an F-8A to serve as a supercritical wing testbed, advanced F-8D and F-8E Crusaders were in combat over North Vietnam.

NASA selected the F-8A because it had an easily removable wing, which technicians could replace with a supercritical-wing (SCW) test installation. It also had landing gear that retracted into the fuselage, which meant that the experimental wing would not need to house the retracted landing gear. The F-8A was readily available from the Navy, could be maintained with relatively little effort, and had genuine transonic performance. NASA acquired three of them. Whitcomb and SCW team members Thomas C. Kelly and Lawrence K. Loftin had decided to use the F-8A at a meeting on 21 March 1967. By mid-May 1968, Langley Research Director Thomas A. Toll was chairing meetings between Langley and Flight Research Center personnel to define the broad responsibilities of each center in running an F-8A SCW proof-of-concept demonstration.[315]

In February 1969, NASA announced that Whitcomb's supercritical-wing concept would be tested on a modified F-8 at the Flight Research Center. NASA Administrator Thomas O. Paine testified before a congressional committee that the tests would probably begin in late 1970. "Because of its potential for enhancing both the cruise performance and the operations economics of subsonic jet aircraft, this new NASA concept has generated widespread interest within the aircraft industry."[316] Whitcomb's team designed a shapely transport-type wing for the F-8 and ran tests in Langley's

[315] Conversation with Thomas C. Kelly, 24 Apr. 1978; memo, Toll to LaRC assistant director, 20 Nov. 1968; L. K. Loftin, SCW meeting minutes, 21 Mar. 1967; Warren C. Wetmore, "New Design for Transonic Wing to be Tested on Modified F-8," *Aviation Week & Space Technology* (17 Feb. 1968) pp. 22-23.

[316] NASA History Office and the Science & Technology Division, Library of Congress, *Astronautics and Aeronautics: 1970–Chronology on Science, Technology, and Policy* (Washington, DC: NASA SP-4017, 1972), p. 58. (This yearly chronology is subsequently cited as *AA*–[year].)

eight-foot tunnel on a model F-8 having such a planform. Military applications of supercritical-wing technology took a different path, that of the Transonic Aircraft Technology (TACT) program, described later in this chapter. The F-8 SCW program was oriented entirely toward civil aviation. Indeed, some observers saw the program as NASA attempting to sell the American aircraft industry on a concept, whereas NACA-NASA's traditional role had been to conduct research, present the results at meetings and symposia, and let industry decide on its own what to do.

The Vought F-8A arrived at the Flight Research Center on 25 May 1969. Center pilots Thomas C. McMurtry and Gary Krier began flying it to gain operational experience in the plane before it was modified. The FRC contracted with North American-Rockwell's Los Angeles division to fabricate the supercritical wing at a cost of $1.8 million. Meantime, North American-Rockwell "gloved" a supercritical airfoil on the wing of a Navy T-2C Buckeye jet trainer at the company's plant in Columbus, Ohio, to gain some preliminary experience with such wings. The Buckeye made its first SCW flight at Columbus on 24 November 1969 without any unusual results. Three weeks earlier, North American had delivered the F-8's supercritical wing to Edwards. NASA planned the first trials of the aircraft in early 1971. By this time, Krier and McMurtry had completed 32 flights in the unmodified Crusader, which received the designation TF-8A.[317] NASA technicians set to work installing the new wing on the plane.[318]

Whitcomb and his Langley team had desired as pure a wing as possible so that the full spectrum of SCW performance could be explored without interference from gaps, flaps, or ailerons. Instead of ailerons on the wings for roll control, he had preferred that the F-8 be modified with a "rolling" tail, such as the one used on the X-15. This proved unworkable, the rolling tail giving inadequate control at low speeds. Whitcomb had to accept an aileron on the supercritical wing. The standard Crusader had a two-position, variable-incidence wing to reduce its landing speed. The test wing was fixed and required a fast landing approach, which made the plane totally unsuited for operations using a conventional runway. Otherwise, Langley could have run the program entirely at Hampton, Virginia. In fact, because the plane touched down at about 195 miles per hour and lacked antiskid provisions, it could not even land on Edwards' 15,000-foot runway without coasting onto the overrun. Takeoffs were from the main runway toward the lakebed, and the craft landed on the dry lakebed.[319]

By early 1971, FRC technicians had installed the shapely wing on the TF-8A. Tom McMurtry was the lead project pilot. Engineer John McTigue, who had earlier shepherded the lifting bodies, was the first program manager. Thomas Kelly acted as

[317] Not to be confused with Vought's abortive two-seat TF-8A Crusader (the F8U-1T), only one of which was completed.

[318] F-8 SCW flight log; *AA–1970*, pp. 359, 377-378; conversation with Thomas C. McMurtry, 20 Apr. 1978.

[319] Interview of Thomas C. McMurtry, 3 Mar. 1977; McMurtry conversation.

The NASA F-8 supercritical-wing (SCW) testbed. (NASA photo EC73-3468)

Langley's project engineer, and Whitcomb took a personal interest in the tests. McMurtry and pilot Gary Krier practiced in an SCW simulator that FRC technicians built, and NASA modified the aircraft to incorporate artificial stability devices. On 9 March 1971, McMurtry took off on the TF-8A's first supercritical-wing flight. During the 50-minute excursion, he evaluated the plane's low-speed handling qualities and stability augmentation system, attaining an altitude of about 10,000 feet and a maximum speed of roughly 350 miles per hour.[320]

The supercritical-wing TF-8A was perhaps the most graceful aircraft flown by NACA-NASA at Edwards. Testing went smoothly as NASA gradually expanded the flight envelope to higher altitudes and higher speeds. During its fourth flight on 13 April 1971, McMurtry took the plane to Mach 0.9 at 36,000 feet. On 26 May, he reached Mach 1.1 at 36,000 feet. The first data-gathering flight came on 18 August, following the installation of special instrumentation, including a network of 250 pressure sensors on the wing's upper surface to locate and measure shock-wave formation. Although the supercritical wing promised great performance improvement at about Mach 0.9, engineers wanted it flown beyond Mach 1 to see if any undesirable trim problems developed there. The early exploratory flights had turned up no surprises, always a pleasant occurrence, and tentative data indicated that the wing's flight performance was close to that expected from tunnel tests at Langley. In fact, the program had already given sufficient encouragement for NASA and the Air Force's

[320] See previous note; *AA–1971*, pp. 65-66; Kelly conversation; F-8 SCW flight log.

Flight Dynamics Laboratory to begin another SCW research program, the military-oriented TACT effort.[321]

Whitcomb envisioned the ideal transonic transport as having both a supercritical wing and transonic area ruling—and, at a later date, winglets—so in May 1972, NASA reworked the F-8's instrumentation and installed new fuselage fairings that gave it pronounced area ruling. It first flew with the fairings on 28 July 1972. By the end of the year, the research utility of the aircraft was nearing an end. Other programs, such as the Blackbirds, demanded funding. Whitcomb was certainly not one to let a program conclude hastily. However, he also recognized that the F-8 effort had reached the point of diminishing returns. Starting in January 1973, the FRC began flying the aircraft on pilot familiarization flights. Ron Gerdes had the honor of making the last flight, on 23 May 1973. As if sensing the end, the plane chose this flight to develop a serious problem: its prime hydraulic system failed. Nevertheless, Gerdes landed the aircraft safely on the lakebed. The plane, as attractive as ever, remains at Dryden to this day.[322]

NASA wasted no time in presenting the results of the SCW F-8 program to the rest of the government and industry in a major symposium at Edwards on 29 February 1972. Richard Whitcomb commented on the good correlation of flight-test and ground-test data. The SCW concept had increased the transonic efficiency of the F-8 by as much as 15 percent, and the tests showed that passenger transports with supercritical wings would increase profits by 2.5 percent over those of conventional aircraft, a total of $78 million per year (in 1974 dollars) for a 280-plane fleet of 200-passenger airliners. Such savings in a fuel-crisis economy were too important to pass by.

Industry rapidly applied the results of supercritical-wing technology to new designs including the Boeing and Douglas YC-14 and YC-15, the Rockwell Sabreliner 65, and the Canadair Challenger. France exploited the concept with an advanced model of the Dassault Falcon business jet. Indeed, foreign interest in employing SCW concepts caused NASA to look closely to determine if NASA-derived data was being used without due consideration of patent law. At NASA Headquarters on 4 June 1974, Administrator James C. Fletcher conferred on Whitcomb the maximum $25,000 prize for invention of the supercritical wing. The National Aeronautic Association awarded him the 1974 Wright Brothers Memorial Trophy.[323]

Before the F-8 had completed its flight program, NASA and Air Force interest in supercritical-wing technology had spawned the Transonic Aircraft Technology (TACT) program. TACT involved modifying a General Dynamics/Convair F-111A to explore how SCW technology could benefit new military aircraft designs. During the 1960s, trying to save the lagging F-111 program, Langley had done a great deal of wind-tunnel work on the F-111. In addition to the transport-type wing tested on the

[321] *AA–1971,* pp. 77, 94, 101, 114, 115, 141, 145, 163, 165, 233; F-8 SCW flight log.

[322] *AA–1971*, pp. 240-241, 252, 256, 270-271, 279, 287, 346, 349; *AA-1972*, pp. 72-73, 94, 104, 109, 113, 123, 155, 163, 184, 186, 193, 274-275, 396; *AA-1973*, pp. 68-69, 143; F-8 SCW flight log; McMurtry interview and conversation.

[323] NASA FRC *X-Press*, 18 Feb. 1972; NASA FRC release 2-72; *AA–1973*, p. 143; *AA–1974*, pp. 44, 112, 173.

NASA's modified F-111A (on loan from the Air Force), the transonic aircraft technology (TACT) testbed, equipped with a Whitcomb supercritical wing, as it descended for a landing after a research flight. (NASA photo ECN-3931)

F-8, Whitcomb had devised a supercritical wing for a transonic maneuvering military aircraft. The F-111 was chosen as the testbed because of its variable-sweep wings. The new wings could be installed easily on the aircraft, with a minimum of other modifications. Indeed, when word of the apparent advantages of supercritical-wing technology reached beyond Hampton, Virginia, the Air Force Flight Dynamics Laboratory began examining the concept. General Dynamics engineers conceived a retrofit program for the entire F-111 fleet, dubbing it the F-111 TIP: Transonic Improvement Program. By mid-1970 General Dynamics had mentioned its retrofit program to the Air Force. The Air Force wanted the F-111 tests as a valuable proof-of-concept evaluation but would not retrofit the entire fleet.

By mid-1971, NASA and General Dynamics had expended over 1,600 hours of wind-tunnel test time on a suitable wing for the F-111. Whitcomb determined its shape, twist, and airfoil coordinates. General Dynamics built the wing, and the Air Force's Flight Dynamics Laboratory furnished the money. On 16 June 1971, NASA and the Air Force signed a joint TACT program agreement to explore the application of supercritical-wing technology to maneuverable military aircraft. The F-111 would be flown at NASA's Flight Research Center, and development of the advanced configuration of the wing would be undertaken by NASA's Ames Research Center. The TACT program, then, affected much of NASA as well as industry and the Air Force. Like the contemporaneous F-8 effort, TACT was far more than just a flight program. Eventually almost as much funding went to support numerous wind-tunnel studies as toward the actual flight program. TACT became primarily a wind-tunnel correlation program, in spite of General Dynamics' earlier hopes that it might spawn an SCW retrofit program for the F-111 fleet. Charles J. Cosenza of the Flight Dynamics Laboratory ran the Air Force's TACT effort. At the Flight Research Center, engineer Weneth D. Painter took over as NASA's TACT project engineer.[324]

The F-111A was an ideal carrier for a supercritical wing. Capable of supersonic speeds above Mach 2, the aircraft had a large volume for fuel and instrumentation. The wings were easily removable. The variable-sweep provision enabled SCW testing over a wide range of wing-sweep angles and aspect ratios. Also the Air Force planned to install pylons under the wings to carry external stores (such as bombs and drop tanks) to evaluate how these shapes interfered with the supercritical flow field.

Fortunately, an F-111A was readily available: the 13th of that first undistinguished and unlucky bunch of F-111A research and development aircraft. NASA signed a loan agreement for the airplane with the Air Force on 3 February 1972, and on 18 February NASA pilot Einar Enevoldson and Air Force pilot Maj. Stu Boyd checked out in the plane. The modified aircraft was ready by the fall of 1973, and on 1 November Enevoldson and Boyd made the first TACT flight, reaching Mach 0.85 at 28,000 feet. During the sixthth flight on 20 March 1974, they exceeded Mach 1. On the twelfth flight, they reached Mach 2.[325]

Thereafter the TACT aircraft flew frequently, with an Air Force/NASA crew. The wing definitely improved the performance of the F-111.[326] At transonic speeds, the wing delayed drag rise and produced twice as much lift as the conventional F-111 wing. The supercritical wing did not impair high-Mach performance either. In fact, the plane spent a great deal of time above Mach 1.3. The external stores tests, with the F-111 carrying drag-inducing multiple bomb shapes on the pylons, came off without a hitch. Fears that the external stores might wipe out any benefits from the supercritical planform proved without foundation. As with the F-8 effort, the correlation between tunnel and flight tests proved close. In November 1975, NASA and the Air Force sent TACT program personnel from Edwards AFB, the Flight Research Center, the Air Force's Flight Dynamics Laboratory, and the Ames and Langley research centers around the country to brief industry and government representatives. The message was simple: TACT, like the earlier F-8 SCW program, had been an unqualified success.

[324] Air Force Systems Command release 188.71; NASA release 71-124; interview of Wen Painter, 8 Aug. 1977; interview of Einar K. Enevoldson, 4 Mar. 1977; General Dynamics *TACT Management Plan and Program Manual*, II, FZP-1124-11, July 1970; General Dynamics, *TACT Technical and Management Proposal*, II, FZP-1260 Revision A, 16 Aug. 1971: General Dynamics, *TACT Manufacturing Plan*, MFGP-595-001, 1 Dec. 1971; General Dynamics, *TACT Aircraft Geometric Characteristics*, MAIR-595-19, 16 Apr. 1973; Air Force-NASA TACT Project Office, *TACT Test Plan*, Oct. 1973; Air Force Flight Dynamics Laboratory, *TACT Data Analysis and Correlation Plan*, July 1975.

[325] Interviews of Painter interview and Enevoldson; Wen Painter, *TACT Project Plan* (NASA FRC, May 1973); F-111 TACT flight chronology, from the files of the TACT program office, DFRC.

[326] Interview of Enevoldson; F-111 TACT flight chronology.

Test results were readily available for the use of industry in developing new and advanced military aircraft.[327]

The F-111 TACT aircraft soon became a workhorse, flying with a variety of aerodynamic experiments, including special shapes to evaluate base drag around the tail, experimental test instrumentation, and equipment destined for use with other airplanes. It was still flying in 1978, the last research flight taking place that September, and it was a highly productive research aircraft. The TACT experience encouraged the Air Force's Flight Dynamics Laboratory to proceed with another research effort: Advanced Fighter Technology Integration (AFTI). Another joint Air Force/NASA effort, it consisted of various "technology sets." AFTI Tech Set II was a direct extension of the TACT program. Like TACT, the AFTI program involved the F-111—in fact, the Flight Dynamics Laboratory examined no fewer than six different F-111 testbed configurations. This second-phase TACT, subsequently called AFTI/F-111, went a step further, with conceptualization of a "mission adaptive wing." This wing would not have the surface irregularities produced by conventional high-lift devices such as flaps and leading edge slats. Instead, an internal mechanism would flex the outer wing skin to produce a high-camber airfoil section for subsonic speeds, a supercritical section for transonic speeds, and a symmetrical section for supersonic speeds—which explains the name "mission adaptive."[328] (See Chapter 12 for coverage of this project in the 1980s.)

By the beginning of the 1980s, a growing number of transonic and high-subsonic aircraft were flying with supercritical-wing planforms. There could be no greater tribute to NASA research and particularly to the work of Richard Whitcomb. A similar situation had happened in the 1950s when his area-rule concept quickly became *de rigueur* for advanced aircraft. It happened yet again, after the full benefits of the Whitcomb winglet were realized, following flight-testing at Dryden of the winglet concept on a modified Air Force KC-135. (See Chapter 12 for details.) During the 1950s, the High-Speed Flight Station had played an important role in validating the area rule. During the 1970s, the center played an equally important role in validating the supercritical wing. Through the efforts of the center, the new, exciting shape of the supercritical wing took its place in the sky.

Radio-Controlled Research

Remotely-controlled aircraft had appeared as early as World War I; by the end of World War II, the major powers were making extensive use of remotely-controlled guided weapons. The technology obviously had great potential. During the 1950s,

[327] See note above; NASA FRC, *TACT: A Briefing to Industry and Government* (Edwards, CA: FRC, Nov. 1975), *passim.*

[328] See Edgar Ulsamer, "On the Threshold of 'Nonclassical' Combat Flying," *Air Force Magazine* (June 1977), pp. 54-58; memo, [DFRC] director of aeronautics (Gene J. Matranga) to [DFRC] director, Aeronautics Update—September 1978, 4 Oct. 1978, in the DFRC Historical Reference Collection.

remotely-controlled Regulus and X-10 missile testbeds were landing on the dry lakebed at Edwards. At the same time, flying radio-controlled model airplanes was becoming a widespread (but expensive) hobby. Electronic advances in the mid-1960s greatly increased the reliability of control systems as tubes gave way to solid-state components. It took the insight of FRC engineer R. Dale Reed to blend this weekend hobby with a professional interest in aeronautical development. The result was a new method of flight-testing and research, using remotely-piloted research vehicles (RPRVs).

The RPRV concept differed appreciably from previous "drone" or remotely-piloted vehicles (RPVs). A limited autopilot had controlled those craft through a restricted number of maneuvers. Some RPVs could be used for military purposes, such as reconnaissance or remotely-controlled strike missions. Drones were used extensively in the Vietnam War and during the 1973 Middle East war. The RPRV, on the other hand, eventually emerged as a study tool capable of versatile applications and of operating in "unexplored engineering territory."[329]

In support of the M2 lifting-body program in the early 1960s, Reed had built a number of little lifting-body shapes and launched them from a twin-engine radio-controlled model called *Mother* that spanned 10.5 feet. By late 1968, *Mother* had made over 120 launch drops. The move to more sophisticated equipment came in late 1968. Following the loss of Mike Adams and the X-15, the FRC installed an X-15-type "eightball" attitude indicator on a TV monitor in the control room. One day, while Reed and test pilot Milt Thompson were monitoring a flight, Reed asked Thompson if he could control an actual research airplane by using the eightball as a reference. Thompson said that he could. Within a month, at a cost of $500, Thompson was flying *Mother* from the ground by reference to the instrument. Next, Reed wanted to see if a pilot could get the same results flying a full-scale research airplane. Because of his interest in lifting-body reentry vehicles, Reed selected the Langley Hyper III configuration, a very slender reentry shape having a flat bottom and flat sides. The Hyper III shape had a lift-to-drag ratio of about 3, and Reed designed it with a fixed wing simulating a "pop-out" wing, such as could be used to improve the low-speed glide ratio of an actual reentry vehicle. Shop personnel built the vehicle at a cost of $6,500. The RPRV weighed about 1,000 pounds, measured 32 feet in length, and spanned 18.5 feet. By December 1969, the FRC was ready for the initial trials. Hyper III was launched from a helicopter at 10,000 feet, glided three miles, reversed course, and glided another three miles to touchdown. As the Hyper III came in for a landing, Thompson transferred control to an experienced model flyer who used standard controls to flare the lifting body and fly it to touchdown. The craft rolled along the lakebed, just like any of the other exotic research aircraft at Edwards.[330]

[329] R. Dale Reed, "RPRV's: The First and Future Flights," *Astronautics & Aeronautics* (Apr. 1974), pp. 26-42.

[330] Ibid. R. Dale Reed, "Can the RC'er Contribute to Aeronautical Research?" *Radio Control Modeler* (Oct. 1968), pp. 28-35; C. M. Plattner, "Lifting Shape for Hypersonic Maneuvers," *Aviation Week & Space Technology* (29 Sept. 1969), pp. 54-58; R. Dale Reed, "Flight Demonstration of a Remote Piloted Concept Utilizing the Unmanned Hyper III Research Vehicle," paper presented at the 26th meeting, Society of Automotive Engineers' Aerospace Vehicle Flight Control Systems Committee, Seattle, WA, 23-25 Sept. 1970.

The Hyper III remotely-piloted research vehicle (RPRV) flew at the Flight Research Center in December 1969. (NASA photo ECN-2304)

Thompson exhibited some surprising reactions during the Hyper III flight: he behaved as if he were in the cockpit of an actual research aircraft. "I was really stimulated emotionally and physically in exactly the same manner that I have been during actual first flights," Thompson recalled afterwards.

> Flying the Hyper III from a ground cockpit was just as dramatic as an actual flight in any of the other [full-scale piloted] vehicles. . . . I, and only I, had to fly the vehicle down to a preselected location for landing. . . . responsibility rather than fear of personal safety is the real emotion driver. I have never come out of a simulator emotionally and physically tired as is often the case after a test flight in a research aircraft. I was emotionally and physically tired after a 3-minute flight of the Hyper III.[331]

Although encouraged by the Hyper III experience, the FRC did not test that shape further since it had a much lower lift-to-drag ratio than predicted. Many other programs—other lifting bodies, the YF-12 Blackbirds, and the SCW F-8—had a more urgent call on the center's time and personnel. Reed and his RPRV team decided to try to control a piloted aircraft by means of a ground pilot, with a backup pilot in the plane. The center selected a Piper PA-30 Twin Comanche, a light, twin-engine airplane already configured as a testbed for general-aviation flight controls. As flown by the FRC, the Twin Comanche had dual controls, one side an electronic fly-by-wire system, the other a conventional system, permitting controls research. That arrangement made

[331] Reed, "Flight Demonstration of a Remote Piloted Concept Utilizing the Unmanned Hyper III Research Vehicle."

The Flight Research Center's Piper PA-30 Twin Comanche, which helped validate the RPRV concept, shown descending to a remotely-controlled landing on Rogers Dry Lake, unassisted by the onboard pilot. (NASA photo ECN-2845)

the aircraft particularly well suited for RPRV research. The FRC already had "downlink" electronics—such as pulsecode modulation telemetering—supported by the center's radar tracking and digital computing equipment. The "uplink" electronics carrying the radio commands to the RPRV came from military research with drones. A forward-pointing television system in the RPRV transmitted images from the aircraft to a ground cockpit, where the operating pilot flew the aircraft by reference to the visual cues. To provide physical cues as well, technicians connected small electronic motors to straps around the pilot's body. During sideslips and stalls, the straps exerted forces on the pilot in proportion to the lateral accelerations being telemetered from the RPRV. The forces on the pilot made it feel natural for the pilot to push rudder pedals to control sideslip.

In October 1971, the FRC began flight trials, with Einar Enevoldson flying the PA-30 from the ground as Tom McMurtry rode as safety pilot. Eventually, Enevoldson flew the airplane unassisted from takeoff through landing, making precise instrument-landing-system approaches, stalls, and stall recoveries.[332]

The next step was applying the RPRV to some meaningful research project. In April 1971, Grant Hansen, assistant secretary of the Air Force for research and development, issued a memorandum calling for a national program to investigate stall and spin phenomena. This area had become critical; many fighter aircraft were being lost in spinning accidents. The Air Force's aeronautical systems division formed

[332] Reed, "RPRV's: The First and Future Flights," pp. 32-33; NASA FRC news release 16-72.

a steering committee that included NASA representatives. It recommended expanding existing programs using radio-controlled free-flight models to evaluate spin entry and post-stall gyrations. The Langley Research Center had made stall-spin studies using small-scale models dropped from helicopters, but the committee recommended using larger models. Scale effects, always significant in model testing, were especially important in stall-spin tests. It was important to verify or refute the Langley tests by examining the results of tests with larger models.

Over the spring and summer of 1971, Reed and other FRC engineers studied the feasibility of stall-spin testing an RPRV model. One advanced Air Force fighter project then under way could benefit from such work—the McDonnell F-15A Eagle, a Mach 2 highly maneuverable dogfighter designed using lessons from air combat over North Vietnam. Maj. Gen. Benjamin Bellis, chief of the F-15 system project office at Wright-Patterson AFB, wanted the Flight Research Center to test an RPRV modeled after the proposed Eagle. In November 1971, the Flight Research Center transmitted a proposal to NASA headquarters for stall-spin testing a 3/8-scale model of the F-15 configuration. OAST's military programs office quickly assented. In April 1972 NASA awarded the McDonnell Douglas Aircraft Corporation a $762,000 contract for the construction of three 3/8-scale F-15 models. NASA placed a variety of contracts with other firms for supporting equipment, including electronic components and parachute recovery equipment.[333]

The first F-15 RPRV arrived at the Flight Research Center on 4 December 1972. The 2,500-pound vehicle, 23.5 feet long, was fabricated from aluminum, hard and soft woods, and fiberglass. It cost a little over $250,000, compared to $6.8 million for a full-scale, F-15 aircraft. McDonnell Douglas built the vehicles, and the Flight Research Center added the avionics, hydraulics, and other subsystems.

The F-15 RPRV was launched from a B-52 mothership at about 50,000 feet, after which an FRC pilot on the ground put the aircraft through its planned research program. Upon reaching 15,000 feet, the RPRV streamed a spin-recovery parachute having a diameter of 12 feet. That chute then extracted two other parachutes, an 18-foot engagement chute and a 79-foot-diameter main chute. As the F-15 RPRV descended, a helicopter snagged the engagement chute with grappling hooks. After a complex series of events, the main chute separated from the F-15, and the helicopter reeled in the RPRV with a winch until the research vehicle was suspended about 15 feet below the helicopter. The helicopter then returned to the Flight Research Center. Should it be impossible to recover the F-15 from a spin or stall, the pilot on the ground could deploy the spin-recovery chute early, initiating the recovery sequence. Similar airborne snatch recoveries were already standard operating procedure for drone aircraft such as the Ryan Firebee. Eventually, NASA landed the F-15 RPRV on the lakebed using skids, just like any other research airplane.[334]

[333] Reed, "RPRV's: The First and Future Flights," pp. 33-37; NASA OAST, "FRC Project Management Report: 3/8 F-15 Remotely Piloted Research Vehicle," internal management document, 15 Jan. 1973.

[334] See note above; interview of Layton.

A 3/8-scale F-15 RPRV being carried to launch altitude under the wing of a B-52 mothership. (NASA photo ECN-3804)

On 12 October 1973, the first F-15 RPRV went aloft under its B-52 mothership for a flawless nine-minute flight, remotely piloted by Einar Enevoldson. He found the task challenging. Researchers monitoring his heart rate found it went from 70-80 beats per minute, normal for a piloted flight test, to 130-140 for the first RPRV flight.[335]

Subsequent testing confirmed the ability of the RPRV to return useful information, encouraging McDonnell Douglas and the Air Force to proceed with piloted spinning trials in the actual F-15 Eagle. The only serious incident in the F-15 RPRV program occurred after pilot Tom McMurtry had remotely flown the aircraft down to parachute deployment, and the helicopter had snagged the parachute. About 3,000 feet above the ground, the lines separated, and the F-15 model was once again in free flight. McMurtry quickly assumed control and guided the plane to an emergency landing in the desert. The plane hit a Joshua tree and a raised roadbank, inflicting some damage, but McMurtry's skill had saved it to fly another day. The incident encouraged those who wished to land the RPRVs using skids. Soon after, NASA did indeed begin landing the F-15 RPRV on the lakebed.[336]

Controversy still surrounds the RPRV concept. Ground researchers have sometimes tended to see the method as a way of relegating piloted flight research to a position of unimportance. More dispassionate champions of the concept recognize that the RPRV complements—but cannot replace—piloted flight research. RPRVs are ideal for use when piloted testing is impossible or unduly dangerous. In some

[335] Reed, "RPRV's: The First and Future Flights," p. 39.

[336] Interview of Layton.

situations, research in RPRVs can be considerably cheaper than research in a piloted aircraft. However, RPRVs cannot match the flexibility of a piloted research airplane. In the words of one Dryden airman, they are "damn limited." Since they cannot fly without a large, complex ground-support system, support costs for RPRV vehicles closely approximate those of piloted aircraft. Nevertheless, the Flight Research Center had proved that the RPRV could make a meaningful contribution to flight research.[337]

Dryden has continued its work with RPRVs since the F-15 RPRV program. In cooperation with Robert Jones and the Ames Research Center, Dryden engineers flew a propeller-driven RPRV having a Jones oblique swingwing. Center engineers have also flown an air-launched Ryan Firebee II in support of advanced RPRV projects. The most ambitious of Dryden's RPRV efforts was the Rockwell/NASA Highly Maneuverable Aircraft Technology (HiMAT). HiMAT is a powered RPRV using an afterburning General Electric J85-21 turbojet engine. It has a wingspan of over 15.6 feet and a length of over 23.5 feet. Designed as a technology demonstrator, the HiMAT aircraft had a sharply sweptwing canard configuration. HiMAT featured a composite structure of glass fibers, graphite composites, and various metals. Following two preliminary study phases, in August 1975, NASA awarded Rockwell International a contract for two HiMAT aircraft, the ultimate cost of which was $17.3 million. After launch from a B-52 mothership, the HiMAT vehicle was flown through a complex series of maneuvers at transonic speeds by a NASA pilot at Dryden's RPRV remote pilot control facility. Then he landed it on Rogers Dry Lake. A chase airplane provided emergency backup control. While HiMAT was Dryden's major RPRV research effort for the first half of the 1980s (see Chapter 12 for further details), Dryden also ran another RPRV project, Mini-Sniffer, beginning in 1975. This was an attempt to develop a propeller-driven RPRV operating on hydrazine monopropellant fuel to altitudes around 80,000 feet to gather air samples from the wakes of high-flying supersonic aircraft. Three Mini-Sniffer configurations have been built. The concept has led to interest by various research facilities, including the Jet Propulsion Laboratory, in using similar vehicles for planetary sampling missions. Such an aircraft could be used on Mars as part of a planetary probe. Clearly, Dryden's RPRV work has been and will continue to be an important aspect of the center's and NASA's research.[338]

Electronic Controls

In the early days of aviation, pilots controlled their aircraft by direct force. They moved a stick or pushed a rudder pedal connected to cables that, in turn, pivoted a control surface. In those days, an on/off switch provided full engine power or none at

[337] Interviews of Manke and Dana.

[338] Interview of Jennifer Baer, 8 Aug. 1977; interview of Reed; Reed, "RPRV's: The First and Future Flights," pp. 39-42; NASA FRC *X-Press*, 15 Jul. 1977; DFRC information sheet "Firebee II," n.d.; Rockwell International, "HiMAT Preliminary Design Review Oral Briefing," 11 May 1976; Jeffrey M. Lenorovitz, "Tooling Begins for Research Vehicles," *Aviation Week & Space Technology* (21 Feb. 1977), pp. 36-39.

all. In time, sets of throttles and fuel mixture controls regulated engine power. As flight speeds rose, control loads increased, eventually reaching a point where pilots could no longer exert sufficient brute strength to control airplanes at high speeds. The next step was hydraulically boosted controls. Control systems now became complex indeed. By the early 1960s, jet aircraft were operating with boosted hydromechanical controls. However, these were very vulnerable to damage. Loss of hydraulic pressure in the control system could spell the end of an airplane even if all other systems functioned smoothly. The necessity for redundant backup systems further complicated aircraft design, while design constraints often minimized the benefit of these backup systems. For example, the Air Force lost many Republic F-105 Thunderchief aircraft over North Vietnam to antiaircraft fire that damaged the plane's hydraulics. "Unfortunately," one "Thud" pilot has written, "a hit that caused loss of one flight control hydraulic system usually got them both."[339] In another case, Grumman lost the first prototype F-14 Tomcat on its maiden flight as a result of hydraulic failure, an accident that delayed the program at a critical time.

Conventional hydraulic-mechanical control systems also imposed design limitations upon aircraft configuration. Designers had to incorporate a degree of inherent stability even if the plane had stability augmentation. During some portions of the flight, the pilot could not be continuously moving the controls. However, the aircraft could not be allowed to go out of control during those moments. Consequently, designers had to use tail surfaces of a certain size and in a certain location; the wing had to be located in a certain position; the fuselage had to be of a certain length. But with electronic controls, in which the pilot's commands go to a computer which sends a signal flashing through a wire to move the controls electronically, all this could be changed. Electronic "fly-by-wire" controls are much less vulnerable to damage than conventional hydromechanical controls, and several wire bundles can be routed through an aircraft with greater flexibility than a maze of pushrods, pulleys, and cables. Furthermore, electronic controls are simpler, smaller, and lighter, advantages that translate directly into improved performance, reliability, payload, and fuel consumption. A fly-by-wire control system could revolutionize the way an airplane looks. No longer do designers have to tailor their configurations a certain way. The electronic controls can provide aircraft stability: a sensing unit can detect any tendency of the aircraft to diverge from its desired flight path and warn the computer to signal corrective control deflection. When the pilot makes a control input, it in fact is a command to the system to relax the stability briefly so that the aircraft moves in the direction the pilot wishes to go. With the electronic control system furnishing stability, designers can reduce the size of some components, such as tail surfaces, or even relocate them. Such changes can reduce the size and weight of aircraft, lessen drag, and permit increases in payload and performance. The primary, and immediate, advantages are in simplicity and maneuverability. "Control-configured vehicles"

[339] Jerry Noel Hoblit, "F-105 Thunderchief" in Robin Higham and Abigail T. Siddall, eds., *Flying Combat Aircraft of the USAAF-USAF* (Ames, IA: Iowa State Univ. Press, 1975), p. 88.

(CCVs) have outstanding maneuvering characteristics. Indeed, with fly-by-wire controls, aircraft can perform such maneuvers as intentional and prolonged yawed flight, which has obvious advantages for military airplanes.

First, however, the fly-by-wire principle had to be proved. Some earlier aircraft had used rudimentary fly-by-wire controls. The Concorde SST used a pseudo fly-by-wire system for primary flight control, but the secondary system was conventionally hydromechanical. At the Flight Research Center, engineers desired a true fly-by-wire testbed having strictly electronic controls. They discussed radically reconfiguring a conventional fighter, such as the Lockheed F-104 or a Vought F-8, with fly-by-wire controls and revised flight-control surfaces, perhaps reducing tail size or incorporating a canard layout. Engineer Melvin Burke was especially interested in flying a digital fly-by-wire testbed.

Considering how important the technology has subsequently become, NASA headquarters expressed little interest in the idea until Neil Armstrong became NASA's deputy associate administrator for aeronautics within the Office of Advanced Research and Technology. During the Apollo program, he had become acquainted with fly-by-wire technology at the controls of the Lunar Module. That vehicle had a digital computer and inertial measuring unit. When Armstrong moved his controls, the computer sent signals to reaction controls that maneuvered the vehicle. Armstrong believed this off-the-shelf system could be readily applied to a testbed airplane, and he supported Burke's project. With OART's approval, the Flight Research Center acquired a Navy Vought F-8C Crusader, disconnected its mechanical flight-control system, including all cables, push rods, and bell cranks, and replaced it with the Apollo-derived digital flight computer and inertial sensing unit. Center engineers and technicians routed sets of wire bundles from the pilot's control stick through linear variable differential transformers to the computer and then through an electrohydraulic servo-actuator and the F-8's hydraulic actuator to the control surfaces. This marked the beginning of FRC's F-8 Digital Fly-by-Wire (DFBW) flight research program.[340] Massachusetts Institute of Technology's Charles Stark Draper Laboratory supported the FRC's effort by reprogramming the Raytheon computer from the Lunar Module. Sperry's flight systems division supplied a backup fly-by-wire system for the aircraft.

On 25 May 1972, FRC research pilot Gary Krier completed the first flight of the F-8 DFBW testbed, which was also the first flight of an airplane completely dependent upon an electronic control system. Using off-the-shelf equipment had enabled NASA to make that flight at least two years earlier than would have been the case starting from scratch. NASA awarded the DFBW project team its Group Achievement Award during headquarters ceremonies in November 1972. By early 1973, after 15 DFBW flights without incident, Krier testified before the House Committee on Science and Astronautics on the benefits the program had already demonstrated. Clearly fly-by-wire-equipped transport aircraft could fly with greater smoothness in turbulence. The

[340] Interview of Krier; NASA FRC news release 8-72; J. P. Sutherland, "Fly-by-Wire Control Systems," Air Force Flight Dynamics Laboratory, 10 Aug. 1967; comments by Gary Krier on the original chapter, 1 Feb. 2000.

Piloted by Gary Krier, the NASA F-8 digital fly-by-wire (DFBW) research testbed cruises on a research mission above the Mojave Desert. (NASA photo ECN-3276)

nearly instantaneous sensing of motion changes, combined with an immediate computer-signaled corrective control response, would rapidly damp any turbulence-induced aircraft motions. "A much larger improvement in performance could be gained by starting from scratch with FBW," Krier testified. "We have been refining aircraft for years now, and the FBW/CCV combination gives us a chance to make a quantum jump in aircraft performance."[341]

Like all trial systems, the F-8's DFBW installation did have some operational quirks. The electronic interface on the Apollo computer was too coarse for the precise pilot stick inputs required to fly the plane. The computer changed control-surface positions in a series of steps, like the small but abrupt movements of a watch's second hand. The pilot felt this as a mild but unpleasant series of nudges, especially when using the all-moving horizontal stabilizer for pitch control. At the FRC's request, the Draper Laboratory changed the computer software, with beneficial results to the handling qualities of the plane. The F-8 flew 42 times without incident, and it was never necessary to resort to the plane's emergency backup flight-control system. Before finishing the test program, the DFBW team tested on the F-8 a prototype version of the electronic side-stick planned for the General Dynamics F-16 fighter, including formation flight and landings. The results lent encouragement to the practicality of using such a stick on the F-16 itself.

The first phase of the F-8 program had only shown that DFBW control was feasible, not that it was practical. The system used much special-purpose hardware

[341] *AA–1973*, pp. 74-75.

and, although it was extremely reliable, it could not operate if the digital computer failed. In a joint program with the Langley Research Center, Dryden received funding to develop and flight-test an advanced redundant digital fly-by-wire system in place of the modified Apollo system. This triplex DFBW system used general-purpose digital computers and would be able to sustain several system failures and still operate. It was flown in August 1976, with a ride-smoothing system, maneuver-driven flaps,[342] and an angle-of-attack limiter that were typical of the characteristics expected on future vehicles employing DFBW control. The F-8 system also demonstrated "fault tolerance" by continuing normal operation after certain computer failures. After the initial development flights, the F-8 was used to test Shuttle computer software and to support the development of the flight-control system of the Shuttle Orbiter.[343] Flying until 1985, it also tested other concepts and helped resolve a pilot-induced oscillation (PIO) on the Shuttle prototype *Enterprise*, resulting in a PIO suppression filter for the Space Shuttle's flight-control system. As at least a partial result of the F-8 DFBW project, the F-16, F-18, Airbus A320, and the Boeing 777 fly with fly-by-wire technology, as do many other aircraft.[344]

Dryden's Digital Fly-by-Wire flight-research program was only one of the electronic control programs that will continue to influence the development of this new technology. Another was the center's Integrated Propulsion Control System (IPCS) evaluated on an Air Force F-111E airplane. This program, run from March 1973 through February 1976, was a cooperative effort of NASA Lewis and the Flight Research Center, the Air Force's Flight Propulsion Laboratory, Boeing, Honeywell, and Pratt & Whitney. In essence, it accomplished for the propulsion system of an airplane what fly-by-wire controls did for flight control. Numerous factors affect engine performance, including throttle position, inlet position for variable-geometry inlets, fuel flow rates, and even the maneuvers that an aircraft is performing at any particular time. As with mechanical aerodynamic controls, the hydromechanical controls used in engine operation grew increasingly complex. Propulsion experts at NASA's Lewis Research Center recognized that future aircraft might demand propulsion control systems capable of controlling a number of variables with much greater accuracy and speed. Digital electronic controls might well provide the answer.

The Air Force's Flight Propulsion Laboratory at Wright-Patterson AFB was willing to fund an experimental effort using a suitable airplane. A twin-engine airplane could be configured so that one engine was electronically controlled. The other engine could remain hydromechanically controlled for flight safety and to provide a comparison with the test engine. One aircraft immediately came to mind, the General Dynamics F-111. The F-111 was a large, two-seat twin-engine aircraft with a complex

[342] In such a system, sensors detect vehicle maneuvers, triggering appropriate flap movement to enhance airplane performance during the maneuver.

[343] *AA–1973*, pp. 196, 266; interview of Krier; information from Kenneth J. Szalai.

[344] James E. Tomayko, *Computers Take Flight: A History of NASA's Pioneering Digital Fly-By-Wire Project* (Washington, DC: NASA SP-2000-4224, 2000), pp. 103-135.

propulsion system. It had a variable position inlet and afterburning fanjet engines, and an internal weapons bay that researchers could use to house the necessary electronic controls. The Air Force had an F-111 available, the first prototype of the General Dynamics F-111E series. Lewis and the Air Force selected Boeing as prime contractor to develop the system, with Honeywell and Pratt & Whitney as subcontractors. NASA awarded the contracts for the Integrated Propulsion Control System program in March 1973.

The program could have been run at Lewis. However, for various reasons, including flight safety, NASA and the Air Force decided to fly the F-111E IPCS testbed from the Flight Research Center at Edwards. Once the FRC became involved, center personnel did far more than just fly and maintain the airplane. Indeed, FRC engineers and pilots initially resented what they saw as an effort by various distant parties to dictate what was to be done, how, and when. "It took a year before we really developed a good working relationship with everybody," one Dryden participant recalled, "so that they trusted us, and we trusted them. And they realized we weren't just being hard to get along with when we wanted changes or said we had a problem. They started believing us."[345] After this initial wariness, the program moved along smoothly.

The Flight Research Center received the F-111E in mid-1974 and before it was modified embarked on a series of 13 flights for acquiring baseline data for comparing with the results of the later IPCS tests. Installation of the IPCS began in March 1975. The system consisted of an instrumentation package, power supply, digital computer, and interface equipment installed in the fuselage weapons bay. The hydromechanical inlet and afterburner controls were replaced by new electronic controls.

Two software programs supported the IPCS evaluation. One of these was a digital representation of a TF30-P-9 afterburning turbofan engine used for assessing the ability of the IPCS system to duplicate the hydromechanical control functions. The other, called the IPCS control mode, integrated the inlet and engine control functions into one operation, exploring the new control concept. All software and the related IPCS control hardware were rigorously bench-tested, installed on a Pratt & Whitney TF30-P-9 engine and run on a test stand. The modified engine was installed in the altitude test chamber of NASA's Lewis Research Center, where engineers ran the engine under planned flight conditions. NASA was especially interested in the operation of the IPCS on high-altitude, low-Mach flights (typically Mach 0.9 at 45,000 feet or Mach 1.4 at 50,000 feet), and flights above Mach 1.9, where the interactions of variable inlet and engine were of critical importance.

NASA had hoped to use the actual IPCS-modified engine tested at Lewis, but this did not prove possible. Instead, another TF30-P-9 was installed. The IPCS

[345] Quote from confidential source; for information on the IPCS system, see Jennifer L. Baer, Jon K. Holzman, and Frank W. Burcham, Jr., "Procedures Used in Flight Tests of an Integrated Propulsion Control System on an F-IIIE Airplane," paper presented to the Aerospace Engineering and Manufacturing Meeting, Society of Automotive Engineers, San Diego, CA, 29 Nov. - 2 Dec. 1976 (hereafter referred to as IPCS procedures); interviews of Wen Painter, Gary Krier, and Einar Enevoldson; planned IPCS schedule as prepared by the research and engineering division of The Boeing Company; NASA FRC IPCS Milestone Schedule, 11 Mar. 1976.

controlled only the F-111E's left engine. Hydromechanical control was available over the left engine for emergency use, and the right engine retained its own hydromechanical system. As a precaution, however, in the event of failure of the manually controlled engine during takeoff and the possibility of simultaneous problems with the experimental IPCS, all takeoffs were made toward Rogers Dry Lake, where an emergency landing could be made.[346]

The F-111E completed its first IPCS flight on 4 September 1975, piloted by NASA's Gary Krier and the Air Force's Stan Boyd.[347] It completed an additional 14 IPCS investigations before the program concluded, making its last IPCS flight on 27 February 1976. NASA returned the F-111E to the Air Force. Restored to its original non-IPCS configuration, it served as a chase aircraft for the B-1 strategic bomber.

The IPCS flights demonstrated that the system worked well. The test crews used rapid throttle manipulation, abrupt aircraft maneuvers such as high-angle-of-attack turns and sideslips, and various inlet positions to evaluate the performance of the IPCS. Because it was not an ideal, best-of-all-possible-worlds system, the gains realized were not spectacular. But at its worst, the IPCS system never performed less efficiently than the hydromechanical system. This was significant, for it indicated that future IPCS technology could be expected to produce major benefits. There were other less visible advantages. Engineers compensated for deficiencies in the hardware used on the IPCS by changing software routines. The project team noted:

> This allowed temporary corrections to be made and verified without the need for extensive design modifications and hardware testing that could have affected the flight scheduling. With this flexibility, the testing and optimization of propulsion systems can be completed without the major hardware modifications that accompany development in hydromechanical systems.[348]

The conclusion of the IPCS program was influenced as much by monetary considerations as by the fact that the system had proved its potential value. Dryden's interest in electronic controls has continued, however, with the similar but more advanced digital electronic engine controls (DEEC) research program using one of the center's F-15s. The advanced aircraft of the 1990s flew with such Dryden-pioneered developments as digital fly-by-wire flight controls and advanced forms of the IPCS. As with the supersonic breakthrough and the dawn of hypersonic flight, Dryden's work on electronic controls had continuing impact in the last two decades of the 20th century. (See Chapters 12 and 13 for further treatment of the Digital Electronic Engine Control (DEEC) program and follow-on work with digital engine controls.)

[346] IPCS procedures; interviews of Painter and Enevoldson.

[347] Not to be confused with Stu Boyd, another Air Force test pilot who by this time had left Edwards.

[348] IPCS procedures.

New Concerns in Aviation Safety

The old NACA did relatively little work in the field of aviation safety, although some of its aerodynamics research had a serendipitous effect on safety. The High-Speed Flight Station undertook virtually no aviation safety projects related to air transportation, the closest being the KC-135 studies supporting the introduction of the 707-generation jetliners into service. Lewis Laboratory had deliberately destroyed surplus military airplanes to study how crash fires propagated. But for the most part, the NACA had left aviation safety to the Civil Aeronautics Administration (the forerunner of the Federal Aviation Agency, later the Federal Aviation Administration), and such organizations as the Flight Safety Foundation and the Cornell-Guggenheim Aviation Safety Center.

This changed in the 1960s and early 1970s. The disconcerting number of general-aviation stall-spin accidents caused NASA to undertake special studies of the spinning characteristics of such aircraft. The agency complemented this work with other studies on the handling qualities of private aircraft. Much of this work was done at Langley, but the Flight Research Center ran a number of flight evaluations on general-aviation airplanes from 1964 through 1966, following these with tests of the center's workhorse Piper PA-30 Twin Comanche. During one test flight of the PA-30, center research pilot Fred Haise encountered severe flutter of the craft's horizontal tail while well within the aircraft's operational limits. Fortunately, this dangerous situation did not cause loss of the tail, and Haise landed safely. A film taken from a chase plane shows the horizontal tail twisting through an alarming arc for what seems an incredibly long time, evidence that the unexpected dangers in flight-testing are not limited to high-performance jets and rocket planes.

Although general aviation was a major research concern, two other problems drew particular attention: wake vortex and clear-air turbulence. In 1907, British aerodynamicist F. W. Lanchester postulated the concept of the tip vortex, a "horizontal tornado," as it were, formed by the flow field around a wing.[349] This whirling column streams around the wing tip and trails in a wake behind the aircraft. Sometimes, under the proper conditions of humidity and temperature, the vortex can be seen. It is easily demonstrable in a wind tunnel or water tank, using injected smoke or dyes. As seen from behind the aircraft, one vortex streams from the right wing tip, rotating counter-clockwise. The vortex from the left wing tip rotates clockwise. These turbulent vortices trailing behind an airplane can affect other aircraft that pass through them. The magnitude of the vortices is directly related to the size and weight of the airplane that generates them: the wake vortex of a light plane such as a Cessna 150 is negligible, while that of a 747 can exceed 150 miles per hour in rotational velocity and can persist for a distance of 20 miles. The vortex of a large transport can easily upset a much smaller aircraft, possibly inducing structural failure or, more likely, throwing

[349] Theodore von Kàrmàn, *Aerodynamics: Selected Topics in Light of Their Historical Development* (Ithaca, NY: Cornell Univ. Press, 1954), pp. 48-50.

the aircraft out of control. If this occurs close to the ground—during a climb-out after takeoff or during a landing approach—the plane might crash. Indeed, many aircraft have been lost in such accidents.

The problems engendered by wake vortices first became a serious concern following the introduction of large jetliners. When the wide-body jumbo jets (the Boeing 747, McDonnell Douglas DC-10, and Lockheed L-1011) entered service, wake vortices became a major hazard. These aircraft trailed vortices powerful enough to roll business jets and even other airliners. Further, their vortices could persist even at high altitudes. In response, the FAA increased minimum separation distances for airplanes from three miles for a small business jet following a wide-body jumbo jet to six miles. Even a wide-body could not follow closer than four miles behind another wide-body aircraft. These separation distances automatically reduced the number of aircraft that could land at an airport in a given time. The FAA undertook the development of sensors that could detect the presence of hazardous vortices in the approach corridor of an airport.[350]

Another method was to attempt to reduce the magnitude of tip vortices. Here is where the Flight Research Center became involved. NASA became interested in vortex research both from the safety aspect and as a matter of aerodynamics. A wing-tip vortex seriously reduces efficiency, causing drag to rise with a consequent penalty in fuel consumption and performance. Minimizing the wake could greatly increase the aerodynamic efficiency of the plane and improve its operating economics, always a vital concern in air transport. This desire for efficiency prompted Richard Whitcomb at Langley to develop the winglet concept: small, nearly vertical wing-like surfaces mounted on the wing tips of an airplane. These winglets reduced the induced drag by four to five percent, offering fuel savings for a 707-class transport of about seven percent. Dryden subsequently tested a Boeing KC-135 equipped with winglets in a proof-of-concept demonstration. Ames engineers experimented with small fins mounted above or below a wing. These fins would generate "good" vortices to break up and disperse the dangerous ones. Langley engineers experimented with a nearer-term solution, deploying an aircraft's spoilers and speed brakes to minimize wake vortex formation. Langley tunnel-tested a 3/100-scale model of a 747. Following up on the Langley work, the FRC flew a 747 on wake-vortex-alleviation studies.[351]

The Flight Research Center had studied wake vortices with a Boeing 727 in November 1973, equipping the plane with smoke generators to trace the patterns and following it with instrumented PA-30 and F-104 chase aircraft to measure the force and effects.[352] The 727 was a small three-engine jetliner, however, not comparable

[350] Philip J. Klass, "Wake Vortex Sensing Efforts Advance," *Aviation Week & Space Technology* (25 Apr. 1977), pp. 92-99.

[351] Ibid.

[352] David Scott, "Today's Research–Tomorrow's Aircraft," Pt. 2, *Aircraft* (Aug. 1974), pp. 30-32.

The Flight Research Center's Aero-Commander, used as a support aircraft and for a variety of general aviation studies, including this study of airflow using tufts on the fuselage. (NASA photo ECN-2119)

even to the 707, let alone to jumbo wide-bodies such as the 747. Fortunately, NASA bought a Boeing 747-100 jetliner from American Airlines for use as the carrier aircraft during the Space Shuttle's approach and landing tests. The FRC petitioned NASA headquarters for use of this aircraft, which had been assigned to the Johnson Space Center. On 16 August 1974, headquarters assented, and the 747 made some 30 flights in a wake vortex research program. Test crews varied the positions of the spoilers and used various spoiler segments in an attempt to determine the optimum method of alleviating wake vortices. Chase aircraft, including a Gates Learjet and a Cessna T-37 trainer (representative of business jets and smaller aircraft) probed the vortices to measure their strength. The results were surprising.[353]

During one test when the 747 crew did not attempt to alleviate the wake vortices by spoiler operation, the T-37 entered a vortex about four miles behind the 747, did two inverted snap-rolls, and developed a roll rate of 200 degrees per second. During another flight, the disturbed vortex flow caused one of the T-37's engines to flame out. The T-37 pilot believed that at least a 10-mile separation was desirable between the T-37 and the 747 when the 747 was in landing configuration, with landing gear and flaps down. Spoiler operation, however, somewhat improved the situation. With

[353] Letter, Robert F. Thompson to FRC Deputy Director, 16 Aug. 1974; "NASA's 747 Shuttle Carrier Aircraft Not New to Heavyweight Ranks," Boeing news release A-0919 (n.d.).

NASA's Boeing 747, used in wake vortex studies, followed by a Gates Learjet (viewer's left) and a Cessna T-37 that penetrated the 747's wake to analyze the turbulence and its strength. (NASA photo ECN-4243)

two spoilers on the outer panels of each wing extended, the vortices were greatly reduced and the T-37 could safely fly three miles behind the larger aircraft.[354]

Although the FRC's 747 wake vortex studies clearly indicated that use of spoilers could reduce the severity of wake vortices, the spoilers were not effective in attenuating vortices when the 747 was in the landing configuration. After the 747 was reassigned to its primary mission—carrying the Space Shuttle Orbiter—the vortex alleviation studies continued, under the direction of program manager Russ Barber. In July 1977, the FRC began a brief series of tests on a Lockheed L-1011 TriStar wide-body to determine if the spoiler fix that worked so well on the 747 could be applied to other wide-body aircraft as well. The test showed that while the spoilers on the TriStar could reduce wake vortices, they were not as effective in doing so as the spoilers on the 747. By the late 1970s, however, fuel price increases from the Organization of Petroleum Exporting Countries (OPEC) sharply reduced projections of high traffic levels at airports and thus decreased interest in reducing separations between airplanes. As a consequence, the work the FRC did in showing that vortices from large airplanes could be attenuated never resulted in design changes on large commercial aircraft.[355]

[354] Klass, "Wake Vortex Sensing Efforts Advance," p. 99.

[355] NASA DFRC *X-Press*, 29 July 1977; interview of Russ Barber by J. D. Hunley, 21 Dec. 1998. See Chapter 12 of this book for a discussion of the OPEC price increase.

Two other research areas for Dryden in the 1970s were clear-air turbulence and pollution of the upper atmosphere. While atmospheric pollution is strictly an environmental problem—and a most serious one—clear-air turbulence can endanger an aircraft by exposing it to sudden and extreme gust loadings possibly exceeding its structural strength. Private researchers, the FAA, and NASA have always had a major interest in turbulence. One of the old NACA's greatest accomplishments was its work on gust-induced flight loads, work that predated World War II. In the late 1950s and 1960s, NASA flight researchers undertook projects on high-altitude clear-air turbulence using Lockheed U-2 aircraft. As concern about pollution of the upper atmosphere became more widespread, NASA sponsored U-2 and Martin WB-57F high-altitude sampling flights, as well as the Dryden Mini-Sniffer RPRV program. Gustiness at high altitudes had caused annoying difficulties during some of the Flight Research Center's work with the XB-70A and YF-12 Blackbird. More seriously, clear-air turbulence had given some commercial aircraft a rough flight, injuring some passengers not using their seat belts and occasionally leading to structural failure.

In response to this interest in atmospheric conditions, Langley and Flight Research Center engineers mapped out joint research to provide "a limited amount of highly accurate measurements associated with mountain waves, jet streams, convective turbulence, and clear-air turbulence near thunderstorms."[356] At the first Langley/FRC meeting on 3-4 June 1969, planners agreed to use a NASA-owned Martin B-57B airplane, a modified medium bomber. In due course, it appeared on FRC's flightline. Difficulties with the data-acquisition system delayed the planned flights,[357] but in time the B-57B supported three atmospheric science programs: measurement of atmospheric turbulence, sponsored by Langley; aerosol-sampling, sponsored by the University of Wyoming; and detection of clear-air turbulence, sponsored by the Department of Transportation.[358] Combining data from these flights with that from many other sources, scientists began to develop a better understanding of the nature—and fragility—of the upper atmosphere.

FRC and the New Generation of Military Aircraft

Because of commitments to the X-15 and other advanced research programs, the FRC lacked the manpower for participating in new military programs such as the F-4 Phantom. Paul Bikle would have preferred to continue the practices of the 1950s, getting involved in as many military-related programs as possible. But the easy days

[356] Memo, J. M. Groen to FRC director, 17 Jan. 1972.

[357] Meanwhile the B-57B was put to work in proof-of-concept testing of the deceleration parachute to be used by the Viking Mars landers. The tests were conducted at the Joint Parachute Test Facility at nearby El Centro. For example, see J. M. Groen, Flight Report, Viking Test #7, 28 April 1972.

[358] See memo, J. M. Groen to De E. Beeler, 3 June 1971; memo, Groen to FRC director, 17 Jan. 1972; letter, James S. Martin to Milton O. Thompson, 4 Jan. 1971.

Two of NASA's Lockheed F-104N Starfighters, used for chase and flight research. (NASA photo ECN-2493)

of the Cook-Craigie procurement plan were long passed, and stronger institutional ties worked to prevent close NASA/Air Force cooperation in flight-testing new military aircraft. Under new procurement policies, if NASA flew an aircraft on loan from one of the military services, it had to pay its operational costs. Bikle nevertheless sought cooperation between the military and the FRC, and because of personal ties dating from his duties as technical director of the AFFTC, he had a great deal of success. Bikle was thwarted in his efforts to acquire military aircraft for the FRC less by the military than by NASA headquarters, which refused several requests for budgetary reasons.

Aside from research, Bikle needed newer aircraft at the FRC so that his pilots could stay current with the latest technology. The FRC acquired three F-104N Starfighters, specially ordered from Lockheed in 1963. Bikle also got the Northrop two-seat T-38 supersonic trainer. This useful and reliable little jet could perform a variety of mission-support chores as well as simulate lifting-body landing approaches. Bikle's managerial philosophy stressed diversity, which helped save the Flight Research Center from those who sought to shut it down during the 1960s.[359]

Following the creation of NASA, the FRC was involved in programs with various military aircraft: the Lockheed F-104A Starfighter, the McDonnell F-4A Phantom II, the General Dynamics F-111A, the Lockheed T-33 Shooting Star, the Northrop F-5A Freedom Fighter, the Vought F-8C Crusader, the Northrop YF-17 Cobra, and the

[359] Interview of Bikle.

McDonnell Douglas F-15A Eagle. NASA had other programs that were military-related, such as the Blackbirds, XB-70A, and the TACT F-111. The FRC also acquired airplanes from abandoned projects, such as the Northrop A-9A, but did not run programs on them. Clearly, then, if the FRC's research using modified military aircraft was not as extensive as it had been in the 1950s, such activity remained substantial—certainly as much as the center could support during a space-conscious era.

During the 1960s and 1970s, NASA continued to fly the workhorse F-104s as testbeds. Aside from using the Starfighters for X-15 mission support, for chase and in support of the lifting-body effort, NASA used them in a number of short programs such as base-drag measurements, sonic-boom measurements in support of Langley research, and tests of "ballute" (balloon-parachute) deceleration devices. In the early 1960s, the FRC flew a brief military-inspired program to determine whether an airplane's sonic boom could be directed; if so, it could possibly be used as a weapon of sorts, or at least an annoyance. In December 1965, the FRC received an ex-Navy McDonnell F-4A Phantom II fighter. It flew briefly in this project before a wing fuel tank burst, producing a large hole in the wing. The pilot landed safely.

The FRC received two early General Dynamics F-111A airplanes. As a result of a poorly thought-out development specification, both the Navy and the Air Force had become committed, much against their will, to a civilian-inspired Tactical Fighter Experimental (TFX) program. This program called for developing a single aircraft—the F-111—to fulfill a Navy fleet-defense interceptor requirement and an Air Force supersonic strike aircraft requirement. This goal was impossible to achieve, especially since planners placed priority upon the Air Force requirement and then tried to tailor this heavy landplane to the constraints of carrier-based naval operations. The naval aircraft, the F-111B, was never placed in production. The Air Force aircraft, which was produced in a variety of models, including the F-111A, F-111D, F-111E, and F-111F, as well as an FB-111A strategic bomber version, had numerous problems. Only the F-111F actually fulfilled the original TFX design specification. The failure was less the fault of General Dynamics than of the civilian planners in the Pentagon whose "cost-effective" inclinations ironically produced the major aeronautical fiasco of the 1960s—and a costly one at that.[360]

The Flight Research Center's F-111A program was the only program of the 1960s that closely followed the earlier pattern of using NACA-NASA flight-test specialists to iron out technical problems with a major new weapon system. The early F-111As had extremely bad engine problems, suffering from compressor surge and stalls. In January 1967, the Air Force sent the sixth production F-111A to the FRC for testing. The plane did not make a favorable impression there. One FRC pilot stated:

> The early ones were rats. . . . It was like flying in a three-dimensional maze. You couldn't sweep the wings beyond a certain point, you couldn't exceed

[360] One FRC wag, noted for his pen-and-ink skills, drew a variety of F-111 "growth" proposals, including a cargo C-111, a helicopter H-111, and an X-111 research airplane. His cartoon was printed in *Aviation Week & Space Technology*, fortunately without credit.

A General Dynamics F-111A Aardvark prototype that the FRC flew in the late 1960s in support of military-testing of the aircraft. (NASA photo ECN-2092)

> so much [angle of attack], you couldn't turn too tight, you couldn't have so much sideslip . . . . It was terrible.[361]

NASA pilots and engineers wrung out the airplane in an attempt to solve its problems, studying the engine inlet dynamics to determine the nature of inlet pressure fluctuations that led to compressor surge and stall. Eventually, as a result of NASA, Air Force, and General Dynamics studies, the engine problems were solved by a major inlet redesign. The FRC's work had been crucial to this effort. The center's second F-111A, the twelfth built, arrived in April 1969 and was flown in a handling-qualities-investigation program. Both aircraft were retired to the boneyard in 1971. The center's experience with its later F-111s (the TACT and IPCS airplanes) was far more pleasant.[362]

The FRC flew numerous brief programs using military airplanes. A Lockheed T-33 Shooting Star jet trainer was flown on a human-factors study to evaluate the effects of visibility restrictions upon a pilot's performance during landing, since many advanced airplanes would have very restricted visibility forward and laterally during landing approach. The center undertook a comprehensive study of high-lift flaps as aids to transonic maneuverability with a series of tests on F-104, Northrop F-5A Freedom Fighter, and Vought F-8C Crusader aircraft during 1970 and 1971. Wind-

[361] Confidential source.

[362] For F-111 information, see the following issues of the NASA FRC *X-Press*: 27 Jan, 1967, 19 Apr. 1968, Apr. 1969, and v. 14 no. 7 (n.d.).

One of the Flight Research Center's F-15A Eagles on an early research flight. (NASA photo ECN 4978)

tunnel results simply were not reliable for this purpose, and the flight-research data would be useful for developing new military aircraft. The FRC's work in this area led to the derivation, by the Department of Defense, of "agility" criteria for fighter turn rate, buffet, maximum lift, and handling qualities. This paid off in the development of a whole new generation of fighter aircraft: the McDonnell Douglas F-15A Eagle, the General Dynamics F-16A, and the Northrop YF-17 Cobra/F-18 Hornet. The center also used the T-33 for evaluating a self-contained liquid-cooled flight garment providing the pilot with heating, cooling, and pressure protection.[363]

The DFRC's most recent exposure to new military aircraft came with the McDonnell Douglas F-15A Eagle and the Northrop YF-17 and F-18. Involvement with the F-15 program came out of earlier work with the F-15 RPRV model and a desire to have a representative of the latest highly maneuverable fighter aircraft. The F-15 Eagle represented a turning point in Air Force doctrine, a return to an airplane designed primarily for agility and air-to-air combat—the first since the old F-86 Sabre.

The opportunity to work on the Eagle came at a time when some NASA engineers and pilots within the agency were grumbling that a return to the military-testing policies of the 1950s was long overdue. Dryden secured approval from NASA headquarters for requesting the transfer of two aircraft from the Air Force's F-15 Joint Test Force. The Joint Task Force's activities were winding down—soon some of its aircraft were refurbished and shipped to Israel—and two specialized prototypes were available:

[363] Harold Walker, "High-Performance Aircraft Research," DFRC, n.d.; "'Short' Program Summary," n.d., transmitted to the author by Milton O. Thompson; Thomas R. Sisk, "Transonic Lift Augmentation Devices," *Basic Research Review* (Aug. 1971); NASA, *Flight Research Center* (Edwards, CA: FRC, 1971), pp. 19, 31.

the second, which had been used for propulsion tests, and the eighth, which had been used for spin-testing. NASA acquired both aircraft on indefinite loan from the Air Force.

The Flight Research Center has flown the two aircraft on a variety of research missions, two of which have been a major propulsion and performance flight-test program and research into high angle-of-attack stall-spin phenomena. At the request of the Air Force, the Dryden F-15 test team also investigated discrepancies between predicted and measured drag values. In flight, the F-15 had greater base drag—drag around the aft end of the plane—than tunnel tests had predicted. This problem has afflicted a range of aircraft—a notable example being the X-15—and remains an area of concern to aerodynamicists. Data from the F-15 full-scale flight tests were also used to validate data taken during testing of the 3/8-scale F-15 RPRV drop model. In another effort to improve wind-tunnel prediction techniques, a small 10-degree cone was installed on the nose boom of one of F-15s. The shape had been tested in 23 wind tunnels, and the data taken in flight up to Mach 2 speeds were compared with wind-tunnel data, furnishing an assessment of the airflow quality and turbulence levels generated in the tunnels.[364]

In the early 1970s, the Air Force pressed for development of a new generation of lightweight fighters—single-seat jet aircraft optimized for agility and air-combat maneuvering, with high thrust-to-weight ratios (above 1 to 1) and good acceleration. Out of this interest came the so-called "Lightweight Fighter" program, which involved construction of two technology demonstrators, the single-engine General Dynamics YF-16 and the twin-engine Northrop YF-17 Cobra. Midway down the development path the stakes changed.What had been a technology demonstration became a Department of Defense competition for a new fighter for both the Air Force and the Navy, and for allied nations as well. Eventually, the YF-16 was declared superior. The Air Force adopted a derivative of it, the production F-16A. The Navy, unhappy with the outcome, proceeded independently with a derivative of the YF-17 Cobra, which evolved into the Navy's Northrop F-18 Hornet fighter program. After sitting briefly in storage, the two YF-17 prototypes flew again, this time as development aircraft for the proposed F-18. At the request of the Navy, Dryden flew the first YF-17 for base drag studies and to evaluate the maneuvering capability and limitations of the aircraft. NASA pilots—all of whom got at least one flight in the plane—and engineers examined the YF-17's buffet, stability-and-control, handling-qualities, and acceleration characteristics.

The YF-17 shocked many of the FRC's pilots, trained on earlier combat aircraft. "I was astounded," one center pilot recalled. "That airplane really is a generation ahead of anything else. It's got twice the performance of current-day airplanes like the F-4 and some of the others. It'll climb twice as fast, and it'll burn half the fuel—just phenomenal."[365] Some at the FRC watched wistfully as the shapely little YF-17

[364] Interview of Kolf; NASA DFRC *X-Press*, 2 June 1978. For further details about this effort that were not readily available at the time this chapter was originally written, see Chapter 12 of this book.

[365] Interview of Krier.

A Northrop YF-17 Lightweight Fighter prototype that Dryden pilots evaluated in 1976 to improve wind-tunnel predictions for future fighter aircraft. (NASA photo EC76-5270)

departed, on its way to help out Northrop and the Navy with the F-18. The greatest shock came when many within NASA realized that the YF-17 typified industry's growing tendency to develop aircraft independent of NASA research. "Now the tail's wagging the dog," one engineer stated. "Industry goes out and builds an airplane like the F-16 and the F-17 . . . . NASA says, 'Let's take a look at it, let's assess the thing.'"[366]

This problem was succinctly summarized in a memo from Milt Thompson, a senior engineering administrator and former X-15 pilot, to Dryden Director David Scott in January 1976, before the YF-17 arrived. The administrator argued for NASA to acquire an F-16, citing the record of the NACA in the 1950s. "We must, however, recognize the fact that we may not have as much to contribute these days as we had in the past." After discussing the NACA station's work on the Century series in the 1950s, he went on:

> NACA was in that . . . period an acknowledged leader in the fields of aerodynamics, stability and control, aerodynamic loads, buffet, flutter, propulsion performance, and possibly others. NASA no longer enjoys that esteemed position in the aeronautics world, largely due to default. NASA was actually unable to provide any substantial guidance or assistance to the designer of the YF-12 and SR-71. Thus, NASA is now in an extremely weak

[366] Confidential source.

> position to bargain for participation in any new aircraft program [, but] NASA should be flight testing new aircraft if for no other reason than to keep abreast of technology.[367]

Certainly NASA occasionally appeared to be playing catch-up to the American aircraft industry. But in many fields including transonic and supersonic aerodynamics, supercritical-wing technology, control-system technology, and aerothermal loads, NASA was well ahead. Those areas where NASA seemed weakest related to the early design of new military aircraft. At worst it was a problem that could be solved by encouraging basic research and involving the centers in new aircraft development programs at an earlier date—that is, before the first flight of an airplane or, better yet, before the design was "fixed" on the drawing board. At best (and this is a view held by many), the problem was fading rapidly as the space program made fewer demands on the time and efforts of the old aeronautics centers, Langley, Ames, Lewis, and Dryden. NASA continued to have much to offer other government agencies and industry. One positive step, coming on the heels of Apollo, was the creation of a Military Aircraft Programs Office within OART in September 1971, charged with overseeing the agency's support of Air Force and Navy aircraft projects.[368]

Dryden's flight-testing of military aircraft continued beyond the end of the period covered by this chapter. From 1978 to 1985, for example, Dryden was involved in a major Navy-sponsored study of the lateral (roll) stability-and-control characteristics of the Grumman F-14A Tomcat fighter in low speed at high angles of attack, in an attempt to develop a better understanding of the spin departure characteristics of the aircraft and then to provide a solution for them. Following their initial deployment to the fleet in October 1972, the Navy's F-14s began to experience out-of-control mishaps. As it turned out, the analog automatic flight-control system on the aircraft had a simple control-law architecture that caused departures from the intended flight path under certain flight conditions. Furthermore, the control system did not provide the pilots full control authority (flight-control-surface deflections) for a recovery from spins and other departures, resulting in the loss of several aircraft and crews. In the course of the project, a NASA/Grumman/Navy team updated the F-14 simulator model since the one the Navy was using was inaccurate. The Navy then used the updated model to upgrade the fleet trainer. In partnership with Grumman and Honeywell, Langley engineers developed new control laws involving what was called an aileron/rudder interconnect (ARI) that succeeded in limiting departures and providing recoveries from spins. The F-14 with the new control laws proved to be "very responsive and maneuverable above 30 degrees angle-of-attack, with no abrupt departure or spin tendencies." The program was an unqualified success, but the Navy did not immediately incorporate the new control laws into its F-14s because of insufficient funding. As a

[367] Memo, Thompson to Scott, 2 Jan. 1976.

[368] OART, *NASA Research and Technology Objectives,* p.1.

result, mishaps with the Tomcats continued. Finally, the Navy contracted with GEC Marconi Avionics of the United Kingdom to incorporate the control laws into a digital flight-control system with minimal changes, and this was deployed on fleet F-14Ds aboard the *USS Kitty Hawk* and *USS Roosevelt* in March of 1999, decreasing the danger of out-of-control flight and greatly increasing the safety of powered approaches to carrier landings. In 1980, Dryden research pilot Einar Enevoldson received the NASA Exceptional Service Medal for his contributions as project pilot on the F-14 stall-and-spin resistance tests.[369]

[369] Joe Renfrow, Naval Air Systems Command, "F-14 Digital Flight Control System," undated briefing provided to the Dryden Historical Reference Collection by Joe Wilson; interview Joe Wilson by J. D. Hunley, 11 Feb. 2000; updates of projects Dec. 1978 through Aug. 1985; quotation from Projects Update for June 1985 (letter, OP/Chief, Dryden Aeronautical Projects Office [Kenneth E. Hodge] to O/Director of Flight Operations and Research, 8 July 1985), p. 2; Lane E. Wallace, *Flights of Discovery: 50 Years at the NASA Dryden Flight Research Center* (Washington, DC: NASA SP-4309, 1996), pp. 148, 185; official biography, Einar K. Enevoldson, BD-1998-01-003-DFRC, on the World Wide Web at URL http://www.dfrc.nasa.gov/PAO/PAIS/HTML/bd-dfrc-p003.html. Dryden engineers developed some simplified control laws for the F-14, but the Navy adopted the Langley or, "Cadillac," version.

11

Shifting Sand 1976-1981

On 26 March 1976, the Flight Research Center opened its doors to hundreds of guests for the dedication of the center in honor of Hugh Latimer Dryden. The evening before, an Air Force Douglas C-9 executive transport had flown into Palmdale, California, with a group of official visitors, including Dryden's widow, other relatives, and prominent NASA officials.

It was a beautiful day, clear and sunny, typical of the Antelope Valley. The dedication was very much a local event. Following Center Director David Scott's opening remarks, the Antelope Valley High School's symphonic band played the national anthem. Then came the invocation, followed by recognition of the invited guests. Dryden, a man of total humility, received praise from all quarters. NASA Administrator James C. Fletcher, Senator Frank Moss, and former NASA Administrator T. Keith Glennan all spoke of his foresight and resourcefulness. Mrs. Dryden unveiled the memorial bust, and with her comments and Scott's closing remarks, the formal ceremonies came to an end. After a buffet lunch, visitors flocked around the center's research aircraft, and official guests returned to Washington.

That night, the center's staff held a more informal celebration in the Longhorn, the ever-popular gathering spot on the outskirts of Lancaster. In contrast to the placid tributes of noon, the conversations in the Longhorn were more questioning. The lifting-body program had ended. The National Hypersonic Flight Research Facility (NHFRF) aircraft faced an uncertain future.[370] The Blackbirds were the only project that seemed to be continuing Dryden's tradition of frontier-probing research. On the horizon loomed the Space Shuttle, but it was less a Dryden project than one for the NASA space centers, notably Johnson. As earlier in the FRC's history, doubts were expressed about the center's future. Could Dryden continue as an independent center in the budget-conscious post-Apollo period? Did headquarters fully appreciate the unique flight-research capabilities of the center? Were non-flight-test-oriented administrators going to homogenize it—turn it into a copy of the other research centers? Many seemed to be celebrating the dedication as an end to what had been, rather than as a promise of what might be.[371]

Whither Dryden?

Dryden's recent administrative history had certainly been unsettled. On 31 May 1971, Paul Bikle had retired from NASA. Bikle had made a major imprint on the center, and

[370] It would be canceled 18 months later (see Chapter 8).

[371] Personal observations.

As Center Director David Scott looks on, Mrs. Hugh L. Dryden unveils the memorial to her husband at the dedication of the Hugh L. Dryden Flight Research Center, 26 March 1976. (NASA photo ECN-5137)

everyone there was aware and appreciative of his role, especially in actively seeking a broad research base. Bikle's deputy director, De E. Beeler, had taken over until October, when Lee R. Scherer became director. His arrival marked a major change in leadership style. Williams and Bikle had been closely attuned to flight-testing and flight research. They were strong managers with a bias toward aeronautics. And they were individualists who favored a great deal of personal and center autonomy. Their immediate successors, however, were more closely in tune with NASA headquarters' management philosophy emphasizing close consultation, coordination, and dependency upon Washington for decision-making. Bikle's successors functioned more as headquarters' agents, in the same sense that project engineers acted at the bidding of a program manager.[372]

Lee R. Scherer, the center's third director, was a graduate of the U.S. Naval Academy with advanced degrees from the Naval Postgraduate School and the California Institute of Technology. A naval aviator, he had served in the 1950s as a special assistant to the assistant secretary of the Navy for research and development, had helped create an antisubmarine warfare center for NATO, and in 1962 had joined

[372] Richard L. Chapman, Robert H. Pontious, and Lewis B. Barnes, *Project Management in NASA: The System and the Men* (National Academy of Public Administration Foundation, Jan. 1973), p. 6.

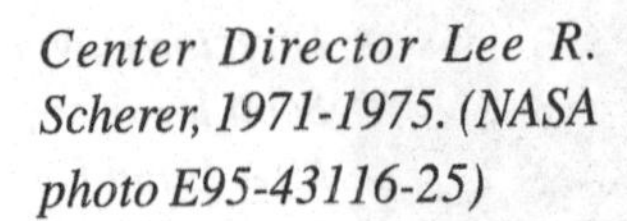

Center Director Lee R. Scherer, 1971-1975. (NASA photo E95-43116-25)

NASA on temporary assignment as manager of the Lunar Orbiter project. After retiring from the Navy in 1964 with the rank of captain, he had risen within NASA to direct Project Apollo's Lunar Exploration Office, where he was responsible for lunar science. A gregarious, athletic individual, Scherer brought to the center a keen awareness of current space and management interests at Headquarters. During his tenure, the Flight Research Center largely continued to run the programs established during the Bikle era. Appointed director of the Kennedy Space Center in 1975, Scherer was replaced at the FRC by his deputy, David R. Scott.[373]

The fourth center director, Scott had joined FRC in August 1973 as deputy director, following the retirement of De Beeler, one of the last of the NACA old-timers. Scott, a West Point graduate and career officer, came to the center as an Air Force colonel. He retired from the Air Force in March 1975. An astronaut of note, Scott had made three flights in Gemini and Apollo. Though a test pilot by training, Scott brought to the center the same orientation and interests as his predecessor, Scherer, for whom Scott had worked for nearly two years. Both sought to bring Dryden more into a line within a standard relationship with other centers and headquarters. Gone was the sometimes paternalistic padrone, Williams or Bikle. In his place was a more tightly structured bureaucracy that rankled many center veterans used to a

[373] Data from Scherer biographical files, DFRC and NASA history office; Interview of Finch.

Center Director Isaac "Ike" Gillam, 1978-1981. (NASA photo E78-34269)

freewheeling style. Some doubted the devotion of the new leaders to atmospheric flight-testing. The cancellation of NHFRF, thwarting of the mini-shuttle research aircraft, termination of the Blackbird effort were seen as symptomatic of this supposed non-aeronautics orientation.[374]

Scott retired in 1977. His deputy Isaac "Ike" Gillam had run the approach and landing tests of the Shuttle at Dryden (discussed later in this chapter). With background in Air Force flight assignments and management of launch vehicles for NASA, a friendly disposition and obvious ability, Gillam had the support of many on the staff who hoped he would become the new director. Until he was appointed in June 1978, Gillam was not inclined to be a mere caretaker.[375] As acting director and then permanent director, his was to be a challenging assignment. With a new administration in Washington pledged to economy, NASA and other agencies would be in a budget squeeze. Dryden would be buffeted by internal wars of institutional assessment, which would determine where NASA's smaller budget would go. At the same time, the Shuttle test flights would bring Dryden massive, unaccustomed publicity.

The Shuttle Comes to Dryden

NASA's thinking on reusable lifting-reentry spacecraft reached fruition in the development of the Space Shuttle. After studying various proposals, NASA awarded

[374] This represents a consensus of statements from numerous interviews with DFRC personnel.

[375] Personal observations and consensus of interview statements; NASA DFRC *X-Press*, 16 June 1978.

study contracts to North American-Rockwell (later Rockwell International, now part of Boeing) and McDonnell Douglas (now also part of Boeing) in July 1970. The design characteristics selected for the craft included a delta wing and a 1,265-mile "cross-range" maneuver on either side of the flight path during reentry. Various designs were submitted, including vehicles launched from the backs of other winged reentry vehicles, vehicles launched on top of boosters, and vehicles attached to large fuel tanks and solid-fuel boosters—the "parallel burn" configuration, in which both liquid-propellant engine and solid-propellant booster would burn during ascent. In March 1972, NASA selected the parallel-burn approach and on 16 July selected Rockwell's proposal for development.

Construction of the first Space Shuttle Orbiter, vehicle OV-101, started at Rockwell's plant in Downey, California, on 4 June 1974. Components were delivered to Rockwell's plant in Palmdale, near Edwards, where final assembly began in August 1975. The OV-101 was rolled out 17 September 1976. The hefty craft, the size of a Douglas DC-9 jet transport, was christened *Enterprise*. In January, the Shuttle was trucked from Palmdale to Dryden. Meanwhile, NASA had bought a Boeing 747 and returned it to the manufacturer for modification so that the Shuttle could be mounted on the back of it. So connected, the Shuttle would be carried aloft for its first flight tests. Later, it would be ferried from one site to another the same way.[376]

The Space Shuttle was an ambitious design. It had a body length of over 122 feet, a height of almost 57 feet, and a wingspread in excess of 78 feet. It combined reaction controls for spaceflight and aerodynamic controls for glide to earth. The reaction controls would not be installed for the approach and landing tests. If the craft went out of control or collided with the 747 after launch, the crew of two would eject. Planning for the approach and landing tests was as complex as for any other research program. And there was the added factor of publicity. Everything that happened at Dryden would be headline news.

In the fall of 1974, the Air Force and NASA executed a joint agreement to establish Space Shuttle facilities at Edwards. Edwards was already designated as the test site for the Shuttle's approach and landing tests and as the prime landing site for the first orbital flights. Within NASA, the Shuttle would be under the overall control of Johnson, with the FRC in a supporting role.[377]

By the time of the Dryden dedication, Shuttle test plans were nearing completion. The Shuttle road was almost ready. In January 1977, *Enterprise* was moved to Dryden. Immediately the center, which had done its most spectacular work under conditions of almost total privacy, was the focal spot of national attention. Ralph Jackson, Dryden's ebullient director of public affairs, had his hands full. Inside the headquarters building, Johnson engineers and technicians roamed the halls. Outside, Johnson astronauts and

[376] E. P. Smith, "Space Shuttle in Perspective: History in the Making," pp. 8-11; NASA release 77-16, "Space Shuttle Orbiter Test Flight Series," pp. 57-58; NASA DFRC *X-Press,* 24 Feb. 1978.

[377] Letter and attachment, Robert F. Thompson to Myron S. Malkin, 16 Oct. 1974; letter, Kraft to Scherer, 23 Oct. 1974.

pilots zipped around in T-38s, and a NASA Grumman Gulfstream II Shuttle trainer simulated Shuttle approaches and landings. The Boeing 747 crews readied themselves for the first flights. Press and television commentators wandered about, interviewing and photographing anything that moved. It was the Cape come to the Mojave, a scene more familiar to Cocoa Beach or Houston than to the high desert–indeed, the reporters who covered the Shuttle were mostly the media veterans of Mercury, Gemini, Apollo, Skylab, and Apollo-Soyuz.

The flight-test program had three phases: captive, captive-active, and free flights. The unpiloted captive flights simply would demonstrate whether the combination—which wags dubbed the world's largest biplane—could fly safely as a mated unit. In the captive-active trials, an astronaut crew would ride in the Shuttle. Finally, in free flights, it would be launched from the back of the 747 and flown down to a landing. During the captive flights and the first of the free flights, the Shuttle's blunt base would be faired over with a tailcone to reduce buffeting on the 747's vertical fin. As another precaution, Boeing had added two more vertical fins to the 747's horizontal stabilizer. Toward the end of the flight trials, NASA hoped to launch the Shuttle without the tailcone, which would reduce the Shuttle's lift-to-drag ratio, resulting in a descent path similar to what it would have upon returning from orbit. A series of high-speed taxi tests by the mated 747 and *Enterprise* in mid-February 1977 went without a hitch.

On 18 February, the first Shuttle flight proved to be a media event unparalleled in the history of Dryden. For the previous week, Johnson and Dryden public affairs officials had been on hand to meet the demands of the hundreds of media representatives who left plusher locales for the sunny but blustery desert. Those who spent the night in motels in Lancaster and Palmdale had to get an early start. At 5 a.m., the sky was still black and clear, the stars as brilliant as always, the temperature in the low 20s. Cars moved along Sierra Highway, down Avenue E, then north on 120th Street East. Despite the urban-sounding names, the surrounding country was bare, scrub desert broken only by an occasional homestead. As the sky began to lighten, Joshua trees and the low hills near Hi Vista were outlined. The revolving beacon at Edwards pulsed brightly on the northern horizon, and the 6 a.m. news on KNX, a radio station in Los Angeles, reported that the Shuttle would fly today. Dryden itself was controlled pandemonium, the public affairs trailer a madhouse. By 6:45 a.m. the sun was spreading a warm glow through the thin fog covering the dry lakebed. Those present were prepared to convoy out to the runway; meanwhile they drank coffee and watched the TV monitors in the public affairs trailer.[378]

By 7 a.m., the Shuttle launch crew–Fitz Fulton, Tom McMurtry, Vic Horton, and Skip Guidry–were in the 747, the inert Shuttle riding on top. No sooner had the reporters journeyed from Dryden to the press site along runway 04-22 than the Air Force staged an impromptu airshow: the YC-14 took off, followed by the B-1, some T-38s, an F-4, and the F-16. Finally those at the site watched the 747/*Enterprise*

[378] Personal recollections.

combination taxi slowly past the Air Force's two large hangars, down to the west end of 04-22. There it held while the test crew completed final checks. Two NASA T-38s flew over, as if impatient to get on with the flight. Camera crews set up their tripods, shivering in the brisk desert morning. It was a beautiful day. Right down the center of 04-22 flew a gaggle of geese—a large V honking along, heading east, unperturbed by the consternation they were causing. Geese and jet engines do not mix, so NASA delayed the departure of the 747 a little longer. Finally, the 747 started to roll down the runway with that peculiar whine so typical of large fan-jet airplanes.

The world's most improbable aerial combination, after a run of 6,000 feet, became airborne, climbing ponderously toward the east, above the dry lake. For 130 minutes, this strange hybrid flew along, anxiously attended by T-38 chase planes, before Fulton and McMurtry returned it gently to earth. First flights are always cautious, and on this one, the test crew held the combination to a maximum altitude of 16,000 feet and a maximum speed of 250 knots. Everything went well. The 747/*Enterprise* flew closer in performance to a standard 747 than simulations had predicted. Nothing serious had happened, a tribute to the test-planning. That afternoon and evening, the dark interior of the Longhorn echoed with the jubilation of Dryden, Rockwell, and Boeing personnel. The Shuttle had taken to the air.

Back at Dryden, over the weekend engineers worked up the data from the flight. At a technical and crew briefing on Monday, the word went forth: "Testing can go on to expand the envelope as planned." The critical concerns of buffeting, flutter, and tail loads proved not to be problems. After five complete successes, NASA abandoned a planned sixth flight, deeming it unnecessary. While the next series of tests was being prepared, the 747 flight crews temporarily returned to more prosaic duties, such as flying the YF-12 Blackbird on its Coldwall tests (see Chapter 9).[379]

Shuttle Summer

NASA had already selected four astronauts for the Shuttle landing tests, placing Fred W. Haise and C. Gordon Fullerton on one crew and Joe H. Engle (the former X-15 pilot) and Richard H. Truly on the other. Haise, a former center pilot, had flown on the ill-fated *Apollo 13* mission. The astronauts prepared for the Shuttle program by practicing in a ground simulator and flying a much-modified Grumman Gulfstream II. Other pilots flew the center's JetStar to test the Shuttle's microwave scanning-beam landing system. In addition, Dryden managers worked closely with their Johnson counterparts to prepare for a most important part of the Shuttle test program: arranging for the Houston center to control the mission while it was in progress at Dryden in the Californian desert. Dryden had controlled the captive-inert flights, but the Mission Control Center at Johnson would have primary responsibility for running subsequent missions, starting when the 747 and *Enterprise* backed away from the Shuttle mate-

[379] Captive-inert flight-test postflight reports, reports of DFRC Shuttle news conference, 18 Feb. 1977; personal recollection.

The Space Shuttle prototype Enterprise *in free flight after separation from its Boeing 747 launch aircraft during the approach and landing tests in 1977. (NASA photo ECN 77-8608)*

demate facility and began taxiing to the runway. By mid-June all was ready, and the Shuttle flight-test program moved into its next phase.[380]

On 18 June 1977, the 747 and *Enterprise* combo went aloft on the first captive-active test. Inside the Shuttle, Fred Haise and Gordon Fullerton had a magnificent view. Not being able to see any portion of the carrier aircraft added to the illusion that they were alone in the sky. The flight lasted nearly an hour and all objectives were achieved. The test data indicated that the Space Shuttle was buffet- and flutter-free up to the maximum speed attained on the flight, over 180 knots. The next captive-active mission, flown by Engle and Truly on 28 June, involved high-speed flutter tests up to 270 knots. It too was successful. NASA concluded that the four flights originally scheduled for the captive-active phase could be safely cut to three. On 26 July, Haise and Fullerton completed the last of the captive-active flights. During this last mission, 747 pilots Fitz Fulton and Tom McMurtry flew a launch separation profile, pushing the 747 over into a shallow dive at 30,250 feet and deploying the speed brakes to make the drag on the 747 about the same as that on the Orbiter.[381] During approach of the 747/*Enterprise* combination to landing, Haise and Fullerton lowered the Shuttle's landing gear to check its operation. It went smoothly.

[380] Donald E. Fink, "Orbiter Flight Plan Expanded," *Aviation Week & Space Technology* (27 June 1977), pp. 12-14; NASA release 77-16, passim.

[381] Comments of Gordon Fullerton upon reading the original chapter.

The Dryden Flight Research Center's Lockheed JetStar, which has been used for a variety of general-purpose airborne simulations, general-aviation research, support of the Space Shuttle approach and landing tests, insect-contamination, ice-particle, and laminar-flow research. (NASA photo ECN-2478)

Indeed, the captive-active phase of the Shuttle testing had gone pleasingly well. Some equipment problems had been experienced: auxiliary power units leaked or overheated, computers were "voted" off-line by other computers, and sometimes a computer tried to take the square root of a negative number. Hoever, these were small concerns that could be remedied by minor fixes or software changes. The important fact was that the Shuttle and the 747 were a safe-flying combination. Now NASA could move to the next phase of the approach and landing tests: the actual free-flight testing of the *Enterprise*. The flock of media reporters, who had left the desert in droves after the first captive flights, now returned.[382]

During the week of 8 August, project officials concluded a two-day Shuttle readiness review and a mission readiness review. All conditions were "go." The most visible—and audible—of the preparations were the Shuttle simulation flights that Fred Haise and Gordon Fullerton made in the Gulfstream II training aircraft. For a few days prior to the flight, the center echoed to the occasional rumble of the Gulfstream and its T-38 chase planes climbing out over the dry lakebed following another approach to landing, or the center's JetStar checking the microwave landing system.

[382] See note 11; Jeffrey M. Lenorovitz, "Shuttle Orbiter Test Phase Trimmed," *Aviation Week & Space Technology* (4 July 1977), pp. 18-19. Captive-active flight-test postflight reports.

The flight plan called for the 747/*Enterprise* to take off at 8 a.m. from runway 22 and climb to the west. The two mated craft would enter a racetrack pattern, flying south toward Los Angeles, turning north over the mountains, and coming up the east side of Rogers Dry Lake. If all went according to plan, the craft would nose into a shallow dive from an altitude of over 30,000 feet. At 24,000 feet and an airspeed of approximately 280 knots, Fred Haise in the Shuttle would press a square white button on the Shuttle's instrument panel, triggering explosive bolts that would separate the *Enterprise* from the 747. If all went well, Fulton would roll the 747 into a descending left turn, and Haise would pitch up the 75-ton *Enterprise* and make a shallow turn to the right. At about 24,000 feet, Haise would initiate a practice landing flare to evaluate the handling qualities of the *Enterprise*. Then, as the Space Shuttle Prototype sank like a rock, its astronaut crew would begin a gradual 180-degree turn to position the *Enterprise* for a 185-knot touchdown on lakebed runway 17–at over 36,000 feet, the longest of the Rogers runways.

The flight attracted massive public attention. Over 1,000 media reporters flocked to Dryden, many from abroad. Parking had to be prepared for the public viewing sites west of Dryden, and the VIP and press sites along runway 04-22 and the west shore of the lakebed, parallel to runway 17. Motels as far away as eastern Los Angeles were booked solid. A wide range of aviation personalities, including NASA Administrator Robert Frosch and such pioneers as Jimmy Doolittle, were present.

As with most test flights, the preparations took days of hard work at all levels. The event came alive in the early morning of 12 August. At 3 a.m., the first media reporters left Lancaster and Palmdale for the lakebed. Once again, the night was perfectly clear. After driving up 120th Street through the base gate, the cars made their way to the FAA radar facility by Hospital Road on old South Base, turned right, and drove through an Air Police checkpoint. Those with authorizations continued on toward the runway site or the lakebed. The lakebed was better, at least for those with a handle on the past and an awareness of the present. Eerily quiet and still, the dry lakebed seemed unconnected with civilization. Further away could be seen the night lights of the mines at Boron and the bustle of activity at Dryden. The Air Force side of the field was still and dark, except for watchlights and the tower and runway lights.

For those interested in omens, the Shuttle's day began with a meteor shower. Looking up toward the Milky Way, clearly visible as a faint patchy white in the crisp desert sky, observers saw a rain of fire, with meteors coming down by the minute. There were fireballs breaking apart in greenish-white trails, streaks of russet, streaks of yellow. Then came the desert dawn, the familiar yellow glow lighting up the eastern sky, shining through high clouds, bathing the rocky outcroppings of Leuhman Ridge in orange, and finally reaching down to illuminate the broad baked expanse of Rogers Dry Lake. The lights on buildings dimmed. Soon an Air Force helicopter clattered noisily over the lakebed, joined by another from Dryden. Far from the dry lake, between 60,000 and 70,000 visitors streamed along Sierra Highway and Rosamond Boulevard into Edwards. At one point, the traffic jam stretched 10 miles. As media personnel whiled

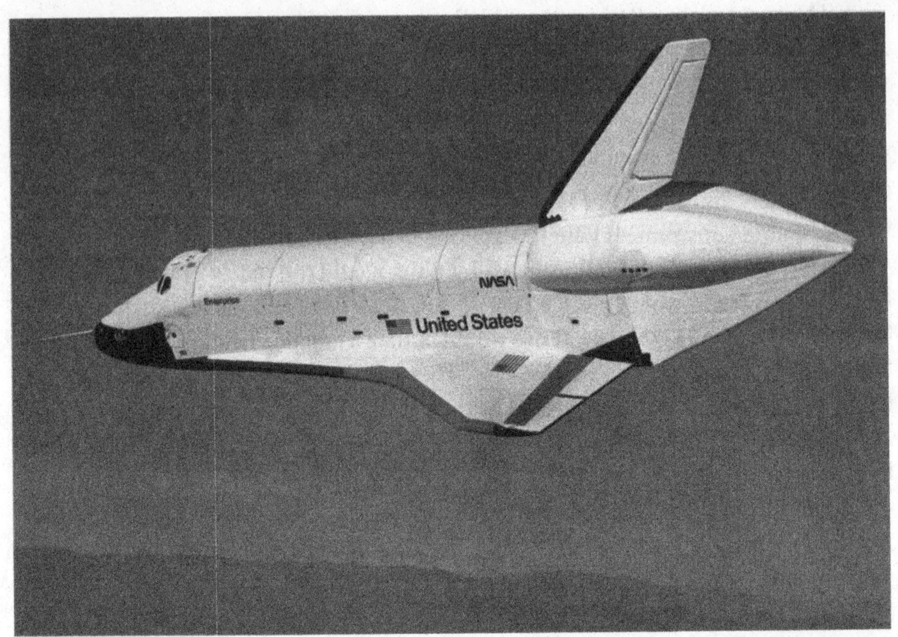

The Space Shuttle prototype Enterprise *glides toward a landing after separation from its carrier aircraft during the approach and landing tests in 1977. (NASA photo ECN-8611)*

away the time setting up equipment and sipping coffee from a Rockwell courtesy van, the technicians, engineers, and flight crews at Dryden readied themselves for the flight.

Finally all was ready, and the 747/*Enterprise* backed out of the mate-demate gantry at Dryden, ran up its engines, and began the long taxi. The Air Force Huey still clattered above. The first of the T-38 chase planes whistled aloft. The 747/Shuttle reached the east end of runway 22, turned, and held for the last checks. At 8 a.m., right on schedule, Fulton called up full power. The 747/*Enterprise* combination, with surprisingly little noise, began to roll and nosed aloft, followed by two T-38s. The aircraft climbed into the prescribed racetrack pattern, joined by the other three chase T-38s. On the ground, media reporters waited for the big moment.

The air launch had been scheduled for 8:45 a.m. In fact, higher-than-normal temperatures at altitude caused the climb to take longer than planned. The 747/Shuttle moved majestically around the racetrack, plainly visible most of the time from the lakebed. The low sun obscured the view of its approach to launch, but video coverage from one of the T-38s outfitted with a portable camera was stunning. The formation continued over Saddleback Butte to the Edwards bombing range. Roughly 48 minutes into the flight, the 747/Shuttle was due east of Rogers Dry Lake, at an altitude of 30,250 feet. Fitz Fulton nosed into a shallow dive and said that he was set. Fred Haise radioed his thanks for the lift, then punched the separation button. Seven explosive bolts detonated, and the Shuttle was flying on its own at 26,000 feet. The 747 pitched down slightly and rolled into a diving left turn, and Haise turned to the right. He initiated a practice landing flare at about 250 knots and made moderate lateral control inputs to evaluate the Shuttle's response. The big delta handled well. Because of the Shuttle's low lift-to-drag ratio, it would remain aloft only for about five minutes. Later, after removing the drag-reducing tailcone, the Shuttle would sink to earth in about two minutes, a descent rate similar to the X-15's.

On the ground, some had seen the separation with binoculars and sun shields. Soon, it became visible to all. The 747 flew alone, trailed by a single T-38, while to the northeast a white speck could be seen growing in size at what seemed a remarkable rate, attended by four T-38s. Media cameras started clicking furiously, and exclamations sounded on all sides. The Shuttle descended over Leuhman Ridge, passed across Highway 58 at Boron, turned west toward California City, swung around over North Edwards, and lined up on runway 17. Houston's Mission Control radioed Haise that the *Enterprise* had a lower lift-to-drag ratio than predicted by tunnel tests. In fact, however, the ratio was just as predicted. Houston had miscalculated. The error caused Haise to fly the final approach at a higher speed, conserving energy to prolong the glide. As a result, the Shuttle was "high and hot" on its final approach. Realizing that the *Enterprise* would land long, Haise deployed the craft's speed brakes from 30 up to 50 percent. At 900 feet above the ground, Haise began the landing flare. As the *Enterprise* leveled out, Fullerton deployed the landing gear. The Shuttle landed long by about 2,000 feet at 185 knots, nearly five and a half minutes after launch. The Shuttle coasted for over two miles before stopping on the south lakebed. As it slowed, its T-38 chase planes streaked by. Soon the 747 and its lone chase plane swept majestically over the landing site. The first Shuttle free flight had been a success. Now all that was left for most at the lakebed was the long trip around the base to Dryden, a quick lunch, and the afternoon press briefing. For the engineers, however, the task of data reduction had just begun.[383]

After the press conference, many called it a day and went to one of the many parties being hosted by mission personnel in and around Lancaster. Most wound up at the main blowout, held at Lancaster's Delta Lady saloon. Others settled for the more tranquil but no less joyous environment of the Desert Inn or Mr. B's Twin Lakes Inn outside Palmdale. The Shuttle obviously flew well, better than the Gulfstream simulator. The major remaining question was how the Shuttle would behave without its tailcone. This actually involved two considerations. One was whether the buffet from disturbed air caused by removal of the cone would cause structural problems for the 747's vertical fin during the climb. The other was whether the Shuttle's low lift-to-drag ratio—made even lower by removal of the tailcone—would present serious piloting problems. After all, the descent rate of the craft would just about double, reducing flight time from over five to just over two minutes. Pending a decision to fly "tailcone

[383] Free-flight test postflight reports; Donald E. Fink, "Orbiter Responsive in Free Flight," *Aviation Week & Space Technology* (22 Aug. 1977), pp. 12-19; George Alexander, "Space Shuttle Sails Through Solo Flight," *Los Angeles Times*, 13 Aug. 1977; Alan Brown, "Fantastic!" Antelope Valley *Ledger-Gazette*, 12 Aug. 1977; Antelope Valley *Ledger-Gazette*, 15 Aug. 1977. Much of this on-site material is from personal recollection.

The Enterprise *on its first tailcone-off flight, 12 October 1977. (NASA photo EC95-43116-26)*

off," Shuttle testing continued with the *Enterprise*'s blunt end still sporting the pointed tailcone.

Rain on the dry lakebed and other delays deferred the next free flight to 13 September, when former X-15 pilot Joe Engle and copilot Dick Truly dropped down to the lakebed, all the while taking data on the craft's longitudinal, lateral, and directional response and lift-to-drag and flutter characteristics during approach and landing. Nothing unusual aloft had occurred, but on the ground, a power surge at Dryden caused a brief loss of all radar data. Fortunately, after a few minutes, everything had come back on-line and the flight had continued. Ten days later, Fred Haise and Gordon Fullerton completed the third Shuttle free flight, and events progressed so smoothly that NASA decided to begin tailcone-off testing with the next flight.[384]

At first, NASA and Rockwell had thought that a series of captive flights with the Shuttle minus its tailcone might be necessary to evaluate whether the buffeting loads on the 747's vertical fin were acceptable. Mission planners soon realized that there was little point in such flights. The 747 could take off with the Shuttle and if the buffeting seemed excessive, the craft could simply abort the mission and land on the

[384] Free-flight test postflight reports.

lakebed. In preparation for the flight, Rockwell and NASA technicians removed the tailcone from the *Enterprise* and replaced it with a configuration identical to what the Shuttle would have during reentry from space, including the three main Shuttle engine nozzles and the much smaller nozzles of the orbital maneuvering subsystem. By this time, the massive media attention that had focused on the earlier Shuttle flights had abated and day-to-day activities at Dryden were more tranquil.

Mission planners decided that during the takeoff roll and liftoff, Fitz Fulton would report any severe buffeting in the cabin. Bill Andrews would monitor the loads on the 747's tail, and if he deemed them excessive, he would call "data abort," terminating the flight. If the 747 was still on the runway, this meant chopping power and stopping. If just airborne, the 747 could land straight ahead on the lakebed. If fully airborne, Fulton and Tom McMurtry would gingerly return the craft to Edwards.

The actual flight on 12 October 1977 came off without difficulty. Again there was the early morning procession to the south lakebed and the long wait until takeoff, while some Air Force Phantoms shot landing approaches. When the 747/Shuttle rolled down the runway this time, observers watched for any indication of an abort. Then it was airborne and climbing out to the east, with no visible problems. At Dryden and Johnson, engineers checked monitors. The tail loads were within acceptable boundaries. After about 40 minutes, the 747/*Enterprise* became visible to the north, approaching the drop. Cameramen peered through telephoto lenses to catch the moment of separation. Fulton pushed into a shallow dive at about 25,000 feet altitude. Thirty-eight seconds later, Joe Engle triggered the explosive bolts. The separation occurred over Peerless Valley. *Enterprise* nosed down sharply, descending over North Edwards on final approach to runway 17. It quickly became apparent that *Enterprise* would land right in the aiming area. The steep diving descent, with the Shuttle plunging to earth followed by its T-38 chase, brought exclamations of surprise even from those who had witnessed the earlier tailcone-on flights. Removing the tailcone certainly made a difference. In what seemed an incredibly brief time, Engle had pulled out of the dive into the landing flare and deployed the gear. There was no excess energy to worry about this time, and *Enterprise* plunked down, streaming a roostertail of playa dust, some two minutes and thrity-five seconds after launch.[385]

During the very brief flight, the Shuttle had flown well, confirming earlier predictions and simulations. It was, in effect, simply a big X-15. The next question was whether the Shuttle could be landed with confidence on a confined runway. It was a critical issue since NASA planned landing the Shuttle on 15,000-foot runways at Vandenberg and Kennedy. For the next tailcone-off flight, NASA planned to land the *Enterprise* on the 15,000-foot runway at Edwards. So far, aside from the high and hot first landing, the Shuttle had had little difficulty in landing at a chosen spot on the lakebed runways, even with the tailcone off. Encouraged, NASA scheduled the fifth Shuttle free flight for 26 October.[386]

[385] Ibid.; personal recollection.

[386] Free-flight test postflight reports.

In that flight, *Enterprise* encountered control problems just at touchdown. The Shuttle had been launched at an altitude of 19,000 feet over the desert (over 19,000 feet above sea level) for a straight-in approach. Mission commander Fred Haise flew a 290-knot approach profile down to the flare maneuver; the *Enterprise* lost speed very rapidly. Haise used the split-rudder speed brake to slow the craft and nosed down to touchdown on the runway at the planned aim point. Instead, the *Enterprise* entered a left roll, which Haise corrected, touched down on its main landing gear, and bounced back into the air. Haise had brought on a pilot-induced oscillation (PIO). Copilot Gordon Fullerton told Haise to relax his grip on the controls, and the *Enterprise* damped out its rolling motions. It touched down again, bounced more shallowly, then touched down for the final time before coasting to a stop. The flight had an important VIP observer: Charles, the Prince of Wales. Prince Charles, a Royal Air Force pilot, was in the United States as part of a goodwill tour. While in Houston, he had flown the Shuttle simulator with Haise and Fullerton. Interestingly enough, during one touchdown in the simulator, the craft had bounced and Prince Charles had encountered the same sort of lateral PIO during the ensuing skip that later occurred during Haise's landing. The rugged arrival prompted NASA briefly to reconsider adding an additional tailcone-off flight, but mission planners decided that it was unnecessary. The astronauts themselves had no reservations about the Shuttle's ability to land on concrete runways at Kennedy and Vandenberg, and their feelings did much to influence the decision not to add an extra flight. Dryden did undertake a landing study of the Shuttle to better understand its low-speed handling and control characteristics. With their usual penchant for thoroughness, center personnel wanted no unresolved questions or doubts when the Shuttle whistled in to land from a Mach 25 reentry sometime in 1981.[387]

The fifth Shuttle free flight concluded *Enterprise*'s flight-testing. Dryden now prepared for the task of ferrying the *Enterprise* aboard the 747 to NASA's Marshall Space Flight Center for a series of ground vibration tests. Technicians reinstalled the tailcone aerodynamic fairing. Fulton and the 747 crew completed a series of test flights with the Shuttle in ferry condition (with its front attachment strut lowered slightly to improve the cruise performance of the two mated vehicles) in mid-November. All indications were that the Shuttle could easily be ferried atop the 747. On 10 March 1978, the *Enterprise* left the runway at Dryden, Fulton and his crew ferrying the Shuttle to Ellington Air Force Base at Houston where, during a weekend stay, it was seen by 240,000 viewers, creating, in the words of Houston police, "the largest traffic jam in Houston's history."

While at Houston, the 747 crew and two other Dryden Shuttle project officers received NASA's Exceptional Service Medal. Nine other Johnson and Kennedy center employees also received the Exceptional Service Medal, and Donald K. Slayton, project director for the approach and landing tests, received NASA's Outstanding Leadership Medal. On 13 March the 747/*Enterprise* departed from Ellington on a short flight to

[387] See note above; Donald E. Fink, "Orbiter Experiences Control Problems," *Aviation Week & Space Technology* (31 Oct. 1977), pp. 16-17; comments of Gordon Fullerton upon reading the original chapter.

Huntsville. Seven thousand NASA and Redstone employees witnessed the arrival of the strange pair. The next day, cranes removed the Shuttle from the 747 preparatory to installing it in a special test rig at Marshall for a series of ground vibration tests simulating the loads a Shuttle would experience in flight.[388]

With the notable exception of flight and other research to solve the PIO problem (see Chapter 10), Dryden's active role with the Shuttle had come to an end, until the time in the future when another Shuttle would reenter from space over the Pacific and glide in for a landing on the Edwards dry lakebed.

Institutional Uncertainties

Despite the fruitful contributions of Dryden to the Space Shuttle program, during the same years in which the DFRC pilots and engineers conducted their Orbiter investigations the entire NASA flight research establishment underwent a period of intense self-analysis and external scrutiny. Because Dryden existed primarily for flight research while Ames, Lewis, and Langley each pursued multiple missions, the drama pivoted on the role of DFRC. The controversy originated in the 1950s. Hugh L. Dryden, director of the NACA, decided early in the decade to allocate a rising proportion of the NACA's budget to space-related and hypervelocity projects. Because the NACA's budgets rose slowly during this period, Dr. Dryden needed to fund the changeover from existing programs, which inevitably resulted in curtailment or slowdown of some aeronautical work. Traditional NACA research suffered a far greater reversal in 1958 when Congress responded to the USSR's space challenge by forming the National Aeronautics and Space Administration from the sinews of the NACA. Then, in 1961, President Kennedy announced an initiative to challenge the Soviet lead in space by sending American astronauts on a round-trip voyage to the moon. Exhilarating as this endeavor may have been, as NASA prepared former NACA research pilot Neil Armstrong to step on the lunar surface, aeronautics spending fell to $65.9 million, about two percent of the agency's research-and-development direct obligations. Meantime, the nation expended roughly $20 *billion* to achieve the first moon walk.[389]

The realization of Kennedy's ambitious objective failed to renew NASA aeronautics. Indeed, even before Armstrong's feat, the space agency found itself immersed in red ink and pressed to articulate new horizons in the post-Apollo era.

[388] NASA DFRC *X-Press,* 24 Mar. 1978; "Space Shuttle Contributions Recognized," *NASA Activities* (Apr. 1978), pp. 25-27; "Shuttle Orbiter Ferried to Huntsville," *Aviation Week & Space Technology* (20 Mar. 1978), p. 15.

[389] Arnold S. Levine, *Managing NASA in the Apollo Era* (Washington, DC: NASA SP-4102, 1982), p. 305; Roger D. Launius, *NASA: A History of the U.S. Civil Space Program* (Malabar, FL: Krieger, 1994), p. 211; Alex Roland, *Model Research: The National Advisory Committee for Aeronautics, 1915-1958,* (Washington, DC: NASA SP-4103, 1985), Vol. 2, p. 475; Jane Van Nimmen and Leonard C. Bruno with Robert L. Rosholt, *NASA Historical Data Book Volume I: NASA Resources, 1958-1968* (Washington, DC: NASA SP-4012, 1988), Table 4-21.

Congress imposed sharp budget reductions, resulting in shortfalls in many programs and in the loss of personnel. The hatchet fell on such ventures as the Nerva II nuclear rocket, and funding fell for the Apollo Applications Project (the initial name for Skylab, a project that NASA conceived as the first step toward a space station). Furthermore, a mere two months after Apollo 11 mesmerized the world, President Richard Nixon's Space Task Group unveiled a set of three options for future space exploration: piloted flight to Mars, launched from space stations orbiting earth and the moon; the Mars mission without the space stations; and a space station served by space trucks or shuttles. The president selected the third and cheapest option, reducing it further by substituting Skylab for the space station. While based on fiscal prudence, the Nixon decision still embraced a space program of grand proportions even as Congress continued to pare down NASA's expenditures. These conflicting forces did not create a climate favorable to the pursuit of flight research. Still, decision-makers recognized that the space program profited greatly from the X-15 data on reaction controls and aerodynamic heating, from the LLRV tests for safe lunar touchdowns, and from the lifting bodies, which paved the way for glide reentries. Consequently, NASA headquarters found itself in a classic bureaucratic conundrum during the 1970s: how to maintain the agency's vigorous flight-research program at a time when budgetary retrenchment and big-ticket space projects like the Shuttle squeezed NASA resources as never before.[390] Neither NASA Administrator James E. Webb nor his successor Thomas O. Paine substantially reduced the aeronautics budget. But when Paine resigned in September 1970, the new administrator, James E. Fletcher, lost no time swinging the efficiency ax. He began by ordering a probe of the agency's flight-research operations after an Office of Management and Budget audit alleged that NASA possessed more aircraft than it required. Accordingly, Roy Jackson, NASA headquarters' associate administrator for aeronautics and space technology (OAST), underook a wide-ranging examination of NASA flight research "with the objective of improving . . . aircraft operations management, modernizing [the] aircraft fleet, and minimizing . . . recurring costs." Jackson named De Elroy Beeler, the Flight Research Center's deputy director, to chair the OAST Committee on Flight Operations. The choice of Beeler seemed to presage a positive outcome for the FRC in the panel's deliberations. Even more encouraging, Jackson closely patterned his directions to Beeler on a memorandum sent by FRC Director Lee Scherer to headquarters in June 1972. Jackson told Beeler to achieve these goals in his final report:

1. Designate the FRC as the lead NASA facility for flight operations and concentrate all of the more dangerous flight-research work at the Flight Research Center.

2. Constitute the FRC as the only center where the agency tested complete flight vehicles (distinct from platforms or testbeds), thus reducing the duplication of flight programs and aircraft.

[390] Levine, *Managing Apollo*, pp. 25, 255-261; Launius, *NASA: The U.S. Civil Space Program*, p. 97.

3. Lower the cost of recurring aircraft operations.

4. Retire obsolete aircraft and replace them with more efficient ones acquired either through military loan or by lease or purchase.[391]

Beeler undoubtedly knew the depth of emotion this realignment would arouse in the three other NASA aeronautics centers. Indeed, at an intercenter meeting in June 1972, only eight weeks before the OAST panel was announced, representatives of Ames, Langley, Lewis, and the FRC evidently displayed "strong parochial feelings. . . . None can be expected to offer more than token changes," wrote a high-ranking observer. "It is clear," he said, "that any significant changes in responsibilities must be directed from OAST after careful consideration of all issues." The fact that flight-research centralization had a long history probably hardened the positions of the combatants. As early as 1960, NASA Associate Administrator Richard E. Horner ordered all four centers to concentrate flight research at Edwards. But the Langley, Lewis, and Ames directors offered successful resistance to the initiative, which collapsed after Horner's brief tenure ended. Although they agreed to transfer token numbers of aircraft to the Flight Research Center, they preserved their main missions, including (but not limited to) helicopter, vertical, and short takeoff and landing research at Ames and Langley, and engine-related high performance flying at Lewis. The question arose again during the period between the Paine and Fletcher administrations. Acting Administrator George M. Low appointed Major General John M. Stevenson (associate administrator for organization and management) to evaluate NASA's aircraft inventory and recommend efficiencies. The subsequent report was unequivocal. It recommended that "all NASA . . . sponsored flight research be centralized at the NASA Flight Research Center." But James Fletcher wanted to review this important matter for himself. He also recognized the necessity of including the center directors in the process. Consequently, he initiated the 1972 OAST investigation.[392]

Beeler's committee gathered data for five months. The four centers inventoried all aircraft—including remotely-piloted vehicles—and estimated the complete annual costs of repairing, maintaining, and fueling them. The seven panelists then took to the road from 23 January to 1 February 1973 to discover the essential programmatic necessities of NASA flight research. A numbing ten days of briefings ensued. They heard Langley, Lewis, Ames, and FRC officials describe almost every existing project,

[391] Launius, *NASA: The U.S. Civil Space Program*, pp. 95-96; Lee Scherer to NASA deputy associate administrator for management, 27 June 1972, DFRC Historical Reference Collection; Roy F. Jackson to OAST Management Council, 10 Oct. 1972, Milt Thompson Collection, DFRC Historical Reference Collection (quoted passage).

[392] Lee Scherer to NASA deputy associate administrator for management, 27 June 1972, Milt Thompson Collection, DFRC Historical Reference Collection (first quoted passage); briefing, "Flight Research Within NASA," n.d. [probably 1977], Milt Thompson Collection, DFRC Historical Reference Collection; George M. Low to associate administrator for organization and management, 27 Apr. 1971, DFRC Historical Reference Collection; Stevenson Report, Fall 1971, DFRC Historical Reference Collection (second quoted passage).

plans for future research, the varying steps required to acquire and decommission aircraft, each center's unique flight-operations procedures and pilot training. Beeler and his associates saw nearly every piece of flight-research equipment owned by NASA. By the first of February, the panelists felt "saturated from the extensive material presented" and made a collective decision to review the documentation individually, rather than as a group. Their findings, announced on 23 April 1973, featured a unique organizational crossbreed: an OAST Aircraft Operations Office (later called the Flight Activities Office). Located at the Flight Research Center, staffed by FRC employees, but responsible to NASA headquarters through the FRC director's office, it made the FRC the focus of flight research but not the master of it. Under this framework the Aircraft Operations Office assumed a decisive role. It controlled the budget of the entire NASA aircraft inventory, evaluated all research proposals, assigned the research projects to the appropriate center, and selected the appropriate aircraft, advised the other centers about flight safety and operations, and planned for the "acquisition, allocation, and disposition" of every aircraft. The Beeler report also recommended realigning the aircraft fleet based on each center's flight-research specialty. Thus, the FRC won control of aircraft and operations suited to its varied missions: experimental, general aviation, proficiency, supersonic, remotely piloted, and high risk, including highly modified flight vehicles. Meanwhile, Langley would assume responsibility for rotary-wing aircraft. The rest would be parceled out among the centers in accordance with such factors as safety, special facilities, program requirements, and the available labor pool.[393]

Despite all of these preparations, between the announcement of these reforms and the naming of the Flight Research Center for Hugh Dryden three years later, the ambitious objectives of the OAST committee all but vanished. Indeed, in 1976 the Flight Activities Office disappeared from the Dryden organization charts, where it had appeared briefly. Robert E. Smylie, the acting OAST associate administrator, attempted to resurrect the Beeler proposals during the same year. The effort ran aground when Smylie failed to win over the directors of Ames, Langley, and Lewis and then left headquarters to become deputy director of the Goddard Space Flight Center. Consequently, few aircraft destined for Dryden under the 1973 plan were actually transferred.

The failure of yet another attempt to consolidate of flight research inspired a climate of self-doubt among many at the DFRC. Some wondered whether the Ames, Langley, and Lewis directors resisted successfully because OAST lacked full confidence in Dryden's competence or because of Dryden's existing stable of projects.

393 Donald Bellman to those concerned, 28 Nov. 1972, Milt Thompson Collection, DFRC Historical Reference Collection; minutes of the OAST Flight Operations Review Committee, 23-24, 25, 26, 29-30 Jan. and 1 Feb. 1973, Ken Szalai Collection, DFRC Historical Reference Collection (first quoted passage from 1 Feb. minutes); De E. Beeler to distribution, 15 Dec. 1972, Milt Thompson Collection, DFRC Historical Reference Collection; Proposed Policy and Implementation Plan for OAST Aircraft Operations Office, 20 Apr. 1973, Milt Thompson Collection, DFRC Historical Reference Collection (second quoted passage); briefing, "Flight Research Activity Within NASA," n.d. [probably 1977], Milt Thompson Collection, DFRC Historical Reference Collection.

Others at Dryden criticized center authorities for assuming projects that bore little relationship to the needs of its main sponsors, the military and the aerospace industry. Requests by headquarters for figures related to DFRC staffing and facilities bred hearsay about personnel reductions and even about an organizational demotion of Dryden to test station status. The center's top legal adviser expressed the feeling of institutional malaise, saying "If we don't come out of our shell [that is, pay attention to the needs of external clients], there is very little chance that DFRC will remain a NASA center for more than another three years. We simply are not a viable and vital part of NASA at this time; and if we don't become so, we leave NASA little choice but to abolish DFRC as the least . . . valuable NASA Center."[394]

This period of self-examination initiated a conversation about Dryden's role, involving individuals both inside and outside the center. Famed D-558 and X-15 pilot Scott Crossfield, who now worked for the House Committee on Science and Technology, offered some suggestions to Gene Matranga, director of aeronautical projects at the DFRC. Crossfield urged Dryden leaders to recreate the Research Airplane Committee in order to rekindle some of the excitement and stature enjoyed during the X-15 program. He recommended specifically that the committee plan for an aerospace plane capable of penetrating "the void between the X-15 envelope and space." Crossfield was joined in his effort to revive past glories by Dryden Chief Engineer Milt Thompson. Thompson sought to renew a beneficial program launched during the early 1950s by Generals Laurence C. Craigie and Orval R. Cook of the U.S. Air Force. (See Chapter 5.) In exchange for its flight research on new fighter aircraft, the center won the right under this program to employ some of the world's most advanced combat planes in whatever experiments its researchers deemed worthwhile. Eventually, a model of nearly every aircraft in the Century Series appeared in the FRC's hangars and on its flight line. During the 1960s, this mutually beneficial relationship faltered as the Air Force produced fewer and fewer front-line fighters. Thereafter, individual military aircraft arrived at the center, but on an ad hoc, rather than a systematic, basis.

During the late 1970s, Milt Thompson, who as a young pilot flew many of these Air Force thoroughbreds, urged DFRC Director David Scott to reestablish the old ties to the Air Force. Thompson set his sights on two new military vehicles: the Air Force's F-16 and the Navy's F/A-18. Scott pressed the request. But rather than make a direct

[394] Not only did Dryden fail to become the focus of aircraft consolidation, but Langley also did not assume preeminence in rotorcraft flight research, despite the Beeler Committee's recommendation. DFRC Organization Chart, May 1976; briefing, "NASA Roles and Missions—DFRC's View," n.d. [probably 1977]; Hallion, *On the Frontier*, p. 235; personal notes, author unknown, "DFRC' [*sic*] Most Significant Problems," n.d. [probably 1976 or 1977]; DFRC Acting Director to Philip Culbertson (draft), n.d., [probably 1976 or 1977]; personal notes, "Potential Manpower Reductions," author unknown, n.d. [probably 1977]; personal notes, "Dryden Management Philosophy," author unknown, n.d., [probably 1976 to 1978]; Isaac T. Gillam to Leonard Jaffe, 10 January 1978; John C. Matthews to DFRC chief engineer, 7 July 1977, DFRC Historical Reference Collection (quoted passage). All unpublished sources are in the DFRC Historical Reference Collection; all but the DFRC Orgonizational Chart are in its Milt Thompson Collection.

appeal to the military services, he used persuasion on his counterparts in NASA headquarters, informing them about the many research possibilities if just two Air Force YF-16s about to be retired were diverted for DFRC use. Scott won the support of Ames Director Hans Mark, who recognized the value of these vehicles for research, but only so long as they were applied to such specific purposes as advanced controls work and high spin tests. Ultimately, Scott's efforts resulted in headquarters forming an intercenter study group to evaluate the utility of military vehicles for NASA flight research. Meantime, Milt Thompson made the case with Vice Admiral Forest S. "Pete" Petersen, commander of Naval Air Systems Command, fellow X-15 pilot, and a longtime friend. Thompson confided that with the onslaught of budget cuts, NASA's high performance flight-research work might have to be abandoned unless military aircraft were procured. Thompson also sent a similar plea to another one-time X-15 pilot, Maj. Gen. Robert Rushworth, now the vice commander of the Air Force's aeronautical systems division. Unfortunately, neither of his former comrades offered immediate help. But the attempt to summon support for this cause did raise awareness about the future of high performance research, clear the path for later (albeit ad hoc) loans of front-line Navy and Air Force airplanes, and offer a sense of positive action in the face of unfulfilled flight-research consolidation.[395]

Indeed, Milt Thompson's campaign encouraged Dryden management to try once more for centralization of their mission. David Scott again contacted his superiors in Washington and raised the subject in the context of coping with NASA's ever-shrinking pool of money and talent. The consolidation of flight research under Dryden, he said, would reduce duplication of projects, make more efficient use of the aircraft inventory, increase the effectiveness of research, and at the same time reduce costs. Scott presented the familiar litany of Dryden's unique assets: a 300-square-mile buffer of government land that prevented noise from disturbing neighboring communities and held ambient air pollution to a minimum; a 15,000-foot Air Force runway and a 44-square-mile dry lakebed; open, unobstructed skies; fine weather; and an unmatched complex of test facilities for weight and balance, heat and loads, rocket engines, and data tracking and acquisition. He also reminded the headquarters staff that the 1973 consolidation plan, if implemented, promised Dryden such important programs as the Ames tilt-rotor project. But once again, these arguments went unheeded. When Associate Administrator of OAST James Kramer paid a visit to Dryden in November 1978, he expressed an unequivocal viewpoint: in essence, Dryden was attempting to "take over the traditional role of the other centers. This approach made the other centers

[395] A. Scott Crossfield, technical consultant, Committee on Science and Technology, U.S. House of Representatives, to Gene Matranga, 30 Nov. 1977, (quoted passage); briefing, "Joint NASA/DOD New Production Aircraft Program"; Milt Thompson to David Scott, 2 Jan. 1976; David Scott to NASA HQ, 5 Jan. 1977; David Scott to Hans Mark, 5 May 1977; Hans Mark to David Scott, 3 June 1977; handwritten note, David Scott to Hans Mark, 27 Sept. 1977; A.M. Lovelace to David Scott, 28 Sept. 1977; Milt Thompson to Forrest Petersen, 29 Mar. 1977; Milt Thompson to Robert Rushworth, 21 Sept. 1977; Robert Rushworth to Milt Thompson, 6 Oct. 1977. These sources are in the Milt Thompson Collection, DFRC Historical Reference Collection.

uncomfortable." Instead, Kramer told his hosts that their "proper role . . . is to provide flight support to the rest of the agency."

With no place else to turn, the center staff looked inward for reform. Once again, Milt Thompson took a leading role. He inaugurated a Dryden Image Committee to recommend steps to raise the center's stature. The subsequent report issued in April 1979 made some candid admissions. It held DFRC managers accountable not just for inattention to such clients as industry and the armed forces, but also for lapses in allocating the center's resources to serve these key patrons. The report detected a disturbing pattern of initiating projects based on internal Dryden dynamics, rather than on the demands of external constituencies. Instead, the Image Committee urged consultation with the manufacturers and the military services before beginning new projects. Additionally, Thompson and his associates felt that once completed, flight-research programs should be terminated punctually, not allowed to linger. The findings also recommended that the DFRC become somewhat less airplane-centered–that is, concentrate more on the technical developments derived from flying the aircraft, than on the flight vehicles themselves. The panelists likewise referred to the absence of strategic planning, an oversight making it difficult to think beyond the day-to-day. Finally, Thompson's group thought the aerospace industry ought to be included in some of Dryden's most important deliberations so the center could better discern its essential needs. One highly-placed engineer underscored the necessity of closer collaborations with the private sector. Gene Matranga felt that by the late 1970s the center focused excessively on "specific configurations" at the expense of the practical design trade-offs often uppermost in the minds of the technical staffs at the nation's aircraft companies.[396] Clearly, by 1980, the Dryden Flight Research Center found itself accustomed to the idea of institutional renewal but not yet certain of its overall direction.

[396] David Scott to James Kramer, 20 Sept. 1977; briefing, "NASA Roles and Missions—DFRC's View," n.d., [probably 1977]; briefing, "Flight Research Activity Within NASA," n.d. [probably 1977 or 1978]; briefing, "Consolidation of Flight Activity Within NASA," [probably 1977 or 1978]; Milt Thompson, "Seventeen Years of Flight Research Consolidation," n.d. [probably 1978]; Berwin Kock to James Kramer, 30 December 1978; (first quoted passages); Berwin Kock to DFRC director, 6 Apr. 1979. All above are in the Milt Thompson Collection, DFRC Historical Reference Collection. Handwritten notes to interview of Gene Matranga by Richard P. Hallion, n.d. [probably 1976], Hallion Collection, DFRC Historical Reference Collection.

12

An Uneasy Marriage 1981-1991

Although NASA Dryden experienced a period of institutional reassessment during the 1970s, the center still reached many significant flight-research milestones. The Langley-sponsored supercritical wing underwent flight research at Dryden. After proving the worth of digital fly-by-wire, the center conducted approach and landing tests on the Shuttle. Yet, by 1980 a sense of disquiet still pervaded NASA flight research. An overall drop in civil service rolls within NASA added to the worry. Between 1977 and 1981, Dryden's share of NASA's reductions dropped staffing from 520 to 450, a fall of nearly 14 percent and a loss of 70 federal employees. Although serious enough, these reductions took on greater significance with the announcement of a reorganization of the center.

Early in 1979, a committee assembled by Dryden Director Ike Gillam informed the workforce about a restructuring and consolidation by which the six center directorates (research, data systems, flight operations, aeronautical projects, shuttle operations, and administration) would be reduced to three. Data systems, shuttle operations, and aeronautical projects disappeared. Administration remained, but all of the other functions were reconstituted into just two directorates: engineering and flight operations.

Intended by the committee to "better carry out [the center's] goals," these changes disheartened not just many Dryden workers but also many in the upper levels of the center's management and its senior technical staff. Chief Engineer Milt Thompson pondered related problems in an informal memo he provocatively entitled, "Why Dryden is not doing more innovative research." He blamed headquarters and his colleagues at Langley for criticizing the DFRC when it proposed long-term, higher-risk ventures, a practice that resulted in the DFRC choosing worthy pursuits having more immediate practical value. According to Thompson, the consequences were significant: Dryden no longer participated in projects of great promise but uncertain application (such as the lifting bodies). He attributed these changes, at least in part, to the weakened position of NASA's center directors in relation to headquarters, especially in regards to the expenditure of discretionary funds. Without adequate seed money to develop good ideas in their formative stages, Paul Bikle would have been unable to start the lifting-body program on local initiative. Milt Thompson concluded that to infuse Dryden with new adventurous projects, NASA headquarters needed to increase the center's discretionary funding to "meaningful levels," around one million dollars. Secondly, Washington needed to tolerate the diversion of some multiyear funding to these more daring, but potentially more rewarding, endeavors.[397]

Thompson's desire to achieve increased institutional autonomy failed outright. Instead of gaining more local control, Dryden lost its independence entirely. A surprise announcement on 27 April 1981 informed the DFRC workforce that the center would be merged with the Ames Research Center on 1 October and assume subordinate status as a NASA facility. So complete was the cloak of silence prior to the notification that even the Dryden public affairs director confessed ignorance, saying "It hit us by surprise. We have no idea what it's going to mean." The reason for the forced wedding of Ames and Dryden soon became apparent. Caught in the vise of congressional budget cuts, NASA headquarters authorities chose to prove their support of fiscal austerity by offering to close two of its field units, an offer that proved to be much less than met the eye. In actuality, these two small field operations would be preserved but consolidated with two larger organizations. Dryden became a constituent part of Ames, and the Wallops Island, Virginia, Flight Center became part of the Goddard Space Flight Center in Greenbelt, Maryland. The motivations presented to the public differed from those of bureaucratic survival. A press release described the new arrangement as an attempt to "focus the resources of each of the installations on what it can do best. The close relationship between Ames' and Dryden's efforts in aeronautical programs as well as the unique facility capabilities and the physical proximity of the installations provides an opportunity to improve overall program effectiveness." Despite this formal statement and Dryden's continuation as a functioning entity, the loss of institutional autonomy hit home. The new words above the main entrance to the Dryden Flight Research Center explained everything: "Ames Research Center, Dryden Flight Research Facility."

To conform to the prevailing conditions, Dryden's 1979 restructuring itself underwent revision. After a gestation period of many weeks, representatives from NASA headquarters' Office of Aeronautics and Space Technology (OAST) and the two merging centers—in addition to Ames Director Clarence Syvertson and Dryden Director Isaac Gillam—announced a new organizational plan. Henceforth, Ames' science platform vehicles as well as its rotorcraft fell under the jurisdiction of the Dryden flight operations office. At the same time, nearly all of Dryden's administrative activities were transferred to Moffett Field (with the exception of a new site manager's office to provide stewardship of Ames interests at Dryden). Moreover, the entire Dryden projects management office received orders to move to Mountain View, California, and integrate itself with the Ames directorate of aeronautics. Although none of the Dryden workforce faced mandatory relocations or terminations, in time the disruptive consequences of some of the reorganization became more apparent. As a result, the

[397] See Appendix B. Handwritten DFRC Manning Chart, Sept. 1977 to Sept. 1986, n.d., Milt Thompson Collection, DFRC Historical Reference Collection; DFRC Organization Charts, May 1976, Aug. 1981, DFRC Historical Reference Collection; "Dryden Research Center Undergoing Reorganization," *Antelope Valley Press*, 4 Feb. 1979; handwritten essay, "Why Dryden is not doing more innovative research," n.d., Milt Thompson Collection, DFRC Historical Reference Collection; comments of an anonymous reviewer on a draft of this section.

project management directorate stayed at Edwards and two marque projects, the Tilt-Rotor and Quiet Short-Haul Research Aircraft, remained at Ames. But the loss of Dryden's administrative capacity did not change. Ames Director Syvertson tried to be positive when he announced these impending realignments to both staffs. "I am firmly convinced that Ames and Dryden can be merged into a single effective and efficient organization for the conduct of advanced aeronautical research," he said. Isaac Gillam, meanwhile, left California to become a special assistant to the administrator at NASA headquarters. To fill the void, Syvertson named John Manke, noted lifting-body pilot and Dryden's director of flight operations, to the dual roles of director of Ames/Dryden flight operations *and* on-site manager of the Dryden Flight Research Facility.[398]

A Joint Project

The first opportunity for the two centers to pool their staffs and their separate technical strengths occurred during a project launched well before the 1981 merger. Ames aerodynamicist Robert T. Jones, celebrated for his sweptwing work at Langley Research Center during the 1940s, arrived at another bold insight some time later. He predicted sharply improved aerodynamic performance among transport aircraft by mounting atop their fuselage a single wing that pivoted at its midpoint. Jones calculated that a vehicle flying at 1,000 miles per hour with its wing swung to an angle of up to 60 degrees would consume half the fuel of a conventional aircraft cruising at the same speed. He also advocated the concept because his research suggested that the pivoting wing would dampen sonic booms and reduce inflight noise.

To test Jones's theory, the Flight Research Center contracted to the Rutan Aircraft Factory in nearby Mojave, California, to design a prototype and to calculate the cost of engineering and fabricating a simple flying prototype, known as the oblique or the yawed wing. In similar fashion to the M2-F1 precedent of building a prototype under a tight-fisted budget using local talent, Rutan made a preliminary estimate of engineering and construction costs of about $87,000 in December 1975. The company presented Dryden with a detailed design nine months later. Meanwhile, in October 1976, Ames Research Center investigators—abetted by their Dryden counterparts—completed three flights of a remotely-piloted version of the oblique-wing aircraft, collecting data on the vehicle's stability and control at pivoting angles up to 45 degrees. The Ames team then correlated these findings in the 12-foot wind tunnel with a model

[398] Essay by Milt Thompson on the Ames/Dryden Consolidation, n.d., pp. 1-2, 6, Milt Thompson Collection, DFRC Historical Reference Collection; "Takeover Surprises Dryden Officials," Public Affairs clipping from unidentified newspaper, 29 Apr. 1981, DFRC Historical Reference Collection (first quoted passage); Dryden news release, "NASA Consolidates Center Operations," n.d., DFRC Historical Reference Collection, (second quoted passage); Donald L. Mallick, unpublished memoir, 1995, p. 302, File No. 001421, NASA HQ Historical Reference Collection; "Ames/Dryden Consolidation to Begin October 1, 1981," Dryden *X-Press*, 7 Aug. 1981, p. 2, DFRC Historical Reference Collection; "Gillam Named to NASA Position," *Bakersfield Californian*, 23 Sept. 1981, p. B6, DFRC Historical Reference Collection; NASA Ames news release, "Manke Named to Head Dryden and Ames Flight Operations," 6 Oct. 1981, DFRC Historical Reference Collection.

Ames-Dryden AD-1 oblique-wing testbed, flying with its adjustable wing in the fully swept (60-degree) position. (NASA photo ECN-13302B)

one-sixth the size of the full-scale vehicle under consideration. The tests suggested the need for re-design of the tail configuration, after which NASA awarded the fabrication contract for the full-scale, piloted vehicle. By this time (1977), the experimental aircraft had received a permanent designation: the Ames-Dryden AD-1.[399]

In a period of budgetary retrenchment, not everyone involved in NASA flight research gave the AD-1 unqualified support. James Kramer, NASA headquarters' acting associate administrator for aeronautics and space technology, expressed reservations about its value. Dryden Director Isaac Gillam replied in January 1978 with two of the classic reasons for conducting flight research. Gillam reminded Kramer that "NASA has a responsibility to explore all concepts regardless of the future applications." While the pivoting wing held promise for transonic flight, the knowledge derived from flying it—whether the project proved to be a success or a failure—held the greater promise. Moreover, despite the extensive model flights and wind-tunnel tests of the yaw-wing, the vehicle still required full-scale, piloted flights "to identify control system and handling qualities problems in conjunction with structural and

[399] Briefing charts, "Oblique Wing Potential Benefits," n.d.; "The AD-1," NASA Dryden Fact Sheet, May 1999; Lane E. Wallace, *Flights of Discovery: 50 Years at the NASA Dryden Flight Research Center* (Washington, DC: NASA SP 4309, 1996), p. 94; Burt Rutan and George Mead, "Feasibility Study of a Small, Low Cost Subsonic, Yawed-Wing Experimental Aircraft," Dec. 1975, pp. 20-22; Dryden Project Highlights, Sept. and Dec. 1976, Feb. 1977; "Annual Report of Research and Technology Accomplishments and Applications, FY 1977," DFRC, p. 7; "Annual Report on Research and Technology Accomplishments, FY 1979," DFRC, p. 7. All unpublished items are in the DFRC Historical Reference Collection.

environmental interactions . . . and [to] establish industry confidence for future application." Headquarters must have been persuaded; construction of the aircraft by the Ames Industrial Corporation of Bohemia, New York, continued throughout 1978 and early 1979. In March 1979, the completed vehicle arrived at Dryden aboard an Air National Guard C-130. A series of ground-vibration tests and static loads tests, the installation of the AD-1's instrumentation, and a complete system checkout followed. The year 1978 may have begun with skepticism about the AD-1, but near the end of 1979 former Ames Director Hans Mark described the concept as "an interesting new idea that should be developed for high subsonic and perhaps low supersonic flight."[400]

Late 1979 also brought the initiation of the AD-1 flight-research program. The aircraft that came off the C-130 was small, light, and inexpensive. Only 38.8 feet long with a wingspan of just 32.3 feet, it weighed a mere 2,100 pounds (with fuel and pilot) and cost under $300,000 to design and construct. For that outlay, NASA bought simplicity itself: two off-the-shelf French Microturbo TRS-18 engines, fixed landing gear, no hydraulics, no altitude indicator, no compass, no rate-of-climb instrument, and manual pilot ejection. It was designed to fly up to 175 knots (far below the transonic range envisioned by R.T. Jones) and up to 15,000 feet. Still, the flight-research program set some ambitious goals: comparing the aircraft's aerodynamics to the predictions of the wind tunnel and the flying model, testing the aeroelastic properties of the airfoil, "learn[ing] the nature and complexity" of yaw-wing flight control, and evaluating its flying and handling characteristics.[401]

After a number of taxi tests on 20 December 1979, project pilot Thomas McMurtry attempted the first piloted flights of the oblique-wing aircraft the following day. First, he flew down the runway just 50 feet above the surface. Then, with the wing still in the perpendicular position, McMurtry achieved an altitude of 10,000 feet and a speed of 140 knots. Six more attempts in January 1980 proved that with the wing unswept the machine operated without incident. Finally, on the ninth flight on 1 February, he pivoted the right side of the wing forward by 15 degrees at five degree increments. At each setting he conducted a series of flutter control inputs at speeds ranging from 110 to 150 knots. McMurtry found the AD-1 handled "extremely well" under these conditions. During the following months, McMurtry and fellow project pilot Fitzhugh Fulton assessed the aircraft's aerodynamics, flight loads, and handling qualities through a series of maneuvers, including pullups and pushovers, descents,

[400] Isaac Gillam to James Kramer, 31 Jan. 1978, (first and second quoted passages); Dryden Project Highlights, Mar. 1978, Apr., June, Aug. 1979; "Annual Report of Research and Technology Accomplishments and Applications, FY 1978," DFRC, p. 15; Hans Mark to Milton Thompson, 31 Oct. 1979 (third quoted passage). All are in the DFRC Historical Reference Collection.

[401] W.H. Andrews, A.G. Sim, R.C. Monaghan, L.R. Felt, T.C. McMurtry, and R.C. Smith, "AD-1 Oblique Wing Aircraft Program," Paper No. 801180, Society of Automotive Engineers, Technical Paper Series, 3 Oct. 1980, Los Angeles, CA, pp. 2-3, 5; Wallace, *Flights of Discovery*, p. 94; T.C. McMurtry, A.G. Sim, and W.H. Andrews, "AD-1 Oblique Wing Aircraft Program," The First Flight Testing Conference of the AIAA, SETP, SFTE, SAE, ITEA, and the IEEE, 11-13 Nov. 1981, Las Vegas, NV, p. 2.

aileron rolls, and 1 g decelerations. These flights began to reveal a profile of a vehicle possessing several distinct characteristics: cross-coupling between lateral-directional and longitudinal movements; lateral-directional trim changes; and pronounced aeroelastic effects.[402]

The first 18 flights of this unusual experimental vehicle went reasonably well. However, the 19th resulted in a pause in the flight-research program lasting more than seven months. On 12 August 1980, as the wing pivoted on the left side between 15 and 16 degrees, McMurtry experienced not wing flutter itself but a narrowing of the margin between safe flight and the onset of this dangerous phenomenon. In itself, this was nothing new. The same phenomenon had occurred before, but at an angle closer to 20 degrees. Still, the 20th flight of AD-1 was postponed pending an investigation. In the meantime, the covering on the wing (a reinforced fiberglass skin) underwent examination, and ground vibration tests–undertaken before the beginning of flight research, were conducted again. The Ames Research Center made another essential contribution to the program. Working under the lead of a Lockheed advanced data processing team, Ames engineers helped prepare a computer model based on the vibration data. The results indicated the phenomenon posed no threat to the structural integrity of the aircraft; it related to the overall structure of the vehicle rather than to the swept wing. The solution lay in revising operational procedures and installing an aileron damper. Similarly, flight research on the AD-1 at the limits of wing sweep raised questions. Past the 50-degree setting and up to the maximum 60 degrees, the NASA pilots found it difficult to control the aircraft. However, the oblique wing's simulator—programmed with the static and dynamic derivatives from the flights—suggested that a control system aided by artificial pitch and roll damping would improve the aircraft's handling qualities at the extremes.

These handling problems surprised no one associated with the project. The AD-1 had been undertaken at minimal cost, unaided by electronic stability augmentation systems. Given these constraints, engineers involved with the oblique wing expected only satisfactory flying qualities, sufficient to allow researchers to achieve the goal of assessing the aircraft's inflight aerodynamics.[403]

Tom McMurtry and Fitz Fulton continued to share AD-1 piloting responsibilities another 15 times. After the 38th flight on 3 September 1981, the oblique wing flew a number of demonstrations (two of them at Ames) and also began a program of pilot familiarization. Beginning in January and ending in May 1982, 13 guest aviators—representing both the Air Force and NASA—took the opportunity to fly this unique

[402] Dryden Project Highlights, Dec. 1979; "AD-1 Swings it's [*sic*] Wings in Flight," Dryden *X-Press*, 8 Feb. 1980, (quoted passage); Tony Landis, compiler, "AD-1 Flight Log," July 1996, p. 1. All of the above are in the DFRC Historical Reference Collection. McMurtry, Sim, and Andrews, "AD-1 Oblique Wing Program," pp. 2-4.

[403] McMurtry, Sim, and Andrews, "AD-1 Oblique Wing Aircraft Program," pp. 1-4; Dryden Project Highlights, Sept., Oct., Nov. 1980, Mar. 1981; "AD-1 Flight Log," p. 1; "Annual Report on Research and Technology, FY 1981," DFRC, pp. 11; comments of Alex Sim on a draft of this section, All are in the DFRC Historical Reference Collection.

machine. After McMurtry and Fulton made four last research flights in the AD-1 during June, it appeared at the Oshkosh air show and went aloft for the 79th and final time on 7 August 1982.

Meanwhile, the project pilots, as well as the guest pilots, evaluated this flying incarnation of Robert T. Jones's design. They found pivot angles below 30 degrees resulted in satisfactory flying qualities in all of the required maneuvers. From 30 to 45 degrees, they encountered a deterioration in their capacity to control the aircraft. The sharpest decline occurred between 45 and 60 degrees. During ailerons rolls, for instance, the machine often failed to roll right without the application of the rudder. However, the project engineer—whose main goals did not include ease of handling—thought that satisfactory flying qualities were attainable even at high-sweep angles with the installation of a simple control augmentation system.[404]

The Shuttle Orbiter Revisited

The approach and landing tests of the Orbiter *Enterprise* during 1977 had established Dryden's *bona fides* as a major participant in the Space Shuttle program. But much more work lay ahead. In an attempt to determine whether the flight loads on the Orbiter altered the shape of its heat-resistant tiles, during the months before the launch of STS-1, Dryden research pilots flew 60 missions simulating the maneuvers of the Shuttle. The ceramic tiles were tested two ways: first, glued to an F-15 fighter and second, mounted on a specially-designed flight-test fixture (FTF) carried aboard an F-104. The fixture, employed originally for panel flutter tests, consisted of a fin-like shape attached to the underside of the aircraft, aligned along the FTF's longitudinal axis and the centerline of the fuselage. The fixture acquired data using its own pulse code modulation (PCM) system, recorded by telemetry and by on-board equipment. The verdict: better methods of adhesion needed to be developed. This finding raised questions about the tiles themselves. Undoubtedly effective in deflecting the torrid temperatures associated with the Orbiter's return from space, they nonetheless needed to be removed, reshaped (to conform to the contours of the spacecraft), and rebonded after every mission.

This time-consuming and expensive process led to a search for a more efficient method of sheathing the Orbiter. Two substitutes appeared promising—one known as felt reusable surface insulation (FRSI) and a second known as advanced flexible reusable surface insulation (AFRSI). Both substances consisted of heat-treated aromatic polyimide encased in white silicon and were highly malleable, assuming the shape of whatever they adhered to. Initially, FRSI and AFRSI appeared to be possible successors

[404] "AD-1 Flight Log," pp. 1-4, DFRC Historical Reference Collection; Dryden Project Highlights, Jan., July 1982, DFRC Historical Reference Collection; Alex G. Sim and Robert E. Curry, "Flight Characteristics of the AD-1 Oblique-Wing Aircraft," (Edwards, CA: NASA TP 2223, 1985), pp. 7-10; W.D. Painter, "AD-1 Oblique Wing Research Aircraft Pilot Evaluation Program," AIAA Aircraft Design, Systems, and Technology Meeting, 17-19 Oct. 1983, Fort Worth, TX, pp. 1-6; comments of Alex Sim on the AD-1, Feb. 2000.

Aerial view of the Hugh L. Dryden Flight Research Center in 1997, looking north to the Rogers Dry Lake. The Shuttle support facilities are at the top of the photo with the main hangar, completed in 1954, near the center of the photo. (NASA photo EC97-44165-35)

to the tiles. Flights aboard the F-104 test fixture showed that even at air loads forty percent higher than those of a Shuttle launch, no failures occurred. However, adhesion tests resulted in less satisfactory findings. On the F-104 tests, the two materials were fastened to flat surfaces. On a subsequent experiment aboard *Challenger* (STS-6, launched 4 April 1983), the impingement of the vortices in the atmosphere caused those portions of AFRSI attached to curved surfaces to break down. As a result, although troublesome, the Shuttle tiles remained in use at least on the windward side of the vehicle.[405]

[405] Bruce Powers and Shafan Sarrafian, "Simulation Studies of Alternate Longitudinal Control Systems for the Space Shuttle Orbiter in the Landing Regime," in A Collection of Technical Papers Presented at the AIAA Atmospheric Flight Mechanics Conference, 18-20 Aug. 1986, Williamsburg, VA, pp. 1, 4; Robert Meyer, "A Unique Test Facility: Description and Results," pre-print of a paper presented at the 13th ICAS Congress/AIAA Aircraft Systems and Technology Conference, 22-27 Aug. 1982, Seattle, WA, pp. 1-2, 5, DFRC Historical Reference Collection; R.R. Meyer, C.R. Jarvis, "In-Flight Aerodynamic Load Testing of the Shuttle Thermal Protection System," The First Flight Testing Conference of the AIAA, SETP, SFTE, SAE, ITEA, IEEE, 11-13 Nov. 1981, Las Vegas, NV, pp. 1-2; Dryden Fact Sheet, "Space Shuttles and the Dryden Flight Research Center," Sept. 1995, p. 3, DFRC Historical Reference Collection; Bianca Trujillo, Robert Meyer, and Paul Sawko, "In-Flight Load-Testing of Advanced Shuttle Thermal Protection Systems" (Washington, DC: NASA TM 86024, 1983), pp. 1, 4; Timothy Moes and Robert Meyer, "In-Flight Investigation of Shuttle Tile Pressure Orifice Installations" (Washington, DC: NASA TM 4219, 1990), pp. 1, 6.

Key members of the Dryden research pilot staff (viewer's left to right): Milt Thompson, Fitz Fulton, Bruce Peterson, Don Mallick, John Manke, Einar Enevoldson, Bill Dana, and Tom McMurtry. (NASA photo ECN 4926)

This conclusion to the tile tests did not end Dryden's contributions to the Shuttle program. In 1990, the Johnson Space Center enlisted the Dryden B-52 aircraft in testing a newly developed drag chute for the Shuttle Orbiter. The project engineers felt that a chute deployed on the Kennedy and the Dryden runways would not only improve the safety of landings but also decrease the extent of wear on the Orbiter's braking system. Research pilots Ed Schneider and Gordon Fullerton had their work cut out: to achieve proper test conditions, the big bomber needed to be landed at 190 knots, well above its regular speed. However, the test program—completed in eight flights during 1991—proved the B-52 to be the perfect testbed, already outfitted for drag chute investigations and capable of applying about the same load on the chute as the Orbiter. As a result of this research, the Shuttle *Endeavour* deployed the system during its first return from space in 1992. It underwent further modifications based on data from subsequent Shuttle landings. Eventually, all of the Orbiters were retrofitted with this device.

The 1990s witnessed another important flight-research project related to Orbiter landings. During 1992, Shuttle authorities at the Johnson Spaceflight Center, concerned about extending Orbiter tire life and hoping to achieve safe landings in crosswinds up to 20 knots, persuaded engineers at Dryden to launch tire and landing tests. The existing limit of 15 knots often inhibited Shuttle operations and the Orbiter's costly tires wore out quickly under high loadings and during high-speed approaches as they roared down the Kennedy Center runway, which had an abrasive surface designed for safety in damp Florida conditions. Although conceived more than a decade after the first Shuttle launch, the tire and landing tests were consequential. They lasted for two

Shuttle Orbiter Discovery *makes a landing at Edwards Air Force Base in 1989. During the 1980s and 1990s, Dryden undertook Orbiter studies relating to heat resistant tiles, tire longevity, and a drag chute. (NASA Photo EC89-0055-2)*

years (155 flights) and enlisted the considerable skills of research pilot Gordon Fullerton, a veteran of the 1977 approach and landing tests and later a Shuttle commander.

The aircraft used to simulate the Orbiter was also a seasoned performer. Once a part of the American Airlines inventory and more recently operated by NASA Ames, the vintage 1962 Convair CV-990 possessed two characteristics necessary to mimic the Orbiter:it weighed about the same and it had roughly the same landing speed (152 knots). To prepare the vehicle for its flights, Dryden technicians made significant modifications inside the fuselage. They mounted at midpoint on its belly a test fixture designed to approximate the vertical tire load and the yaw angle on the Orbiter. Capable of operating with tires loaded to as much as 150,000 pounds, the apparatus included parts actually used on an Orbiter, although for safety the CV-990's original landing gear remained down during the tests.

Once the CV-990 touched down on its own gear, a high-pressure hydraulic system lowered the test gear according to the commands of an on-board computer, which caused the test gear to approximate Orbiter landing forces. A computer-controlled steering system turned the tire to match the skid angles that the Shuttle encountered during crosswinds. Simulated Orbiter braking was usually applied toward the end of the landing roll. As soon as an individual landing test ended, the loads on the Shuttle tire were reduced, and the landing roll was completed on the CV-990's own landing gear.

In its incarnation as the Landing Systems Research Aircraft (or LSRA), the CV-990 increased in unloaded weight from 115,000 to 177,000 pounds. Gordon Fullerton's standard flight maneuver both at Dryden and on the rougher Kennedy runways consisted of a landing approach, touchdown, deceleration to the required speed, and deployment of the test fixture. Instrumentation yielded data about vertical loads, tire braking, and slip angles. The findings altered the flight operations of the Shuttle in several respects. The Dryden landings provided valuable data about touchdowns in higher crosswinds, and the information about vertical loads resulted in revisions in the Orbiter simulator's tire drag model. Finally, the Kennedy flights persuaded authorities there to repave the runways for a smoother surface, a decision supported by the LSRA's tire-wear data.[406]

With the last of the LSRA tests, Dryden concluded a quarter century of specific Shuttle flight research. However, Dryden's broader commitment did not end in 1995. Since the initial Shuttle launch 14 years earlier, the Dryden staff—and the institution itself—had received the notoriety associated with supporting the landing site of the nation's banner space program. As each Orbiter neared Edwards Air Force Base on final approach, a team of Dryden employees staffed the Shuttle mission control room. Technicians from Dryden and other NASA centers readied equipment needed to service the incoming spacecraft, medical personnel prepared to examine the astronauts, and public affairs specialists handled the international press.

Then, after landing and being prepared for ferrying, the Orbiter was mated at Dryden to the 747 carrier aircraft and flown to its next destination. Gradually, however, Dryden's role diminished. From 1981 to 1996, nearly 60 percent of Shuttle missions (45 of 76 flights) ended at Edwards. But this period of intense activity lasted only through the 1980s, when Dryden managed 80 percent of all landings. Once Kennedy became the primary and Dryden the alternate location, Dryden's contribution receded to 45 percent during the early 1990s and a mere 30 percent after 1992. Still, even Dryden's limited participation yielded important rewards. It resulted in an expanded budget to satisfy Shuttle requirements, enhanced national and international recognition, and (as a consequence of this affiliation) wider acknowledgment of Dryden's flight research role. Yet, with all the advantages came a disadvantage. When the mammoth X-15 program wound down during the 1960s, some felt Dryden's reason for being ended with it. Similarly, as the Shuttle became associated less and less with Dryden,

[406] Wallace, *Flights of Discovery*, p. 142; Roy Bryant, B-52 program manager, "History of the B-52 at Dryden," completed 8 Oct. 1999, p. 2; Dryden Project Highlights for Oct. 1990; Don Nolan, "CV-990 Expands Orbiter Crosswind Limits," Dryden *X-Press*, Sept. 1994, p. 1; Dryden Fact Sheet, "Landing Systems Research Aircraft," Nov. 1994; Cheryl Heathcock, "CV-990 Completes Orbiter Wheel and Tire Tests," Dryden *X-Press*, Sept. 1995, p. 1. All are in the DFRC Historical Reference Collection. C.D. Michalopoulos and David Hamilton, "Orbiter Tire Traction and Wear," AIAA Report No. 95-1256-CP, 1995, pp. 851, 859; John Carter and Christopher Nagy, "The NASA Landing Gear Test Airplane" (Washington, DC: NASA TM 4703, 1995), pp. 1-3, 5-7; SAE Briefing, "CV-990 Landing Systems Research Aircraft," 9 April 1990, and SAE briefing, "LSRA Status Update" May 1994, both in the DFRC Historical Reference Collection; comments of anonymous reader of this chapter, Feb. 2000.

many at Dryden worried about the future. As a result, during the 1990s Dryden officials undertook two initiatives deemed essential to Dryden's institutional survival: making significant contributions to the next generation of space launch vehicles and achieving autonomy from the Ames Research Center.[407]

Aerodynamic Efficiency

Both initiatives awaited events partly beyond Dryden's control. In the meantime, numerous important projects were conceived and pursued, among the most significant being a series of investigations relating to aerodynamic efficiency. The first of these involved a simple device ultimately tested in 23 different wind tunnels: an instrumented, precision-made, 10-degree-sharp cone. Between 1970 and 1978, air-flow measurements taken with this cone at subsonic, transonic, and supersonic speeds defined the extent of turbulence generated inside each tunnel. But even after the testing, a fundamental question remained: how did the tunnel data compare with actual flight conditions? Eager to know how well their equipment approximated reality, researchers at the Air Force's Arnold Engineering Development Center (AEDC), who developed the device, sponsored flight research at Dryden to answer this question. The Air Force supplied $250,000 and NASA Headquarters $169,000 for the total flight program, lasting (with ground preparation) most of 1978. Using the same cone and instrumentation devised by the AEDC, engineers mounted the cone on the nose of an F-15 aircraft on loan from the Air Force. NASA research pilots flew 30 missions ranging in elevation from 5,000 to 50,000 feet and in speed from Mach 0.5 to 2.0. All data were recorded at zero angle of attack. Originally, AEDC engineers hoped to derive from the flight experiment a mathematical model by which to correct the inaccuracies in the wind-tunnel data. But the results provided a pleasant surprise. Below Mach 1.4 (Mach 1.2 in some analyses), only negligible differences existed between turbulence encountered in the tunnels and turbulence measured on the F-15. Above that speed, however, the differences widened. Three Dryden participants concluded,

> . . . this experiment is valuable because of the thousands of hours of wind-tunnel testing time . . . which will be more useful, more conclusive and more meaningful because of the opportunity to evaluate the respective wind tunnels against a common standard which was obtained under unusually controlled conditions. It is fair to say

[407] While Dryden's role in Shuttle operations declined during the 1990s, Orbiter flights continued to have engineering and scientific value to DFRF and DFRC researchers. For instance, a small number of engineers at Dryden—obtaining data from Shuttle missions as if they were research flights—continued to publish papers about Orbiter aerodynamics. For examples of these papers, see David F. Fisher, compiler, *Fifty Years of Flight Research: Annotated Bibliography of Technical Publications of the NASA Dryden Flight Research Center, 1946-1996*, (Edwards, CA: NASA TP 1999-206568, 1999), pp. 378, 384, 417. Dryden Fact Sheet, "Space Shuttles and the Dryden Flight Research Center," Sept. 1995, p. 6; Dryden Fact Sheet, "Completed Space Shuttle Missions," May 1996; interviewed of Richard Day by Richard P. Hallion, 24 Feb. 1977, pp. 56-58; these sources are in the DFRC Historical Reference Collection.

Mounted on the nose of an F-15A fighter, the 10-degree-sharp cone helped aerodynamicists determine the accuracy of their wind-tunnel readings by comparing the data from the tunnels to that gleaned from actual flight. (NASA Photo ECN-9810)

that this experiment has the "leverage" . . . to increase the value of . . . literally hundreds of other experiments and tests.[408]

The protean mind of Richard Whitcomb also contributed to the drive for aerodynamic efficiency during this general period. He proposed to update an old idea. At the end of the nineteenth century, celebrated British aerodynamicist F.W. Lanchester received a patent for a new concept: a vertical surface mounted at the wingtip, designed to reduce drag. Seventy-five years later, Whitcomb took up where Lanchester left off. Wind-tunnel tests and analyses convinced him that Lanchester's idea would succeed only if the vertical plane (called a winglet) produced side forces, thus diminishing induced drag at the wing tips. He predicted improved cruise efficiencies of six to nine

[408] David Scott to James Kramer, 24 Jan. 1977, DFRC Historical Reference Collection; Edwin Saltzman and Theodore Ayers, *Selected Examples of NACA/NASA Supersonic Flight Research,* (Edwards, CA: NASA SP-513, 1995), p. 31; N. Sam Dougherty, Jr., "Tunnel/Flight Res Correlation," Project P32A-BOA, presented at a USAF/NASA Planning Conference, AEDC; "10-Deg Transition Cone Flight Test," Edwards AFB, CA, 27 May 1976, DFRC Historical Reference Collection; N.S. Dougherty, Jr. and D.F. Fisher, "Boundary Layer transition on a 10-Degree Cone: Wind Tunnel/Flight Data Correlation," AIAA 18th Aerospace Sciences Meeting, 14-16 Jan. 1980, Pasadena, CA, pp. 1-2, 6-7; David F. Fisher and N. Sam Dougherty, Jr., "In-Flight Transition Measurement on a 10 Degree Cone at Mach Numbers from 0.5 to 2.0," (Edwards, CA: NASA TP 1971, 1982), pp. 1-2, 6-7, 14-15; Dryden Project Highlights, Jan. and Nov. 1978, DFRC Historical Reference Collection; David Fisher, Edwin Saltzman, and Theodore Ayers to DFRC director of flight operations, 30 Nov. 1978, DFRC Historical Reference Collection (block quote).

An Air Force KC-135 tanker equipped with Richard Whitcomb's innovative winglets, adopted by jetliners worldwide to improve fuel efficiency. Winglet flight research lasted from July 1979 to January 1981. (NASA Photo EC79-11484)

percent in commercial aircraft employing winglets. Conceived during the 1970s—the Organization of Petroleum Exporting Countries (OPEC) quadrupled petroleum prices in 1973 and tripled them between 1974 and 1980—Whitcomb's winglets soon assumed practical significance. The airline industry paid strict attention to any technology that could be retrofitted to existing jetliners and yield substantial fuel savings. Toward the end of the 1970s, NASA responded officially to the OPEC crisis with a program called Aircraft Energy Efficiency (ACEE).

The ACEE managers linked Whitcomb's design to the Air Force's objective of upgrading its KC-135 tanker fleet. In a memorandum of understanding, the Air Force agreed to provide a KC-135 and money to construct the winglets; NASA agreed to contract for and install the new structures and to undertake an extensive program of flight research at Dryden. A KC-135A model arrived at Dryden at the end of 1977 and was instrumented for the research effort. Starting in July 1979 and ending in January 1981, the flight-research program tested several different winglet configurations fabricated by Boeing, subjecting them to envelope expansion, flutter tests, and performance measurements. The research pilots flew the KC-135 at Mach 0.70, 0.75, and 0.80 at a nominal altitude of 36,000 feet. Small in comparison to a big commercial aircraft, the winglets were nonetheless imposing. They measured more than nine feet tall, were over six feet at the base, and weighed some 215 pounds each. Despite delays

The Variable Sweep Transition Flight Experiment (VSTFE), flown on a Navy F-14A aircraft, entailed the installation of two composite wing "gloves" to evaluate the extent of natural laminar airflow during swept-wing flight. (NASA Photo EC86 33491-09)

due to fuel leaks in the tanker caused by a broken wing spar cap, the flight research yielded impressive results. Its data deviated little from Whitcomb's wind-tunnel predictions. The aircraft's flying qualities and flight loads behaved as expected. Finally, the winglet with a 15-degree cant angle and minus-four-degree-incidence configuration proved to be the best performer, capable of increasing the efficiency of the aircraft by about six percent with acceptable flutter effects. In due course, much of the airline industry embraced the winglets. The result is still plain to see on airport ramps throughout the world on Boeing 747-400s, MD-11s, and other airliners.[409]

Quick on the heels of the winglet flight research, Dryden inaugurated a series of aerodynamic investigations to achieve smoother airflow over airfoils. Dryden researchers experimented both with *passive* and *active* laminar-flow control (LFC) techniques. They tried to achieve active LFC by cooling the airfoil surface or by

[409] Richard T. Whitcomb, "Research Methods for Reducing the Aerodynamic Drag at Transonic Speeds," the ICASE/LaRC Eastman Jacobs Lectureship Award, the Inaugural Eastman Jacobs Lecture, NASA LaRC, 14 Nov. 1994, pp. 8-9; interview of Russ Barber by J. D. Hunley, NASA DFRC, 21 Dec. 1988, pp. 3-4, DFRC Historical Reference Collection; "Winglet Flight Testing to Begin at Dryden," Dryden *X-Press*, 27 July 1979, p. 2, DFRC Historical Reference Collection; Dryden Project Highlights, Feb. and June 1980, DFRC Historical Reference Collection; "Annual Report on Research and Technology, FY 1981," NASA TM 81367, 1981, pp. 19-20.

mechanically bleeding a small portion of the boundary layer through holes or slots on the wing surface. In passive or *natural* laminar flow, the researchers avoided mechanical contrivances, instead designing airfoil surfaces so that air pressure fell sharply near the leading edge and rose gradually toward the trailing edge. Active LFC projects included the JetStar Leading-Edge Flight Test and the F-16XL supersonic LFC project. Both the F-111 natural laminar-flow research and the F-14 Variable Sweep Transition Flight Experiment (VSTFE) were examples of passive LFC.

Like winglets, the initial LFC project of the 1980s fell under the rubric of NASA's Aircraft Energy Efficiency program, initiated to foster fuel savings in aeronautics. Two manufacturers offered technical proposals. Working with Langley researchers, Lockheed and Douglas engineering teams each proposed new wing leading edges that combined three essential and integrated features necessary for successful, practical laminar flow: suction, de-icing, and insect protection. The Lockheed system envisioned a network of ducts incorporated into the wing structure, consisting mainly of graphite epoxy composites. Laminar suction resulted from 27 spanwise slots cut into the upper and lower wing surfaces. Insect and ice abatement occurred by means of fluid fed through eight of the slots on the leading edge. Douglas, on the other hand, offered less ducting and a configuration consisting of perforated suction strips. A retractable Krueger flap at the front portion of the wing served as a barrier to insects, behind which spray nozzles were mounted to wet the leading edge.

This flight-research program, known as the Leading-Edge Flight Test, required an aircraft capable of simulating the big transports and airliners for which the designs were intended. A Dryden JetStar served as the testbed. Langley authorities considered the JetStar experiments to constitute a "very high priority item," so they advanced fiscal year 1982 funding to Dryden, then pressed Dryden to launch its investigation without delay. Yet, after a meeting among the principals (Langley, Douglas, Lockheed, and Dryden) held at Dryden in February 1982, speed became less of an issue. The timetable anticipated a flight program beginning during summer 1983. The resulting interval allowed time for the manufacturers to fabricate the test sections–the Douglas apparatus on the right wing, the Lockheed on the left. Afterwards, the Dryden technicians and mechanics attached the new wing sections and made other necessary preparations, such as installing a turbo-compressor to draw air through the wing perforations or slots, generating smooth flow over the wings. The first two flights occurred on 30 November 1983, the last four during October 1987. During this period, the small jet flew simulated airline missions around the country to compare the two competing systems.

While Langley did not achieve the quick results originally desired, the four-year flight program did discover some important insights about laminar flow. After installation of leading edge notch bumps to counteract contamination along the attachment lines of both test sections, the Douglas wing produced 95 percent laminar flow from Mach 0.72 to 0.78 at altitudes from 32,000 to 38,000 feet. The Lockheed device was less consistent, but near the design point yielded 80 to 94 percent laminar flow. Unfortunately, inclement weather became the nemesis of the experiment. Clouds and ice disrupted the smooth motion of the air above the wings, but laminar flow returned immediately after the JetStar passed through these conditions. Perhaps more seriously, the weight of the machinery required for active LFC proved too heavy for efficient operation on subsonic aircraft.[410]

Despite this drawback, the efficiencies possible with laminar flow continued to inspire research during the 1980s. Roughly during the same period in which Dryden engineers, pilots, and technicians investigated LFC for airliners and transports, fighter aircraft also underwent scrutiny. Starting in late 1979 and early 1980 with the modification of an F-111 bearing the acronym TACT (Transonic Aircraft Technology), flight researchers installed natural laminar-flow gloves—specially designed test sections fitted over the existing wing structure—to learn how wing sweeping affected air flow over airfoils. Research flights occurred during May through August 1980. Regrettably, the inquiry was curtailed because the Air Force required the F-111, on loan to NASA, for other purposes. Other factors inhibited obtaining good data. The gloves covered only part of the F-111's wingspan, and the aircraft flew with only partial instrumentation. Still, the interim results were promising and provided a basis for future research.

About two years passed before a follow-on program, called the Variable Sweep Transition Flight Experiment (VSTFE), carried the F-111 TACT research forward. This time, a borrowed Navy F-14A served as the testbed vehicle. These investigations required the construction of one glove for each wing, both composed of fiberglass, foam, and resin. There was a full-span, upper-surface glove on the left wing designed to "clean-up" the aircraft's original airfoil shape and provide baseline data. Meanwhile, researchers at Langley conceived of a more experimental glove on the right wing, tailored for the demands of specific flight conditions. After a natural laminar-flow program review at Langley in April 1984 at which Dryden and Langley representatives agreed upon the flight program, in October 1984 the Composite Development Corporation of Mojave, California, won the glove fabrication contract.

As in many flight-research projects, instrumentation played an important role in this process. It provided the capacity to measure surface static pressures, boundary layer profiles, and acoustic phenomena. Checkouts of the instrumentation and the aircraft systems reached their final stages at the start of 1986. The gloved F-14 first flew on 25 February 1986. After 68 flights, the program ended in July 1987. It made an important contribution. As a result of these and the F-111 glove tests (complemented by earlier wind-tunnel research in the Langley Low-Turbulence Pressure Tunnel and

[410] Albert L. Braslow, *A History of Suction-Type Laminar-Flow Control with Emphasis on Flight Research* (Washington, DC: NASA Monographs in Aerospace History No. 13, 1999), pp. 1-2; Roy V. Harris, Jr., and Jerry N. Hefner, "NASA Laminar-Flow Program—Past, Present, Future," in *Research in Natural Laminar Flow Control and Laminar-Flow Control* (Washington, DC: NASA Conference Publication 2487, Part I, 1987), pp. 12-13; David F. Fisher and Michael C. Fischer, "Development Flight Tests Of JetStar Leading-Edge Flight Test Experiment," in *Research in Natural Laminar Flow Control and Laminar-Flow Control* (Washington, DC; NASA Conference Publication 2487, Part I, 1987), pp. 118-119, 121, 138; Wallace, *Flights of Discovery*, p. 96; Dryden Project Highlights, Jan., Feb., and Apr., 1982, Nov. 1983, and Oct. 1987, DFRC Historical Reference Collection.

Designed to test the concept of changing wing camber during flight, the AFTI/F-111 MAW aircraft is shown aloft in February 1986. (NASA Photo EC86-33385-002)

in the Ames 12-Foot Tunnel), laminar flow on sweptwings appeared to be unaffected until about 18 degrees. At higher angles, engineers found cross-flow turbulence sufficient to reduce laminar flow significantly. However, they concluded that such disturbances could be overcome by combining suction at the leading edge of swept wings with an appropriate pressure gradient behind it.[411]

In a separate project not related significantly to laminar flow but associated closely with aerodynamic efficiency and using the same F-111 aircraft employed in the TACT experiments, researchers at Dryden reconfigured the aircraft to test a device referred to as the Mission Adaptive Wing (MAW). The program was part of an Air Force program known as Advanced Fighter Technology Integration (AFTI), designed to evaluate a number of promising aeronautical advances for possible use in military vehicles. The timetable of the AFTI/F-111 MAW flight program roughly paralleled that of the F-14 natural laminar-flow project, although the AFTI/F-111 MAW program germinated for a longer period of time.

Pursued jointly by Dryden and the Air Force Flight Dynamics Laboratory (AFFDL), the AFTI/F-111 MAW project actually began as early as 1978, with the AFFDL responsible for overall program management. The Langley and Ames research

[411] Braslow, *A History of Suction-Type Laminar-Flow Control*, pp. 17-18, 35; Statement of Work, F-14 VSTFE Glove Fabrication/Modification, n.d., DFRC Historical Reference Collection; Marta R. Bohn-Meyer, "Constructing Gloved Wings for Aerodynamic Studies" (Edwards, CA: NASA TM 100440, 1988), pp. 1-2; Dryden Project Highlights, May, June, and Aug. 1980, Jan. and Nov. 1983, April, July, Oct., and Nov. 1984, Nov. 1985, Feb. 1986, and July 1987, DFRC Historical Reference Collection.

centers along with the Air Force's Arnold Engineering Development Center at Tullahoma, Tennessee, provided wind-tunnel testing, and the General Dynamics Company of Fort Worth, Texas, consulted in the area of airplane design interfaces and supporting systems maintenance. The Boeing Military Airplane Company won the Air Force contract to modify the F-111's wings in February 1979. Of course, Dryden—with the support of the Air Force Flight Test Center at Edwards AFB—conducted the flight research.

But a long delay elapsed before the first flight test, in part due to the complexity of the endeavor. The MAW attempted nothing less than a smooth, flexible airfoil whose camber could be modified to best suit the aircraft's intended mission, thus eliminating drag-inducing wing structures such as flaps, slats, and spoilers to provide improved aerodynamic efficiency over a range of flight conditions from subsonic to supersonic. The wing itself contained mechanisms for changing the wing camber.

Pilots could change the wing leading edge and each trailing-edge segment to any position using a manual control system. They could also select one of four wing modes, each one activated and controlled by two digital computers: *cruise camber control,* which adjusted the trailing edge for maximum speed at a given altitude and throttle setting; *maneuver camber control,* which automatically moved the leading and trailing portions of the wing to achieve optimal lift-to-drag ratios; *maneuver enhancement-gust alleviation,* which improved the aircraft's response to pilot commands while reducing sensitivity to strong winds (gusts); or *maneuver load control,* which mitigated wing bending during intense maneuvering.

Perhaps not surprisingly, delays in development occurred, principally involving hardware and software integration between the modified wings and the F-111, which was old and hard to maintain. Moreover, the MAW system involved a great deal of complexity inside the wing. When the AFTI/F-111 MAW program entered its sixth year without a research flight, some people started to express reservations about it. They feared having insufficient funds to sustain the project and insufficient time to transfer its lessons to the next-generation Advanced Tactical Fighter (ATF). Eventually, the exhaustive ground preparations did lead to flight research.

On 18 October 1985, Lieutenant Colonel Frank Birk of the Air Force Flight Test Center and co-pilot Rogers Smith of NASA Dryden took off over the Mojave Desert and flew to 15,000 feet and 350 knots. During the ensuing flights, the MAW system "was exercised extensively and performed without a glitch." Almost nine months later, it conducted its maiden supersonic flight, flying up to Mach 1.2 at 27,500 feet. After 26 missions in manual MAW mode only, the first phase of flight research ended late in 1986. The major thrust of this phase was simply to make the overall system work. It was reasonably successful in this effort. Data analysis revealed some of the system's potential value: 25 percent longer range than a conventional F-111 in supersonic low altitude missions and 30 percent more range during conventional high-altitude flight.

The second phase (33 flights) concentrated on flying the MAW aircraft in all four modes using an automatic flight-control system. It took nine months to install the automatic flight-control system and to calibrate the sensors, verify and validate the system, perform the ground testing, and complete the modifications the testing indicated were necessary. The first flight of the second phase occurred in August 1987, with the final flight in December 1988.

Although this segment of the MAW flight research yielded a lot of data, only two of the four modes—maneuver camber control (also described as maneuver cruise control) and maneuver load control—worked as intended. Gordon Fullerton joined the project as co-project and then project pilot, but the team could never get the maneuver enhancement-gust alleviation (MEGA) mode to work properly, in part because the full evaluation of the MEGA mode was limited to the last flight of the program. Researchers planned to fly through as much turbulence as practical during that flight. Unfortunately, it was one of the calmest flying days of the program, with turbulence difficult to find. The team would have liked to have used the data obtained from this flight to modify the MEGA system. Had this been possible, the MEGA mode might have had more positive results.

The second problematic mode—cruise camber control (CCC)—was supposed to optimize the camber, but it would drift off the optimized setting. The flight tests showed the CCC concept was feasible, identifying factors to consider. Again, modifications to the CCC mode might, if implemented, have led to more positive results. The team at least had learned a lot about the internal complexity of a variable camber wing. Team members had brought the MAW concept to the execution phase, but funding ran out before they could make it all functional. Thus, the project yielded a lot of data that evaluated "the total airplane performance, wing performance, buffet characteristics, aerodynamic loads, handling qualities, stability, and control." While it did not result in a perfected system, the program did accumulate 145 hours of flight and 1,524 hours of ground operation of the MAW system over the course of a program of 59 flights.[412]

[412] Albert E. Preyss, Welbourne G. Williams, and Charles Cosenza, "AFTI—Advanced Fighter Technology Integration," AIAA Aircraft Systems and Technology Meeting, Dallas, TX, 27-29 Sept. 1976, p. 1; Dryden Project Highlights, Feb. 1979, Aug. 1987, and Dec. 1988, DFRC Historical Reference Collection; "Mission Adaptive Wing Program Shaping Up," Dryden *X-Press*, 2 Mar. 1984, DFRC Historical Reference Collection; "NASA to Flight Test Advanced Technology Wing," Dryden *X-Press*, 27 Sept. 1985, DFRC Historical Reference Collection; "AFTI-F-111 Project Team Readies Aircraft for Initial Flights," *Aviation Week and Space Technology*, (7 Oct. 1985), p. 70; "Mission Adaptive Wing Research Aircraft First Flight A Success!" Dryden *X-Press*, 25 Oct. 1985, DFRC Historical Reference Collection (first quoted passage); "Mission Adaptive Wing Flies Supersonic," Dryden *X-Press*, 18 July 1986, DFRC Historical Reference Collection; Bruce A. Smith, "AFTI/F-111 Flight Tests Demonstrate Potential of Mission Adaptive Wing," *Aviation Week and Space Technology*, (24 Nov. 1986), p. 40; "Second-Phase Flight Tests Begin on Mission Adaptive Wing," Dryden *X-Press*, 14 Aug. 1987, DFRC Historical Reference Collection; William B. Scott, "Performance Gains Confirmed in Mission Adaptive Wing Tests," *Aviation Week and Space Technology*, (3 Oct. 1988), pp. 77; Charles F. Bostwick, "Air Force Decides to Retire Experimental Wing Aircraft," *Antelope Valley Press*, 29 Mar. 1989; John W. Smith, Wilton P. Lock, and Gordon A. Payne, *Variable-Camber Systems Integration and Operational Performance of the AFTI/F-111 Mission Adaptive Wing* (Washington, DC: NASA TM 4370, 1992); Sheryll Goecke Powers, Lannie D. Webb, Edward L. Friend, and William A. Lokos, *Flight Test Results from a Supercritical Mission Adaptive Wing with Smooth Variable Camber* (Washington, DC: NASA TM 4415, 1992); comments of Rogers Smith on a draft of this section, 13 Sept. 2000; S. G. Powers, "Wing," *McGraw-Hill Yearbook of Science and Technology for 1995* (New York: McGraw-Hill Book Co., 1995), pp. 462-465 (second quotation from p. 465); comments of Sheryll Goecke Powers on a draft of this section, 6 Oct. 2000.

Researchers then pursued the next logical step in laminar-flow research. At the end of the 1980s, Langley Research Center's laminar-flow control project office revived an initiative to apply LFC techniques to a proposed generation of supersonic airliners. If successful, this innovation promised benefits similar to those for subsonic flight, including increased range, better fuel economy, and reduced aircraft weight. But laminar-flow control also offered additional advantages at supersonic speeds. Lower fuel expenditure at higher altitudes resulted in less ozone depletion, and lower weight diminished the intensity of sonic booms and takeoff noise. By definition, LFC also resulted in smooth (rather than turbulent) flow over wings, which diminished aerodynamic heating and drag. Under contract to Langley, Boeing and Douglas engineers reviewed the supersonic LFC concept and, although favorably inclined, suggested a flight-research program before undertaking any additional work. Meanwhile, Rockwell International conducted its own inquiries into the subject and also concluded that flight research was required.

At this point, Langley assumed the role of program management, and Dryden became responsible for flight tests. During 1988, the project sought an appropriate testbed. The solution: not one, but two. General Dynamics of Fort Worth, Texas, had in storage two F-16XL fighter aircraft owned by the Air Force. Those involved in the project selected this vehicle because of its highly swept, cranked-arrow wing design, one that closely approximated the High Speed Civil Transport (HSCT) planforms being discussed by aircraft manufacturers. (See Chapter 9 for further information about the HSCT.) The Air Force loaned the two F-16XLs to Dryden, and after refurbishment at Fort Worth, they arrived at Dryden to engage in two essential experiments. The first experiment, using F-16XL-1, examined the degree of laminar flow at supersonic speeds with and without suction. The second experiment, flown aboard the F-16XL-2, employed data from the first experiment and explored means to maximize laminar flow at high speeds. By the end of the investigation, the Langley engineers hoped to achieve laminar flow over 50 to 60 percent of the chord on a swept wing at supersonic velocities, compile a set of computational fluid dynamics (CFD) codes and design data for industry to fabricate the wings, and produce design criteria for the suction system.[413]

The Phase 1flights began after a Rockwell/Boeing team designed and delivered an LFC wing glove to Dryden in late July 1989 and Dryden technicians installed instrumentation, checked out all systems, and conducted a series of suction-system ground tests. The supersonic laminar-flow experiment on F-16XL-1 began in May 1990 and ended in September 1992. On the F-16XL-2, initial flight research started in January 1992 and ended in July 1993. Then, after more than two years of modifications, the F-16XL-2 took wing again in October 1995. In its final incarnation, the F-16XL wing glove proved to be a remarkable technical achievement. It consisted of a titanium

[413] Braslow, *A History of Suction-Type Laminar-Flow Control*, p. 32-33; Wallace, *Flights of Discovery*, p. 96; Statement of Work, "F-16XL Supersonic Laminar Flow Control Experiment," Langley Research Center, July 1991, p. 1, DFRC Historical Reference Collection; Dryden Project Highlights, June, July, Aug., and Sept. 1988.

An overhead view of the F-16XL-2 aircraft revealing its asymmetrical wing planform and the glove on the left wing. In association with NASA Langley, Dryden engineers used the vehicle to test laminar-flow control during supersonic flight. (NASA EC96-43831-5)

sheet with over 10 million tiny holes drilled by laser, arrayed at the density of 2,500 to 3,000 per square inch. Smaller at the surface than below it, these orifices were designed to minimize the amount of dirt flowing through the openings and clogging the system. Beneath the titanium panel, a matrix of ducts—connected to a vacuum pump stored in the aircraft's ammunition bay and fed by air bled from the F-16XL's engine—sucked turbulent air through the millions of holes. The flight research that tested this unique system lasted until November 1996. By its conclusion, supersonic LFC yielded impressive results, achieving laminar flow to 46 percent of chord, only slightly less than the original objective of the Langley researchers.[414]

Maneuverability

Essentially, the many aerodynamic flight experiments conducted at Dryden from the late 1970s to the early 1990s represented a sustained attempt to improve the operating

[414] Briefing, "F-16XL-2 Supersonic Laminar Flow Control Experiment," Flight Readiness Review, Aug. 1995, DFRC Historical Reference Collection; Dryden Project Highlights, July 1989, Mar. 1990, May 1990, Nov. 1991, Sept. 1992, 13 Oct. 1995, 7 June 1996, DFRC Historical Reference Collection; Mark A. Gottschalk, "Going with the Flow: Developments in Laminar-Flow Technology May Dramatically Reduce Drag on Next-Generation Aircraft," *Design News*, (9 Sept. 1996), pp. 23-24; Laurie A. Marshall, "Boundary-Layer Transition Results from the F-16XL-2 Supersonic Laminar Flow Control Experiment," (Edwards, CA: NASA TM-1999-209013, 1999).

efficiency of aircraft. Almost simultaneously, an entirely different set of investigations opened the possibility of drastic improvements in the maneuverability of military aircraft. These investigations had their origins in the Cold War. During the early 1970s, Western military planners assumed that the 1980s and 1990s would witness an increased threat of Soviet air attack. Consequently, the U.S. armed forces, acting cooperatively with NASA, surveyed a number of promising aeronautical technologies. They realized that combining them in a single technology demonstrator might pay greater dividends and speed their maturation than would developing them individually.

Linking such new discoveries as the supercritical wing, structural composites, digital fly-by-wire, variable geometry, and active controls might yield great gains in maneuverability. But since trying several untested approaches at once might endanger the lives of pilots and others, a consensus emerged to use a remotely-piloted vehicle as a testbed. In May 1972, Roy Jackson, the NASA associate administrator for aeronautics, instructed the directors of the Flight Research Center, Ames, Langley, and Lewis to formulate NASA's role in the highly maneuverable aircraft demonstrator. Four months later, Jackson sent FRC director Lee Scherer a proposed project plan for a Highly Maneuverable (Combat) Aircraft Technology program (later given the acronym HiMAT). Headquarters announced in November 1972 its intention to release a request for proposals to industry for studies of this project.

In August 1975, Rockwell International won the competition (over Grumman and Douglas) to build the aircraft. For payments totaling $17.3 million, Rockwell fabricated two remotely-piloted research vehicles (RPRVs) scaled at 44 percent of the original full-size design. The HiMAT vehicle measured a mere 23.5 feet in length and 4.3 feet in height at the tail, with a wingspan of 15.6 feet. It featured a close-coupled canard, winglets, digital fly-by-wire controls, aeroelastic tailoring, supercritical airfoil, and a composite structure. In March 1978, during a rollout ceremony at Rockwell's Los Angeles plant, 500 spectators watched as the first HiMAT demonstrator was unveiled for the media. Both models arrived at Dryden, the lead center for HiMAT, before the end of 1978.

Dryden engineers actually initiated a flight program for HiMAT during the two years prior to the Rockwell delivery, using a general aviation PA-30 aircraft to simulate the HiMAT's approaches and landings, guided by a television camera in the cockpit. Since landing the vehicle using only television would deny the pilot any peripheral vision, flight safety required such a simulation. When the HiMAT did arrive, a fully developed set of flight-research objectives awaited it. These included a demonstration of transonic maneuverability, proof of supersonic endurance, and the collection of data to evaluate the design tools and the RPRV concept.

The HiMAT's operational procedure involved launching from a B-52 aircraft, flying a set of maneuvers controlled by a pilot from a ground cockpit, and ending with a horizontal landing on Rogers Dry Lake. One of the more novel innovations in the flight-research program, which began on 27 July 1979 and ended 25 flights later on 12 January 1983, augmented the pilot's controls with an automatic flight-control system called a Flight Test Maneuver Autopilot (FTMAP). Designed to help the pilot

One of two HiMAT demonstrators produced by Rockwell for NASA and the Air Force, this 44-percent-scale vehicle flew maneuvers—some too hazardous for humans—using a ground cockpit. (NASA Photo EC80-14281)

execute maneuvers requiring extraordinary precision, the FTMAP was linked to two ground-based computers that guided the aircraft through the three essential HiMAT maneuvers: pushover/pullups, excess thrust windup turns, and thrust-limited turns. At the end of the program, researchers found the control so exacting that "the data collected with the FTMAP . . . were of a quality unequaled at Dryden."

Dryden's overall experience with HiMAT, while impressive, proved more ambiguous than this quotation suggested, although it was valid as stated. The aircraft's systems exhibited more complexity than predicted, its flight operation was more labor-intensive than expected, and its engineers found the subscale size of HiMAT restrictive. Still, its accomplishments could not be denied. Near the speed of sound, flying at 25,000 feet, HiMAT made sustained 8 g turns, compared to the F-16's maximum turning capability of about 4.5 g. Moreover, its supersonic endurance surpassed the design goals significantly. At its design points, the HiMAT's aerodynamics were as good as, or better than, expected.

On the other hand, research pilot Bill Dana expressed some reservations about the system. He observed that cockpit workload in the HiMAT—which exceeded that of a piloted aircraft because all "flying" occurred on instruments—required the assistance of a head-up display. While Dana acknowledged the HiMAT's uncommon maneuverability, admitting the 8 g turns produced data impossible aboard a piloted aircraft, he questioned the value of technical information devoid of direct pilot input.

Despite these misgivings, he recognized the value of subscale, remotely-piloted aircraft that were still capable of collecting data through maneuvers too hazardous or strenuous for pilots to attempt.[415]

While Dryden and NASA grappled with the complexities of HiMAT, the Air Force undertook two maneuverability-related projects of its own during the 1980s. The first—part of the AFTI program discussed earlier in this chapter in connection with the F-111 MAW—included a number of technical innovations designed to perfect and uncomplicate cockpit management. Flown aboard an F-16 aircraft, these technologies allowed pilots to pursue combat maneuvers more effectively, although maneuver was not the only purpose of the research. The project began after a number of meetings in 1979 among the Air Force Flight Dynamics Laboratory, the Air Force Flight Test Center, and General Dynamics representatives. In the end, the parties agreed to an ambitious flight-research program using an F-16A fighter (ship number 6 off the assembly line). General Dynamics started modifications of the vehicle in April 1980 and conducted the first contractor flights in July 1982. The overall objectives embraced a cluster of technologies that, in combination, improved the survival rates of combat aircraft, lessened pilot workloads, and increased the accuracy of weapons.

The flight research consisted of three phases. The first, from 1982 to 1983, involved the testing of a digital flight-control system. The second, lasting from August 1984 to June 1987, emphasized automated air-to-air and air-to-ground attack systems, voice control systems, digital map displays, and software that automatically extricated an aircraft from hazardous situations. Finally, from February 1988 to January 1993, the F-16 acted as a testbed for advanced and interconnected close air support technologies, including night vision systems, pilot/vehicle interface, integrated flight/fire control, and digital navigation. While the Air Force assumed the leading role in this research, Dryden contributed significantly by installing instrumentation, offering data processing services and mission control facilities as well as providing staff support. Moreover, seasoned Dryden research pilots Bill Dana and Steve Ishmael shared the AFTI F-16 flying duties with their Flight Test Center counterparts, including Dana Purifoy who later became a Dryden pilot. In addition, Dryden researchers sat on the planning meetings with the Air Force and General Dynamics from the beginning of the AFTI F-16 project. Although, as one participant admitted during the second phase

[415] Henry R. Arnaiz, "Overview of the Highly Maneuverable Aircraft Technology (HiMAT) Program," n.d., DFRC Historical Reference Collection; Roy P. Jackson to directors of Ames, Flight Research Center, Langley, and Lewis, 11 May 1972, DFRC Historical Reference Collection; Roy P. Jackson to Lee R. Scherer, 20 Sept. 1972, with attached Project Plan, DFRC Historical Reference Collection; "L.A. Division Unveils the Future: Top Officials, News Media at HiMAT Rollout," *Rockwell News*, (Mar. 1978), p.13, DFRC Historical Reference Collection; Project Plan for Highly Maneuverable Aircraft Technology Program (HiMAT), DFRC, 13 Jan. 1976, DFRC Historical Reference Collection; E.L. Duke and D.P. Lux, "The Application and Results of a New Flight Test Technique," AIAA Atmospheric Flight Mechanics Conference, 15-17 Aug. 1983, Gatlinburg, TN, pp. 1-5 (quoted passage, p. 5); "HiMAT Flight Operation Summary," n.d., DFRC Historical Reference Collection; NASA Facts, "HiMAT," 1998, pp. 1-2, DFRC Historical Reference Collection; interview of William H. Dana by Michael Gorn, 9 Mar. 1999, pp. 4-5, DFRC Historical Reference Collection.

NASA participated in an Air Force program known as Advanced Fighter Technology Integration, flown aboard the F-16A aircraft. The AFTI/F-16 evaluated technologies for improving survival rates on combat aircraft. (NASA Photo EC89-0016-20)

of the project, "development has taken longer and has been more difficult than originally envisioned," the broad technical umbrella of AFTI F-16 opened new prospects for simplifying military cockpits, enabling pilots to maneuver and to fight more effectively under combat conditions.[416]

The second Air Force project devoted to better agility and handling actually arose as a Defense Advanced Projects Agency (DARPA) initiative. For decades, the concept of forward-swept wings had held high appeal to designers of fighter aircraft. They hoped to benefit from the presumed advantages of reversing the traditional swept-back wings. Such expected improvements included lower compressibility at transonic speeds, higher lift at subsonic speeds, and greater maneuverability. The German firm Junkers had experimented with the concept during World War II. But a technical barrier existed. To be effective, forward-swept wings required airfoils of greater structural stiffness than were necessary for conventional swept wings. The reinforced construction necessary for metal wings to withstand the flight loads inherent in a

[416] Briefing Chart, HQ NASA Office of Aeronautics and Space Technology Annual Program Review, DFRF, Oct. 1986, DFRC Historical Reference Collection; Briefing Chart, HQ NASA OAET Code RX, Ames Research Center Program Review, 6 Dec. 1990, DFRC Historical Reference Collection; Dana Purifoy and Peter F. Demitry, "AFTI F-16 Night Close Air Support Testing," 36th Symposium *Proceedings of the Society of Experimental Test Pilots*, Beverly Hills, CA, Sept. 1992, pp. 111-117; Dryden Project Highlights, July 1979, Mar. 1980, July 1982, June 1987, Feb. 1988, Jan. 1993, DFRC Historical Reference Collection.

forward-swept design resulted in much added weight. Here the matter stood until composites and laminates came of age during the 1970s. Once materials science provided reliable, lighter substitutes for metal wings, DARPA acted. In cooperation with the Air Force Flight Dynamics Laboratory, DARPA in 1977 issued study contracts to three aircraft manufacturers to investigate the feasibility of the forward-swept wing. Their positive replies persuaded DARPA and the Air Force to issue requests for proposals to the aerospace industry to build actual testbeds. Finally, at the end of 1981, Grumman Aircraft won an $87-million contract to construct two full-scale, piloted, forward-sweptwing demonstrators.

Although much larger than the subscale HiMAT, the X-29A (as the Air Force designated it) measured about 48 feet long, nearly 15 feet tall at the tail, and a little more than 27 feet at the wingspan, slightly smaller than most fighter aircraft. It bore important similarities to the HiMAT: the close-coupled canard, the thin supercritical wings, digital fly-by-wire, and aeroelastically tailored wing covers. Yet the X-29 differed in some crucial respects. It featured three control surfaces (canards, wing flaps, and strakes) to the HiMAT's two (canards and elevons). The X-29's canards were movable while HiMAT's were fixed. The X-29 was much more unstable in pitch than other known aircraft, roughly 35 percent unstable compared with about 10 percent of instability for the Shuttle and the F-16.

In contrast to the HiMAT program, Dryden assumed a supporting, rather than a leading role in X-29A flight research. The X-29A advanced program office at Wright-Patterson Air Force Base managed the overall program, while Dryden served as the responsible test organization with the Air Force Flight Test Center as the participating test organization. The Air Force's overriding concern was to fly the X-29A and its breakthrough technologies without delay. Optimal aerodynamic design, for instance, was not the highest priority. Indeed, the front part of the fuselage and the canopy came directly from an F-5A fighter. Despite the pressures imposed on the project engineers to develop the control laws for an aircraft with more negative static margin than any ever flown, the Air Force succeeded in readying the X-29 quickly. Even allowing for the necessary checkouts and installation of instrumentation after the first X-29A arrived at Dryden, X-29A-1 flew over Edwards on 14 December 1984, just three years after Grumman won the demonstrator competition. Nevertheless, this aircraft had a long flight-research career. Its final flight on 5 August 1990 represented its 254th time aloft.[417]

During these missions, the Air Force, Grumman, and NASA pilots performed six essential maneuvers. They conducted stick raps to evaluate flutter and aeroelasticity, wind-up turns for divergence, buffet, loads, and performance, and pushover/pullups for performance. They also undertook accelerations and decelerations in level flight for airspeed calibrations and performance, stability-and-control maneuvers to extract

[417] Edwin J. Saltzman and John W. Hicks, "In-Flight Lift-Drag Characteristics for a Forward-Swept Wing Aircraft (and Comparisons with Contemporary Aircraft)," (Edwards, CA: NASA TP 3414, 1994), p. 3; NASA Facts, "The X-29," Apr. 1998, DFRC Historical Reference Collection; Flight Log, "X-29A-1 Test Program," DFRC Historical Reference Collection.

Pursuing an old aeronautical concept, NASA, the USAF, and the Defense Advanced Research Projects Agency (DARPA) collaborated on the forward-swept-wing X-29A aircraft, conceived to test high maneuverability and high-angle-of-attack flight. (NASA Photo EC91-491-6)

aerodynamic derivatives, and formation flying in trail to simulate refuellings. While slowly expanding the flight envelope, Dryden research pilots Rogers Smith and Steve Ishmael along with their Air Force and Grumman counterparts, concentrated on a wide range of data collection. They flew maneuvers designed to yield information on handling qualities, loads, angles of attack, and buffeting. In the end, this massive effort yielded important findings. The Number 1 aircraft's tailored aeroelastic wing prevented structural divergences, and its digital flight-control system provided sufficient artificial stability and predictable handling qualities in an otherwise unstable aircraft. Moreover, its supercritical wing contributed to good maneuvering and cruise characteristics in the transonic range, and its forward-swept wings inhibited stalling at the wing tips during moderate angles of attack.

While this aircraft still plied the skies, the X-29A-2 began its flight career on 23 May 1989. Although it flew only about half the flights (120) of its predecessor and remained active for only two years during its initial program, its research nonetheless revealed important results for flying at high angles of attack. Dryden pilots Ishmael and Smith returned to the X-29 cockpit for these flights, designed to test the demonstrator's capacity in "maneuvering, control system, flying qualities, and military utility aspects . . . in the high AOA [angle-of-attack] envelope extending to approximately 70 degrees." In actual flights, the vehicle achieved up to 45 degrees angle of attack with excellent control and maneuverability, much better than suggested by the mathematical models and simulators. Even at 67 degrees (the ultimate limit), the X-29A-2 displayed limited control.

Engineers at Langley and Dryden achieved high angles of attack without recourse to thrust vectoring (see Chapter 13). Instead, they reprogrammed the Number 2 ship's computer based on tests conducted at Langley on a small-scale radio-controlled X-29. But this was not the end of X-29 high angle-of-attack flight research. Between May and August 1992, the Air Force initiated a project to fly the aircraft with two small-nozzle jets implanted on the forward upper portion of the nose. Known as vortex flow control, the system operated by emitting high-pressure nitrogen from these jets, allowing pilots to "steer" the aircraft left and right at very high angles of attack even as the rudder lost effectiveness. Pilots made 60 flights using this technique. Despite these undeniable achievements for the X-29A program, one predicted advantage was not realized: the anticipated higher ratio of lift to drag at subsonic speeds did not materialize in these tests. In fact, the lift/drag ratio of the X-29A about equaled that of three of the front-rank combat aircraft of the day, the F-15C, the F-16C, and the F/A-18.[418]

Computerized Propulsion

During this period in which computerized flight controls proved themselves in the Digital Fly-By-Wire F-8 and were later used in HiMAT and the X-29A, aircraft engines also experienced the influence of on-board electronic brains. The initial attempt to marry engines to digital computer control occurred during the early 1970s in a program called the Integrated Propulsion Control System (IPCS). Using the large cumbersome computers of the day, Dryden and NASA Lewis research engineers—in an Air Force-led program carried out in partnership with Boeing, Pratt & Whitney, and Honeywell—succeeded in controlling one engine and one inlet on an F-111 aircraft. This enabled engine and inlet combinations to perform successfully in flight. (See Chapter 10.)

In a more recent incorporation of this sort of research a preproduction F-15A fighter on loan since 1976 from the Air Force to NASA served as the testbed for a joint NASA/USAF/Pratt and Whitney program known as Digital Electronic Engine Control (DEEC). In its long career at Dryden, this flight-research war horse would carry some 25 separate experiments. During the DEEC project, designed to test the contractor's prototype system, interest centered not on the airframe but on the vehicle's F100-PW-100 powerplant. It came equipped not only with hydromechanical controls (such as levers, bellows, governors) but also with an added supervisory electronic

[418] Flight Log, "X-29A-1 Test Program" (see Appendix W, this history); NASA Facts, "X-29,"; T.W. Putnam, "X-29 Flight Research Program," The 2nd Flight Testing Conference of the AIAA, AHS, IES, SETP, SFTE, and DGLR, Las Vegas, NV, 16-18 Nov. 1983; Flight Log, "X-29A-2 Test Program," DFRC Historical Reference Collection and Appendix W; Integrated Research and Flight Test Plan, X-29A-2 High Angle of Attack, 8 Mar. 1989, p. 1, DFRC Historical Reference Collection (quoted passage); Saltzman and Hicks, "In-Flight Lift-Drag Characteristics," p. 16.

During the early 1970s, Dryden engineers pioneered digital engine control by flight-testing an F-111 aircraft equipped with the Integrated Propulsion Control System (IPCS). (NASA Photo ECN 4359)

engine control. Flights began during mid-1981 to determine whether the Pratt & Whitney F100 engine with the added DEEC lived up to expectations as a full-authority single-channel control, backed by a redundant hydromechanical secondary control. After 75 flights, the Dryden team completed its analysis of the flight data in January 1984. The researchers found the DEEC system yielded major "improvements in engine efficiency, performance, and operations." By superceding the traditional machinery required to govern aircraft engines with an electronic computer control system, project engineers found better precision and response, reduced weight and cost, and an enhanced capacity to process data as well as to detect system faults. Moreover, an engine-pressure-ratio mode made engine retrim unnecessary. This change saved time for the ground crew, conserved fuel, improved engine life, and cut noise. In spite of two sensor failures during the flight-research program (neither of which necessitated the use of the mechanical backup), project engineers proclaimed the DEEC a success, responsible for

> unrestricted and stall-free throttle operation; an 11,000 ft increase in afterburning envelope; a 100 knot improvement in airstart capability; 3 second faster idle-to-max throttle transients; and major improvements in R & M

> [reliability and maintainability] (220-875 percent). The DEEC is now [March 1989] in production for the F100-PW-220 engines. The USAF and P & W credit the early NASA flight evaluation for advancing the introduction of DEEC by 1 1/2 years and saving $42M.[419]

As the DEEC program concluded, the NASA F-15 equipped with the new digital engine control again served as the testbed for the computerized integration of engines and airframes. During the first half of 1984, McDonnell Douglas modified the F-15 for the Highly Integrated Digital Engine Control (HIDEC) program, a follow-on to DEEC. The aircraft started flying on behalf of the new initiative in July. Flight research on HIDEC included four phases, beginning with installation and flight of the F100 engine model derivative (EMD) powerplant and the replacement of the analog control augmentation system. Secondly, technicians installed and tested the HIDEC control panel. Thirdly, the team incorporated into the F-15 system an auxiliary general-purpose airborne computer for control modes. Finally, team members reprogrammed the DEEC computer to accept commands either from the auxiliary airborne computer or from ground uplinks. By the time the HIDEC program wound down in winter 1989, its impressive accomplishments had become well known. The first fully integrated propulsion system, HIDEC counted many achievements (in comparison to the F-15 without HIDEC). It increased engine thrust by 10 percent, lowered fuel flow by up to 14 percent at maximum power, reduced climbing time as much as 12 percent, resulted in accelerations as high as 15 percent faster, and operated without a single stall during extreme maneuvers.[420]

[419] "Ames-Dryden NASA F-15 Flight Research Facility," brochure, n.d., pp. 1, 7, DFRC Historical Reference Collection; comments of Bill Burcham and Trindel Maine on a draft of this section; Dryden Project Highlights, Jan. 1984, DFRC Historical Reference Collection; Frank W. Burcham, Jr., et al., "Recent Propulsion System Flight Tests at the NASA Dryden Flight Research Center," 1st Flight Testing Conference of the AIAA, SETP, SPTE, and SAE, Las Vegas, NV, 11-13 Nov. 1983, pp. 3-4, DFRC Historical Reference Collection; "DEEC Program Starts at Dryden," Dryden *X-Press*, 10 July 1981, DFRC Historical Reference Collection; Lawrence P. Myers, Karen G. Mackall, and Frank W. Burcham, Jr., "Flight Evaluation of a Digital Electronic Engine Control System in an F-15 Airplane," AIAA, SAE, ASME 18th Joint Propulsion Conference, 21-23 June 1982, Cleveland, OH, p. 1 (first quoted passage); F.W. Burcham, Jr., L.P. Myers, K.R. Walsh, "Flight Evaluation of a Digital Electronic Engine Control in an F-15 Airplane," *Journal of Aircraft* (Dec. 1985): p. 1077; David C. Spencer, "NASA Highly Integrated Digital Electronic Control (HIDEC)/Adaptive Engine Control System (ADECS) Program," Mar. 1989, DFRC Historical Reference Collection (block quote, p. 1); interview of Bill Burcham by J. D. Hunley at the DFRC, 23 Feb. 1999, DFRC Historical Reference Collection.

[420] Dryden Project Highlights, Feb. and July 1984, Feb. 1989, DFRC Historical Reference Collection; E.L. Duke and K.L. Petersen, "Systems Development and Controls Research Plan for the Highly Integrated Digital Engine Control (HIDEC) Program," 30 Nov./1 Dec. 1983, DRFC Historical Reference Collection; Spencer, "NASA HIDEC Program," Mar. 1989, DFRC Historical Reference Collection; Frank W. Burcham, "Briefing for the Aeronautical Advisory Committee on HIDEC and Performance Seeking Control," 17 Dec. 1986, DFRC Historical Reference Collection.

Dryden's F-15 Highly Integrated Digital Engine Control (HIDEC) flight research program yielded significant increases in engine power and reductions in fuel flow. (NASA Photo EC91-677-1)

An Intentional Crash

Normally, Dryden research pilots, engineers, technicians, and mechanics dreaded nothing more than losing an aircraft in a flight mishap. Through a combination of careful preflight planning, skilled flying technique, deliberate and incremental envelope expansion, and a little good fortune, comparatively few such events occurred in the more than 50 year history of the center. But one crash, which happened in 1984, saddened and surprised no one. Early in the 1980s, Federal Aviation Administration (FAA) authorities contacted Dryden with a novel project. They wanted to intentionally crash a deactivated Boeing 720 airliner in an effort to document by extensive instrumentation the extent to which a four-engine transport aircraft and a cabin full of dummies serving as mock passengers withstood an accident. The test also sought to discover whether flame-retardant/anti-misting fuel additives conceived in the United Kingdom and developed by the FAA might prevent commercial aircraft from becoming roaring infernos on impact. Research on such mixtures stemmed from a ground collision between two Boeing 747s on the island of Tenerife in 1977. The event resulted in a deadly fireball that the FAA and other aviation safety organizations hoped to avert in the future.

Acting on a FAA request, Dryden researchers conducted a Controlled Impact Demonstration (CID) of an remotely-piloted Boeing 720 airliner. The 1984 crash determined the (in)efficacy of anti-misting kerosene (AMK) in preventing intense aviation fires. (NASA Photo EC84-31805)

The FAA specified the particulars of the staged incident. It had to replicate not just any crash but one in which passengers had a chance of surviving if the aircraft did not burst into flames. To achieve this objective, FAA officials asked Dryden to simulate either a final approach or landing, a missed approach, or an aborted take-off with the landing gear retracted. The jetliner—loaded with anti-misting kerosene (AMK) fuel—needed to follow a glide path between 3.3 and 4 degrees with the nose up one degree. Finally, the test designers were told to recreate a wing tank rupture of 20 to 100 gallons per second, yet at the same time to preserve the fuselage's integrity. The FAA delivered the doomed 720 to Dryden in summer 1981 and the preparations began.

Naturally, the final landing of the 720 would be flown by remote control. But Dryden researchers planned twelve piloted checkout and rehearsal flights for spring, summer, and fall 1984. The trial flights of this Controlled Impact Demonstration (CID), as it was called, were made by pilot Tom McMurtry. Fitzhugh Fulton and Ed Schneider alternated between the 720's second seat and the ground cockpit. Each flight focused on a few essential parts of the final mission, such as experiencing the CID approach not just from the cockpit but by remote control. On other occasions, remotely-controlled takeoffs and ground steering were practiced. For instance, during the eighth flight on 28 August 1984, the old airliner—a veteran of some 50,000 landings—underwent a number of procedural checks. By this time, some of the aircraft's tanks had been purged of regular jet fuel and filled with anti-misting kerosene, one tank at a time, until the actual demonstration, when the aircraft took off filled only with AMK. In addition, Langley engineers, at work on airworthiness studies, instrumented the ship's cabin and placed dummies aboard to approximate human responses to the impact, complete with motion pictures to record the aftermath. When the twelfth flight was completed in the first week of November (a piloted one for a final systems check), the Dryden staff readied the jet for its last mission. Finally, on 1 December, under the gaze of some 400 technical personnel and about 100 journalists, the big plane flew at approximately 150 knots on approach to the Rogers Dry Lake runway. Upon impacting on a specially prepared patch of desert not far from Mercury Boulevard, its belly and fuel tanks were punctured by iron posts embedded in the landing strip.

Unfortunately, the FAA's additives did not spare the jetliner. It burst into flame immediately, and the ensuing holocaust destroyed it. Still, the investigation proved successful in its own way. The Langley researchers collected excellent crashworthiness data from their instruments, and their cameras faithfully captured the scene inside. More importantly, the CID test proved once more a pivotal fact about flight research. Before this event, the FAA had run extensive ground tests of its anti-misting kerosene at Lakehurst, New Jersey, towing aircraft into various sources of fire and finding their formula highly effective against ignition. Although the Boeing 720 did not land exactly as planned and some declared the demonstration a failure due to the plane's destruction, only realistic flight research showed what might happen in an actual incident. The results proved to be persuasive enough for the FAA to discontinue its investigations of AMK.[421]

[421] "FAA Plans Largest Destructive Test Ever," *Industrial Research and Development* (Nov. 1981): p. 113; FAA Impact Scenario, "Full-Scale Transport Controlled Impact Demonstration Program," DFRC Historical Reference Collection; M.R. Barber to all concerned (at DFRF), 20 Jan. 1984, with attachment: CID landing zone, DFRC Historical Reference Collection; Eugene Kozicharow, "Planned 720 Crash to Benefit Safety," *Aviation Week & Space Technology* (11 July 1983): pp. 28-29; Statement of Capability, NASA CID, 18 May 1983, DFRC Historical Reference Collection; "720 Prepares for CID," Dryden *X-Press*, 16 Mar. 1984, DFRC Historical Reference Collection; "Fuel for the Fire," *FLIGHT International* (7 Apr. 1984): p. 933; CID Pilot Flight Reports, 13 July, 8 Aug., 17 Aug., and 28 Aug. 1984, DFRC Historical Reference Collection; Dryden Project Highlights, Oct. and Nov. 1984, DFRC Historical Reference Collection; interview of Russ Barber by J. D. Hunley, DFRC, 21 Dec. 1998, pp. 5-7, DFRC Historical Reference Collection; comments of L. Dean Webb on a draft of this section.

13

A Fresh Start 1991-1999

Initiated when Dryden operated under the administration of the Ames Research Center, the Walter C. Williams Research Aircraft Integration Facility all but eliminated manual checkout of aircraft and enabled researchers to assess the interactions of major components prior to first flight. (NASA Photo EC96-43393-1)

Despite the unwieldy union between the Dryden Flight Research Facility and Ames after the amalgamation in 1981, Dryden's program of flight research achieved many milestones during the 1980s and early 1990s. Even though the marriage of the two partners never ripened into an enduring relationship (see below), under Ames leadership important projects continued to be pursued at Dryden. Moreover, the infra-structure to support these endeavors underwent improvements. Ames officials approved plans at Dryden for the design and construction of a ground-based laboratory for testing aeronautical vehicles that was unlike any other in the United States. About five years after its groundbreaking, the Integrated Test Facility (ITF) opened on 24 October 1992, at a total cost of $22.5 million. The ITF not only curtailed much of the time-consuming process of manual checkouts of new aircraft and components, but its computer simulations and diagnostics also reported the *interactions* of such systems as flight controls, electrical circuitry, and avionics before an aircraft ever took to the air. An imposing, 120,000-square-foot structure, the ITF possessed the floor space and the height to shelter an entire transport aircraft under its roof. Most of the banner research vehicles of the 1990s—the CV-990, X-29, X-31, SR-71, F-15, F-16XL, F/A-18,

Kenneth J. Szalai served as director of Dryden during the last years of the amalgamation with Ames (1990 to 1994) and the first years after Dryden won its independence (1994 to 1998). Szalai inaugurated two major initiatives: Access to Space and Uninhabited Aerial Vehicles. (NASA Photo EC91-601-1)

Hyper-X, and the X-33—benefited from the comprehensive preflight analyses possible only in the ITF. After November 1995, the ITF became known as the Walter C. Williams Research Aircraft Integration Facility, or RAIF, in honor of the center's first director.

In the midst of a full calendar of flight research projects, a change of leadership occurred at Dryden. In mid-1984, John Manke retired from his dual role as Ames director of flight operations and as Dryden site manager. His replacement—one-time U-2 pilot Martin Knutson, a familiar figure at Ames since 1971—at first raised some concern at Dryden. Unlike Manke, Knutson assumed his job with no ties to Dryden, and he lacked his predecessor's intimate familiarity with its mission and its methods. Still, he won converts. He assumed a low profile, gave wide discretion to his division chiefs to manage day-to-day details, seemed to adapt well to the new surroundings, and developed an appreciation for Dryden's approach, as well as for its technical capabilities. In addition, conscious of his awkward role as an Ames emissary, Knutson prudently avoided the path of energetic reform and instead contented himself with competent, if somewhat aloof, management. At the same time, as mediators in the tug-of-war between Ames' requirements and Dryden's preferences, Knutson and his deputy Ted Ayers urged NASA headquarters to make an organizational distinction between the two organizations.

Knutson and Ayers were not alone in desiring greater autonomy for Dryden. Ken Szalai, who gained recognition during the F-8 digital fly-by-wire investigation, also pressed for a change in Dryden's status. Szalai held a series of important positions during the 1980s and early 1990s. In 1982, he was the director of research engineering, overseeing such essential Dryden activities as simulations, vehicle technology, fluid and flight mechanics, flight systems, and aerostructures. Szalai's position enabled him to become more familiar with a number of different flight-research specialties and to practice his talents as a multi-disciplinary manager. Recognizing these experiences and talents, Ames Director Dale Compton selected him in 1989 to be the acting associate administrator of Ames.

Despite gaining a fresh appreciation for the Ames point of view, Szalai still believed the existing institutional framework failed to provide Dryden with the administrative support it needed. Szalai and others believed that Ames represented a bureaucratic obstacle between NASA headquarters and Dryden, one which hindered Dryden's capacity to present itself in Washington, DC. Former Dryden research pilot and Chief Engineer Bill Dana pointed out another irritant. "To some extent," he said, "Dryden was a rubber band in the Ames budget process, and if Ames got cut big money, [it] could take it out of Dryden. So we were kind of a whipping boy for Ames during bad times."

Apparently, Szalai found some sympathetic listeners to his ideas. After Marty Knutson left Dryden in December 1990 to become the chief of a newly constituted flight operations directorate at Ames, Szalai succeeded him at Dryden, although not in the same role. Instead, he assumed the dual responsibilities of deputy director of Ames *and* director of Dryden. In this way, Dryden under Szalai ceased to be administered as one of many directorates of the Northern California center and instead received autonomy within the Ames framework. With this change in status, the new director gained oversight of Dryden's budget and its administrative apparatus. He also won the authority to make internal policies. Nevertheless, partial independence seemed little better than none at all. Szalai referred to his new task as "probably the most difficult job I ever had."[422]

Both before and after he took charge at Dryden, Szalai worked quietly to achieve Dryden's complete separation from Ames. He argued that during the 10-year marriage no true sense of cooperation had developed between the two. Ames engineers and scientists, for instance, did not select wind-tunnel experiments with a view toward their propensity for later flight-testing at Dryden. Each organization continued to follow its traditional research interests with little reference to the other. Moreover, few staff members served on job details from Ames to Dryden (or the reverse), and the tools of research were not often shared. Indeed, out of all the high angle-of-attack projects

[422] Because this chapter covers some projects (such as Hyper-X) that were still underway at the time of publication, their final outcomes could not be determined. Also, this chapter owes much improvement to the careful reading of an anonymous commentator and Rogers Smith, acting director of flight operations. Wallace, *Flights of Discovery*, p. 28; NASA Facts, "Integrated Test Facility," Aug. 1995, DFRC Historical Reference Collection; NASA Facts, "Walter C. Williams Research Aircraft Integration Facility," May 1998, DFRC Historical Reference Collection; Ames news release, "New NASA Ames Research Center Director of Flight Operations, Ames Site Manager Named," May 1984, DFRC Historical Reference Collection; Ames Biographical Data, "Martin A. Knutson," Dec. 1990, DFRC Historical Reference Collection; Theodore Ayers to J.D. Hunley, (approximately) 9 Mar. 1999, DFRC Historical Reference Collection; Dryden Biographical Data, "Kenneth J. Szalai," Mar. 1994, DFRC Historical Reference Collection; Organization Chart, Research Engineering Division, Oct. 1985, DFRC Historical Reference Collection; Martin Knutson, "Knutson Bids A Fond Farewell to Dryden Employees," Dryden *X-Press*, 30 Nov. 1990, DFRC Historical Reference Collection; Ken Szalai to Bill [Ballhaus?] and Dale [Compton?], 7 May 1989, with attached comparison between Dryden arguments and those encountered by Szalai during his assignment at Ames, Milt Thompson Collection, DFRC Historical Reference Collection; interview of Bill Dana by Michael Gorn, 9 Mar. 1999, p. 11 (first quoted passage); Ames news release, "New Ames Deputy Director to Head Ames-Dryden," 5 Nov. 1990, DFRC Historical Reference Collection; "New Dryden Director: Challenges Ahead," Dryden *X-Press*, 20 Nov. 1990, DFRC Historical Reference Collection (second quoted passage); Dale Compton to staff of Ames Research Center, 3 Dec. 1990, DFRC Historical Reference Collection.

conducted at Dryden, just one Ames researcher acted as a full-time contributor, and then not down in the desert, but from his office at Moffett Field. Even regular NASA air service between Mountain View and Edwards failed to span the gulf between the two facilities. Not only did the merger fail to save money, but Szalai suggested it actually wasted the precious asset of time.

> What is indisputable is that much valuable time is being used by senior managers to implement the 1981 decision, to travel between the sites, to solve intersite problems, to coordinate administrative and financial activities, and to attempt to advocate and formulate Center wide programs over two geographically widely separated sites. Also, it is clear that the principal DFRF mission and capabilities are significantly different than the Ames-Moffett mission. A great deal of effort has been expended to overcome this gap, by managers, technical staff, and administrative staff. It is time to admit that the 1981 reorganization did not produce the desired resource savings and has placed an excessive burden on Ames management at both sites which is hurting the Agency's aeronautical research program. It is time to reestablish Ames and Dryden as separately managed facilities in [Headquarters] OAST.[423]

Before long, thoughts such as these—at first spoken discreetly only at Dryden—were expressed more openly by Szalai. Arnold Aldrich, headquarters associate administrator for aeronautics, invited the Dryden director to Washington, DC, to argue his case for detaching Dryden from Ames. In a written exchange that followed, Szalai told Aldrich candidly that he persevered in the "current arrangement . . . [out of] loyalty to NASA. It is," he added, more than likely thinking of himself in particular, "a stressful situation for several people at DFRF; this probably filters down to the rank and file inadvertently." He therefore urged Aldrich to separate Dryden and Ames, telling him such a step would be welcome in both organizations. He even advised Aldrich to consult with Martin Knutson—a man more associated with Ames than Dryden—to substantiate the inefficacy of the existing situation. Through the good offices of Milt Thompson, Szalai also enlisted the help of Walt Williams. Williams corresponded in October 1989 with NASA Administrator Richard Truly (familiar at Dryden for his role as a crew member during the *Enterprise*'s approach and landing tests and later as a Shuttle pilot) and pressed him to end Dryden's connection to Ames.

Among those who agreed with Szalai, Milt Thompson showed the greatest persistence and forcefulness. Thompson bemoaned the drop in morale since 1981:

> After nine years of consolidation, none of the older [Dryden] employees really accepted consolidation. They have no sense of belonging to the Ames Research Center. They have no loyalty to Ames. Dryden employees are still proud of and loyal to NASA, but they still hope for deconsolidation. They have seen no benefits of consolidation. On the contrary, they have witnessed a decline in the quality of life at Dryden. Ames management is seen as another superfluous layer of management that unlawfully [*sic*] taxes Dryden funding

[423] Ken Szalai to Bill [Ballhaus?] and Dale [Compton?], 5 July 1989, with attached essay: "Should NASA Spin-Off Dryden as a Separate Entity?" Milt Thompson Collection, DFRC Historical Reference Collection.

> in a somewhat arbitrary manner. Dryden has for example been assessed to support a number of activities at Ames-Moffett due to shortfalls in funding, but this never seems to work in reverse. Dryden shortfalls are Dryden problems. The younger Dryden employees, those hired after consolidation, are not as emotionally effected [*sic*] by consolidation, and yet they wonder why Dryden lost its center status after thirty-five years of independence. There is no good answer. There was no obvious benefit of consolidation. The two sites represent two different cultures.[424]

The end of this institutional pairing occurred after Daniel Goldin succeeded Richard Truly as NASA administrator. On 1 March 1994, Goldin made public Dryden's return to full independence. In doing so, he may have responded to the advice of Szalai, Williams, and Thompson, or he may have been persuaded by reports that Ames Director Dale Compton supported Dryden's bid for reinstatement as a NASA center. Whatever his motivations, Goldin started a six-month preparatory period, at the end of which Ken Szalai became the center director. "This change," remarked Goldin via satellite from Washington, DC, "reflects the commitment on the part of NASA to reduce layers of management and empower operating organizations to carry out their missions with maximum benefit to the country."

Szalai expressed gratification at the news and found it reassuring that NASA once more "trust[ed] us" to conduct flight research without supervision. He then moved without delay to erect a fully-realized organizational framework composed of five essential functional parts: research facilities, research engineering, flight operations, aerospace projects, and intercenter aircraft operations. He subsequently directed his staff to revise the Dryden Basic Operations Manual (BOM). Its publication in February 1995 underscored safety and reminded its readers of the valuable lessons learned in flight research "through tears, sweat, and worse. . . ." Finally, in 1996 Szalai presided over a year of special observations to honor the center on its 50th anniversary.[425]

However, the new center director wanted to look to the future. As early as 1993, he and a group of close advisors concentrated on the probable reduction in military sponsorship of Dryden research as the Cold War wound down. Since military projects constituted a large proportion of Dryden's overall workload, Szalai decided to analyze the implications fully, appointing Robert Meyer (former chief of the Dryden

[424] Letter, Ken Szalai to [?] Geastman, n.d., Milt Thompson Collection, DFRC Historical Reference Collection (first quoted passage); Milt Thompson essay on Dryden/Ames consolidation, n.d., DFRC Historical Reference Collection (second quoted passage and block quote).

[425] "An Important Briefing," 4 June 1993, Milt Thompson Collection, DFRC Historical Reference Collection; NASA News, "NASA Administrator Announces Management Changes," 6 Jan. 1994, DFRC Historical Reference Collection (first quoted passage); Jim Skeen, "Dryden Instructed to Split from Ames," *Los Angeles Daily News*, 8 Jan. 1994, DFRC Historical Reference Collection (second quoted passage); Sharon Moeser, "NASA Facility Will Gain its Independence," *Los Angeles Times*, 7 Jan. 1994, DFRC Historical Reference Collection; Nancy Lovato, "Up Front With Dryden Director Ken Szalai," Dryden *X-Press*, Mar. 1994, DFRC Historical Reference Collection; DFRC organization chart, Oct. 1995, DFRC Historical Reference Collection; DFRC Basic Operations Manual, Mar. 1995, introductory letter by Ken Szalai, inside cover, 17 Mar. 1995, DFRC Historical Reference Collection; Ken Szalai, essay on Basic Operations Manual entitled "Back to Basics," n.d., DFRC Historical Reference Collection (third quoted passage).

aerodynamics branch) to serve as his assistant for strategic planning and new program development. Subsequently, in collaboration with some top staff, Meyer initiated a survey of the current programs in order to begin planning for the future. If the military services required less flight research, what other avenues should Dryden pursue? The initial conclusions were outlined in November 1993 at an off-site meeting attended by the senior Dryden leadership. A candid discussion ensued. Then, fresh from these deliberations, Meyer and Ken Szalai presented their recommendations in January 1994 to the NASA center directors, as well as to headquarters aeronautics representatives.

Meyer drew a troubling picture. He predicted that most of Dryden's present projects—including the military work—would be completed during mid-1995. Unfortunately, no candidates existed to fill the void between the programs about to vanish and the relatively meager resources committed to the activities not canceled or terminated, such as civil and hypersonic projects. Szalai and Meyer proposed a way to bridge the gap. They identified for their listeners the main resources enjoyed by Dryden as a tenant on Edwards Air Force Base: Rogers Dry Lake, the huge runways, the fine climate for flight research, the protected airspace, and the modern ground facilities. They also pointed to the high technical competence of employees. Then they enumerated the kinds of ventures likely to be attracted to such facilities. First, they listed such historically proven ventures as subsonics and high-speed, high-performance flight, including hypersonics.

The dramatic moment of the presentation centered on a new initiative. Known as Access to Space, it promised several appealing prospects. It offered a mission that complemented most of Dryden's strengths; a chance to become active in one of Administrator Goldin's fondest wishes, low-cost launch to orbit; the opportunity to participate fully in the burgeoning fields of communications and satellites, both at the leading edge of technological innovation; and the advantage of involving the Dryden leadership not only with the satellite and communications manufacturers, but also with the rocketeers at NASA's Marshall Space Flight Center (MSFC). Additionally, Access to Space carried the cachet of history. America awoke to the possibilities of space travel in part through two marque Dryden programs of the past, the X-15 and the lifting bodies. Moreover, the center's involvement in the Orbiter approach and landing tests and the operational flight test of the Space Shuttle programs contributed significantly to the U.S. space program. The Access to Space initiative also enjoyed the advantage of timeliness. During the mid-1990s, NASA, Congress, and the White House teemed with talk about a new generation of X-airplanes designed for space launch, in itself a powerful incentive for Dryden's participation in this venture.[426]

[426] Late in Nov. 1994, John McCarthy, a NASA HQ official highly familiar with long-range planning, joined Robert Meyer's strategic initiative. In the end, they added another marquee project to complement Access to Space known as Environmental Research and Sensor Technology, or ERAST (see below, this chapter). Like Access to Space, ERAST claimed a close affinity to the resources available at Dryden. However, ERAST also raised controversy because it gathered data without the human pilot, who until this point epitomized Dryden's historic mission. See interview of Robert Meyer by Michael Gorn (by telephone), 18 Mar. 1999; Robert Meyer, notes of a meeting with Michael Gorn, 30 Mar. 1999; interview of Robert Meyer by Michael Gorn (by telephone), 28 July 1999; briefing, "Dryden Strategic Planning Brief for Associate Administrator and Division Directors," 21 Jan. 1994. All are in the DFRC Historical Reference Collection.

Access to Space

Upon assuming his responsibilities as director of Dryden, Ken Szalai expressed interest in restoring the experimental (X-series) airplanes—perhaps the most notable programs of Dryden's first 50 years—to a position of prominence. An opportunity presented itself in the aftermath of the Shuttle *Challenger*'s explosion after launch in January 1986. Concepts for alternate means of space transportation began to circulate in Washington, DC. The Shuttle's technology, dating from the 1960s, appeared outmoded in comparison to the space launchers offered by Arianespace, a consortium of European nations. But the will to modernize the American program clashed with Congress' desire to slash the federal budget at the end of the Cold War. Republican President George H. Bush and his administration considered these realities and decided, nonetheless, to press for an STS replacement. In his capacity as chairman of the National Space Council, Vice President Dan Quayle became the government's chief proponent of this policy and endorsed a generation of new launch vehicles. Toward the end of his term, the president appointed Daniel Goldin, vice president of satellite systems at TRW, to succeed Richard Truly as NASA administrator. Goldin retained his position after the election of Democratic President William Clinton in November 1992, perhaps because of Goldin's determination to pursue reforms at NASA.

Confronted by shrinking budgets, Goldin chose a two-tiered approach to space exploration: continue to fund a few high-cost, high-visibility programs, but devote far greater resources to many smaller projects under the banner "faster, better, cheaper." Consequently, during the month in which the new president took office, Goldin initiated a top-to-bottom reassessment of NASA's long-term objectives, presented in a report called *Access to Space*. Unlike some earlier studies, this one proposed a *fully* reusable launch vehicle (RLV), powered neither by expendable fuel tanks nor by jettisoned boosters. The concept of a single-stage-to-orbit (SSTO) spacecraft capable of delivering 25,000 pounds of payload to the International Space Station led to exploration of the requisite technologies via a subscale technology demonstrator designated the X-33.[427]

Three bidders responded to a request for proposals for phase I of the X-33: Lockheed Martin, McDonnell Douglas Aerospace, and Rockwell International (the Shuttle prime contractor, later part of Boeing). The competitors began work in March 1995 on concept definitions and technical designs. Upon review, NASA would decide whether any one, or any combination of the three, presented sufficient evidence of probable success to be achieved in phase II, the fabrication and flight of the X-33 demonstrator. During 15 months in which NASA officials—including Dryden

[427] Comments of an anonymous individual on a draft of this chapter; Nancy Lovato, "Up Front with Dryden Director Ken Szalai," Dryden *X-Press*, DFRC Historical Reference Collection; Andrew Butrica, "X-33 Fact Sheet #1: Part I: The Policy Origins of the X-33," 7 Dec. 1997, pp. 1-8, X-33 Home Page on the World Wide Web (http://www1.msfc.nasa.gov/NEWSROOM/background/facts/x33.htm), printout in DFRC Historical Reference Collection; Bill Sweetman, "VentureStar: 21st Century Space Shuttle," *Popular Science* (Oct. 1996): pp. 43-47; "RLV Overview: About the Reusable Launch Vehicle Technology Program," 6 Dec. 1995, p. 1, X-33 Home Page on the World Wide Web, DFRC Historical Reference Collection. The narrative here summarizes developments that are covered in detail in Richard W. Powell, Mary Kae Lockwood, and Stephen A. Cook, "The Road from the NASA Access-to-Space Study to a Reusable Launch Vehicle," IAF paper 98-V.4.02 presented at the 49th International Astronautical Congress, 28 Sept.-2 Oct., 1998, Melbourne, Australia.

An air drop model of the X-33 lands on the dry lakebed at Edwards Air Force Base in 1998. (NASA Photo EC98-44814-11)

representatives—studied the candidate systems, the space agency divided future X-33 responsibilities among its centers. Headquarters selected the Marshall Space Flight Center as the lead center for the X-33.

From the outset, the structure of the program aroused almost as much attention as its daunting technical demands. The industries involved in the X-33 agreed not only to develop a concept for a single-stage vehicle able to demonstrate suborbitally the technologies needed for an orbital vehicle, but also to do so quickly and with minimal government funding. Even if a viable X-33 did materialize, some expressed concern about the short timespan allowed for flight research and about the decision to fabricate only one prototype. Despite such skepticism, the competition continued. Meanwhile, Dryden simulations engineer John Bresina—relying on preliminary wind-tunnel data from the three alternative planforms—devised piloted simulations based on aerodynamic characteristics and proposed flight-control systems. Two Dryden researchers also contributed to concept evaluations and suggestions for the software integration of the X-33. At the same time, Langley engineers played a major role. Following work on the Space Shuttle lasting into the 1980s, they had done computer analyses and wind-tunnel testing on follow-on RLV concepts. The consequent design and analysis tools that resulted enabled them to do analyses and testing at the request of the three competitors for the X-33 design. At last, on 2 July 1996, the Lockheed Martin design won the competition for a cooperative agreement under which Lockheed Martin would construct the X-33 demonstrator, the potential precursor of a hoped-for full-sized vehicle Lockheed Martin called VentureStar. Following that decision, Langley researchers continued to perform wind-tunnel tests and analyses at the request of Lockheed Martin.

Initial plans called for NASA to spend $941 million on the project through 1999 and for Lockheed Martin to invest $220 million in the X-33 design. This relied on three main technical concepts, two with long histories: a wedge-shaped lifting-body airframe borrowed directly from Dryden flight research of the 1960s and 1970s, a linear aerospike powerplant conceived originally by Rocketdyne (a division of the firm that became Rockwell International) during the 1960s but never flown, and a metal thermal protection system. While the Lockheed Martin Skunk Works of Palmdale, California, assembled the airframe, Boeing/Rocketdyne (as it now became) in Canoga Park, California, fabricated the engines as an industry partner. Lockheed Martin and its industry partners then agreed to construct the half-size prototype of VentureStar known as the X-33, a vehicle 67 feet long, and 68 feet at the widest part of the wedge, that weighed 64,000 pounds empty, and was equipped with a five-by-ten-foot cargo bay. NASA Administrator Daniel Goldin hailed the announcement as the first step in a journey to slice space payload costs from the $10,000 per pound on existing launchers to just $1,000 per pound on the operational RLV.[428]

Although Dryden had only a supporting role in the X-33 drama, it took advantage of the opportunity to serve as a partner on a program of national importance. Dryden initially contributed to the program in the following ways:

1. Advising Marshall about the safety of the vehicle.
2. Conceiving and implementing the flight-research program (beginning with a staff of about 25 and increasing it as needed).
3. Designing the test range and supervising construction required to extend communications, radar, and Global Positioning System (GPS) tracking for the X-33's flight trajectory.

[428] More complete accounts of the origins of the three X-33 designs are available in a handwritten schematic by an individual who wishes to remain anonymous entitled "30 Minute Draft" [of Reusable Launch Vehicle History], 2 Mar. 1999, DFRC Historical Reference Collection, and interview of Ken Iliff by Michael Gorn, 22 Mar. 1999, DFRC Historical Reference Collection. Also, see "X-33 Advanced Technology Demonstrator," Marshall Space Flight Center Fact Sheets, as of 27 July 1999, X-33 Home Page on the World Wide Web http://www1.msfc.nasa.gov/NEWSROOM/background/facts/x33.htm), hard copy in DFRC Historical Reference Collection; "X-33 Program Risks Unnecessarily High,"*Aviation Week & Space Technology* (29 Jan. 1996), p. 74; briefing charts, "X-33" (two pages showing X-33 work years, accomplishments, and major issues), 8 Feb. 1996, DFRC Historical Reference Collection; NASA news release, "Lockheed Martin Selected to Build X-33," 2 July 1996, DFRC Historical Reference Collection; Lockheed Martin information release, "Lockheed Martin VentureStar Wins X-33 Competition: Program Valued at More than $1 Billion through 2000," July 1996, DFRC Historical Reference Collection; specifications sheet, "X-33 Advanced Technology Demonstrator," n.d., DFRC Historical Reference Collection; Gray Creech, "Dryden Plays Major Role in X-33," Dryden *X-Press*, Aug. 1996, DFRC Historical Reference Collection; Powell, Lockwood, and Cook, "The Road from . . . Access-to-Space to a Reusable Launch Vehicle," pp. 1-9. On the Langley studies and contributions, see Douglas O. Stanley, Theodore A. Talay, Roger A. Lepsch, W. Douglas Morris, and Kathryn E. Wurster, "Conceptual Design of a Fully Reusable Manned Launch System," *Journal of Spacecraft and Rockets*, Vol. 29, No. 4 (Jul.-Aug. 1992), pp. 529-536; Delma C. Freeman, Theodore A. Talay, Douglas O. Stanley, Roger A. Lepsch, and Alan W. Wilhite, "Design Options for Advanced Manned Launch Systems," *Journal of Spacecraft and Rockets*, Vol. 32, No. 2 (Mar.-Apr. 1995), pp. 241-249; and Delma C. Freeman, Jr., Theodore A. Talay, and Robert Eugene Austin, *Single-Stage-to-Orbit—Meeting the Challenge* (Washington, DC: NASA TM-111127, 1995), pp. 1-10.

4. Cooperating with Marshall in developing flight-control components and real-time computer simulations.

5. Supporting the X-33 launches and collaborating with Edwards Air Force Base in selecting launch and landing locations.

6. Flight-testing an SR-71 aircraft equipped with a 20-percent-scale, half-span model of the X-33 (minus the fins) that was rotated 90 degrees and equipped with eight thrust cells of an aerospike engine.

7. Calculating X-33 support costs for Dryden's program planning in such a way as to protect funding for Dryden's on-going flight-research activities.

Although still limited, Dryden's collaboration in this project increased over time. While the document signed between NASA and Lockheed essentially consigned Dryden to a subcontracting role, Ken Szalai pushed hard to broaden the center's participation. He and his associates realized that the fullest possible participation offered Dryden an opportunity to recover at least part of the stature lost during the 13-year amalgamation with Ames. Consequently, not satisfied merely to be passive supporters, Szalai, Dryden project manager Gary Trippensee, and the Dryden engineers attending X-33 technical meetings actually volunteered *additional* services that Dryden could render to the contractors and to the other NASA centers. In fact, Dryden engineers did join forces with their Lockheed Martin counterparts on sensor development. They also provided the contractor with system configuration and with operational experience. At first, it appeared that Dryden might take responsibility for transforming the Shuttle Carrier Aircraft (SCA) from its Orbiter-ferrying configuration to serve as an X-33 transport. This project also entailed flight research on the combined 747/X-33 vehicles, envelope expansion tests, and eventual supervision of flight operations and maintenance. However, program planners eliminated air transportation of the X-33 in favor of ground travel.[429]

Meanwhile, the scaled 20-percent-scale linear aerospike engine arrived at Dryden for testing on the SR-71 Blackbird, an important event because the success of the entire X-33 project rested in some degree on the flight research of this unusual powerplant. It differed significantly from traditional rockets. The system expelled the exhaust produced by liquid hydrogen and liquid oxygen (propellants with the highest known performance) from eight small combustors (20 in the full-sized X-33). The combustors were arrayed in parallel rows and positioned to fire along curved, rectangular plates. On the Shuttle, bell-like nozzles possessing a *fixed* shape and unchanging expansion ratios—based on a compromise configuration that operated at peak efficiency only during critical parts of the launch—were attached to a central combustion chamber. By contrast, the angle of exhaust flow in the linear aerospike was not constrained on the outside, but instead expanded and contracted according to the density of the atmosphere encountered in flight. If proven satisfactory in actual flight, this system promised significant reductions in weight and fuel consumption for a given level of thrust. It also offered the advantages of simple design, durable construction, high thrust, and ideal expansion ratios at all altitudes, among other characteristics demonstrated through a long history of ground-testing.

[429] Interview of Robert Meyer by Michael Gorn, 30 Mar. 1999, DFRC Historical Reference Collection; briefing charts, "X-33: Director's Weekly Update," 14 Feb. 1997, DFRC Historical Reference Collection; Gray Creech, "Dryden Plays Major Role in X-33."

The Linear Aerospike SR-71 Experiment (LASRE) flew during 1997 and 1998. Oxygen leaks discovered during cold flow tests conducted as part of LASRE flight research resulted in the cancellation of hot firings of the engine. (NASA Photo EC 97-44295-108)

Dryden first contributed to the trials of this engine when its technicians began attaching the 20-percent-scale linear aerospike powerplant to NASA SR-71 Blackbird 844. This mating, completed in August 1997, began the Linear Aerospike SR-71 Experiment, or LASRE. After completion of a series of ground tests, the Blackbird flew on 31 October under its own power, with the aerospike disabled. This flight lasted almost two hours at speeds up to Mach 1.2 and altitudes up to 33,000 feet. It collected data relative to the SR-71's aerodynamics, stability and control, and structural integrity as it carried the aerospike mock-up. The following year, on 4 March 1998, the first cold-flow tests occurred, cycling gases through the engine during flight.

Unfortunately, oxygen leaks were detected during the three flights that followed in spring and summer. In the same period, two engine hot firings took place on the ground. Project managers, fearing the consequences of the liquid oxygen leaks in the presence of hydrogen, decided that enough data existed from the four cold-flow flights and the ground tests to make reasonable extrapolations about the behavior of the engine operating with hot gases in flight. As a result, the LASRE researchers chose not to try a hot firing aboard the SR-71. Flying operations ended in November 1998 with some data but significantly less than anticipated.[430]

As LASRE prepared to undergo testing, industry and NASA representatives convened at Dryden during November and December 1996 for the X-33 Preliminary Design Review. The attendees agreed that sufficient progress had been made to allow more detailed design and fabrication, including a subject of keen interest to DFRC, the X-33 launch and ground facilities. During 1997, Dryden officials hosted a delegation from the Sverdrup Corporation which surveyed the base and surrounding terrain for a suitable launch site. Ultimately, a six-member panel comprised of Edwards and Dryden staff evaluated seven possible locations suggested by Lockheed Martin. The panel eventually selected a 25-acre parcel a few hundred yards north of Haystack Butte, on the eastern side of the base in an angle formed by Highways 58 and 395. It appealed to the group because the X-33 launch path projected from this area offered the least disruption to the main base and to the local population, yet lay just 30 miles away from Dryden by road. Once the Air Force officially accepted the site, a 30-day period of public review of the site's Environmental Impact Statement (EIS) ensued. This document explained the project's probable affect on the people and the land in question. The plan passed muster at a series of regional town meetings, and NASA headquarters then declared Haystack Butte the preferred option. Sverdrup began construction of the $30 million facility on 14 November 1997, and during the next year a launch pad, X-33 rolling shelter, fuel storage tanks, water storage tank (for a sound baffling system), and concrete flame trench materialized on the arid ground.[431]

[430] "Linear Aerospike Engine—Propulsion for the X-33 Vehicle," Marshall Space Flight Center fact sheets, as of 27 July 1999, X-33 Home Page on the World Wide Web (http://www1.msfc.nasa.gov/NEWSROOM/background/facts/aerospike.htm), DFRC Historical Reference Collection; Summary Paper, "X-33 Year in Review: Accomplishments and Challenges," 1996, DFRC Historical Reference Collection; Kathy Sawyer, "Bargain-Hunting NASA Picks Blast From Past," *Washington Post*, 3 Feb. 1997, p. A03; Warren Leary, "Novel Rocket to Power Shuttle Successor," *New York Times*, 30 July 1996, p. C1, C8; Cheryl Agin-Heathcock, "Linear Aerospike Engine Fitted to SR-71 #844," Dryden *X-Press*, April 1996, DFRC Historical Reference Collection; briefing charts, "X-33 Status as of 20 June 1997," DFRC Historical Reference Collection; "Linear Aerospike SR-71 Experiment Talking Points," and "LASRE Project Information Summary," 4 Oct. 1997, DFRC Historical Reference Collection; Dryden press release, "Linear SR-71 Experiment Completes First Cold Flow Flight," 5 Mar. 1998, DFRC Historical Reference Collection; Draft DFRC press release on the LASRE Project, July 1999, pp. 1-2, DFRC Historical Reference Collection.

[431] Dryden news release, "X-33 Launch Facility Site Survey Underway," 6 Mar. 1997, DFRC Historical Reference Collection; Gray Creech, "1st Phase Ends for X-33 Launch Site Survey," Dryden *X-Press*, 21 Mar. 1997, DFRC Historical Reference Collection; "Preferred X-33 Sites Chosen," Dryden *X-Press*, 3 Oct. 1997, DFRC Historical Reference Collection; briefing charts, "X-33 Director's Weekly Update," 14 Feb. 1997, DFRC Historical Reference Collection; NASA news release, "X-33 Launch Facility Ground Breaking Held," 14 Nov. 1997, DFRC Historical Reference Collection; briefing charts, "Striving for Affordable Access to Space," 17 Oct. 1996, DFRC Historical Reference Collection; NASA news release, "X-33 Program Completes Operations Review," 18 Dec. 1996, DFRC Historical Reference Collection.

The prototype demonstration of the X-33 actually received the go-ahead as late as 1997. The decision followed a long and tortured path. During the year, delegates representing government and private institutions presented no fewer than 51 detailed briefings about the craft's subsystems and components. At the close of October, the process reached its climax in a conference sponsored by Edwards Air Force Base. For five days, roughly 600 persons met for an X-33 Critical Design Review. In the end, they approved the fabrication of the final pieces of the demonstrator and agreed to its final assembly. Meanwhile, Dryden released some important technical findings and pursued a second flight-research project in support of the X-33. One paper announced the results of the center's investigation of a flush air data sensing (FADS) system, a group of pressure orifices located on the X-33's nose to collect data on airflow in flight. Another published the results of an analysis by Dryden engineers of the immense archive of flight data from six lifting-body and the Shuttle Orbiter flight programs. The authors contrasted this data with the X-33 wind-tunnel tests and determined models of uncertainty for the Lockheed Martin demonstrator's subsonic and supersonic flights. Finally, in the skies over Edwards, the X-33 thermal protection system (TPS) was subjected to a thorough flight-research program, similar to ones performed for the Shuttle during the 1980s.

Mounted to a test fixture on an F-15B aircraft were three substances to be tested for shear and shock loading: metallic Inconel tiles, soft advanced flexible reusable surface insulation tiles, and sealing materials. The project pilot flew the aircraft through several vigorous flights, maneuvering it at speeds up to Mach 1.4 and at altitudes to 33,000 feet. Close inspections of heat-resistant items on the test fixture revealed no wear or damage caused by air loads below Mach 1 or from shock waves at the transonic range. F-15B project manager Roy Bryant expressed justifiable satisfaction at the speed and the frugality with which his project personnel compiled the resultant data for the X-33 designers.[432]

Nonetheless, programmatic and technical factors beyond the control of the Dryden flight-research team rendered the X-33 a vehicle with an uncertain future as early as 1999. Speaking before the House Subcommittee on Space and Aeronautics in September 1999, NASA Deputy Associate Administrator (Space Transportation Technology) Gary Payton conceded that the X-33's first flight, originally scheduled for summer 1999, had to be postponed. This delay implied serious consequences for the program. Lockheed and its industry partners had by then invested $356 million in the X-33 over and above the $942 million expended by NASA. Moreover, Lockheed Martin had at stake the opportunity to develop VentureStar or whatever commercial launch vehicle might result from X-33 research and testing. Since NASA considered its share of the costs to be fixed, Lockheed Martin faced some difficult choices.

[432] NASA Ames participated in the X-33 thermal-protection-system tests by conducting wind-tunnel analyses and by offering advice about the TPS flight research program. "X-33 Program Completes CDR," Dryden *X-Press*, 21 Nov. 1997, DFRC Historical Reference Collection; Stephen Whitmore, Brent Cobleigh, and Edward Haering, "Design and Calibration of the X-33 Flush Air Data Sensing (FADS) System," (Washington, D.C.: NASA TM 206540, 1998), p. 1; Brent Cobleigh, "Development of the X-33 Aerodynamic Uncertainty Model," Washington, DC: NASA TP 206544, 1998), p. 1; Dryden Press Release, "X-33 Thermal Protection System Materials Fly on F-15B," 18 May 1998, DFRC Historical Reference Collection; NASA news release, "X-33 Thermal Protection System Tests Complete," 30 June 1998, DFRC Historical Reference Collection.

Payton attributed the delays to two technical hurdles: the linear aerospike engine and the X-33's fuel tanks. During the first week of November 1999, one of the X-33's liquid hydrogen tanks—fabricated from composites to reduce weight—failed during a validation test at the Marshall Space Flight Center. A difficult decision followed. The program abandoned the composite materials in favor of more traditional aluminum lithium construction. But the change would take time. The X-33 with the new tanks would not be ready to resume flight research until 2002, according to projections at the end of 1999.

Meanwhile, the linear aerospike powerplant had encountered better fortune but still required more development time. An engine mounted on its A-1 test stand at NASA's Stennis Space Center fired successfully for 18 seconds in December 1999 and for 290 seconds in May 2000. The second time far exceeded the necessary duration for an X-33 flight. But the flight-research program still awaited the time-consuming process of preparing and testing (at Stennis) two linear aerospike powerplants in the side-by-side configuration to be used aboard the X-33.

Given the delays these developments entailed, Lockheed Martin hesitated in June 2000 about making a further financial commitment to X-33 and withdrew an offer to add funding. Since both government and industry would benefit from the technical data gleaned from the effort to solve the single-stage-to-orbit conundrum, an agreement between NASA and Lockheed Martin in September 2000 broke the deadlock temporarily. The manufacturer pledged to fund the program through March 2001, after which it could compete for more support through the NASA Space Launch Initiative. The revised plan scheduled the X-33 to fly in 2003. As further events unfolded and the administration of President George W. Bush assumed office, NASA finally announced on 1 March 2001 that it would no longer fund the X-33. Art Stephenson, director of Marshall Space Flight Center, stated that NASA had gained "a tremendous amount of knowledge" from X-33 and X-34 (see below) "but one of the things we have learned is that our technology has not yet advanced to the point that we can successfully develop a new reusable launch vehicle that substantially improves safety, reliability and affordability."[433]

Meanwhile, Dryden involved itself with a number of other projects affiliated with the Access to Space initiative. During the early 1990s, it contributed to a novel space launch endeavor conceived during the late 1980s by the Orbital Sciences Corporation (OSC). The firm conducted feasibility studies under the sponsorship of

433. J. Sumrall, C. Lane, and R. Cusic, "VentureStar—Reaping the Benefits of the X-33 Program," *Acta Astronautica*, Vol. 44 (1999), P. 727; Statement of Gary E. Payton, Hearing on NASA's X-33 Program Before the Space and Astronautics Subcommittee of the Committee on Science, House of Representatives, 29 Sept. 1999 (NASA website: http//www.hq.nasa.gov/office/legaff/payton9-29.html), copy in the Dryden Historical Reference Collection; Craig Covault, "Aerospike, RBCC Address RLV Goals," *Aviation Week & Space Technology*, (13 Dec. 1999), pp. 76-78; "X-33 Program Status," 4 and 15 Nov., and 21 Dec. 1999, 16 May 2000 (Marshall Space Flight Center website: http://x33.msfc.nasa.gov/status/991104.html, .../991115.html, .../991221.html, and http://x33.msfc.nasa.gov/x33 status.html); Frank Morring, "NASA, Lockheed Martin Eye X-33 Restructuring," *Aerospace Daily*, 22 Mar. 2000, copy in the DFRC Historical Reference Collection; Aerospace Projects Weekly Highlights, 2 and 16 June 2000, DFRC Historical Reference Collection; telephone interview of Gary Trippensee, DFRC X-33 Project Manager, by Michael Gorn, 31 May 2000; Dryden news release 00-77, "NASA, Lockheed Martin Agree on X-33 Plan," 29 Sept. 2000 (Dryden website: http://www.dfrc.nasa.gov/PAO/PressReleases/2000/00-77.html); NASA Press Release 01-31, "NASA Reaches Milestone in Space Launch Initiative Program; Also Announces No SLI Funding for X-33 or X-34," the source for the Stephenson quotation in the last paragraph, gives a total investment of $912 million for NASA and $357 for Lockheed, differing from other sources

Mounted under the wing of a B-52 aircraft, the Pegasus rocket consisted of one booster that launched the vehicle from the bomber and a second one that propelled payloads (up to 1,500 pounds) into orbit. It flew from beneath the B-52 between 1990 and 1994, after which time it was launched from an Orbital Sciences Corporation L-1011, including a flight with the Pegasus Hypersonic Experiment (PHYSX). A steel glove fitted over one of Pegasus' wings enabled PHYSX to collect data about hypersonic flight between Mach 5 and 8. (NASA Photo EC94-42690-7)

the Defense Advanced Research Projects Agency (DARPA) to define an economical method of placing small payloads into orbit, a high priority issue after the *Challenger* catastrophe imperiled American access to space. The contractor proposed a system in which a small rocket, launched from a large transport aircraft, propelled a satellite into space. The aircraft served the purpose of the traditional first stage rocket, from which a winged booster took off, ascended, and launched a second booster that lifted the payload (weighing about 1,500 pounds) out of the atmosphere. Representatives of OSC approached Dryden officials about using the venerable B-52 #008—already outfitted with a launch pylon—as a platform for this rocket, which they called Pegasus.

As a result, Dryden's staff arranged for three inert captive flights during December 1989 and January 1990. After some minor difficulties with telemetry had been corrected, DARPA called two prelaunch meetings: a Mission Readiness Review on 6 February 1990 in Fairfax, Virginia, and a Consolidated Safety Review at Dryden one week later. These sessions generated nineteen items requiring the attention of the participants before Pegasus could fly on its own. Meantime, as B-52 project manager Roy Bryant and his associates checked off these items, in March the bomber's research pilot, Gordon Fullerton, took a proficiency flight during which he tested the hook and safety pin system on the pylon adapter used to secure Pegasus. Clouds between Edwards and the launching point over the Pacific Ocean prevented the first launch, planned for 4 April 1990. The next day, Fullerton initiated the launch sequence by which the vehicle separated from the B-52 pylon and the Pegasus igniters fired, in turn activating the first-stage booster motor. After about twelve minutes the satellite entered a satisfactory (but not perfect) orbit. Five more successful Pegasus launches followed between July 1991 and August 1994. The third one, in February 1993 from Kennedy

Space Center, marked the first time a commercial satellite had been sent into orbit by a launch vehicle dropped from an aircraft.

But Pegasus was not yet retired from service as a Dryden research vehicle—a role it fulfilled in addition to its capacity as a small launch vehicle dropped, after the sixth flight under the B-52, by an Orbital Sciences L-1011. Dryden engineers collaborated with OSC in what came to be known as the Pegasus Hypersonic Experiment, or PHYSX. Designed to shed light on the aerodynamics of air-breathing vehicles traveling from Mach 5 to Mach 8, the investigation involved the bonding of a steel "glove" to the right wing of Pegasus's first stage. A similar device underwent ground tests at Dryden in 1996. Then, riding along as a commercial satellite was boosted into orbit, the glove, arrayed heavily with sensors, was launched over the Atlantic from Orbital Science's L-1011 mothership after taking off from Cape Canaveral Air Force Station, Florida, on 22 October 1998. PHYSX succeeded in collecting valuable information about the transition from laminar to turbulent airflow as the wing accelerated with stage one to Mach 8 and reached altitudes of 250,000 feet.[434]

The Pegasus legacy of launch to orbit from an aircraft lived on in another Orbital Sciences Corporation project to which the Dryden lent support. The firm won a $85.7 million contract from NASA headquarters' office of aerospace technology to construct three prototypes of an unpiloted technology demonstrator known as the X-34. Like Pegasus, the X-34 would be launched from an L-1011 jetliner, but rather than boost satellites into orbit, it would fly suborbital missions to conduct flight research on a variety of technologies essential to reusable launch vehicles. Some of the important test possibilities included composite airframe structures, new thermal protection systems, automated flight controls, and inexpensive avionics. About 58 feet in length, 28 feet in wingspan, and almost 12 feet tall at the tail, the sweptwing vehicle was expected to complete some 25 flights a year once it became operational. It featured a single-stage Fastrac engine propelled by a mixture of liquid oxygen and kerosene, a powerplant being developed by the Marshall Space Flight Center, the project's lead organization. Flying to Mach 8 and 250,000 feet, the X-34, if successful, would provide a testbed for a number of improvements, including quick relaunch capability, low-cost operation, safe flight through bad weather, and reliable landing in 20-knot crosswinds.

Dryden's role—as with the X-33, subsidiary to the contractor—began during spring 1999 when OSC trucked one of the X-34 airframes to Edwards. After reassembling the vehicle, the Dryden staff performed ground vibration tests on the X-34 and the L-1011 carrier aircraft. Then the OSC/Dryden team inaugurated a series of seven planned captive-carry flights, necessary for FAA certification of the paired vehicles. The first outing on 29 June uncovered a problem: the airliner developed aft panel vibration in flight. Apparently, cracks had developed in the L-1011's skin,

[434] Comments of Roy Bryant, B-52 project manager, on a draft of this section, 26 Jan. 2000; Wallace, *Flights of Discovery*, pp. 142-144; Roy Bryant, essay on the Dryden B-52 aircraft, 8 Oct. 1999, DFRC Historical Reference Collection; Dryden Project Highlights for Jan.-Apr. 1990, DFRC Historical Reference Collection; NASA Facts, "PHYSX: Pegasus Hypersonic Experiment," Oct. 1998, available on the Dryden website at http://www.dfrc.nasa.gov/PAO/PAIS/HTML/FS-053-DFRC.html and in hard copy in the DFRC Historical Reference Collection; Dryden Project Highlights, 26 Oct. 1998, DFRC Historical Reference Collection; Arild Bertelrud et al., "Pegasus Wing-Glove Experiment to Document Hypersonic Crossflow Transition—Measurement System and Selected Flight Results" (Edwards, CA: NASA TM-2000-209016, Jan. 2000), pp. 1-4, 9, 10.

Orbital Sciences Corporation—the firm that designed and fabricated the Pegasus—also produced the X-34, shown here on the ramp at Dryden in 1999. Launched from a L-1011, this pilotless aircraft, if successful, would showcase technologies for the new reusable launch vehicles. (NASA Photo EC99-44976-4)

requiring extensive repairs in Dryden's hangars. The mated aircraft underwent the second captive flight on 3 September, flying for more than four hours, with a third and final flight on 14 September. As with the X-33, NASA decided in 2001 that the benefits of continuing the X-34 program failed to justify the costs, so it began the process of terminating its X-34 contract with OSC. NASA still hoped to benefit from technologies developed under both the X-33 and X-34, according to Marshall Director Stephenson. The technologies in question included the linear aerospike engine for the X-33, advanced thermal protection systems, lightweight materials, advanced aerodynamic designs, and self-healing avionics.[435]

The X-38 advanced technology demonstrator for a Crew Return Vehicle (CRV) carved out a much different niche in the continuum of access to and return from

.[435] Before its termination, the X-34 underwent significant changes before it ever flew. One modification that was discussed involved placing a pilot in the system so that in future flights—planned not just over water but over land as well—a research pilot on the ground could enhance safety. In addition, the idea of an inexpensive, "single-string" flight-control system had given way to an X-34 possessing redundant systems for greater safety. Redundant systems also offered the hope that the vehicle might be recovered if one system failed. Andrew Blankstein, "X-34 Rocket Craft to be Tested at Dryden Center," *Los Angeles Times*, 26 Aug. 1999; Kirsten Williams, "Dryden Assists with X-34," Dryden *X-Press*, 12 Mar. 1999, DFRC Historical Reference Collection; Dryden Project Highlights, 2 July, 3 Sept., and 17 Sept. 1999, DFRC Historical Reference Collection; telephonic interview of John Bosworth, X-34 project engineer, by Michael Gorn, 31 May 2000; telephonic interview of John McTigue, 1 June 2000; Gregory J. Brauckmann, "X-34 Vehicle Aerodynamic Characteristics," *Journal of Spacecraft and Rockets*, (Mar.-Apr. 1999), p. 229; NASA Press Release 01-31; Paul Hoversten, "Citing Costs, NASA kills futuristic X-33, X-34," *Aerospace Daily*, 2 Mar. 2001, item 3, copy in the files of the NASA Dryden Historical Reference Collection.

The X-38 Crew Return Vehicle (CRV) made unpiloted flights beginning in 1998 (with a captive flight in 1997). Shaped like the X-24A, it underwent tests as a "lifeboat" for the International Space Station, capable of returning up to seven astronauts to Earth by combining a lifting-body glide with a parafoil landing. (NASA Photo EC99-45080-25)

space. In its initial incarnation–designers hoped it would later serve as an international spacecraft boosted into orbit aboard an Ariane 5 rocket–the CRV was intended as a space "lifeboat," a vehicle by which as many as seven people inhabiting the International Space Station could escape to earth. The first two test aircraft–about 25 feet long, 15 feet wide, and 80 percent scale of the expected final version[436]—had a familiar look. The X-38 closely resembled the highly successful X-24A lifting body that last flew in 1971. The two prototypes (designated Vehicles 131 and 132) matured under somewhat unusual circumstances. Scaled Composites of Mojave, California, fabricated the airframe while engineers at the Johnson Space Center (which managed and originated the project) designed the avionics and computer software. The spacecraft's planned descent from the Space Station also broke new ground. On-board engines would be fired (and then jettisoned) to remove the CRV from orbit; then it would glide unpowered in lifting-body fashion through the atmosphere until, at final descent, a parafoil parachute opened. Originally, the engineers designed the parafoil's control to be totally autonomous. Subsequently, they added pilot-steering as a backup mode.

[436] The original vehicles were expected to be the right size for the CRV. They became 80-percent versions when the Space Station program changed the requirement from two four-person vehicles to one seven-person vehicle.

The joint Johnson Space Center/Dryden Flight Research Center team, responsible for the flight-test part of the program, set four objectives. They would fly and evaluate the transition from lifting-body to parafoil flight, assess the flight-control systems of the X-38 and the parafoil, demonstrate the subsystems, and then record and analyze the lifting-body and parafoil flight dynamics. Dryden actually began research on the X-38 by conducting thirteen drop tests using a 1/6-scale model. Then, in July 1997, a series of captive-carry flights began with the B-52 008 and the 80-percent-sized X-38s. These trials lasted through that year and into 1998. During March 1998, the first launch of the 80-percent X-38 revealed some problems with the parafoil. After further research on this critical part of the descent system at the Army's Proving Grounds in Yuma, Arizona, free flights of the X-38 resumed at Dryden on 6 February 1999, and continued in March and July with the two vehicles. Number 132 was equipped with (and Number 131 flew without) flight-control surfaces. Following these successes, additional captive-carry flights of Vehicle 132 occurred in September and November 1999. These flights culminated in a final free flight for Vehicle 132 on 30 March 2000. This was the highest, fastest, and longest X-38 flight to date. Released from an altitude of 39,000 feet, Vehicle 132 flew for 45 seconds and reached a speed of over 500 miles per hour before deploying its parachutes for a landing on Rogers Dry Lake. Due to budgetary considerations, the X-38 program ended after a final flight in December 2001.[437]

Finally, among the initiatives that belonged generically to Access to Space, the Hyper-X project represented perhaps the most distinct break from present means of high-speed cruise or escape from the atmosphere. (Technically, Hyper-X fell under the rubric of NASA headquarters' office of aerospace technology's Revolutionary Technology Leaps rather than Access to Space, but it was developing a technology with potential application to space.) A technology demonstrator, the Hyper-X research vehicle—designated the X-43A—represented a hypersonic aircraft scheduled to undertake "the first free-flight ever of an airframe-integrated supersonic combustion ramjet (scramjet)." Unlike conventional turbine engines, which compress air by means of a rotating mechanical compressor, the scramjet uses air compressed by shock waves created as an aircraft flies at hypersonic speeds (above Mach 5). Internal airflow and combustion processes in a scramjet occur at supersonic speeds, in contrast to the ramjet whose internal airflow and combustion processes happen subsonically. Since its powerplant "breathes" air like a traditional jet engine, the scramjet has the potential to transport heavier payloads than rockets because it needs to carry no oxidizer. The X-43's fuel, consisting of gaseous hydrogen, will be injected initially into a silane-hydrogen mixture to start combustion. Ultimately, designers expected the vehicle would fly at seven to ten times the speed of sound.

[437] Telephonic interview of Chris Nagy, X-38 project engineer, by Michael Gorn, 31 May 2000 and comments by Nagy on a draft of this section, 14 Feb. 2000; Ricardo Machin, Jenny Stein, and John Muratore, "An Overview of the X-38 Prototype Crew Return Vehicle Development and Test Program," AIAA Paper 99-1703, presented in Toulouse, France, at the CEAS/AIAA Aerodynamic Decelerator Systems Technology Conference, 8-11 June 1999, pp. 20-21; (Draft) Dryden fact sheet, "Sub-Scale Flight Research Vehicles," 26 Feb. 1999, DFRC Historical Reference Collection; Dryden fact sheet, "X-38: Back to the Future for a Spacecraft Design," from the Dryden website (http://www.dfrc.nasa.gov/PAO/PAIS/HTML/FS-038-DFRC.html), Mar. 1998; briefing, "X-38 Crew Return Vehicle," 26 Jan. 1999, DFRC Historical Reference Collection; Dryden Project Highlights, 9 July and 23 Aug. 1999, DFRC Historical Reference Collection.

The Hyper-X (or X-43A) vehicle required a boost from the Pegasus launcher, after which its own supersonic combustion ramjet (scramjet) engine would operate. (NASA Photo ED99-45243-01)

Langley Research Center and Dryden Flight Research Center embarked on the project jointly in October 1996, under the sponsorship of NASA headquarters' office of aeronautics and space transportation (since redesignated the office of aerospace technology). Dryden's role entailed contractor management, flight operations, and flight research. Langley's contributions included the conceptual design of the vehicles, wind-tunnel testing, and comparison of flight data (when available) with predictions of performance.

During the flight-research phase, the B-52 at Dryden was scheduled to make three drops of the expendable, autonomous X-43A, a vehicle roughly 12 feet long and five feet wide that would be attached to a modified Orbital Sciences Corporation first-stage booster from a Pegasus launch vehicle. Two drops were projected to lead to Mach 7 flights and one to Mach 10. Each flight would consist of two vehicle separations: the first, when the Pegasus "stack" dropped from the B-52; the second, when the X-43A was ejected from the booster at the desired Mach number. During the Mach 7 and Mach 10 flights, the initial separation would occur near San Nicholas Island off the California coast, followed respectively by flights of 400 and 900 miles over the Pacific ocean.

For this project, Dryden's management of both the flight-research program and the building of the test vehicles was no small task with manufacturers scattered from Seal Beach, California (system design by Boeing), to Chandler, Arizona (integration and construction of three Pegasus first-stage launch vehicles by Orbital Sciences), to Long Island, New York (engine fabrication by GASL-Ronkoncoma), and to Tullahoma, Tennessee (assembly and integration of three X-43As). Gradually, during 1998 and

1999, the various elements of this strange-looking system—featuring a long, tapered booster with a triangular wing mounted atop the fuselage and the slender X-43A itself—rolled off their respective assembly lines. Significantly, the first scramjet engine arrived at Langley in November 1998 for ground tests, and by September of the following year, engineers there reported "successful engine light and burns have been accomplished" at a range of dynamic pressures from 650 pounds per square foot (psf) to 1,000 psf (concentrating on the flight condition of 1,000 psf). The X-43A flew for the first time on 2 June 2001. Unfortunately, after the Pegasus stack separated from the B-52, its motor ignited and it flew for only a few seconds, when a malfunction caused the X-43A vehicle and booster to depart from controlled flight, leading to their destruction within the range area set aside for the flight. NASA appointed an investigation board to determine the cause of the malfunction.[438]

No One Aboard: Remotely Piloted Vehicles

While the series of projects involving Access to Space became one of the most visible new initiatives pursued by Dryden after it resumed full independence, two others identified by Ken Szalai and Bob Meyer also assumed prominence. Uninhabited Aerial Vehicles (UAVs), including a group of airplanes developed under the Environmental Research Aircraft and Sensor Technology (ERAST) program, received considerable attention during the mid-1990s, although their precursors had been associated with Dryden for decades. As discussed in Chapter 10, the center used the PA-30 Twin Comanche to research and practice flying an airplane from a ground cockpit. It then flew the 3/8-scale F-15 as an RPRV to conduct spin research in the 1970s. Moreover, while the remotely-piloted HiMAT plied the skies over Edwards, a modified Teledyne drone called Firebee II also flew.

Langley and Dryden engineers, hoping to validate theoretical aerodynamic predictions through flight research, gave the reconfigured Firebees the name Drones for Aerodynamic and Structural Testing (DAST). First used as gunnery targets, these 28-foot-long, missile-shaped Firebees were obtained from the Air Force as surplus and outfitted with supercritical wings, themselves more than 14 feet in span. During their flight program, the DAST vehicles were launched first from a B-52 but produced most of their data following subsequent drops from a DC-130. Pilots controlled the DAST remotely from the ground, putting it through a series of maneuvers (including 5 g turns), and ending each mission with a descent by parachute. The program tackled a complicated problem: to determine the speed at which wing flutter occurred and to design and test a digital flutter suppression system. By controlling flutter electronically, wing structures could be fabricated with lower-than-normal stiffness and less weight.

[438] Briefing, "Hyper-X Flight Project Overview," approximately 6 Oct. 1998, DFRC Historical Reference Collection (first quoted passage); Dryden fact sheet, "Dryden and the Hyper-X Program," 1998, DFRC Historical Reference Collection; "Sub-Scale Flight Research Vehicles" (draft), Feb. 1999, DFRC Historical Reference Collection; Dryden Project Highlights for 1999 (second quoted passage from 24 Sept. 1999), DFRC Historical Reference Collection; C. R. McClinton, J. L. Hunt, R. H. Rickette, P[aul] Reukauf, and C. L. Peddie, "Airbreathing Hypersonic Technology Vision Vehicles and Technology Dreams," AIAA Paper 99-4978 presented at the National Space Planes and Hypersonic Systems and Technologies Conference in Norfolk, VA, 1-5 Nov. 1999; comments of Paul Reukauf, X-43A deputy project manager, on a draft of this section, 3 Mar. 2000; Dryden Flight Research Center news release 01-40, 2 June 2001.

Dryden engineers converted Firebee II drones into Drones for Aerodynamic and Structural Testing (DAST), shown here. DAST flew from 1977 to 1983. (NASA Photo EC 80-14090)

The project began in December 1973 when the Air Force agreed to turn over the Firebees to NASA. After eighteen flights and five program managers, it ended almost a decade later, in October 1983, with the publication of a report on the loss of a DAST in June of that year. Of the eighteen flights, just ten involved actual launches (the others were captive flights), and of these ten, two crashed and four were aborted. But the low cost of the DAST flights gave engineers the freedom to experiment boldly, and much of the data derived from the program found its way into other flight programs.[439]

The ERAST family of remotely-piloted aircraft—for which Dryden acted as the lead NASA center—also offered data-collection of an entirely different kind at low cost. Discussion of ERAST-like aircraft began in NASA about 1990. The program's estimated annual budget of $12 million between its proposed initiation in 1994 and the year 1998 was a small price to pay for flight research on several new vehicles and a winnowing process designed to arrive at those best suited to the intended mission. Their requirements and modes of operation clearly satisfied the Szalai/Meyer definition of a top-priority program. They needed open and secure terrain, modern test facilities, and the particular talents of the Dryden staff.

The project was designed to test platforms capable of flying very high, very slowly, for very long periods of time. It involved a collaboration between private

[439] "Sub-Scale Flight Research Vehicles" (draft), DFRC Historical Reference Collection; Donald Gatlin to Director of Flight Operations, 15 Nov. 1983, DFRC Historical Reference Collection; Don Gatlin, "DAST Project Management History," 16 Sept. 1983, DFRC Historical Reference Collection; Peter Merlin, "DAST Flight History," 25 Nov. 1975-1 June 83, DFRC Historical Reference Collection; Wallace, *Flights of Discovery*, pp. 137, 148.

firms that designed and flew the ERAST prototypes and the Dryden team that oversaw the flight research. The ERAST planners wanted to pursue the prospect not just of scientific work in the upper atmosphere but of commercial applications as well. They ultimately foresaw at least two roles for their aircraft: as collectors of environmental information relative to the global climate and as telecommunications platforms (much like satellites, but without the associated high launch costs). To address these objectives, scientists with projects to send aloft proposed two distinct flight envelopes: one of extreme high altitudes (80,000 to 100,000 feet), short duration (about two hours), and small payload capacity (100 to 200 pounds); the other to achieve altitudes between 50,000 and 75,000 feet, last at lease 96 hours, and carry a cargo up to about 1,000 pounds. In pursuit of these requirements, the center's commercial partners—assisted by Dryden's engineering staff—concentrated on testing such advanced aeronautical concepts as lightweight structures, engine innovations, and fault-tolerant flight-control systems.

The ERAST effort envisioned and ultimately resulted in a group of small companies working in alliance with NASA to meet the requirements of the atmospheric-science community, which needed to take measurements and validate atmospheric models in the 80,000-foot range where the High Speed Research program intended to fly the High Speed Civil Transport (see Chapter 9). Big companies had initially been invited to participate in the program, but they declined because of the small potential market for such aircraft. Program managers then came to believe that they could help little companies establish a new industry for producing high-altitude UAVs. If NASA helped such companies find legitimate commercial markets for high-altitude, slow-flying aircraft, it could eventually purchase them at comparatively low cost. It could then use them to collect data from such dangerous locations as the polar regions or from areas in which it could not currently fly for extended durations.

One of the first ERAST models was Perseus A, conceived and built by Aurora Flight Sciences Corporation of Manassas, Virginia. It started flight research at Dryden in 1993 under what was then called the Small High Altitude Science Aircraft (SHASA) Project, which later evolved into ERAST. Aurora Flight Sciences built two Perseus As designed to fly up to 82,000 feet and stay aloft for five hours. Their mode of propulsion consisted of an experimental, closed-system, four-cylinder piston engine that recycled exhaust gases and obtained oxygen for combustion from stored liquid oxygen. This design allowed the engine to operate in the thin air found at high altitudes. Since it used a propeller that was too long to operate on the ground, it had to be disengaged from the propulsion system in the horizontal position while a truck towed the aircraft until it was airborne. Then the propeller could be operated once the tow line was dropped. Unfortunately, the engine never functioned as well as expected, and the aircraft did not reach the anticipated altitude before one of them came apart in flight late in 1994 when an autopilot gyro failed. The aircraft then descended on a flight-termination-system parachute. With more developmental time and funding, the engine might have become much more successful. As it was, Perseus A reached an altitude of some 50,000 feet.

Perseus B, also constructed by Aurora Flight Sciences Corporation, attempted to increase the time aloft to *24 hours* by employing a triple-turbocharged engine to provide sea-level air pressure up to 60,000 feet. Perseus B had its maiden flight in 1994 and was damaged in a hard landing. It experienced another hard landing in 1996

Like the ERAST aircraft that succeeded them, Perseus A and B offered long duration flight at low speed. Conceived by Aurora Flight Sciences Corporation, Perseus also typified ERAST in its program structure. Privately developed, it required Dryden's technical knowledge and range facilities for flight research. (NASA Photo EC91-0623-17)

for a variety of reasons, including failure to feather the propellers after a drive-train failure forced the aircraft to return to base. The high drag from the unfeathered propeller caused a sharp increase in the descent rate. The airplane resumed flying in 1998 with numerous improvements and upgrades, including a wingspan extended from 58.5 feet to 71.5 feet. The upgraded Perseus B reached an altitude of 60,280 feet on 27 June 1998, an unofficial altitude record for a single-engine, remotely-piloted, propeller-driven aircraft. Unfortunately it crashed near Barstow, California, on 1 October 1999. Fluctuations in the electrical system damaged the control-surface actuators, and the throttle cable fouled a safety parachute. In addition, the manufacturer-operator failed to perform integrated system tests as a follow-on to testing the parachute extraction system and conducted a less than fully adequate "independent design review of the complete final configuration." Despite the mishaps, Perseus's research flights yielded useful data about selecting instrumentation for a remotely-piloted vehicle, identifying potential failures from feedback that can be made available in a ground cockpit, and correcting those failures.

The creators of the Pathfinder flying wing approached the technical challenges of ERAST in a much different manner. Designed and built in the early 1980s by AeroVironment of Monrovia and Simi Valley, California (a company founded by Paul MacCready, the designer of the Gossamer Albatross and Gossamer Condor), Pathfinder had nearly the same wingspan as a Boeing 737 (98.4 versus 94.9 feet), yet it weighed a mere 530 pounds. Powered by solar panels stretching the length of its wings, which

Unlike Perseus, which used a piston engine, Pathfinder and Pathfinder-Plus, shown here in 1998 over the Hawaiian island of N'ihau, flew by solar power. (NASA Photo EC98-44621-252)

in turn powered six electric motors and propellers, it carried just a 50-pound payload. Originally battery-powered, Pathfinder first flew by solar power at Dryden in 1993. On 11 September 1995, it set a world record for altitude by solar-powered aircraft (50,567 feet). Then after modifications, it completed a checkout flight at Dryden in November 1996. Pathfinder received international media coverage on 7 July 1997 when it broke its own altitude mark with a flight reaching 71,500 feet in the skies over Kauai, Hawaii. It also carried two newly-developed imaging instruments conceived especially for ERAST. With them, it mapped some of the Hawaiian terrain, forest, and coastal waters.

Pathfinder Plus, an expanded version of Pathfinder, appeared over Kauai in June 1998. There, a larger range than the one available near Dryden made ascent to higher altitudes less problematical than it would have been in the California desert, where the angle of the Sun's rays was also less favorable than it was further south. The new vehicle's wingspan (121 feet) exceeded Pathfinder's by about 22 feet, and its upper surface was equipped with more efficient silicon solar cells. Although Pathfinder Plus attained an altitude of 58,000 feet during its June flight, it experienced a cyclic vibration at high altitudes called "whirl flutter". The research team returned to Dryden to analyze the phenomenon. The project engineers decided to return to using the original Pathfinder propellers rather than the experimental propellers with larger chord that had been used on the previous flight. Flying again on 6 August, Pathfinder Plus experienced no significant vibration, achieved an altitude of 80,200 feet, and subsequently won recognition as one of NASA's ten most memorable accomplishments in 1998. It also sustained flight at 70,000 feet for three hours.

Pathfinder Plus' successor, known as Centurion, measured 206 feet across its wingspan and came equipped with 14 seven-foot propellers. On its last flight late in 1998, it carried a dummy payload weighing a substantial 605 pounds. (NASA Photo EC98-44803-110)

Subsequently, a bigger successor to Pathfinder Plus, bearing the name Centurion, rolled out of the AeroVironment shops. With a 206-foot wingspan and 14 seven-foot propellers, it lumbered out of the Dryden hangars on 10 November 1998 for its first flight. Two more preliminary flights followed, and none experienced problems. Indeed, during the last one, flown in December, the aircraft carried a 605-pound simulated payload. Centurion engineers designed the vehicle for flight at 100,000 feet, but this altitude awaited a different ERAST aircraft.

The largest AeroVironment aircraft to date, known as the Helios Prototype, flew twice over the Dryden area during September 1999. Budget constraints required that the objectives of Centurion (100,000-foot altitudes) and those of Helios (96 hours at 50,000 feet) be accomplished by a single aircraft. Consequently, engineers at AeroVironment modified the Centurion by adding a sixth wing section or panel (thus increasing the capacity for solar cells) and a fifth landing gear pod. The resulting addition to the ERAST fleet featured a massive 247-foot wingspan, larger than that of a 747 airliner. Powered by fourteen motors for short- and eight motors for long-duration flight, the Helios Prototype prepared to assume the mission of Centurion as well as its own 96-hour mission, although doing both required some design tradeoffs that made the resultant aircraft somewhat less capable than two separate vehicles would have been. The first four flights of the giant flying wing in the fall of 1999 employed the "extreme altitude" configuration using all 14 electric motors and carrying only minimal ballast. The final two flights, the last on 8 December, used only eight motors in the configuration designed for extreme-duration flights. Also on board for these two trials were several hundred pounds of ballast to simulate the weight of a planned energy storage system that would be needed for a four-day flight. These flights, all under

AeroVironment, builder of the Pathfinders and Centurion, also fabricated the even bigger Helios Prototype, pictured here. The solar-powered Helios received a dual assignment: fly to 100,000 feet (a goal cut from Centurion's flight program for budgetary reasons), and fly continuously up to 96 hours. (NASA Photo EC99-45285-6)

battery power, completed a series of low-altitude flights (less than 1,000 feet) at Dryden designed to evaluate such technical factors as the aircraft's handling qualities, stability and control, flight in mild turbulence, and use of differential motor thrust to control turns and changes in pitch. After these flights, the craft returned to AeroVironment's Design Development Center in Simi Valley, California, for upgrades to its systems and installation of solar arrays in preparation for flights of higher altitude and longer duration at the Navy's Pacific Missile Range Facility in Hawaii.

If the Helios Prototype succeeded in sustaining an altitude of 100,000 feet for shorter durations and 50,000 feet for four days in the ensuing years, as projected, one ERAST official envisioned a follow-on Helios vehicle able to fly continuously not just for hours or days but for months at a time. Such capabilities brought within the grasp of researchers and commercial users of the technology an "atmospheric satellite," a vehicle supplement to existing telecommunications systems and actual satellites. The new technology could be used for a variety of commercial and other applications such as mapping; oceanographic and atmospheric research; monitoring and early warning for hurricanes, flooding, and other natural disasters; and hurricane tracking.

The ERAST program sponsored several other aircraft as well. Among them was Altus II, a dual-turbocharged vehicle with a rear-mounted pusher propeller driven by a small, four-cylinder Rotax piston engine. Constructed by General Atomics-Aeronautical Systems Inc. of Rancho Bernardo near San Diego, California, Altus II

Altus II, seen over the California desert, flew using a single pusher propeller powered by a piston engine. (NASA Photo EC98-44684-1)

was descended from a single-turbocharged Altus I. The Altus II vehicle began flight-testing with its new powerplant over the Mojave Desert during the second half of 1998. It moved to the Hawaiian island of Kauai in the late spring of 1999 and flew nine scientific flights supporting the Atmospheric Radiation Measurement studies conducted by Sandia National Laboratories for the Department of Energy, flying up to an altitude of 55,000 feet in an extremely successful effort. In general, Altus II validated such technologies as propulsion, command and control, and avionics for the ERAST Program.

As these lines were being written, General Atomics-Aeronautical Systems was developing the Predator B aircraft to meet the requirements of NASA's Earth Science Enterprise for an unpiloted aerial vehicle using consumable fuel to carry out scientific missions. Under a cost-sharing agreement with NASA, General Atomics was building three prototypes, the final version to be powered by a 700-horsepower turboprop engine. It was to be capable of carrying a 660-pound payload for as many as 32 hours at altitudes up to 52,000 feet. To be successful, the aircraft would have to demonstrate its ability to fly safely in civil airspace controlled by the FAA.

There were other aircraft in the ERAST Program, including the Proteus tandem-wing design built by Scaled Composites Inc. of Mojave, California, primarily as a piloted aircraft for relaying commercial telecommunications. It could also be flown by remote control. At the time this section of the book was being written, ERAST seemed on the verge of transferring at least some of its technologies to the commercial sector for use in scientific, telecommunications, early-warning, or other applications, the results of the program yet to be determined.[440]

More Than a Wing and a Prayer

In addition to the extra workload imposed by the Szalai initiatives, Dryden's traditional responsibilities—serving the flight-research needs of civilian and military aeronautics—continued to require time and attention. Consequently, during the 1990s, the Dryden staff found itself stretched to capacity with new and old commitments. One major investigation conducted during this period had the potential to serve the interests not only of the airlines and the aircraft manufacturers but of the flying public as well.

An air disaster in 1989 precipitated the research. As United Airlines Captain Al Haynes cruised over Iowa farm country on 19 July 1989, nothing suggested trouble. All systems appeared to be functioning normally on his McDonnell Douglas DC-10 airliner, carrying a complement of 296 passengers and crew. Then the tail-mounted rear engine shut down with a roar. As it quit, a burst of shrapnel tore through the aft

[440] *Annual Report, FY 1998*, NASA DFRC, p. 10, and *Annual Report, FY 1999*, p. 6, DFRC Historical Reference Collection; ERAST program description, from Ames Research Center website (http://erast.arc.nasa.gov/erast/erast2.html), 27 Feb. 1997; Don Haley, "Dryden is Lead Center in ERAST Role," Dryden *X-Press*, Oct. 1994, DFRC Historical Reference Collection; John H. Del Frate and Gary B. Cosentino, *Recent Flight Test Experience with Uninhabited Aerial Vehicles at the NASA Dryden Flight Research Center* (Edwards, CA: NASA TM-1998-206546, 1998), pp. 1-9; Jennifer Baer-Riedhart, "ERAST Program Marks Early Successes," *Aerospace America* (Jan. 1998), pp. 38, 41-42; NASA news release, "Perseus B Damaged in Crash on California Highway," 1 Oct. 1999, DFRC Historical Reference Collection; NASA Facts, "Pathfinder," Dec. 1996, DFRC Historical Reference Collection; John H. Del Frate, "Four Remotely Piloted Aircraft Mishaps—Some Lessons Learned," in Association for Unmanned Aerial Vehicles International '96 Proceedings, Orlando, FL, 15-19 July 1996, pp. 202-204, 205-207; "Blocking and Tackling," Remarks of NASA Administrator Daniel S. Goldin, AIAA Global Air & Space Conference, 11 May 2000, for quotation on Perseus B crash; Alan Brown, "Pathfinder Returns to Skies; Hawaii Next," Dryden *X-Press*, 20 Dec. 1996, DFRC Historical Reference Collection; "Pathfinder Sets World Record," Dryden *X-Press*, 4 July 1997, DFRC Historical Reference Collection; "Solar-Powered Pathfinder Sets New Record; Prepares to Monitor Deforestation of Hawaiian Island," Ames Research Center website (http://erast.arc.nasa.gov/erast/erast2/html), 30 Oct. 1997; "NASA ERAST—Solar Electric Aircraft," AeroVironment, Inc., n.d., DFRC Historical Reference Collection; NASA Facts, "Solar-Power Research at Dryden," 1998, DFRC Historical Reference Collection; NASA Facts, "Pathfinder," Dec. 1997, DFRC Historical Reference Collection; NASA Facts, "ERAST," June 2000; Dwain A. Deets and Dana Purifoy, *Operational Concepts for Uninhabited Tactical Aircraft* (Edwards, CA: NASA TM-1998-206549, 1998), pp. 1-3; Dryden News Release, "Upgraded Pathfinder Solar-Powered Aircraft Begins New Flight Season," June 1998, DFRC Historical Reference Collection; Dryden Project Highlights, 2 July, 7 Aug., 13 Nov., 20 Nov., 4 Dec. 1998, DFRC Historical Reference Collection; interview of John H. Del Frate by Lane Wallace, 1 Sept. 1995, p. 42, DFRC Historical Reference Collection, and informal interview of Del Frate by J. D. Hunley, 21 July 2000; Wallace, *Flights of Discovery*, pp. 81-84; W. Ray Morgan, Robert F. Curtin, and John Del Frate, "NASA ERAST Solar Aircraft, On the Road to New Heights," unpublished paper provided by John Del Frate electronically, July 2000; *ERAST: Scientific Applications and Technology Commercialization*, compiled by John D. Hunley and Yvonne Kellogg (Edwards, CA: NASA CP-2000-209031, 2000); Proceedings of addresses, sessions, and workshops of the NASA ERAST Exclusive Preview, esp. remarks by Rich Christiansen, John Sharkey, and Basil Papadales; Alan Brown, "Helios Completes Research Series," Dryden *X-Press*, 17 Dec. 1999; comments of John Del Frate on a draft of this section, 14 Sept. 2000.

portion of the jetliner, disabling the engine and the hydraulic systems that actuated the aircraft's flight-control surfaces. Douglas engineers had designed the jumbo to fly without the rear powerplant, so this problem alone did not pose a grave danger. However, lacking the means of moving the flaps, ailerons, and rudder that gave him control of the aircraft, Captain Haynes faced a catastrophic situation. As air-traffic control diverted him to Sioux City Airport for a crash landing, Haynes received crucial assistance from the regular crew of the aircraft and especially from check pilot Dennis Fitch, who was off duty that day and riding in a first-class seat. After Fitch volunteered his assistance, Haynes asked him to take control of the throttles in an attempt to control pitch and roll through engine power. Fitch discovered that the jetliner tended to pull right, making it difficult to maintain a stable attitude in pitch, but the crew managed to coax it towards the desired flight path. Haynes chose not to attempt a landing on runway 31, the one designated by air-traffic control, since it would have required some difficult left turns. Instead, bearing down on runway 22, the widebody reached Sioux City at 4 p.m. It struck the runway, upended, and caught fire. Yet because the captain had managed to land at an airport, where emergency procedures awaited the broken vehicle, nearly two-thirds of those aboard (184 people) survived the impact and the aftermath. Those familiar with this type of incident knew that far more fatalities usually accompanied crashes of this kind. In fact, between the mid-1970s and the late 1990s, at least ten aircraft—including a B-747, L-1011, DC-10, B-52, and C-5A—encountered major flight-control losses. In a number of these cases, the crews turned to propulsion control as a desperation measure, with some success. However, in the B-747, DC-10, and C-5A disasters alone, a total of more than 1,000 passengers died.[441]

As details of Haynes' remarkable landing reached the media, Dryden researchers, led by propulsion engineer Bill Burcham, began to discuss the application of some of Dryden's past investigations to solve such problems of total control failure. Identified for years with the DEEC and HIDEC programs, Burcham wondered whether the digital devices that performed such functions as governing fuel consumption and improving performance might also be programmed to enhance pilot control in case of complete hydraulic failure. He began to consider a project later known as Propulsion Controlled Aircraft (PCA). In emergency situations, computer regulation of engine thrust offered an opportunity to give pilots more accurate handling of a plane's trajectory than was possible with manual controls. This might prove especially helpful during an oscillation called Dutch Roll, a "slithering" effect produced as the aircraft flew forward, each wing alternately rising and falling, moving the machine left and right in heading. These motions assumed critical importance during landings. Computer-aided controls improved longitudinal handling and lessened the pivoting of the aircraft's nose during Dutch roll from three degrees to one. After months of indifference and even hostility to this concept by many on Dryden's staff who doubted its value, Ken Szalai sent Burcham a handwritten note in March 1990, encouraging him to proceed with the PCA project.

[441] Interview of Robert Meyer by Michael Gorn (by telephone), 18 Mar. 1999, DFRC Historical Reference Collection; Tom Tucker, *Touchdown: The Development of Propulsion Controlled Aircraft at NASA Dryden* (Washington, DC: Monographs in Aerospace History No. 16, 1999), pp. 1-2, 12, 36-37; Frank W. Burcham, Jr., Trindel A. Maine, John J. Burken, and John Bull, *Using Engine Thrust for Emergency Flight Control: MD-11 and B-747 Results* (Edwards, CA: NASA TM 1998-206552, 1998), pp. 1-3; comments of Bill Burcham and Trindel Maine on a draft of this section.

Burcham and his team began by programming a simulator to mimic a Boeing 720 flying under propulsion control. Spurred by some positive initial results, Burcham approached Gordon Fullerton, one of Dryden's most experienced research pilots as well as a former Shuttle commander, to try the simulator and to consider flying a real aircraft under such conditions. Fullerton accepted the unusual challenge. The project gained further momentum when Capt. Al Haynes visited Dryden in early 1991 and spoke to the staff about the Sioux City incident. Fullerton took the opportunity to introduce Haynes to the electronic version of propulsion control, and the captain landed the phantom plane in the Boeing 720 simulation on his first attempt.

Burcham's group then designed a propulsion-controlled simulation for the F-15 HIDEC aircraft. Gordon Fullerton reported that landing only by throttle adjustment "was a very demanding task, but with practice it could be done repeatedly." By contrast, he observed that the "use of the PCA system dramatically reduced pilot workload, and performance, in terms of touchdown position, and sink rate was greatly improved . . . with little pilot training." However, this case proved an old fact about flight research: flying an actual machine through the air proved quite different from flying in a simulator. Fullerton discovered that because of the unaugmented fighter's poor stability, low natural rate damping, and some dynamics not present in the electronic model, it was "much harder to control than expected from simulation," rendering a safe landing impossible. Yet, after only nine flights using the PCA control logic developed by McDonnell Douglas, Fullerton made a safe propulsion-only landing with the former Air Force fighter in April 1993, the first ever achieved.[442]

The PCA program reached its zenith with the decision to flight-test the system on an airliner—specifically, a McDonnell Douglas MD-11. Initially, McDonnell Douglas failed to find one to lend to Dryden. Many still doubted the capacity of PCA for emergency landings, and whoever loaned Dryden such a testbed risked losing a vehicle valued at about $120 million. Finally, McDonnell Douglas PCA project manager Drew Pappas found one being used as a flight testbed at the corporation's airstrip in Yuma, Arizona. Dryden engineers arrived on the scene and changed nothing on the MD-11 but its software. Research pilot Gordon Fullerton arrived in Yuma on 29 August 1995 to fly the jumbo to Dryden and land it using the PCA system. Once over Edwards, he tried approaches at 100, 50, and 10 feet. Then, as light winds, light turbulence, and occasional thermals played over the area, Fullerton lined up with the runway, making small track changes as the altitude decreased. Sinking at four feet per second and flying at a speed of 175 knots, the plane "touched down smoothly on the center line . . . without [Fullerton] making flight control inputs." In November, he returned to the cockpit and flew high-altitude approaches (at about 10,000 feet) in the widebody, simulating landings at that altitude while in other respects emulating the circumstances faced by Al Haynes–that is, flying with the control surfaces floating, the rear engine nearly idling, and with no hydraulics available. Between these two flights, some 20 pilots chosen from the airlines, the aircraft industry, the military

[442] Frank [Bill] Burcham, Ronald Ray, Timothy Conners, and Kevin Walsh, *Propulsion Flight Research at NASA Dryden From 1967 to 1997* (Edwards, CA: NASA TP-1998-206554, July 1998), p. 15; Tucker, *Touchdown,* pp. 9-22; Wallace, *Flights of Discovery*, p. 122; C. Gordon Fullerton, "Propulsion Controlled Research Aircraft," *Proceedings of the 37th Symposium of the Society for Experimental Test Pilots*, Beverly Hills, CA, Sept. 1993, 78-88, DFRC Historical Reference Collection (quoted passages).

After Captain Al Haynes landed a crippled DC-10 solely by engine thrust in July 1989, Dryden and industry engineers developed software to enable propulsion-controlled landings. Research pilot Gordon Fullerton subsequently touched down in an MD-11 (depicted here) equipped with this system. (NASA Photo EC95-43355-1)

services, and the Federal Aviation Administration tried the PCA system in low-level approaches. They declared it a success after more than 30 total hours of flight research.

But the airline manufacturers did not readily adopt the PCA system for their jetliners. Boeing chose to concentrate on measures to prevent the loss of the standard flight-control system. Airbus likewise seemed reluctant, relying on its aircraft's redundant electronic and mechanical backups. Perhaps wishing to avoid the issue of possible, but unlikely catastrophic air incidents, the industry instead devoted more time to training pilots for problems more likely to occur than total control failure. Meanwhile, in partnership with others at Ames Research Center and the private sector, Dryden engineers developed modified approaches dubbed PCA Lite and PCA Ultralite. Both could be installed on older aircraft not yet equipped with fly-by-wire systems because they could operate by using existing engine-control computers. Both also required less pilot training than the original PCA system. Even so, there had been no manufacturers adopting either of the two systems as of June 2000.[443]

Paddles and Nozzles: Thrust Vectoring

In addition to its many other projects, Dryden continued to pursue research related to the requirements of combat aircraft. Since their inception, fighters had depended for their survival not merely on speed, but also on their capacity to outmaneuver the enemy. Groundbreaking work on high maneuverability began at Dryden during the 1980s with the HiMAT and the X-29 projects. During the same decade, planning began for a third high-profile initiative that reached full flower during the 1990s. Using as a testbed a specially outfitted Navy F-18 Hornet, which had been trucked to Dryden in need of repair and restoration, researchers christened their investigation the High Angle of Attack Research Aircraft (HARV) program. Among their more important objectives, the HARV engineers sought to determine whether this particular aircraft performed more effectively at high angles of attack using specially designed thrust-vectoring engine paddles. They also wanted to determine whether thrust vectoring held the potential to improve performance for other vehicles and whether wind-tunnel and computer predictions in this flight regime might achieve better accuracy as a result of the data gleaned from flight research.

The HARV program developed slowly and, in its different embodiments, lasted a long time. Planning began in 1984 among Dryden, Langley, Ames, and Lewis, at which time the F-18 was transferred from the Navy to Dryden and the process of refurbishment started. The first flight-research phase began in mid-1987, during which the aircraft—although in full working order—had not yet been equipped with the engine paddles. During this phase, researchers developed a base of information by positioning film and video cameras aboard the HARV to record airflows over the nose and other portions of the aircraft. They made these flow patterns visible using such techniques as oil-based dye sprayed from holes on the plane's nose, pieces of yarn (tufts) attached at various points, and tracer smoke expelled from ports ahead of the wings' leading edges. Sensors located around the nose also monitored air pressure. Once they completed these tests, investigators turned to wind tunnels and computational fluid dynamics (CFD) to analyze the flight-research data. The resulting analyses allowed them to revise their forecasts and to improve their predictive abilities in this

[443] Tucker, *Touchdown*, pp. 22-29; Burcham, Maine, Burken, and Bull, *Engine Thrust for Emergency Flight Control*, pp. 12-14 (quoted passage); Phill Scott, "Throttled: Manufacturers balk at steering and landing with engine thrust alone," *Scientific American* (June 2000) on the Internet at http://www.sciam.com/2000/0600issue/0600scicit4.html.

A Navy F-18 Hornet seen in its role as Dryden's High Angle of Attack Research Vehicle (HARV). The HARV flight research project (1987 to 1996) increased our understanding of flight at high angles of attack. (NASA Photo EC94-42645-9)

particular area of inquiry. In reality, however, the data derived from the early HARV flights represented only a fraction of the high-angle-of-attack work undertaken since the mid-1980s. Led by Langley researchers and abetted by their colleagues at Ames, Lewis (later renamed Glenn), and Dryden, the High-Angle-of-Attack Technology Program (HATP) collected and digested data using computational fluid dynamics, wind-tunnel tests, and flight research.

As a consequence of the broad base of knowledge acquired by the HATP team—as well as the Phase 1HARV flight-research data—between late 1990 and mid- 1991 the F-18 underwent modification for phase-two flying. Technicians installed three paddle-like vanes (fabricated from Inconel metal) at the exhaust nozzles of each engine. Unfortunately, the entire apparatus added quite a lot of weight (2,100 pounds) to the airframe. But if the concept could be shown effective in flight, it offered a technique to enhance low-speed maneuverability, prevent the loss of control, and delay the onset of stalls common at very high angles of attack. By using these vanes to vector engine thrust, pilots gained auxiliary yaw and pitch control. Indeed, at the cockpit's command, a specially programmed PACE 1750A computer selected the most effective combination of aerodynamic control and guided thrust for each desired maneuver. Flights of this system started during summer 1991 under the overall supervision of project manager Don Gatlin, with project pilots Bill Dana and Ed Schneider at the HARV's controls. Ultimately, the aircraft proved capable of achieving significantly improved roll rates at 65-degrees angle of attack (also known as alpha) and controllable flight at 70 degrees. During 1995 and 1996, the HARV appeared in its last role. Technicians added two nose strakes to the Hornet in an effort to modify the flow of vortices over this part of the aircraft. Schneider and research pilot Jim Smolka reported that at high alpha, the strakes also improved control substantially. In recognition of his contributions to HARV, Ed Schneider won the prestigious Octave Chanute Award from the American Institute of Aeronautics and Astronautics in 1996.[444]

Perhaps more daring than the HARV in some technical respects, the X-31 project certainly operated under a more demanding programmatic structure. After German aerodynamicist Wolfgang Herbst conceived it, the aircraft's prohibitive design and production costs persuaded German defense officials to seek partners for X-31 development and flight research. Consequently, five separate aeronautical entities formed an alliance to fabricate the first international X-series aircraft. An X-31 International Test Organization consisted of representatives from DARPA, the US Navy, the German Ministry of Defense, Rockwell International, and Deutsche Aerospace. As the complexities of the flight research program became more apparent, NASA and the US Air Force joined the team. Dryden became the institutional home for the project in its later stages and served as host for the International Test Organization's staff. Despite its many participants, the multilateral group worked together well, in part because it was clustered in the Integrated Test Facility and also because there was little turnover in staff. Indeed, despite the presence of no fewer than six research pilots from six of the partners, dividing the workload never posed real problems.

The X-31 aircraft itself not only combined the research breakthroughs of a number of its predecessors—and even some of its contemporaries—but in a number of ways it also improved upon them. Like the HARV, it employed three thrust-vectoring paddles; but unlike those on the F-18, these vanes were composed of graphite epoxy, rather than Inconel metal. Like the X-29, the X-31 had small canards forward of the wings; unlike the forward-sweptwing vehicle, it also featured fixed nose strakes. Like many aircraft since the F-8 DFBW, it flew using a digital flight-control system; but its four computers operated without analog or mechanical backups.

The program unfolded in three stages. First, the many parties agreed upon a concept that enlisted those technologies of probable utility to future air conflicts. Second, the three government agencies evaluated the design of the demonstrator. Finally, two aircraft were constructed on Rockwell International's assembly line in Air Force Plant 42, located at Palmdale, California. Meanwhile, German manufacturers conceived and fabricated the wings and paddles. Contractor flight research followed once the two vehicles emerged from the factory in 1990. After 108 flights at Rockwell's Palmdale facility (in which pilots achieved angles of attack up to 40 degrees), testing

[444] Interview of Ed Schneider by Lane Wallace, n.d., p. 22, DFRC Historical Reference Collection; NASA Facts, "F-18 High Angle-of-Attack (Alpha) Research Vehicle," 5 May 1999, on the Dryden website (www.dfrc.nasa.gov/PAO/PAIS/HTML/FS-002-DFRC.html) and filed in the DFRC Historical Reference Collection; briefing, "F-18 High Angle of Attack Flight Research," n.d., DFRC Historical Reference Collection; approved baseline schedule, "F-18 HARV Project," 30 Aug. 1990, DFRC Historical Reference Collection; Wallace, *Flights of Discovery*, pp. 103-106; Dryden Project Highlights for 1995 and 1996, DFRC Historical Reference Collection; comments of Ronald "Joe" Wilson, DFRC aerospace engineer, on a draft of this section, 4 Feb. 2000.

Another thrust-vectoring pioneer flown at DFRC, the X-31 aircraft flew angles of attack up to 70 degrees. Remarkably, this success involved two countries and seven partners: NASA, the US Air Force, DARPA, the US Navy, the German Ministry of Defense, Deutsche Aerospace, and Rockwell International. (NASA Photo EC94-42478-4)

activities moved to Dryden in February 1992. Two months later, the international consortium began post-stall envelope expansion flights at Dryden. The development process encountered some difficulties. The aircraft computers required frequent reprogramming as the idiosyncracies of these vehicles became known. In some respects, the inflight aerodynamic qualities of the two planes differed somewhat from those forecasted in wind tunnels. Moreover, at angles of attack around 50 degrees, pilots discovered lurches, or "kicks," caused by vortices striking the sides of the aircraft.

Engineers mitigated this problem by placing narrow strips of grit on the noseboom and radome, disrupting the vortex flow.

Despite such headaches, the program enjoyed some impressive successes. In November 1992, the number two aircraft not only achieved 70-degrees angle of attack but performed a controlled roll at that same high alpha. At the end of April the following year, this aircraft flew a post-stall 180-degree turn at a minimum radius, surpassing the capabilities of any conventional combat vehicle. (The HARV had almost equaled the feat, flying in a post-stall 180-degree turn at 60-degrees angle of attack in October 1992.) The X-31 proved that in the flight regime past normal high-alpha stalling, it continued to perform in a completely flyable and controllable manner. Then, it engaged in a long series of mock aerial battles to demonstrate its dogfighting capacities against an F-18 (winning over the Navy fighter roughly 30 times for each time it lost, but losing to an F-16 equipped with a newer F100-PW-229 engine that gave it a better thrust-to-weight ratio than the X-31). This unique multinational project ended in January 1995 after well over 500 research flights. Unfortunately, in the final flight of the original program, number one X-31 crashed due to a malfunction of the noseboom airspeed system. The pilot ejected and parachuted to the ground. Despite a gloomy ending to this vehicle's flying career, the X-31 demonstrated the potential of thrust vectoring. It subsequently caused quite a stir when the other X-31 vehicle demonstrated post-stall maneuvering at the Paris Airshow that year. "The general consensus [among pilots]," concluded a report on the X-31 flight-research program, boiled down to two observations: "the [X-31] aircraft had good handling qualities. . .in this [high alpha] flight regime" and offered promise in the war-fighting role.[445]

Dryden engineers participated during the 1990s in one more project devoted to the study of thrust vectoring and advanced engine control. Aboard a specially configured F-15 Eagle on loan from the Air Force for this project, researchers pursued an investigation known as Advanced Control Technology for Integrated Vehicles, or F-15 ACTIVE. Engine designers at Pratt & Whitney, impressed by the capabilities of thrust vectoring proven on the HARV and the X-31, conceived of a more operationally viable system of controlling the deflection of engine exhaust than that demonstrated in these two earlier vehicles. The manufacturer's axisymmetric nozzle—mounted on the standard F100 powerplant—eliminated the heavy paddles common to both of the other demonstrators and replaced them with a gimballing nozzle capable of swiveling 360 degrees at angles up to 20 degrees. It provided thrust control in pitch and yaw or any combination of the two axes. Yet the ACTIVE flight-research project intended not merely to validate the new nozzle but also to further perfect the computer programming that determined (in real time) the most beneficial engine settings, to investigate the aerodynamic details of the airflow around the nozzles, and to explore ways to dampen the noise produced by the jet flow in the nozzles. Pursued

[445] Al Groves, Fred Knox, Rogers Smith, and Jim Wisneski, "X-31 Flight Test Update," P*roceedings of the 37th Symposium of the Society for Experimental Test Pilots*, Beverly Hills, CA, Sept. 1993, pp. 100-101; Dave Canter, "X-31 Post-Stall Envelope Expansion," *Fourth High Alpha Conference,* NASA DFRC, 12-14 July 1994 (Edwards, CA: NASA Conference Publication 10143, vol. 2, 1995); Draft NASA Fact Sheet, "X-31 Enhanced Fighter Maneuverability Demonstrator," 20 October 1999, DFRC Historical Reference Collection; Patrick C. Stoliker and John T. Bosworth, *Evaluation of High-Angle-of-Attack Handling Qualities for X-31A Using Standard Evaluation Maneuvers* (Edwards, CA: NASA TM 104322, 1996), p. 5; Wallace, *Flights of Discovery*, pp. 106-111.

Dryden and Pratt & Whitney engineers conducted tests on an engine nozzle capable of swiveling 360 degrees, thus eliminating the paddles previously used for thrust vectoring. This research occurred on the F-15 ACTIVE (Advanced Controls Technology for Integrated Vehicles), shown here. (NASA Photo EC96-43485-5)

cooperatively by Dryden, the Air Force Wright Laboratories, and F-15 manufacturer McDonnell Douglas, the initial ACTIVE ground demonstrations occurred from 1990 to 1992. The program began officially late in 1992. By November 1995, the engine had passed all requirements for safe operation in the Air Force Flight Test Center's horizontal thrust stand.

During the flight-research phase, Dryden made use of previous experience; HARV project manager Don Gatlin initially ran the ACTIVE program as well. After the first flight on 27 March 1996, gradual envelope expansion led to the initial supersonic flight in summer 1997 (Mach 1.2 at 30,000 feet). By mid- to late 1998 ACTIVE's flying characteristics became apparent in a process of inquiry that once again proved the value of flight research, pointing up the limits in the predictive value both of wind tunnels and of computational fluid dynamics. Even though both of these research tools arrived at similar conclusions about ACTIVE performance, actual flight experience showed "large differences" from ground-based analyses (perhaps because the real nozzles differed significantly from those on the models used in the CFD and the wind-tunnel tests). Relying on strain gages installed in engine mounts to record thrust and vector forces, those involved in the project decided that the nozzle's performance database needed to be broadened to include a greater variety of test conditions. On the other hand, pilots assessing the F-15 ACTIVE's handling qualities found them to be "outstanding," a consequence of the computer system that successfully integrated engine thrust controls with standard aerodynamic flight surface controls. Such strong endorsements encouraged those in the ACTIVE program to continue thrust-vectoring flights until July 1999. Dryden engineers also explored ways in which thrust-vectoring technology might be adapted to those production aircraft requiring advanced maneuvering qualities. The possibilities for harnessing the additional propulsive forces of vectoring seemed promising as the project wound to a close. Meanwhile, the project had gathered a great deal of data about thrust vectoring and had learned that direct measurement of thrust and vector forces provided an effective way to determine the turning efficiencies of vectored flow.

In a further project involving the F-15 ACTIVE, during the spring of 1999 Dryden conducted flights to test new "smart" software designed to enable pilots to control and safely land disabled aircraft. Called the Intelligent Flight Control System, the technology employed experimental "neural network" software developed by computer scientists at NASA Ames Research Center and the Boeing Company's Phantom Works division, St. Louis, Missouri. When fully developed, the software promised a significant margin of safety for future military and commercial aircraft equipped with the system. Neural network software is distinguished by its ability to "learn" by observing patterns in the data it receives and processes and then performing different tasks in response to new patterns. Simple neural network software has been in use for decades. Beginning in the 1960s, it operated through computer modems, enabling them to receive error-free data over often-noisy telephone lines. But until the ACTIVE tests, it had never before been demonstrated in such a complex safety-related environment. Dryden research pilots made about a dozen flights over a three-week period. They demonstrated how a preliminary version of the neural network software that was pretrained to the F-15's aerodynamic database, operating with an adaptive controller, could correctly identify aircraft stability-and-control characteristics and immediately adjust the control system to maintain the best possible flight performance.[446]

A Flying Smorgasbord

For projects of limited duration requiring data quickly and inexpensively, Dryden pilots flew an F/A-18B testbed designated the Systems Research Aircraft (SRA). A former chase plane, it won its SRA role when some Dryden researchers decided to

[446] Wallace, *Flights of Discovery*, pp. 122-124, 158-159; Roger Bursey, "The F-15 Aircraft 'Next Step,'" draft paper, 6 July 1995, pp. 1-3, DFRC Historical Reference Collection; NASA news release, "NASA Will Test New Nozzle to Improve Performance," 22 Nov. 1995, DFRC Historical Reference Collection; United Technologies/Pratt and Whitney news release (draft), "Pratt and Whitney's Multi-Directional Vectoring Nozzle System Completes First Vectored Flight," Feb. 1996, DFRC Historical Reference Collection; "First Flight Imminent for F-15 ACTIVE Aircraft," *Aviation Week & Space Technology* (29 Jan. 1996), p. 54; NASA news release, "NASA Tests New Nozzle to Improve Performance," 27 Mar. 1996, DFRC Historical Reference Collection; "First Supersonic Yaw-Vectoring Flight for the ACTIVE Program," *NASA Tech Briefs*, July 1997, p. 74; John S. Orme, Ross Hathaway, and Michael D. Ferguson, *Initial Flight Test Evaluation of the F-15 ACTIVE Axisymmetric Vectoring Nozzle Performance* (Edwards, CA: NASA TM-1998-206558, 1998), p. 18 (first quoted passage); NASA news release, "Flight Research Reveals Outstanding Flying Qualities Result From Integrating Thrust Vectoring with Flight Controls," 29 Dec. 1998, DFRC Historical Reference Collection (second quoted passage); Dryden Project Highlights, 30 July , 16 Aug., and 20 Aug. 1999, DFRC Historical Reference Collection; comments of Joe Wilson, an anonymous reviewer, and Rogers Smith (DFRC acting chief of flight operations) on a draft of this chapter; *Dryden Flight Research Center, FY 1999 Annual Report*, p. 9; John S. Orme and Robert L. Sims, "Selected Performance Measurements of the F-15 ACTIVE Axisymmetric Thrust-Vectoring Nozzle," paper no. IS 166 presented at the 14th Annual Symposium of the International Society for Airbreathing Engines, 5-10 Sept. 1999 in Florence, Italy.

Originally used for chase at Dryden, during the 1990s this F/A-18B served many testbed assignments as the Systems Research Aircraft (SRA). Some of these tests involved the replacement of mechanical flight-control linkages with electronic ones. (NASA Photo EC98-44672-1)

mount a few experiments on it while it continued its chase duties. Gradually, as engineers in aerospace industries also became aware of this asset, the research and platform responsibilities became paramount. Eventually, Langley and Lewis [later, Glenn] Research Centers also participated in the SRA program. Several practical reasons conspired to promote the use of the F/A-18. About five of these vehicles flew chase at Dryden at any one time, making for easy access to spare parts, to knowledge of their repair and upkeep, and to technical documentation. The vehicle also possessed full-authority digital fly-by-wire capability, a dual front-rear cockpit configuration, two engines, and two vertical stabilizers. Its duplicative features allowed engineers to try new flight techniques with the assurance of conventional backups. From the early days of the SRA testbed, program managers recognized its unusual value:

> A primary goal is to identify and flight-test high-leverage technologies beneficial to subsonic, supersonic, hypersonic, or space applications. Demonstrating new system concepts in flight will greatly promote the transition of research and development technology from widespread, highly specialized ground-based laboratories to cost-effective flight research and production applications. The SRA flight test facility will enable government and industry to focus the integration, ground test, and flight validation

> of breakthrough technologies. The intent of flight testing new technologies is to eliminate perceived and real technical barriers.[447]

A number of worthy technologies presented themselves to the SRA staff during the early 1990s. The most promising ones included active flutter suppression, onboard envelope expansion techniques, pilot associate systems, advanced displays, massive parallel processing architectures, vehicle management techniques, automated vehicle checkout techniques, and advanced vehicle system interfaces. The first project to win a spot on the SRA—so-called "smart" actuator technology—featured an innovation designed to make a generation of actuators less vulnerable to failure than any to date. Under the umbrella of the Electronically Powered Actuation Design (EPAD) validation program, the existing left aileron actuator was replaced with one that actually *evaluated* the aileron's performance in flight to determine whether it obeyed pilot commands. In this system, loop closure occurred at the actuator rather than at the flight-control computer. The smart actuator system also contained a fiber-optic interface capable of transmitting commands from the flight-control computer to the actuator, thus eliminating the need for bundles of wires.

Later in the 1990s, another EPAD development, the Electro-Hydrostatic Actuator (EHA) underwent flight research. The EHA incorporated a fluid reservoir and a motor-driven pump into a piston-cylinder assembly. When the EHA received specific commands from the SRA's flight-control computer, it responded by powering the pump, which contained small amounts of hydraulic fluid. In this way, the system transformed electrical power to mechanical power. If successful, the EHA held the promise of eliminating heavy and costly hydraulic equipment from an aircraft, reducing fuel consumption, making combat aircraft less vulnerable to disasters resulting from severed hydraulic lines, and lowering maintenance costs. The EHA flew first in early 1996 and continued to be tested over the entire SRA envelope during 1997. It performed as well as the standard actuator.

Yet a third EPAD system, called the Electro-Mechanical Actuator (EMA), although powered itself by electrical current, operated the SRA's ailerons by a mechanical ball-and-screw shaft. Flight tests of EMA began during 1997 and by March 1999 had advanced sufficiently to be the subject of a paper at a meeting of the Society of Automotive Engineers, where it engendered "widespread interest in EMA applications and flight results."

Many other SRA investigations occurred, including nine in 1998 alone. One in November 1998 was noteworthy: the SRA team succeeded in demonstrating a *wireless* flight-control system (using fiber-optic cable) on the F-18 iron bird, a simulator in an actual F-18 inside the research aircraft integration facility. Commands fed into a radio-frequency interface between the left aileron input connector and the iron bird's flight-control computers resulted in aileron motions equal to those delivered by wire. The SRA projects proceeded at a similar pace during 1999, with 38 research flights involving six different experiments, and continued into the year 2000. Among other projects pursued during those two years was one validating the performance of a

[447] Wallace, *Flights of Discovery*, pp. 124-126; Joel R. Sitz, *F-18 Systems Research Aircraft* (Edwards, CA: NASA TM 4433, 1992), pp. 2-3 (block quote, p. 2); comments of Robert Navarro, SRA project engineer, on a draft of this section.

vehicle health monitoring (VHM) system for the X-33. A series of flights in late fall 1999 and early spring in 2000 successfully tested a VHM computer, remote health nodes, and advanced fiber-optic cables. It also identified problems with fiber-optic distributed strain sensors integral to the VHM system that still remain to be resolved.[448]

Aircraft Consolidation, Part II

After the collapse of the Beeler committee initiative recommending the consolidation of NASA aircraft under Dryden control, the issue lay dormant for 20 years. Then, budgetary constraints raised the practicality of the idea once again. During 1995, NASA headquarters underwent a far-reaching zero base review (ZBR), targeting bureaucratic reorganization, infrastructure reductions, and revisions in the agency's roles and missions as areas in which savings might be realized. During this analysis, the logic of aircraft consolidation under one center seemed compelling, as it had a generation earlier during the drawdown from the Apollo program. This time, however, rather than canvas the centers for advice and suggestions, headquarters acted on its own initiative.

Accordingly, on 23 January 1996, Associate Deputy Administrator General John Dailey told Ken Szalai and the Ames, Langley, and Lewis directors that the ZBR showed aircraft consolidation to be a valid approach to paring overhead expenses. By restructuring the fiscal-year-1997 budget to reflect these anticipated savings, headquarters, in effect, planned to impose the decision on the affected centers. General Dailey instructed Robert Whitehead, the associate administrator for aeronautics, to assign all NASA flight research and platform aircraft to Dryden. The headquarters implementation schedule left no time for debate. The directors of Ames, Langley, and Lewis had just a week to inform the manager of the consolidation project, former research pilot Gary Krier, now serving as head of the Dryden intercenter aircraft operations directorate, appointed to oversee the transfer of airplanes and staff to Dryden.

Whitehead underscored the need for speedy compliance in order to "demonstrate our commitment to . . . this decision." He delegated to Dryden the task of drafting an implementation plan and asked that it announce the movement of aircraft at "the

[448] Wallace, *Flights of Discovery*, pp. 124-126, 187; briefing, "Systems Research Aircraft (SRA)" by Joel R. Sitz, 1 May 1992, DFRC Historical Reference Collection; NASA news release, Don Nolan, "NASA System Research Aircraft Makes First Test Flight," 24 May 1993, DFRC Historical Reference Collection; briefing charts, "Systems Research Aircraft (SRA)," 16 Sept. [1994 (?)], DFRC Historical Reference Collection; NASA Facts, "F-18 Systems Research Aircraft," Mar. 1995, DFRC Historical Reference Collection; NASA news release, "NASA Flight Tests New Flight Control Actuator," 18 Jan. 1996, DFRC Historical Reference Collection; Lynn Corzine, "F-18 SRA Test-Bed Research Proves Itself," Dryden *X-Press*, Sept. 1996, DFRC Historical Reference Collection; briefing, "SRA 1998," DFRC Historical Reference Collection; Robert Navarro, "SRA 1999 Experiment Accomplishments Summary for Code R," DFRC Historical Reference Collection; Dryden Project Highlights, 17 and 31 Jan. 1997, 6 Nov. 1998, 12 Mar. 1999 (quoted passage), 5 Nov. 1999, 12 Nov. 1999, 26 Nov. 1999, 6 Dec. 1999, 17 Dec. 1999, 17 Mar. 2000, and 14 Apr. 2000, DFRC Historical Reference Collection; briefing charts, Gerard Schkolnik, "System Research Aircraft," for the meeting of the Flight Research Subcommittee, Aero-Space Technology Advisory Committee at DFRC, 22 Aug. 2000, DFRC Historical Reference Collection; interview of Keith Schweikhard by J. D. Hunley, 28 Aug. 2000.

earliest possible date." This would give the affected civil servants the longest possible interval for deciding whether to transfer to Dryden or exercise other options. However, during September 1996, Congress intervened to modify some of the headquarters directives. Apparently, a member of Congress–who wanted to shield Wallops Island, Virginia, from the loss of its air fleet–teamed up with representatives from Cleveland, Ohio, and Hampton, Virginia, to protect the flight operations of Lewis and Langley as well as those of Wallops. As a result, clauses added to HR 103-812, the bill that appropriated funds for NASA, prohibited the permanent reassignment of the agency's flying assets east of the Mississippi River to the West Coast. This legislation left just Dryden and Ames subject to consolidation.[449]

A number of Ames aircraft arrived at Dryden during 1997 and 1998. Most had served as platforms in the Moffett Field Airborne Science program, a prestigious and well-publicized activity conducted by Ames for 25 years. Airborne Science undertook flying investigations of the earth's ecosystems and atmosphere, made celestial observations, and contributed to sensor development and satellite sensor verification. The program also helped contain forest fires by penetrating smoke and locating hot spots.

The first aircraft, a Lockheed C-130B Hercules, arrived at Dryden in June 1997. It represented one of Ames' two Earth Resources and Applications Laboratories. The other—a McDonnell Douglas DC-8 Super 72—landed at Edwards at the end of 1997. That November, two high-flying Lockheed ER-2s (newer, larger versions of the famed U-2) transferred from Moffett Field to Dryden. These Earth Resources Survey Aircraft, along with the DC-8, served as platforms from which scientists surveyed Antarctica and the Arctic from 1989 to 1992. The resulting reports—that chlorofluorocarbons had depleted the planet's ozone layer—made headlines throughout the world. Finally, DFRC received a smaller vehicle in February 1998, a Learjet 24 outfitted as an Airborne Sensing Laboratory. These aircraft constituting the Airborne Science program—and the accompanying personnel who followed them from Ames and other NASA centers—launched operations at Dryden in Building 1623 early in 1998. The complete workforce numbered about 92: some 68 contract employees who supported the ER-2s and the DC-8, and 24 civil servants who acted as permanent staff and flight crew.[450]

[449] John Dailey to directors of Dryden, Ames, Langley and Lewis Research Centers, 23 Jan. 1996, DFRC Historical Reference Collection; Robert Whitehead to directors of Dryden, Ames, Langley, and Lewis Research Centers, n.d., DFRC Historical Reference Collection (quoted passages); Mitzi Peterson to Ken Szalai, Kevin Peterson, Chuck Brown, Dwain Deets, Gary Krier, Tom McMurtry, Bob Meyer, and Joe Ramos, 24 Sept. 1996, DFRC Historical Reference Collection; briefing, "Aircraft Consolidation," n.d., (probably Feb. 1996), DFRC Historical Reference Collection.

[450] The Dryden inventory was also increased by the addition of two general-purpose aircraft from Ames: a Lockheed YO-3A and a Beechcraft Model 200 Super Air King. *Ames Research Center: The Future Begins Here*, n.d., (about 1993), "Science and Earth Science," pp. 18-19, DFRC Historical Reference Collection; Inventory of Ames Aircraft, compiled by Robert Burns, 3 Mar. 1992, DFRC Historical Reference Collection; J.D. Hunley, compiler, "Aircraft Transfers from Ames Research Center," 31 July 1998, DFRC Historical Reference Collection; NASA Facts, "Dryden Historical Milestones," Apr. 1998, p. 11, DFRC Historical Reference Collection; J.D. Hunley, compiler, "Personnel Support Airborne Science at DFRC as of July 29, 1998," DFRC Historical Reference Collection; anon., "Around Center," Dryden *X-Press*, 16 Jan. 1999, DFRC Historical Reference Collection; Dryden news release, "Airborne Science Flights Begin at NASA Dryden," n.d., DFRC Historical Reference Collection.

Transferred from NASA Ames as a result of aircraft consolidation, the DC-8 Earth Resources and Applications Laboratory first flew for Dryden in April 1998. (NASA Photo EC98-44444-6)

In its early days, the new operation was assigned to a branch in Dryden's aerospace projects directorate under the supervision of Randy Reynolds, who came to Dryden with the transferred program. But on 30 August 1998, the Airborne Science branch became a separate directorate. Gary Krier, who had supervised the aircraft consolidation as head of the intercenter aircraft operations directorate and later led the aerospace projects directorate, became the chief of the new Airborne Science Directorate. Gary Shelton, the deputy director for Airborne Science since December 1997, became the director on 1 August 1999 when Gary Krier assumed the duties of chief engineer at Dryden.

Despite the institutional and personnel changes required to accommodate the new mission, platform aircraft research did not begin at Dryden with the arrival of the Ames air armada. Perhaps the most famous examples from the past were the later flights of the X-15, which concentrated on exposing experiments to, or collecting samples from, the upper atmosphere. Similarly, during the mid-1990s, Dryden's SR-71 Blackbirds served as platforms to measure the impact of sonic booms on the earth. But Airborne Science differed from these earlier investigations. It represented a permanent program with long-range objectives and operated aboard aircraft configured especially for the mission—not simply performing it as a part of some other project.

Airborne Science began work from its new base in winter and spring 1998. Specifically, the DC-8 Laboratory became operational in April 1998 after undergoing

One of the two ER-2s received by Dryden from Ames in November 1997 is pictured outside a hangar in Kiruna, Sweden, during ozone depletion tests. The ER-2 team also participated in flights over Brazil in 1999, during which researchers observed weather patterns associated with tropical rainfall. (NASA Photo EC00-0037-5)

some maintenance, as well as improvements in its satellite communications apparatus. Jet Propulsion Laboratory (JPL) scientists reserved time on the aircraft to assess the data collection and imaging capacities of a piece of sensing equipment known as the Airborne Synthetic Aperture Radar. The JPL team flew above the Missouri River, the Texas Gulf Coast, and the Pacific Northwest. Team members trained their device on forest canopies and cloud cover, seeing how well it portrayed such topographic, oceanographic, and geologic features as the vegetation, ocean currents, glaciers, and soils far below.

The DC-8 team also turned its attention once more to the measurement of ozone in the earth's atmosphere. During October 1999, technicians installed aboard the DC-8 some 29,000 pounds of payload, containing experimental equipment devised by JPL, Langley, and the Goddard Space Flight Center. Ames Research Center provided project management. Late the following month, the big airliner flew north of the Arctic Circle to Kiruna, Sweden, where eighteen ozone-detection flights were scheduled during the winter of 1999-2000. Under the rubric of the Stratospheric Aerosol and Gas Experiment (SAGE), Dryden supported these wintertime flights known as the SAGE III Ozone Loss and Validation Experiment (SOLVE). Researchers hoped to learn the processes affecting ozone volume in cold weather at mid to high latitudes. Eventually, the SAGE III instrumentation would orbit the earth aboard

Russia's Meteor-3 satellite, during which time the data collected by the DC-8 over Sweden would be used to validate the readings taken from space.

The ER-2s first undertook projects under Dryden's auspices in January and February 1998. One responded to a request from the Johnson Space Center to capture high-altitude particulate material. Meantime, the Anderson Group, sponsored by Harvard University, employed the services of the other ER-2 to test an instrument called an atmospheric thermal radiometer to determine whether it detected heat radiation more accurately than existing sensing devices aboard National Oceanic and Atmospheric Administration (NOAA) satellites. During 1999, the high-flight capabilities took the ER-2s farther afield. In early January, one was ferried to Brasilia, Brazil, where for six weeks it tracked the same orbit (at lower altitude) as NASA's Tropical Rainfall Measuring Mission satellite. Scientists wanted to complement the data collected from space with information gathered by the ER-2 about convection currents and moisture associated with tropical rainfall. Meantime, Gary Krier and his staff basked in an unexpected event that added still more luster to the ER-2 platforms recently acquired from Ames. The National Aeronautic Association announced in April 1999 that for the year 1998 it had bestowed the historic Collier Trophy on the ER-2 and U-2 aircraft.[451]

Into the New Millenium

Dryden entered the year 2000 under new leadership. Kevin L. Petersen became center director on 8 February 1999. Petersen had been acting deputy director and then deputy director of Dryden since 1994, serving previously as chief of Dryden's National AeroSpace Plane Projects office after lengthy assignments in the research engineering division. Petersen described five essential Dryden roles in a state-of-the-center address delivered on 18 April 2000. As well as traditional aeronautical research, he emphasized

[451] Alan Brown, "Shelton Leads Airborne Sciences Program," Dryden *X-Press*, 3 Apr. 1998, DFRC Historical Reference Collection; Alan Brown, "Dryden Launches Airborne Science Program," Dryden *X-Press*, 20 Mar. 1998, DFRC Historical Reference Collection; NASA Biographical Data, "Gary E. Krier," Aug. 1999, taken from DFRC website (http://www.dfrc.nasa.gov/PAO/PAIS/HTML/bd-dfrc-p022.html), filed in DFRC Historical Reference Collection; Randy Albertson, Dryden airborne science directorate, telephone conversations with Michael Gorn, 13, 14, and 18 Jan. 2000, DFRC Historical Reference Collection; Dryden news release, "Airborne Science Flights Begin from NASA Dryden," n.d., DFRC Historical Reference Collection; Alan Brown, "DC-8 Studies the Earth," Dryden *X-Press*, 1 May 1998, DFRC Historical Reference Collection. For sonic boom research aboard the SR-71, see Domenic J. Maglieri, "Sonic Boom Ground Pressure Measurements for Flights at Altitudes in Excess of 70,000 Feet and at Mach Numbers up to 3.0," *2nd Conference on Sonic Boom Research* (Washington, DC: NASA SP-180, 1968), pp. 19-27; Domenic J. Maglieri, Vera Huckel, and Herbert R. Henderson, *Sonic Boom Measurements for SR-71 Aircraft Operating at Mach Numbers to 3.0 and Altitudes to 24 384 Meters,* (Washington, DC: NASA TN D-6823, 1972), pp. 1-28; Edward A. Haering, Jr., L.J. Ehernberger, and Stephan A. Whitmore, *Preliminary Airborne Measurements for the SR-71 Sonic Boom Propagation Experiment* (Edwards, CA: NASA TM 104307, 1995), pp. 1-23; Beth Hagenauer, "DC-8 Begins SOLVE Mission," Dryden *X-Press*, 26 Nov. 1999, taken from the DFRC website (http://www.dfrc.nasa.gov/PAO/X-Press/1999/Nov26/frontfull0.html), filed in the DFRC Historical Reference Collection; NASA news release, "NASA ER-2 Deploys for Brazil to Study Tropical Weather Systems," 24 Dec. 1998, DFRC Historical Reference Collection; notes from DFRC director's meeting, 11 Jan. 1999, DFRC Historical Reference Collection (filed with Dryden Project Highlights); NASA news release, "Dryden's Airborne Science Plane, ER-2, Selected for Multiple Awards," Apr. 1999, DFRC Historical Reference Collection.

Dryden Director Kevin L. Petersen succeeded Kenneth Szalai in February 1999. (NASA Photo EC99-44890-1)

access to space flight, scientific missions aboard platform aircraft, Space Shuttle support, and development of a crew return vehicle for the International Space Station. Petersen's comments echoed those expressed the previous month in a briefing to NASA Administrator Daniel Goldin and the recently appointed associate administrator in the Office of Aerospace Technology, Samuel L. Venneri. This presentation anticipated an increase in experimental research aircraft as well as access-to-space flight projects in the future.

For his part, Goldin regarded Dryden as "the place to test X-Planes," an important remark that recognized the continued value of experimental vehicles in the overall NASA agenda and Dryden's role therein. Goldin also wanted to eliminate any ambiguity about Dryden's primacy in flight research. "If it involves flight," he said, "Dryden needs to be involved . . . from the beginning."[452]

[452] NASA Biographical Data, "Kevin L. Petersen," Feb. 1999; [Kevin L. Petersen], "State of the Center Address," 18 Apr. 2000, DFRC Historical Reference Collection (quoted passages from p. 32).

Epilogue: "I Hadn't Missed it After All"

Even the most informed observers of events may be misled occasionally. No less an authority than NASA research pilot Milt Thompson recounted how he had misjudged the unfolding of aeronautical research.

> When the Space Shuttle *Columbia* came out of the rentry blackout April 14, 1981, my eyes filled with tears. NASA had accomplished a lifting reentry and a fly-back to a precise landing on Rogers Dry Lake at Edwards Air Force Base. When I arrived at Edwards in 1956, I didn't foresee what I was about to be part of. The golden age of flight test was nearing its end. We were almost finished with the X-1 program; the X-3, X-4, and X-5 programs had ended: and the X-2 was in trouble. I thought to myself, 'The X programs are all over, I've missed it all.' Then came the X-15, the Dyna-Soar program, the paresev, and the lifting bodies. I hadn't missed it after all.[1]

As Thompson's testimony illustrates, flight research at Dryden did not unfold in a predetermined fashion, one project following another predictably. Rather, like most episodes in the history of technology, the process of discovery began, progressed, ended, and renewed itself for a variety of reasons. To some extent, the internal dynamics of technical progress shaped the direction of research. To those participating in the first epoch in Dryden flight research–one in which speed and altitude records fell in rapid succession during the 1940s and1950s–faster and higher flying seemed to be a predetermined destiny. But equally, various factors such as fiscal, political, and military considerations governed these early investigations at Dryden. Pursued at the height of the Cold War, the initial X-planes became investments in national security, not merely the instruments of advanced aeronautical research. As such, they usually won the funding they required.

By the time Milt Thompson arrived on Edwards Air Force Base, however, flight research had undergone a transformation and entered its second phase. The necessity of winning the high battleground of space caused a re-direction of the energies of the High Speed Flight Station (as Dryden was then designated). The hypersonic X-15 aircraft initiated the new period and its frequent leaps into space resulted in an invaluable pool of data for future spaceflight. While the X-15 flew, the Lunar Landing Research Vehicle gave former Dryden research pilot Neil Armstrong and the other Apollo astronauts a preview of the complexities of touching down on the moon. Then, during the 1960s and 1970s, first the paresev and then the lifting body programs enlisted the services of engineers like Dale Reed and pilots like Thompson to prove that wingless aircraft could indeed return crews safely from space.

During the third phase of research, Dryden Flight Research Center engineers embarked on some pivotal experiments on the Space Shuttle Orbiter, testing its basic

[1] Thompson and Peebles, *Flying Without Wings*, p. 1.

aerodynamics in flight and compensating for its deficiencies. In order to fly reliably, the Shuttle Orbiter required digital fly-by-wire, a system of flight control pioneered at Dryden. Indeed, the third period of Dryden history began with the application of DFBW. After its first successful flight during the 1970s and subsequent adoption by the armed forces and industry, DFBW inaugurated at Dryden a time in which practical improvements in civil and military aircraft dominated the calendar. Rather than launching headline-making projects, Dryden undertook a series of more modest endeavors designed to improve the safety, aerodynamic efficiency, maneuverability, range, and fuel economy of air travel. They included (among many others) such notable projects as the Controlled Impact Demonstration, digital engine control, laminar flow control research, the Highly Maneuverable Aircraft Technology demonstrator, several high angle-of-attack investigations, the supercritical wing, and winglets.

Finally, at the end of the twentieth century, the center re-focused its attention on such space-related subjects as hypersonics, return from space, and reusable launch vehicles. The exact direction of the Dryden Flight Research Center during the twenty-first century is still being determined. But this much is clear: the three main traditions that have sustained it for nearly 60 years—high speed, high altitude, and hypersonic flight research; space-related flight research; and flight research that pursues practical discoveries—will continue to head the agenda, either singly or in combination.

A Note on Sources: First Edition

The majority of sources cited in this study are records such as memoranda, policy statements, progress reports, and planning documents found within Record Group 255, Records of the National Advisory Committee for Aeronautics and the National Aeronautics and Space Administration, maintained by the National Archives and Records Administration. Many other documents are in the files of Dryden Flight Research Facility [now once again, the Dryden Flight Research Center]; at other NASA centers such as Ames, Langley, and Johnson; and in the possession of present and former employees. Other sources include documents held by the USAF Systems Command Historical Office [since merged with the Air Force Logistics Command to form the Air Force Materiel Command], and the Historical Office of the United States Navy's Naval Air Systems Command. Finally, some documentation was provided by individuals at major aircraft and aerospace manufacturing concerns such as McDonnell Douglas, Rockwell, and Lockheed [now parts of Boeing and Lockheed Martin, respectively].

As with other aspects of aviation history, there are few useful published sources dealing with the subjects discussed in this work; indeed, only one previous book had dealt with the origins of supersonic flight in America, and that by the author of this study. The lack of published works dealing with the development of postwar aerospace technology and the management and utilization of that technology can be taken as a general indication of the future studies that are required to enhance our understanding of the impact and role that aviation and aerospace science have had upon modern society.

The following discussion of sources useful to this study should indicate the scope of records available to a researcher examining other aspects of NACA-NASA history as well.

My research began in the NASA History Office at NASA Headquarters, Washington, DC. The History Office provided great assistance over a number of years, both on this topic and others. The History Office's Historical Records Collection consists of 500 cubic feet of records [now expanded to some 2,000 linear feet], with another 395 cubic feet stored at the Federal Records Center at Suitland, Maryland. The series that contain information useful to this study are: Congressional Documents, Industry, Organization and Management, Budget Documentation, NASA Headquarters, NASA Centers, Human Space Flight, Tracking and Data Acquisition, Biography File, and Aeronautics.

The Dryden Flight Research Center records are retired to the Federal Records Center, Laguna Niguel, California. (Previously, they were at the Federal Records Center at Bell, California.) Record Group 255 boxes that proved of particular value at Bell are 310 and 312 (X-1 reports), 321 (X-2 documentation), and 361, 362, and 366, all on the D-558 program [box numbers have since changed, and copies of many documents are presently available in the Dryden Historical Reference Collection]. The files of the NASA Langley Research Center contain much useful material on the early days of Dryden when it operated as a satellite of Langley, including the annual and semiannual reports of the NACA Research Airplane Projects Panel and the NACA High-Speed Flight Station. [Many of these reports are now available in the Hallion Collection at the Dryden Historical Reference Collection; others may have been retired to the National Archives and Records Administration.] The retired records of Ames Research Center at the Federal Records Center at San Bruno, California, contain documentation pertaining to Ames's role in Dryden's affairs during the 1940s and 1950s. The Lyndon B. Johnson Space Center's History Office maintains a large collection of records on human spaceflight. This was particularly useful for the Space Shuttle project and its origins.

Since Dryden operates as a tenant on Edwards Air Force Base (AFB), Air Force records reveal the nature of the relationship between the service and this civilian agency. Particularly pertinent are the semiannual and annual historical reports submitted to higher command concerning the activities of the Air Force base. These reports, including primary documents in appendixes, are on file at Edwards, but copies are also held by the Air Force systems Command History Office at Andrews AFB, Maryland [since moved in large part to the Air Force Materiel Command headquartered at Wright-Patterson AFB in Dayton, Ohio] and at the Albert F. Simpson Historical Research Center, Maxwell AFB, Alabama [now called the Air Force Historical Research Agency]. Though Dryden's connections with naval aviation never equaled those with the Air Force, there was a significant interchange of ideas between the Navy and Dryden especially in the 1940s and 1950s. The files of the Naval Air Systems command contain memorandums and reports of the Bureau of Aeronautics.

This history could not have been prepared without the cooperation and assistance of a number of persons, especially the staff of the Dryden center itself. Many preserved key documents over the years, and these were made available to the author. More important, however, many of the staff consented to interviews, as did many former NACA, NASA, Air Force, Navy, and industry officials. These interviews provided insight into the human story of Dryden—the working of the staff, viewpoints and goals, what individuals felt they had accomplished. While oral history is no substitute for written records, it is a supplement that can flesh out and elaborate on the more traditional sources.

What remains to be done? Dryden's story offers a view of a unique research center operating at the forefront of technological change in a critical period of aeronautical and astronautical development. We have seen the emergence of aerospace technology. In many sections of the manuscript, the fulfillment of that technological promise is alluded to. Further study is indicated on the supersonic breakthrough and the turbojet revolution and their impact on modern business, industry, transportation, military affairs, and society as a whole. We are an aerospace society and need to understand what this means to us and to our descendants.

A Note on Sources: Second Edition

Because they treat the recent story of flight research at DFRC, the two chapters (12 and 13) added to the second edition of *On the Frontier* were based essentially on documents collected at Dryden, rather than on outside sources. A good many of these items materialized from present or former center employees who held copies of important documents in their personal files. The bulk of the documentation, however, exists in the Dryden Historical Reference Collection, a growing and efficiently organized resource founded during the mid-1990s and consisting of such varied materials as program documents, correspondence, technical papers, briefings, and symposia proceedings. Particularly valuable are the many transcribed interviews conducted by the center historian and others with important Dryden figures of the past and present. Finally, in bringing the story of the flight research center up to the year 2000, the research for this new edition entailed specialized interviews with Dryden personalities, as well as frequent recourse to the DFRC Library and Web site for the most recent NASA technical publications and other published material.

For minor revisions to the original chapters, publications that have appeared since the first edition have shed some new light on various topics and have been incorporated into footnotes. In some cases, sources that Richard Hallion used are now available in the Dryden Historical Reference Collection and this has been noted in the footnotes. In other cases, present locations of his sources are not known, and his footnotes have been left unchanged.

About the Authors

Richard P. Hallion earned his PhD in aerospace history from the University of Maryland in 1975 and served as a curator at the Smithsonian Institution's National Air and Space Museum from 1974 through 1980 before becoming an Air Force historian at the Air Force Flight Test Center on Edwards Air Force Base (AFB) in California from 1982 to 1986. Later in 1986 he became the Director, of the Special Staff Office in Aeronautical Systems Division at Wright-Patterson AFB in Ohio. Two years later he became the executive historian for advanced programs in the Directorate of Advanced Programs at Headquarters Air Force Systems Command at Andrews AFB, Maryland. He was a senior issues and policy analyst for the Secretary of the Air Force during 1991 and became the Historian of the United States Air Force the same year. He has written more than a dozen books in aerospace history including *Supersonic Flight: Breaking the Sound Barrier and Beyond* (originally published in 1972 and reprinted in a revised edition in 1997) about the D-558 and X-1 programs; *Test Pilots: The Frontiersmen of Flight* (1981), and the original version of *On the Frontier*. He is a recognized expert on the history of aviation and space.

Michael H. Gorn received a doctorate in history from the University of Southern California in 1978. He served in the Air Force history program as a staff historian, command historian, and Deputy Air Force Historian. He also served as Chief Historian for the U.S. Environmental Protection Agency (EPA). He is presently the chief historian of the NASA Dryden Flight Research Center, Edwards, California. Gorn is a recipient of the National Air and Space Museum's Alfred V. Verville Fellowship in Aerospace History and was selected for the Fellowship in Aerospace History, awarded by the American Historical Association. He is the author of *Hugh L. Dryden's Career in Aviation and Space* (1996), *The Universal Man: Theodore von Kàrmàn's Life in Aeronautics* (1992) and *Harnessing the Genie: Science and Technology Forecasting for the Air Force, 1944-1986* (1988), among other works. His latest book—*Expanding the Envelope: Flight Research at NACA and NASA*—has been published in fall 2001 by the University Press of Kentucky. In addition, he has completed research on a biography of Dr. Hugh L. Dryden, whose publication is forthcoming.

About the Authors

Appendices,
A Note on Sources
Index

Appendix A

Organizational Charts, 1948-1999

Unlike other aeronautical centers in NASA, the Flight Research Center, later Dryden Flight Research Center, has always been small, with a single overriding purpose: flight research. Thus, changes of major significance in its administrative organization have been comparatively few. The 1948 chart reflects the close identity of the Muroc Flight Test Unit with specific aircraft programs. As the unit expanded to station and eventually to center size, gaining autonomy along the way, its administrative organization of necessity became broader and more in line with that of other NACA/NASA research facilities. As indicated in chapters 3 and 6, the center's organization remained largely unchanged in 1960 from the days of the station in 1954; there was strong continuity from the period of HSFS Chief Walter C. Williams to the period of FRC Director Paul F. Bikle, although increasingly during Bikle's tenure a matrix organization emerged in which many individuals worked on a variety of projects, not just for their functional supervisors. Despite this change, organization charts for 1960 through 1966 also suggest continuity. Even into the 1970s, under Director Lee R. Scherer (1975 chart), the FRC's formal organization remained strongly oriented along previous lines, mainly structured around four key directorates: Research, Data Systems, Flight Operations, and Administration. Under Center Director David Scott (1976 chart) came the first significant formal departure from the previous structural framework, with the addition of two new directorates, Aeronautical Projects and Shuttle Operations.

On the eve of consolidation with Ames Research Center (ARC), the DFRC's structure showed further changes introduced by Director Isaac Gillam. By 1979, the traditional directorate structure had disappeared, replaced by a more centralized and integrated one built around a strong executive-staff support network and three major directorates: Engineering, Flight Operations and Support, and Administration (see chart for 1979). These changes reflected the wishes of NASA as a whole to consolidate the functions of the centers within the contexts of both the centers themselves and that of the agency as a whole. These same wishes were responsible for the eventual decision to merge DFRC and ARC into a single operating administrative unit (see chapter 11). The August 1981 chart shows the center on the eve of the merger. Those for 1982 and 1990 reveal the results of the merger, with that for 1992 indicating how the Ames-Dryden Flight Research Facility by itself looked toward the end of the merger period.

The chart for 1996 reveals the organization shortly after DFRC re-emerged as a separate center. At that time, it had five major directorates: for Research Facilities, Research Engineering, Flight Operations, Aerospace Projects, and Intercenter Aircraft Operations. The last of these disappeared in 1997 with the congressional mandate not to transfer research aircraft from east of the Mississippi River to Dryden. In 1998, the Airborne Science Directorate joined the other four, and in October 1999 Gary Krier became head of the new System Management Office in addition to his duties as Chief Engineer (a position not shown on the December 1999 chart).

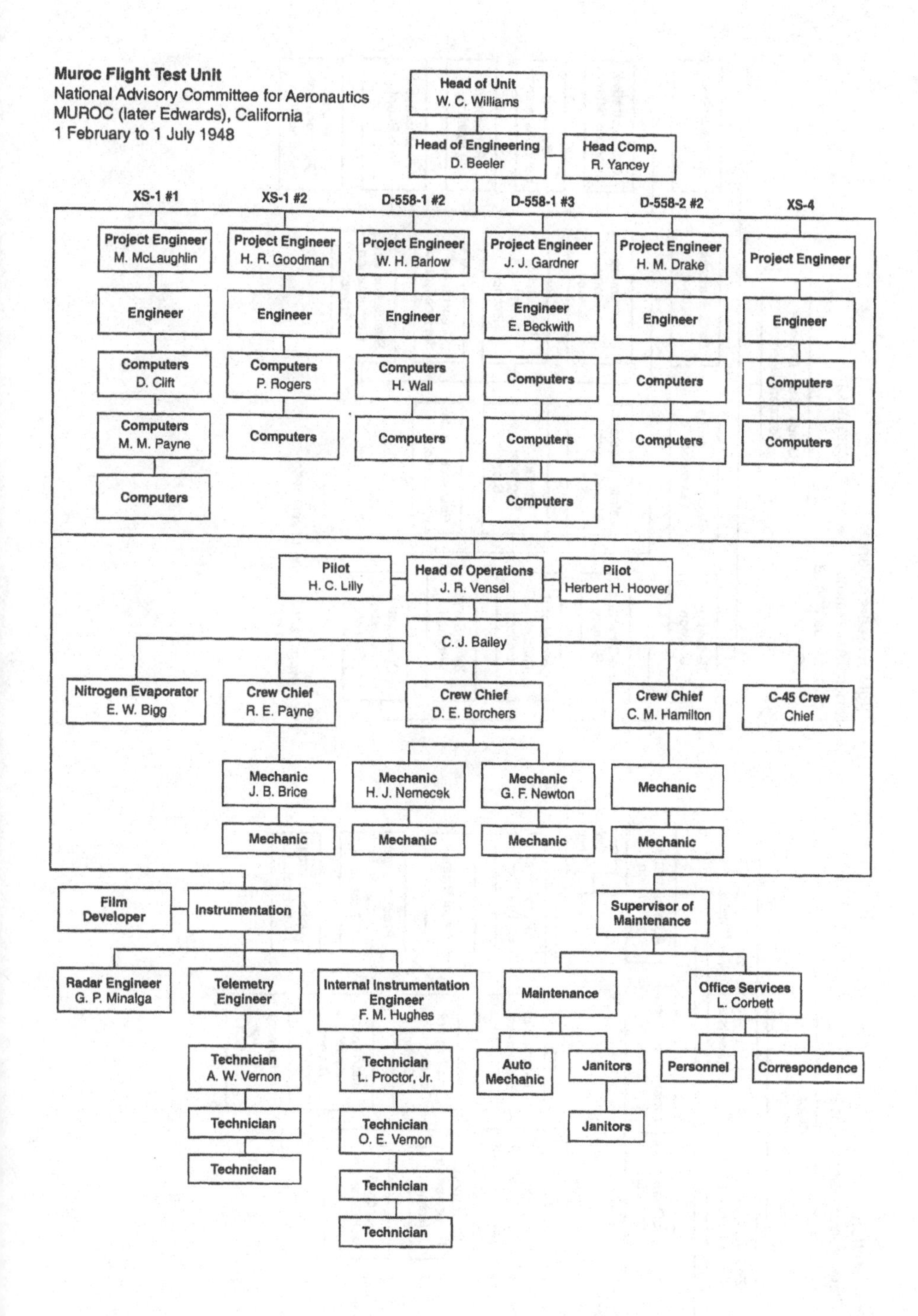
Muroc Flight Test Unit
National Advisory Committee for Aeronautics
MUROC (later Edwards), California
1 February to 1 July 1948
Head of Unit
W. C. Williams
Head of Engineering
D. Beeler
Head Comp.
R. Yancey
XS-1 #1
XS-1 #2
D-558-1 #2
D-558-1 #3
D-558-2 #2
XS-4
Project Engineer
M. McLaughlin
Project Engineer
H. R. Goodman
Project Engineer
W. H. Barlow
Project Engineer
J. J. Gardner
Project Engineer
H. M. Drake
Project Engineer
Engineer
Engineer
E. Beckwith
Computers
D. Clift
Computers
P. Rogers
Computers
H. Wall
Computers
M. M. Payne
Pilot
H. C. Lilly
Head of Operations
J. R. Vensel
Pilot
Herbert H. Hoover
C. J. Bailey
Nitrogen Evaporator
E. W. Bigg
Crew Chief
R. E. Payne
Crew Chief
D. E. Borchers
Crew Chief
C. M. Hamilton
C-45 Crew Chief
Mechanic
J. B. Brice
Mechanic
H. J. Nemecek
Mechanic
G. F. Newton
Mechanic
Film Developer
Instrumentation
Supervisor of Maintenance
Radar Engineer
G. P. Minalga
Telemetry Engineer
Internal Instrumentation Engineer
F. M. Hughes
Maintenance
Office Services
L. Corbett
Technician
A. W. Vernon
Technician
L. Proctor, Jr.
Auto Mechanic
Janitors
Personnel
Correspondence
Technician
O. E. Vernon
Technician

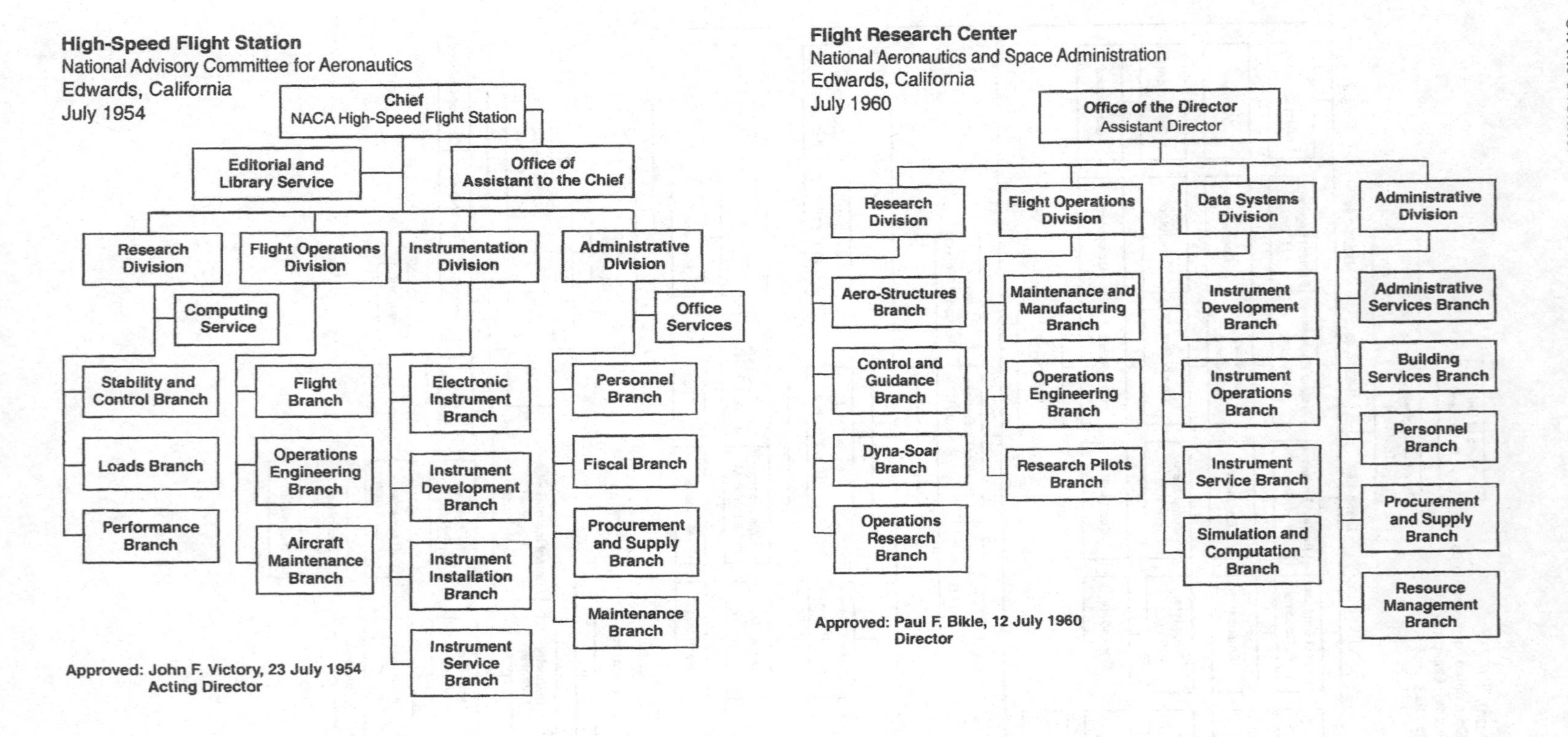
High-Speed Flight Station
National Advisory Committee for Aeronautics
Edwards, California
July 1954
Chief
NACA High-Speed Flight Station
Editorial and
Library Service
Office of
Assistant to the Chief
Research
Division
Flight Operations
Division
Instrumentation
Division
Administrative
Division
Computing
Service
Office
Services
Stability and
Control Branch
Loads Branch
Performance
Branch
Flight
Branch
Operations
Engineering
Branch
Aircraft
Maintenance
Branch
Electronic
Instrument
Branch
Instrument
Development
Branch
Instrument
Installation
Branch
Instrument
Service
Branch
Personnel
Branch
Fiscal Branch
Procurement
and Supply
Branch
Maintenance
Branch
Approved: John F. Victory, 23 July 1954
Acting Director
Flight Research Center
National Aeronautics and Space Administration
Edwards, California
July 1960
Office of the Director
Assistant Director
Research
Division
Flight Operations
Division
Data Systems
Division
Administrative
Division
Aero-Structures
Branch
Control and
Guidance
Branch
Dyna-Soar
Branch
Operations
Research
Branch
Maintenance and
Manufacturing
Branch
Operations
Engineering
Branch
Research Pilots
Branch
Instrument
Development
Branch
Instrument
Operations
Branch
Instrument
Service Branch
Simulation and
Computation
Branch
Administrative
Services Branch
Building
Services Branch
Personnel
Branch
Procurement
and Supply
Branch
Resource
Management
Branch
Approved: Paul F. Bikle, 12 July 1960
Director

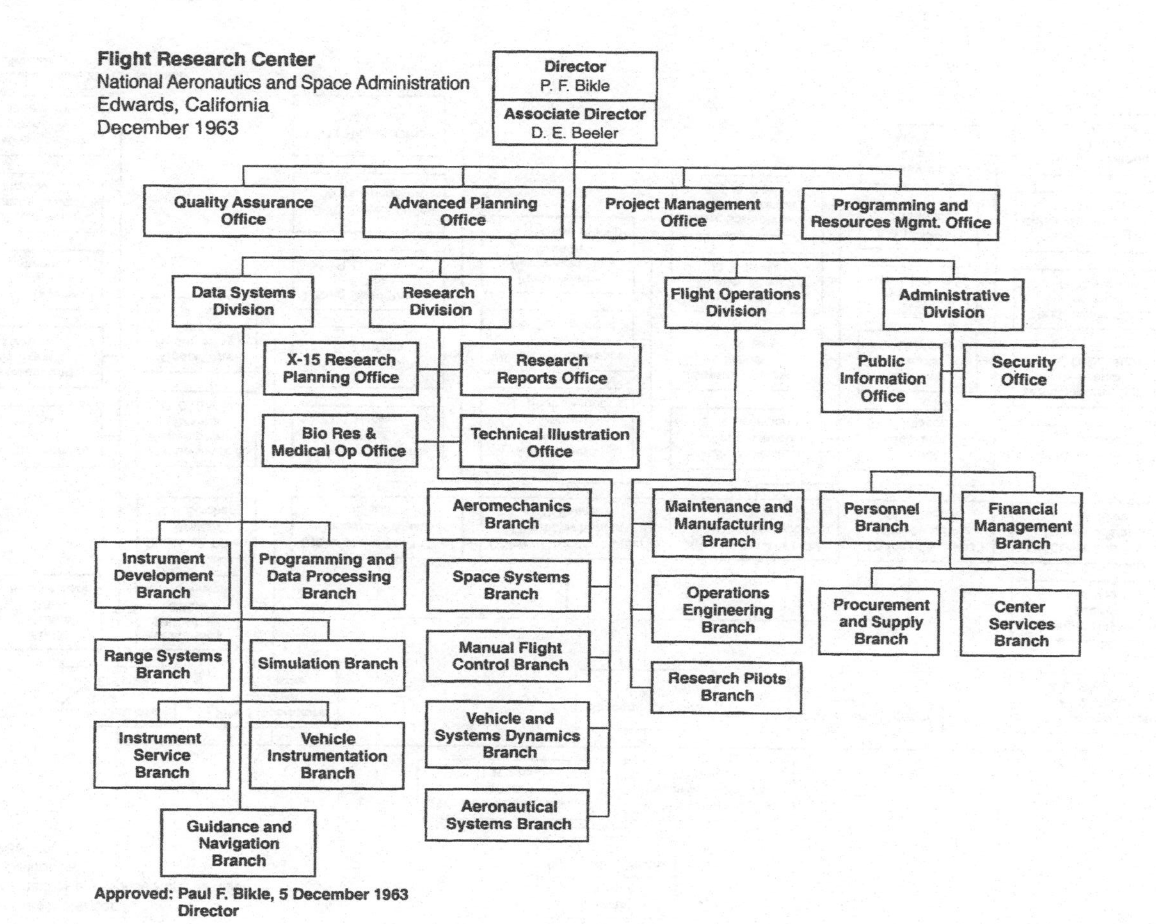

Flight Research Center
National Aeronautics and Space Administration
Edwards, California
December 1963
Director
P. F. Bikle
Associate Director
D. E. Beeler
Quality Assurance Office
Advanced Planning Office
Project Management Office
Programming and Resources Mgmt. Office
Data Systems Division
Research Division
Flight Operations Division
Administrative Division
X-15 Research Planning Office
Research Reports Office
Bio Res & Medical Op Office
Technical Illustration Office
Public Information Office
Security Office
Aeromechanics Branch
Maintenance and Manufacturing Branch
Personnel Branch
Financial Management Branch
Instrument Development Branch
Programming and Data Processing Branch
Space Systems Branch
Operations Engineering Branch
Procurement and Supply Branch
Center Services Branch
Range Systems Branch
Simulation Branch
Manual Flight Control Branch
Research Pilots Branch
Instrument Service Branch
Vehicle Instrumentation Branch
Vehicle and Systems Dynamics Branch
Guidance and Navigation Branch
Aeronautical Systems Branch
Approved: Paul F. Bikle, 5 December 1963
Director

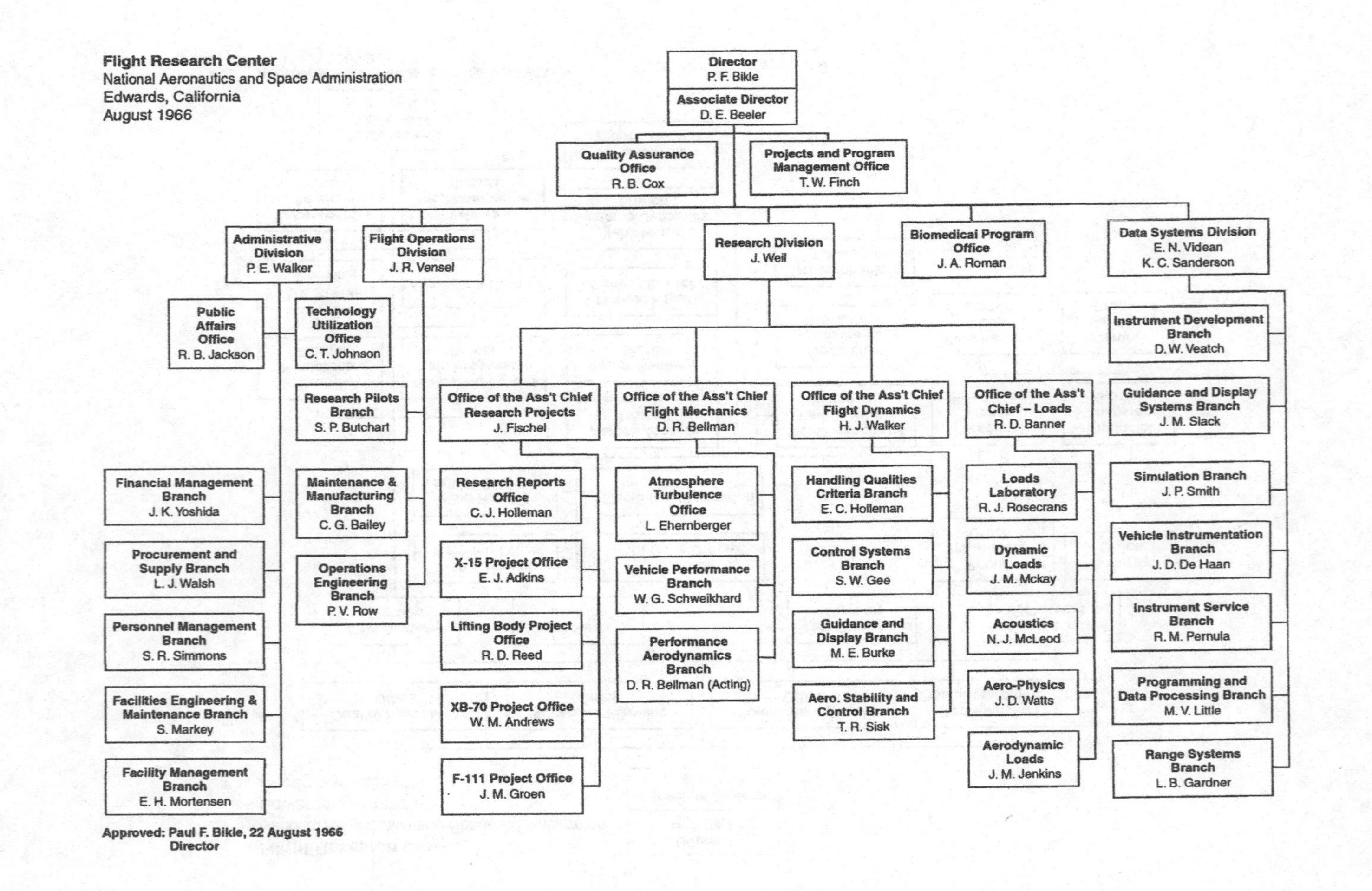

Flight Research Center
National Aeronautics and Space Administration
Edwards, California
August 1966
Director
P. F. Bikle
Associate Director
D. E. Beeler
Quality Assurance Office
R. B. Cox
Projects and Program Management Office
T. W. Finch
Administrative Division
P. E. Walker
Flight Operations Division
J. R. Vensel
Research Division
J. Weil
Biomedical Program Office
J. A. Roman
Data Systems Division
E. N. Videan
K. C. Sanderson
Public Affairs Office
R. B. Jackson
Technology Utilization Office
C. T. Johnson
Financial Management Branch
J. K. Yoshida
Procurement and Supply Branch
L. J. Walsh
Personnel Management Branch
S. R. Simmons
Facilities Engineering & Maintenance Branch
S. Markey
Facility Management Branch
E. H. Mortensen
Research Pilots Branch
S. P. Butchart
Maintenance & Manufacturing Branch
C. G. Bailey
Operations Engineering Branch
P. V. Row
Office of the Ass't Chief Research Projects
J. Fischel
Research Reports Office
C. J. Holleman
X-15 Project Office
E. J. Adkins
Lifting Body Project Office
R. D. Reed
XB-70 Project Office
W. M. Andrews
F-111 Project Office
J. M. Groen
Office of the Ass't Chief Flight Mechanics
D. R. Bellman
Atmosphere Turbulence Office
L. Ehernberger
Vehicle Performance Branch
W. G. Schweikhard
Performance Aerodynamics Branch
D. R. Bellman (Acting)
Office of the Ass't Chief Flight Dynamics
H. J. Walker
Handling Qualities Criteria Branch
E. C. Holleman
Control Systems Branch
S. W. Gee
Guidance and Display Branch
M. E. Burke
Aero. Stability and Control Branch
T. R. Sisk
Office of the Ass't Chief – Loads
R. D. Banner
Loads Laboratory
R. J. Rosecrans
Dynamic Loads
J. M. Mckay
Acoustics
N. J. McLeod
Aero-Physics
J. D. Watts
Aerodynamic Loads
J. M. Jenkins
Instrument Development Branch
D. W. Veatch
Guidance and Display Systems Branch
J. M. Slack
Simulation Branch
J. P. Smith
Vehicle Instrumentation Branch
J. D. De Haan
Instrument Service Branch
R. M. Pernula
Programming and Data Processing Branch
M. V. Little
Range Systems Branch
L. B. Gardner
Approved: Paul F. Bikle, 22 August 1966
Director

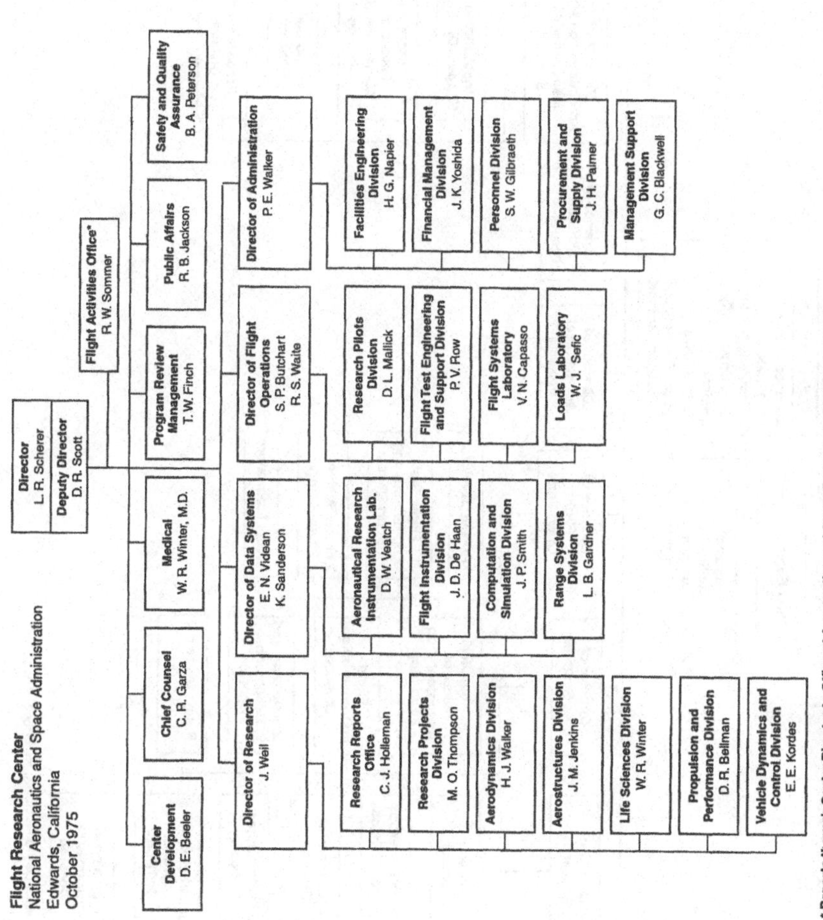

Flight Research Center
National Aeronautics and Space Administration
Edwards, California
October 1975
Director
L. R. Scherer
Deputy Director
D. R. Scott
Flight Activities Office*
R. W. Sommer
Center Development
D. E. Beeler
Chief Counsel
C. R. Garza
Medical
W. R. Winter, M.D.
Program Review Management
T. W. Finch
Public Affairs
R. B. Jackson
Safety and Quality Assurance
B. A. Peterson
Director of Research
J. Weil
Research Reports Office
C. J. Holleman
Research Projects Division
M. O. Thompson
Aerodynamics Division
H. J. Walker
Aerostructures Division
J. M. Jenkins
Life Sciences Division
W. R. Winter
Propulsion and Performance Division
D. R. Bellman
Vehicle Dynamics and Control Division
E. E. Kordes
Director of Data Systems
E. N. Videan
K. Sanderson
Aeronautical Research Instrumentation Lab.
D. W. Veatch
Flight Instrumentation Division
J. D. De Haan
Computation and Simulation Division
J. P. Smith
Range Systems Division
L. B. Gardner
Director of Flight Operations
S. P. Butchart
R. S. Waite
Research Pilots Division
D. L. Mallick
Flight Test Engineering and Support Division
P. V. Row
Flight Systems Laboratory
V. N. Capasso
Loads Laboratory
W. J. Sefic
Director of Administration
P. E. Walker
Facilities Engineering Division
H. G. Napier
Financial Management Division
J. K. Yoshida
Personnel Division
S. W. Gilbraeth
Procurement and Supply Division
J. H. Palmer
Management Support Division
G. C. Blackwell
* Reports through Center Director to Office of Aeronautics and Space Technology, NASA Headquarters
Approved: George M. Low, 03 October 1975

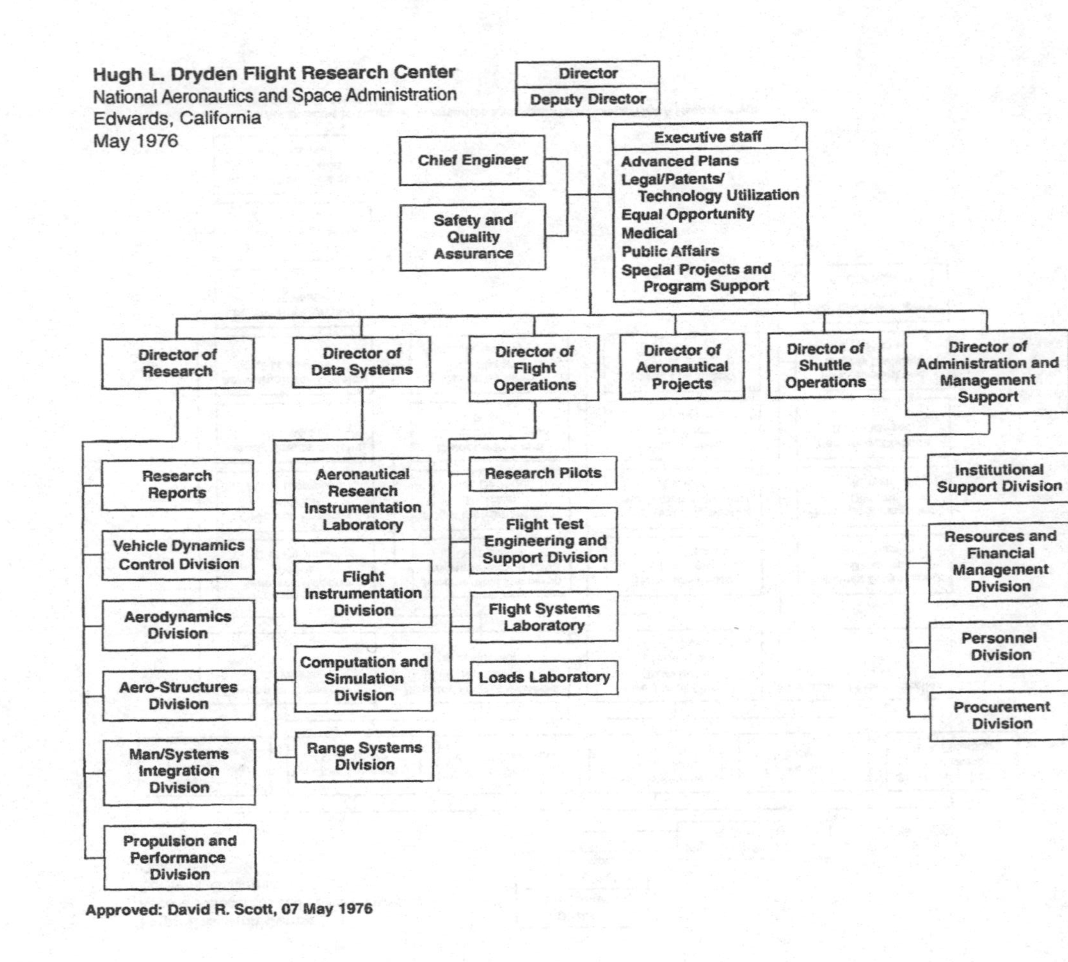
Hugh L. Dryden Flight Research Center
National Aeronautics and Space Administration
Edwards, California
May 1976
Director
Deputy Director
Chief Engineer
Safety and Quality Assurance
Executive staff
Advanced Plans
Legal/Patents/
Technology Utilization
Equal Opportunity
Medical
Public Affairs
Special Projects and
Program Support
Director of Research
Research Reports
Vehicle Dynamics Control Division
Aerodynamics Division
Aero-Structures Division
Man/Systems Integration Division
Propulsion and Performance Division
Director of Data Systems
Aeronautical Research Instrumentation Laboratory
Flight Instrumentation Division
Computation and Simulation Division
Range Systems Division
Director of Flight Operations
Research Pilots
Flight Test Engineering and Support Division
Flight Systems Laboratory
Loads Laboratory
Director of Aeronautical Projects
Director of Shuttle Operations
Director of Administration and Management Support
Institutional Support Division
Resources and Financial Management Division
Personnel Division
Procurement Division
Approved: David R. Scott, 07 May 1976

Hugh L. Dryden Flight Research Center
National Aeronautics and Space Administration
Edwards, California
March 1979

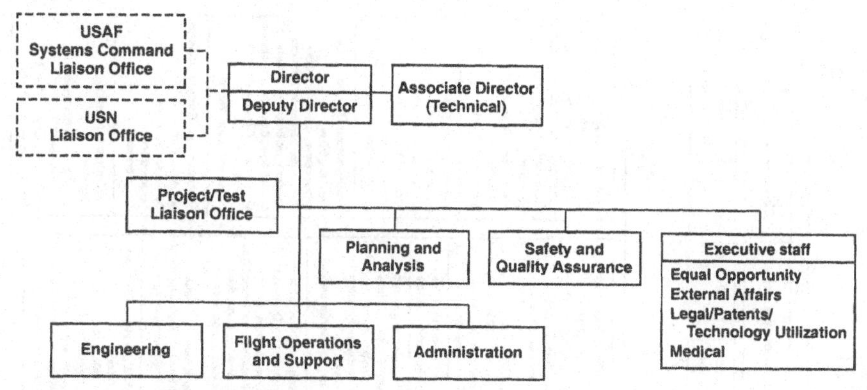

Approved: A. M. Lovelace, 13 March 1979

Hugh L. Dryden Flight Research Center
National Aeronautics and Space Administration
Edwards, California
August 1981

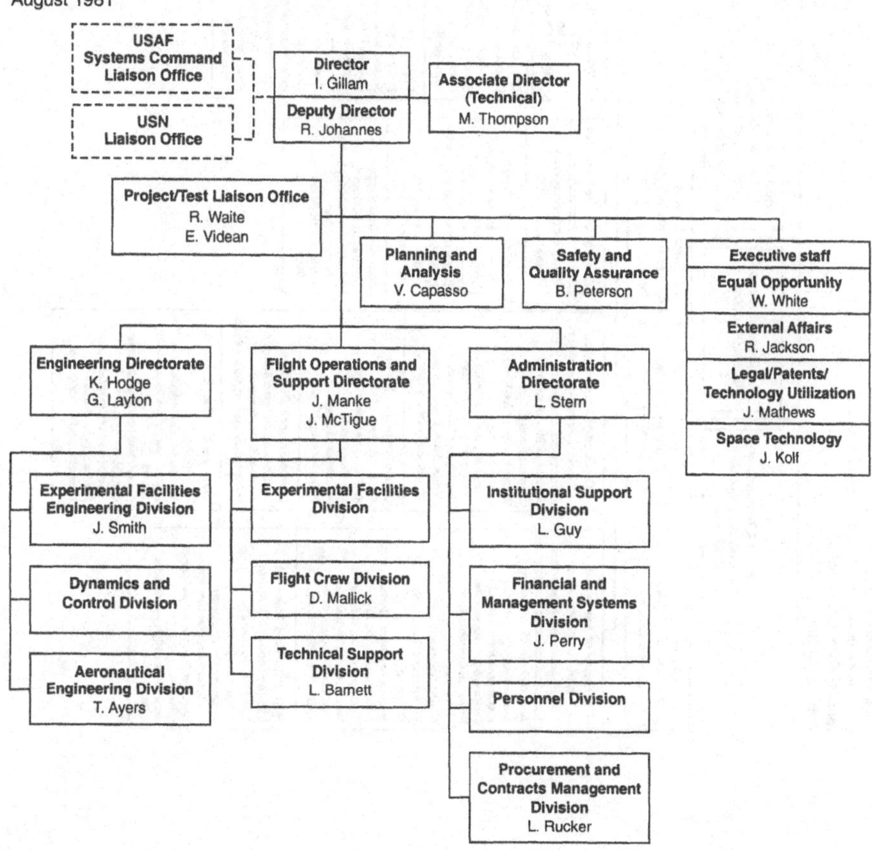

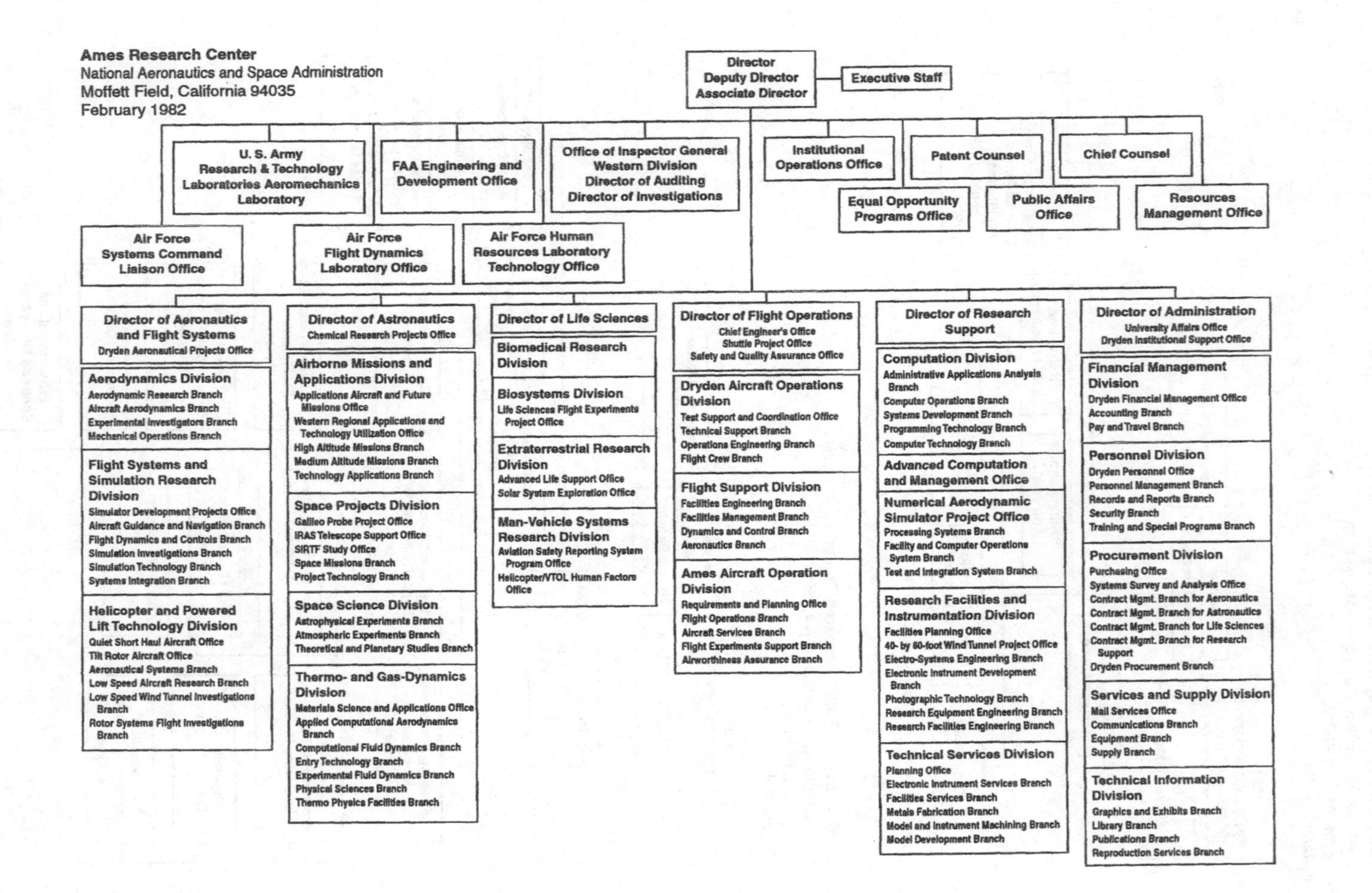
Ames Research Center
National Aeronautics and Space Administration
Moffett Field, California 94035
February 1982
Director
Deputy Director
Associate Director
Executive Staff
U. S. Army Research & Technology Laboratories Aeromechanics Laboratory
FAA Engineering and Development Office
Office of Inspector General Western Division Director of Auditing Director of Investigations
Institutional Operations Office
Patent Counsel
Chief Counsel
Equal Opportunity Programs Office
Public Affairs Office
Resources Management Office
Air Force Systems Command Liaison Office
Air Force Flight Dynamics Laboratory Office
Air Force Human Resources Laboratory Technology Office
Director of Aeronautics and Flight Systems
Dryden Aeronautical Projects Office
Aerodynamics Division
Aerodynamic Research Branch
Aircraft Aerodynamics Branch
Experimental Investigators Branch
Mechanical Operations Branch
Flight Systems and Simulation Research Division
Simulator Development Projects Office
Aircraft Guidance and Navigation Branch
Flight Dynamics and Controls Branch
Simulation Investigations Branch
Simulation Technology Branch
Systems Integration Branch
Helicopter and Powered Lift Technology Division
Quiet Short Haul Aircraft Office
Tilt Rotor Aircraft Office
Aeronautical Systems Branch
Low Speed Aircraft Research Branch
Low Speed Wind Tunnel Investigations Branch
Rotor Systems Flight Investigations Branch
Director of Astronautics
Chemical Research Projects Office
Airborne Missions and Applications Division
Applications Aircraft and Future Missions Office
Western Regional Applications and Technology Utilization Office
High Altitude Missions Branch
Medium Altitude Missions Branch
Technology Applications Branch
Space Projects Division
Galileo Probe Project Office
IRAS Telescope Support Office
SIRTF Study Office
Space Missions Branch
Project Technology Branch
Space Science Division
Astrophysical Experiments Branch
Atmospheric Experiments Branch
Theoretical and Planetary Studies Branch
Thermo- and Gas-Dynamics Division
Materials Science and Applications Office
Applied Computational Aerodynamics Branch
Computational Fluid Dynamics Branch
Entry Technology Branch
Experimental Fluid Dynamics Branch
Physical Sciences Branch
Thermo Physics Facilities Branch
Director of Life Sciences
Biomedical Research Division
Biosystems Division
Life Sciences Flight Experiments Project Office
Extraterrestrial Research Division
Advanced Life Support Office
Solar System Exploration Office
Man-Vehicle Systems Research Division
Aviation Safety Reporting System Program Office
Helicopter/VTOL Human Factors Office
Director of Flight Operations
Chief Engineer's Office
Shuttle Project Office
Safety and Quality Assurance Office
Dryden Aircraft Operations Division
Test Support and Coordination Office
Technical Support Branch
Operations Engineering Branch
Flight Crew Branch
Flight Support Division
Facilities Engineering Branch
Facilities Management Branch
Dynamics and Control Branch
Aeronautics Branch
Ames Aircraft Operation Division
Requirements and Planning Office
Flight Operations Branch
Aircraft Services Branch
Flight Experiments Support Branch
Airworthiness Assurance Branch
Director of Research Support
Computation Division
Administrative Applications Analysis Branch
Computer Operations Branch
Systems Development Branch
Programming Technology Branch
Computer Technology Branch
Advanced Computation and Management Office
Numerical Aerodynamic Simulator Project Office
Processing Systems Branch
Facility and Computer Operations System Branch
Test and Integration System Branch
Research Facilities and Instrumentation Division
Facilities Planning Office
40- by 80-foot Wind Tunnel Project Office
Electro-Systems Engineering Branch
Electronic Instrument Development Branch
Photographic Technology Branch
Research Equipment Engineering Branch
Research Facilities Engineering Branch
Technical Services Division
Planning Office
Electronic Instrument Services Branch
Facilities Services Branch
Metals Fabrication Branch
Model and Instrument Machining Branch
Model Development Branch
Director of Administration
University Affairs Office
Dryden Institutional Support Office
Financial Management Division
Dryden Financial Management Office
Accounting Branch
Pay and Travel Branch
Personnel Division
Dryden Personnel Office
Personnel Management Branch
Records and Reports Branch
Security Branch
Training and Special Programs Branch
Procurement Division
Purchasing Office
Systems Survey and Analysis Office
Contract Mgmt. Branch for Aeronautics
Contract Mgmt. Branch for Astronautics
Contract Mgmt. Branch for Life Sciences
Contract Mgmt. Branch for Research Support
Dryden Procurement Branch
Services and Supply Division
Mail Services Office
Communications Branch
Equipment Branch
Supply Branch
Technical Information Division
Graphics and Exhibits Branch
Library Branch
Publications Branch
Reproduction Services Branch

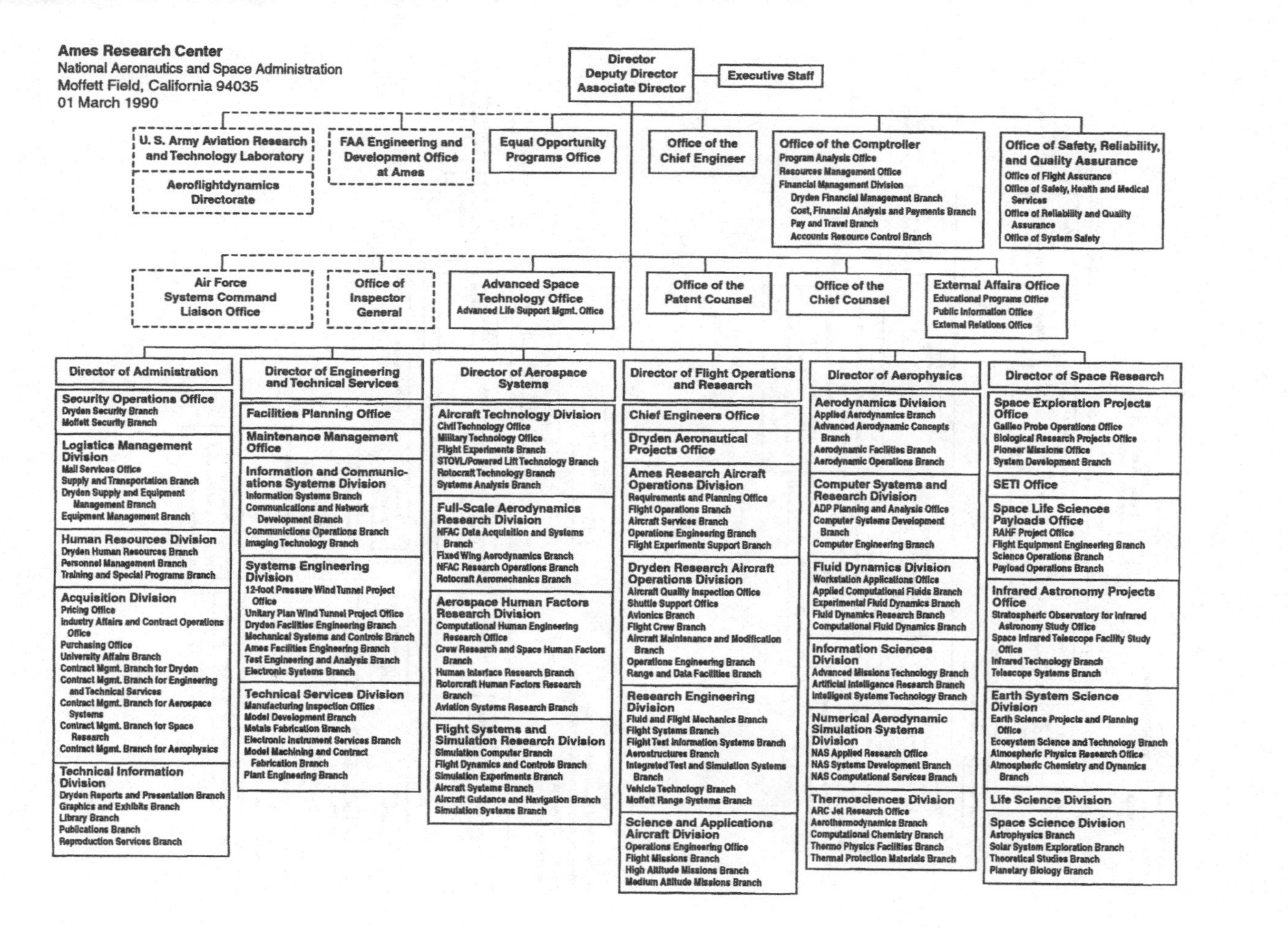
Ames Research Center
National Aeronautics and Space Administration
Moffett Field, California 94035
01 March 1990
Director
Deputy Director
Associate Director
Executive Staff
U. S. Army Aviation Research and Technology Laboratory
Aeroflightdynamics Directorate
FAA Engineering and Development Office at Ames
Equal Opportunity Programs Office
Office of the Chief Engineer
Office of the Comptroller
Program Analysis Office
Resources Management Office
Financial Management Division
Dryden Financial Management Branch
Cost, Financial Analysis and Payments Branch
Pay and Travel Branch
Accounts Resource Control Branch
Office of Safety, Reliability, and Quality Assurance
Office of Flight Assurance
Office of Safety, Health and Medical Services
Office of Reliability and Quality Assurance
Office of System Safety
Air Force Systems Command Liaison Office
Office of Inspector General
Advanced Space Technology Office
Advanced Life Support Mgmt. Office
Office of the Patent Counsel
Office of the Chief Counsel
External Affairs Office
Educational Programs Office
Public Information Office
External Relations Office
Director of Administration
Security Operations Office
Dryden Security Branch
Moffett Security Branch
Logistics Management Division
Mail Services Office
Supply and Transportation Branch
Dryden Supply and Equipment Management Branch
Equipment Management Branch
Human Resources Division
Dryden Human Resources Branch
Personnel Management Branch
Training and Special Programs Branch
Acquisition Division
Pricing Office
Industry Affairs and Contract Operations Office
Purchasing Office
University Affairs Branch
Contract Mgmt. Branch for Dryden
Contract Mgmt. Branch for Engineering and Technical Services
Contract Mgmt. Branch for Aerospace Systems
Contract Mgmt. Branch for Space Research
Contract Mgmt. Branch for Aerophysics
Technical Information Division
Dryden Reports and Presentation Branch
Graphics and Exhibits Branch
Library Branch
Publications Branch
Reproduction Services Branch
Director of Engineering and Technical Services
Facilities Planning Office
Maintenance Management Office
Information and Communications Systems Division
Information Systems Branch
Communications and Network Development Branch
Communictions Operations Branch
Imaging Technology Branch
Systems Engineering Division
12-foot Pressure Wind Tunnel Project Office
Unitary Plan Wind Tunnel Project Office
Dryden Facilities Engineering Branch
Mechanical Systems and Controls Branch
Ames Facilities Engineering Branch
Test Engineering and Analysis Branch
Electronic Systems Branch
Technical Services Division
Manufacturing Inspection Office
Model Development Branch
Metals Fabrication Branch
Electronic Instrument Services Branch
Model Machining and Contract Fabrication Branch
Plant Engineering Branch
Director of Aerospace Systems
Aircraft Technology Division
Civil Technology Office
Military Technology Office
Flight Experiments Branch
STOVL/Powered Lift Technology Branch
Rotocraft Technology Branch
Systems Analysis Branch
Full-Scale Aerodynamics Research Division
NFAC Data Acquisition and Systems Branch
Fixed Wing Aerodynamics Branch
NFAC Research Operations Branch
Rotocraft Aeromechanics Branch
Aerospace Human Factors Research Division
Computational Human Engineering Research Office
Crew Research and Space Human Factors Branch
Human Interface Research Branch
Rotorcraft Human Factors Research Branch
Aviation Systems Research Branch
Flight Systems and Simulation Research Division
Simulation Computer Branch
Flight Dynamics and Controls Branch
Simulation Experiments Branch
Aircraft Systems Branch
Aircraft Guidance and Navigation Branch
Simulation Systems Branch
Director of Flight Operations and Research
Chief Engineers Office
Dryden Aeronautical Projects Office
Ames Research Aircraft Operations Division
Requirements and Planning Office
Flight Operations Branch
Aircraft Services Branch
Operations Engineering Branch
Flight Experiments Support Branch
Dryden Research Aircraft Operations Division
Aircraft Quality Inspection Office
Shuttle Support Office
Avionics Branch
Flight Crew Branch
Aircraft Maintenance and Modification Branch
Operations Engineering Branch
Range and Data Facilities Branch
Research Engineering Division
Fluid and Flight Mechanics Branch
Flight Systems Branch
Flight Test Information Systems Branch
Aerostructures Branch
Integrated Test and Simulation Systems Branch
Vehicle Technology Branch
Moffett Range Systems Branch
Science and Applications Aircraft Division
Operations Engineering Office
Flight Missions Branch
High Altitude Missions Branch
Medium Altitude Missions Branch
Director of Aerophysics
Aerodynamics Division
Applied Aerodynamics Branch
Advanced Aerodynamic Concepts Branch
Aerodynamic Facilities Branch
Aerodynamic Operations Branch
Computer Systems and Research Division
ADP Planning and Analysis Office
Computer Systems Development Branch
Computer Engineering Branch
Fluid Dynamics Division
Workstation Applications Office
Applied Computational Fluids Branch
Experimental Fluid Dynamics Branch
Fluid Dynamics Research Branch
Computational Fluid Dynamics Branch
Information Sciences Division
Advanced Missions Technology Branch
Artificial Intelligence Research Branch
Intelligent Systems Technology Branch
Numerical Aerodynamic Simulation Systems Division
NAS Applied Research Office
NAS Systems Development Branch
NAS Computational Services Branch
Thermosciences Division
ARC Jet Research Office
Aerothermodynamics Branch
Computational Chemistry Branch
Thermo Physics Facilities Branch
Thermal Protection Materials Branch
Director of Space Research
Space Exploration Projects Office
Galileo Probe Operations Office
Biological Research Projects Office
Pioneer Missions Office
System Development Branch
SETI Office
Space Life Sciences Payloads Office
RAHF Project Office
Flight Equipment Engineering Branch
Science Operations Branch
Payload Operations Branch
Infrared Astronomy Projects Office
Stratospheric Observatory for Infrared Astronomy Study Office
Space Infrared Telescope Facility Study Office
Infrared Technology Branch
Telescope Systems Branch
Earth System Science Division
Earth Science Projects and Planning Office
Ecosystem Science and Technology Branch
Atmospheric Physics Research Office
Atmospheric Chemistry and Dynamics Branch
Life Science Division
Space Science Division
Astrophysics Branch
Solar System Exploration Branch
Theoretical Studies Branch
Planetary Biology Branch

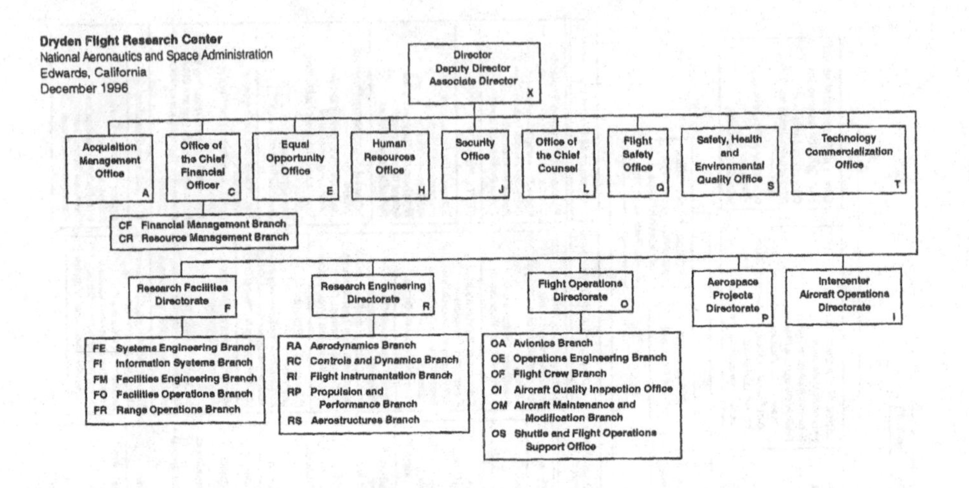
Dryden Flight Research Center
National Aeronautics and Space Administration
Edwards, California
December 1996
Director
Deputy Director
Associate Director X
Acquisition Management Office A
Office of the Chief Financial Officer C
Equal Opportunity Office E
Human Resources Office H
Security Office J
Office of the Chief Counsel L
Flight Safety Office Q
Safety, Health and Environmental Quality Office S
Technology Commercialization Office T
CF Financial Management Branch
CR Resource Management Branch
Research Facilities Directorate F
Research Engineering Directorate R
Flight Operations Directorate O
Aerospace Projects Directorate P
Intercenter Aircraft Operations Directorate I
FE Systems Engineering Branch
FI Information Systems Branch
FM Facilities Engineering Branch
FO Facilities Operations Branch
FR Range Operations Branch
RA Aerodynamics Branch
RC Controls and Dynamics Branch
RI Flight Instrumentation Branch
RP Propulsion and Performance Branch
RS Aerostructures Branch
OA Avionics Branch
OE Operations Engineering Branch
OF Flight Crew Branch
OI Aircraft Quality Inspection Office
OM Aircraft Maintenance and Modification Branch
OS Shuttle and Flight Operations Support Office

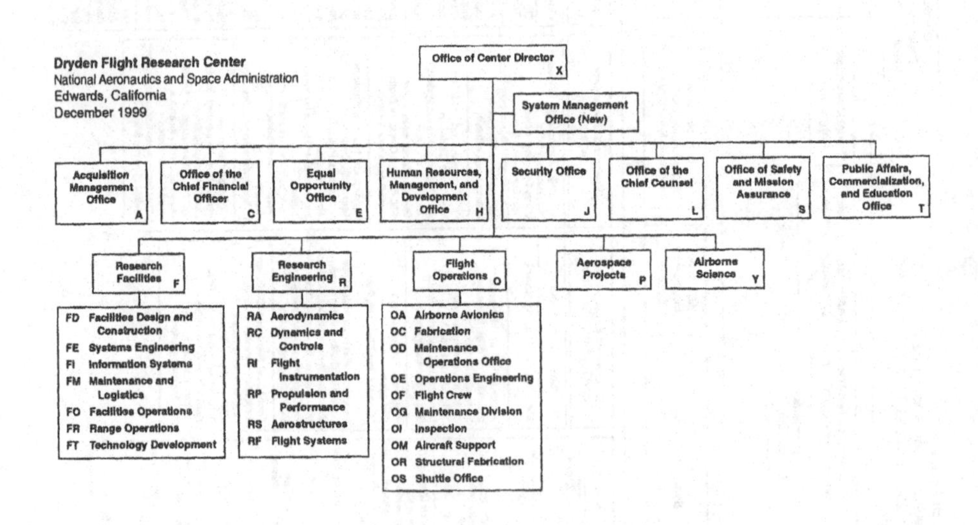
Dryden Flight Research Center
National Aeronautics and Space Administration
Edwards, California
December 1999
Office of Center Director X
System Management Office (New)
Acquisition Management Office A
Office of the Chief Financial Officer C
Equal Opportunity Office E
Human Resources, Management, and Development Office H
Security Office J
Office of the Chief Counsel L
Office of Safety and Mission Assurance S
Public Affairs, Commercialization, and Education Office T
Research Facilities F
Research Engineering R
Flight Operations O
Aerospace Projects P
Airborne Science Y
FD Facilities Design and Construction
FE Systems Engineering
FI Information Systems
FM Maintenance and Logistics
FO Facilities Operations
FR Range Operations
FT Technology Development
RA Aerodynamics
RC Dynamics and Controls
RI Flight Instrumentation
RP Propulsion and Performance
RS Aerostructures
RF Flight Systems
OA Airborne Avionics
OC Fabrication
OD Maintenance Operations Office
OE Operations Engineering
OF Flight Crew
OG Maintenance Division
OI Inspection
OM Aircraft Support
OR Structural Fabrication
OS Shuttle Office

Appendix B

Personnel Summary for FRC/DFRC, other aeronautical Centers, and NASA as a whole as of the end of the fiscal year[a]

Year	FRC/DFRC	AMES RC	Langley RC	Lewis/ Glenn RC	NASA Total
1959	340	1,464	3,624	2,809	9,235
1960	408	1,421	3,203	2,722	10,232
1961	447	1,471	3,338	2,773	17,471
1962	538	1,658	3,894	3,800	23,686
1963	616	2,116	4,220	4,697	29,934
1964	619	2,204	4,330	4,859	32,499
1965	669	2,270	4,371	4,897	34,049
1966	662	2,310	4,485	5,047	35,708
1967	642	2,264	4,405	4,956	35,860
1968	622	2,197	4,219	4,583	34,641
1969	601	2,117	4,087	4,399	33,929
1970	583	2,033	3,970	4,240	32,548
1971	579	1,968	3,830	4,083	30,506
1972	539	1,844	3,592	3,866	28,382
1973	509	1,740	3,389	3,368	26,777
1974	531	1,776	3,504	3,172	26,007
1975	544	1,754	3,472	3,181	25,638
1976	566	1,724	3,407	3,168	25,426
1977	546	1,645	3,207	3,067	24,188
1978	514	1,691	3,167	2,964	23,779
1979	498	1,713	3,125	2,907	23,360

[a] NASA civil servants as of 30 June until 1976, thereafter as of 30 September. The NASA totals include space centers as well as the former NACA laboratories, so the figures do not add up horizontally to the NASA totals.

1980	499	1,713	3,094	2,901	23,470
1981	491	1,652[d]	3,028	2,782	22,736
1982	434[b]	2,041	2,801	2,485	21,620
1983	Not available until 1994.	2,033	2,904	2,632	21,505
1984	"	2,043	2,821	2,624	21,050
1985	"	2,052	2,827	2,715	21,423
1986	"	2,072	2,814	2,598	21,228
1987	"	2,079	2,851	2,663	22,646
1988	"	2,101	2,840	2,649	22,823
1989	"	2,151	2,864	2,749	23,893
1990	"	2,205	2,961	2,728	24,566
1991	"	2,263	2,969	2,835	25,741
1992	"	2,243	2,953	2,799	25,421
1993	"	2,173	2,859	2,731	25,062
1994	434[c]	1,696	2,789	2,457	23,097
1995	428	1,559	2,504	2,258	20,563
1996	445	1,484	2,468	2,200	20,638
1997	474	1,392	2,406	2,060	18,970
1998	504	1,312	2,245	1,933	17,901
1999	593	1,404	2,291	1,909	17,939

SOURCES: NASA Office of Management Operations data; NASA Pocket Statistics, Jan. 1997, p. C-26; the [NASA] Civil Service Workforce, Past, Present, Future, on the Internet at http://www.hq.nasa.gov/office/HR-Education/workforce/.

NOTE: These figures are for civil service employees and do not include contractors. A ready source for the data on contractors from other centers and for NASA as a whole is not available, but in a State-of-the-Center Address on 18 April 2000, Dryden Center Director Kevin Petersen revealed that in FY96, Dryden had 450 contractors and that in FY00 (not, obviously, the end of the fiscal year) there were 550 contractors and 634 civil servants.

---

[b] Ames-Dryden Flight Research Facility; following this year, personnel figures for Dryden were not available until Dryden again became a separate center.
[c] Dryden Flight Research Center once again.
[d] Including Ames-Dryden Flight Research Facility from FY82 through FY93.

Appendix C
HSFS/FRC/DFRC Technical Facilities

Obligations for Facilities Construction at FRC/DFRC

Year	Amount(in millions)
1959	$0
1960	1.8
1961	0
1962	0
1963	1.8
1964	2.5
1965	0
1966	0
1967	0
1968	0
1969	0
1970	0.9
1971	0
1972	0
1973	0
1974	0
1975	0
1976	0
1977	0.8
1978	0.4
1979	0.245
1980	0.325
1981	0.245
1982	0.0
1983	4.7
1984	0.796
1985	0.2
1986	0.48
1987	22.5
1988	0.0
1989	1.1
1990	0.48
1991	4.49
1992	0.65
1993	0.3
1994	0.0
1995	8.08

1996	0.0
1997	8.383
1998	2.7
1999	0.0
2000	2.9

Note: The data in this appendix from 1959 through 1978 appeared in the original version of this history and apparently is given for calendar years. The data for 1979 through 2000 is fiscal year data covering from 1 October of the calendar year before the year listed to 30 September of that calendar year. For example, Fiscal Year 1979 ran from 1 October 1978 to 30 September 1979. The appendix in the original history showed a 0 for 1979 and also for 1980 through 30 September, so the information is somewhat suspect. Incidentally, it does not include the Mate-Demate Device and the Shuttle Hangar built in or around 1975 (and *not*, in any event, funded by the FRC), and it shows only construction of facilities (major) funding, not minor construction. Unfortunately, the records to check or correct the data before 1979 apparently no longer exist.

In any event, it is worth remembering that the FRC/DFRC's greatest research resources have been the experimental aircraft themselves, which are analogous to the wind tunnels, shock tubes, etc., at other centers. The DFRC's need for specialized facilities on the ground for testing and research has thus traditionally been far less than other NACA-NASA centers. In a 1973 OAST research evaluation, DFRC's major facilities listed only 1 ground laboratory (for high-temperature loads calibration) and 10 specialized flight research aircraft, an appropriate example of the importance attached to research aircraft. It should be added that two important facilities available to the FRC/DFRC but not owned by it have been (1) the natural "facility" of Rogers Dry Lake and (2) the taxiways, runways, and control tower of Edwards Air Force Base.

Sources: NASA Office of Management Operations data for the period through 1978; records of the Dryden Facilities Design and Construction Branch (including the NASA Budget Requests to Congress for Fiscal Year 1979 to the present), Fiscal Year 1979 to Fiscal Year 2000

Major HSFS/FRC/DFRC Technical Facilities

(Not including specialized flight research aircraft, for which see Appendix E.)

Facility Name	Bldg No.	First Year	Capitalized Investment 12/31/99 ($M)
Research, Development, and Test Building	4800	1954	21.4
Aircraft Hangar (originally called the Loads Hangar and Then the Calibration Hangar)	4801	1954	8.0
Main Hangar	4802	1954	3.8
Aeronautical Tracking Facility #1	4982	1964	1.7
Communications Building	4824	1964	0.4
Flight Loads Laboratory	4820	1966	4.1
Aircraft Maintenance Hangar (originally, the YF-12 Hangar)	4826	1968	3.7
Structural Fabrication Facility (originally the Sheet Metal Shop and F-104 Dock)	4837	1975	2.8
Shuttle Mate/De-Mate Device	4860	1976	3.6
Shuttle Hangar and Shops	4833	1976	3.1
Integrated Support Facility	4825	1976	1.8
Data Analysis Facility	4838	1985	6.1
Aeronautical Tracking Facility #2	4720	1985	0.7
Post-Flight Science Support Facility	4822	1986	0.9
Consolidated Warehouse	4876	1990	1.2
Research Aircraft Integration Facility (so redesignated in honor of Walt Williams in 1995 but originally called the Integrated Test Facility)	4840	1991	32.2
Audio/Video Support Center	4851	1992	0.6

SOURCE: Records of the Dryden Facilities Design and Construction Branch (including the NASA Budget Requests to Congress for Fiscal Year 1979 to the present)

Thanks to Richard W. Rieder for compiling the data Appendix C for the period after 1978.

Appendix D
Authorized Funding for Research and Program Management at FRC/DFRC, other OART/OAST/OAT Centers, and NASA

(in millions by fiscal year)

Year	FRC/DFRC	Ames RC	Langley RC	Lewis RC	NASA TOTAL[a]
1959	$3.3	$16.3	$31.4	$27.8	$87.6
1960	4.3	17.8	33.0	31.2	118.6
1961	5.1	19.9	39.1	35.8	222.7
1962	7.2	22.9	46.6	45.2	315.6
1963	7.5	25.6	51.8	53.4	438.7
1964	9.4	26.9	52.1	58.5	496.8
1965	10.5	31.8	59.0	69.3	623.3
1966	9.4	33.2	63.5	66.4	611.2
1967	9.5	33.8	64.3	66.3	646.6
1968	9.5	33.8	62.2	66.2	639.3
1969	9.7	34.0	63.0	67.9	648.0
1970	10.3	37.6	69.8	73.9	702.2
1971	11.1	40.6	75.3	78.0	730.2
1972	11.7	42.2	80.2	82.5	732.3
1973	11.7	42.4	78.6	81.2	721.8
1974	12.2	46.4	83.3	79.6	744.0
1975	13.2	48.6	88.6	80.3	764.7
1976	14.5/19.7[b]	50.9/63.9	93.1/115.7	80.7/102.4	792.3/1,012.5
1977	17.2	53.1	94.7	83.3	844.4

[a] The NASA totals include space centers as well as the former NACA laboratories shown here, so that the figures do not add up horizontally to the NASA totals.

[b] Second figure in this row indicates the fiscal year including the transition quarter to a redefined FY ending Sept. 30 instead of June 30, as previous fiscal years had done.

1978	18.2	57.8	100.7	84.7	889.5
1979	18.9	62.8	106.6	87.5	933.8
1980	20.2	67.4	113.8	94.8	996.0
1981	22.6	72.2	120.8	99.9	1,071.1
1982	24.4	76.6	126.6	106.4	1,183.1
1983	--	107.2[c]	132.7	118.8	1,197.4
1994		165.0	184.0	181.8	1,673.5

SOURCES: NASA Office of Management Operations data; NASA Pocket Statistics, Jan. 1994, p. C-26; 1995, p. C-25. Research and Program Management Funding is not shown in the 1996 or 1997 Pocket Statistics.

[c] From 1981 until 1994, the Ames-Dryden Flight Research Facility was part of Ames Research Center, and separate Research and Program Management Funding is not readily available for DFRF.

Appendix E

Muroc Flight Test Unit-DFRC Research Aircraft, 1947-1999
Research Aircraft operated by the NACA and NASA

Piloted Experimental Research Aircraft

AIRCRAFT	SERIAL NUMBER	NACA/NASA CODE	RECEIVED DATE[1]	TRANSFER DATE	REMARKS
*Bell XS-1 #1	46-062	—	Oct. 7, 1946	May 1950	Not flown by, but supported by NACA. In Smithsonian National Air and Space Museum. (59 AF flights).
Bell XS-1 #2	46-063	—	21 Sept. 1947	—	Last flight 23 Oct. 1951. (54 NACA flights) Converted to X-1E in 1955.
Bell XS-1 #3	46-064	—	June 1951 glide flight.	—	Destroyed 09 Nov. 1951 in a ground explosion. Called "Queenie" (1 glide flight).
Bell X-1A	48-1384	—	23 Feb. 1955	—	Destroyed in crash 08 Aug. 1955. (1 NACA flight.)
Bell X-1B	48-1385	—	01 Aug. 1956	26 Jan. 1959	To USAF Museum, WPAFB. (17 NACA flights)
*Bell X-1D	48-1386	—	July 1951	—	Jettisoned following onboard explosion on 22 Aug. 1951.
Bell X-1E	46-063	—	03 Dec. 1955	—	Last Flight 06 Nov. 1958. Mounted as display at NASA DFRC 29 Apr. 1960. (26 NACA flights.)
*Bell X-2 #1	46-674	—	July 1954	—	Crashed 27 Sept. 1956.
Douglas X-3	49-2892	—	8 Dec. 1953	28 Sept. 1956	To USAF Museum, WPAFB. (27 NACA flights)
Northrop X-4 #1	46-676	—	08 May 1950	22 Mar. 1954	Acquired for spare parts. To the Air Force Academy and then Edwards Air Force Base Museum.
Northrop X-4 #2	46-677	—	08 May 1950	22 Mar. 1954	To USAF Museum, WPAFB. (79 NACA flights.)
Bell X-5 #1	50-1838	—	23 Nov. 1951	05 Apr. 1957	To USAF Museum, WPAFB. (133 FRC flights).
*Bell X-5 #2	50-1839	—	—	—	Fatal crash 13 Oct. 1953.
Convair XF-92A	46-682	—	18 May 1951	12 Mar. 1954	25 NACA flights.
Douglas D-558-1 #1	37970	NACA 140	11 Apr. 1949	—	Acquired for spare parts.
Douglas D-558-1 #2	37971	NACA 141	23 Oct. 1947	—	Destroyed in crash, 03 May 1948. 19 NACA flights.

[1] Dates refer only to period actually operated by the NACA or NASA at the Test Unit, Station, or Center.

* Supported, instrumented, and sometimes maintained by NACA or NASA personnel.

Douglas D-558-1 #3	37972	NACA 142	24 Jan. 1949	04 Oct. 1954	78 NACA flights.
Douglas D-558-2 #1	37973	NACA 143	31 Aug. 1951	07 Apr. 1958	Converted to pure rocket, 29 June 1954. To Los Alamitos (1 NACA flight)
Douglas D-558-2 #2	37974	NACA 144	31 Aug. 1951	07 Apr. 1958	Converted to pure rocket. (75 NACA flights)
Douglas D-558-2 #3	37975	NACA 145	15 Dec. 1950	21 May 1958	To Antelope Valley College. (66 NACA flights)
North American X-15 #1	56-6670	—	03 Feb. 1960[2]	May 1969	To National Air and Space Museum. Last flight 24 Oct. 1968. (81 flights.)
North American X-15 #2	56-6671	—	08 Feb. 1961	09 Nov. 1963	Damaged in landing on 9 Nov. 1959.
North American X-15A-2	56-6671	—	18 Feb. 1964	Oct. 1969	To USAF Museum, WPAFB. Returned to FRC from NAA in June 1968. (50 flights in the A-2 configuration.)
North American X-15 #3	56-6672	—	04 Oct. 1961	—	Fatal crash, 15 Nov. 1967. (65 flights.)
Northrop X-21	55-0410	—	17 Oct. 1966	18 Jan. 1967	No NASA flights; program cancelled. On Edwards AFB photo range as of 2000.
Grumman X-29-1	82-0003	—	10 Oct. 1984	04 Nov. 1994	Last research flight 08 Dec. 1988. Displayed at USAF Museum, WPAFB.
Grumman X-29-2	82-0049	—	05 Nov. 1988	—	Static display-NASA DFRC. Last flight 1992.
Rockwell X-31-1	164584	—	10 Feb. 1992	—	Crashed 19 Jan. 1995.
Rockwell X-31-2	164585	—	10 Feb. 1992	23 Feb. 1999	Moved to Boeing hangar, Palmdale, and then to Patuxent River, MD.
NASA FRC Paresev I	N9765C	—	20 Jan. 1962	—	Large wing, 150 sq. ft. area. Damaged in rollover. Rebuilt as Paresev 1A.
NASA FRC Paresev 1A	N9765C	—	08 May 1962	—	Large wing, 150 sq. ft. area. Converted to 1B.
NASA FRC Paresev 1B	N9765C	—	27 July 1962	—	Small wing, 100 sq. ft. area. Converted to 1C.
NASA FRC Paresev 1C	N9765C	—	04 Mar. 1963	14 Apr. 1964	Wing 179 sq. ft. area. Inflatable wing spars. Held as of 2001 by the Smithsonian National Air & Space Museum at the Garber Facility.
NASA FRC M2-F1	N86652	—	22 Jan. 1963	—	Lightweight. Last flight, 16 Aug. 1966. Static display at NASA DFRC.

[2] This and the three following dates in this column are the dates North American Aviation deeded the aircraft to NASA. The #1 X-15 arrived at the North American (NAA) hangar at Edwards approximately 17 Oct. 1958 and was delivered to NASA about 3 Feb. 1960. The #2 X-15 arrived at the NAA hangar in late Oct. of 1958. It was delivered to NASA about 8 Feb. 1961. The # 3 airplane arrived at the NAA hangar about 29 June 1959 and was delivered to NASA about 29 Sept. 1961 after blowing up on the test stand 8 June 1960 and being sent back to NAA in Los Angeles for repair.

Northrop M2-F2	N803NA	NASA 803	16 June 1965	—	Landing accident 10 May 1967. Rebuilt as M2-F3.
Northrop M2-F3	N803NA	NASA 803	July 1969	—	Permanent display at National Air & Space Museum. Last flight 20 Dec. 1972.
Northrop HL-10	N804NA	NASA 804	31 Mar. 1966	—	Last flight, 17 July 1970. Static display at NASA DFRC.
Martin X-24A	66-13551	—	14 Sept. 1967	—	Last flight, 04 June 1971. Rebuilt as X-24B.
Martin X-24B	66-13551	—	24 Oct. 1972	25 May 1976	Last flight, 26 Nov. 1975. To USAF Museum, WPAFB.
Bell LLRV #1	—	—	14 Apr. 1964	Dec. 1966	Last flight, 12 Dec. 1966. Sent to JSC, crashed in 1968.
Bell LLRV #2	—	—	May 1964	—	Last flight, 17 Jan. 1967. Sent to JSC, returned to FRC. On static display at NASA DFRC.
Ames Industrial Corporation AD-1	001	NASA 805	12 Mar. 1979	27 Aug. 1982	Skewed wing research. On loan to Langley.

Military Models Flown on Experimental Research Flights

AIRCRAFT	SERIAL NUMBER	NACA/NASA CODE	RECEIVED DATE	TRANSFER DATE	REMARKS
North American A-5A	147858	—	19 Dec. 1962	20 Dec. 1963	Used in SST landing approach study. Returned to the Navy. Total of 48 flights.
McDonnell F-4A	145313	—	03 Dec. 1965	—	Damaged by in-flight explosion, 25 July 1967.
McDonnell F-4C	63-7424	—	05 Oct. 1983	29 Aug. 1985	Span wise blowing study. Returned to USAF.
Douglas F5D-1	139208	NASA 708	16 Jan 1961	04 Mar. 1963	SST studies. Transferred to Ames. Loaned to Victor Valley College 12-19-75.
McDonnell F5D-1	142350	NASA 802	15 June 1961	19 May 1970	To WPAFB, Dayton OH. Donated to the Neil A. Armstrong Museum in Wapakoneta, OH.
North American ETF-51D	44-84958	NACA 148	05 Sept. 1950	09 July 1959	Dives to 0.8 Mach; also used for proficiency flights. Damaged in ground mishap with T-37A 15 Apr. 1959.
Republic YF-84A	45-59488	—	18 Dec. 1950	21 Apr. 1954	Used primarily for proficiency.
Republic YF-84A	45-59490	NACA 134	28 Nov. 1949	21 Apr. 1954	Vortex generator research.
Republic YRF-84F	51-1828	NACA 154	27 Apr. 1954	29 Oct. 1956	Pitch-up research. Swept wing A/C.
North American F-86A	48-291	—	1951	---	To investigate maneuver characteristics and buffet characteristics. On loan from Ames.
North American F-86F	52-5426	—	23 June 1954	10 Sept. 1954	Pitch-up research. Chase for the D-558-2.
North American F-86D	50-577	—	Aug. 1958	Dec. 1958	Handling Quality data. Chase for X-1-1. On loan from Ames.
North American F-86E	50-606	—	29 Dec. 1958	30 Dec. 1958	Handling quality data. On loan from Ames. Two flights 29 Dec., 30 Dec. 1958.

*Northrop YF-89D	49-2463	—	30 Oct. 1951	---	Wing instrumented. Fatal crash on 20 Oct. 1953.
*Republic XF-91	46-681	—	28 Aug. 1952	---	Handling qualities reports.
North American JF-100A	52-5778	—	31 July 1954	10 June 1960	Define directional-stability problems and roll-coupling boundaries. To Davis-Monthan AFB for storage.
North American F-100C	53-1712	—	18 Sept. 1956	26 Mar. 1957	Behavior of the pitching motion damper.
North American F-100C	53-1717	—	27 Sept. 1957	23 Feb. 1961	Chase and pilot proficiency.
North American JF-100C	53-1709	—	02 Nov. 1960	03 Feb. 1964	Variable stability studies. Airborne simulation studies in support of X-15 & SST.
McDonnell F-101A	53-2432	—	10 May 1956	22 Aug. 1956	Research on inlet-flow distortion studies; pilot familiarization; total 20 flights.
Convair YF-102	53-1785	—	20 Sept. 1954	27 June 1958	104 research flights; to EAFB salvage.
Convair JF-102A	54-1374	—	03 Apr. 1956	04 May 1959	48 research flights. To Davis-Monthan AFB.
Lockheed YF-104A	55-2961	NASA 818	23 Aug. 1956	18 Nov. 1975	Aero. data, devel. of reaction control attitude thruster system for X-15. Redesignated JF-104A, to National Air & Space Museum for permanent display. 1,444 total flights.
Lockheed F-104A	56-0734	—	07 Oct. 1957	10 Jan. 1961	Returned to USAF for conversion to QF-104A drone. 164 total flights
Lockheed JF-104A	56-0749	—	13 Apr. 1959	—	Configured with a center-line launcher system. Crashed 20 Dec. 1962. 249 total flights.
Lockheed F-104B	57-1303	NASA 819	16 Dec. 1959	16 June 1983	Carried bio-medical experiments. 1,731 total flights.
Lockheed F-104N	N811NA	NASA 011/811	19 Aug. 1963	Oct. 1990	Pilot proficiency, low L/D landings. 4,370 total flights. Sent to Embry Riddle Aeronautical University, Prescott, AZ.
Lockheed F-104N	N812NA	NASA 012/812	30 Sept. 1963	Jan. 1987	Pilot proficiency, low L/D landings. 4,442 total flights. Static display Lockheed, Palmdale, CA.
Lockheed F-104N	N813NA	NASA 013/813	22 Oct. 1963	08 June 1966	Pilot proficiency, low L/D landings. Mid-air collision 08 Jun. 1966. 409 total flights.
Lockheed F-104A/G	56-0790	NASA 820	27 Dec. 1966	01 June 1977	Replacement for 013/813. Support; investigated wake vortices of jumbo jets. 1,022 total flights. EAFB Museum display.
Lockheed TF-104G	61-3065	NASA 824	27 June 1975	09 Sept. 1995	HiMAT control ship, pilot proficiency, support aircraft. 1,127 total flights. Sent to Estrella Warbird Museum, Paso Robles, CA

Lockheed TF-104G	66-13628	NASA 825	25 June 1975	07 Sept. 1995	"0" g experiment for Shuttle. 1,890 total flights. Sent to Moffet Field static display
Lockheed F-104G	8213	NASA 826	27 June 1975	03 Feb. 1994 last flight	Thermal Protection System (TPS) shuttle tile tests. On display at NASA DFRC. 1,415 total flights.
Republic F-105B	54-0102	—	01 Oct. 1959	07 Jan. 1960	Pilot familiarization, 8 flights. Last flight 28 Dec. 1959.
*Convair NF-106B	57-2516	NASA 816	27 Nov. 1978	29 Jan. 1979	Installation of wing leading edge vortex flaps. To Langley. All shop work done at NASA FRC.
*Convair QF-106A	60-0010	—	17 June 1997	30 Apr. 1998	Define wake turbulence. To Davis-Monthan AFB.
*Convair QF-106A (EXD 01)	60-0130	—	08 Oct. 1997	01 May 1998	Modified EXD-01. Eclipse program, 6 tows behind C-141A 61-2775. To Davis-Monthan AFB.
North American F-107A	55-5118	—	06 Nov. 1957	—	Grounded for spares on 25 Nov. 1957.
North American JF-107A	55-5120	—	10 Feb. 1958	01 Sept. 1959	Evaluated sidestick flight control system. Last flight, fire in forward wheel well. 1 Sept. 1959.
Gen. Dynamics F-111A	63-9771	—	20 Jan. 1967	18 Dec. 1969	Returned to AF.
Gen. Dynamics F-111A	63-9777	—	17 Mar. 1969	28 July 1971	Sent to Davis Monthan AFB for storage.
Gen. Dynamics NF-111A TACT	63-9778	NASA 831	18 Feb. 1972	—	Transonic Aircraft Technology and AFTI/F-111 mission adaptive wing. Last flight, 22 Dec. 1988.
Gen. Dynamics F-111E IPCS	67-0115	—	24 Sept. 1974	18 June 1976	Ferry to Sacramento to return to AF. Retired.
Vought F-8A	145385	NASA 816	11 Jan. 1969	03 Apr. 1970	Placed in storage 03 Apr. 1970, spare parts.
Vought TF-8A	141353	NASA 810	25 May 1969	—	Last flight, 23 May 1973. Supercritical wing. On static display at NASA DFRC.
Vought YTF-8A	143710	—	1977	—	Spare parts.
Vought F-8C	145546	NASA 802	25 Sept. 1969	—	DFBW program. Last flight 16 Dec. 1985. On display at NASA-DFRC.
Lockheed YF-12A	60-6935	NASA 832	01 Feb. 1970	07 Nov. 1979	To USAF Museum, WPAFB.
*Lockheed YF-12A	60-6936	—	Mar. 1970	—	Destroyed (Air Force flight), 24 June 1971.
Lockheed "YF-12C"	61-7951	NASA 833	16 July 1971	17 Oct. 1978	Actually an SR-71A. Given a A-12 tail number 06937. To Lockheed Palmdale, CA, and then to the Pima Air Museum, AZ.
Grumman F-14A	157991	NASA 991	09 Aug. 1979	06 Sept. 1985	ARI, asymmetric thrust, angle of attack. Spin control and recovery tests. Returned to Patuxent River.

Grumman F-14A	158613	NASA 834	08 Apr. 1984	20 July 1987	Laminar flow, gloved wing. Variable Sweep Transition Flight Experiment (VSTFE). Returned to Navy.
McDonnell Douglas F-15A	71-0281	NASA 834	08 Jan. 1976	12 Aug.1983	Propulsion, buffet, base drag, and Shuttle tile studies. Last flight 07 Jan.1981. On display at NASA Langley.
McDonnell Douglas F-15A	71-0287	NASA 835	24 Feb. 1976	---	DEEC, Laminar HiDEC flow, 10-deg. cone. Last flight 27 Oct. 1993. In storage at NASA DFRC.
McDonnell Douglas F-15B	74-0141	NASA 836	24 Jan. 1993	On-going[3]	Flown with flight test fixture for various experiments.
McDonnell Douglas NF-15B	71-0290	NASA 837	11 June 1993	On-going	Called the ACTIVE through 1999.
Gen. Dynamics F-16A	75-0746	—	04 Nov. 1983	---	Decoupler pylon tests.
Gen. Dynamics NF-16A	75-0750	—	15 July 1982	04 Nov. 1997	AMAS, AFTI. To General Dynamics, Fort Worth, TX.
Gen. Dynamics F-16A	82-0976	NASA 816	23 Jan. 1995	—	In storage at NASA DFRC.
Gen. Dynamics F-16XL-1	75-0749	NASA 849	10 Mar. 1989	—	Laminar flow, CAWAP. In storage at NASA DFRC.
Gen. Dynamics F-16XL-2	75-0747	NASA 848	13 Feb. 1991	—	Supersonic laminar flow control studies. In storage at NASA DFRC. Last flight of program April 1996.
*Northrop YF-17	70-1569	—	27 May 1976	14 July 1976	To test base drag and maneuverability at transonic speeds. Returned to Navy, then to Northrop.
McDonnell Douglas F-18A	160775	—	3 Jan. 1985	03 Nov. 1986	Flying Qualities, flutter, spare parts. Airlifted to China Lake and returned to Navy. On display in U.S. Navy Museum, China Lake, CA.
McDonnell Douglas F-18A	160780	NASA 840	22 Oct. 1984	—	High angle-of-attack and spin tests. HARV Storage at DFRC. Last flt. 17 Nov. 1998.
McDonnell Douglas F-18A	160785	—	07 Nov. 1985	03 Nov. 1986	Engine maintenance, used for spare parts. Airlifted to China Lake and returned to Navy.
McDonnell Douglas F-18A	161214	NASA 842	24 Aug. 1987	03 Mar. 1997	On display in front of Jethawks' stadium in Lancaster, CA. Last flight 09 Feb. 1994.
McDonnell Douglas F-18A	161216	NASA 841	01 Oct. 1985	28 June 1993	Support, retired to NAWC, China Lake, CA.
McDonnell Douglas F-18A	161213	NASA 844	23 July 1987	—	Destroyed in crash, 07 Oct. 1988.
McDonnell Douglas F-18A	161250	NASA 843	06 Nov. 1987	17 June 1991	Returned to the Navy.
McDonnell Douglas F-18A	161251	—	13 Dec. 1985	On-going.	Arrived by truck. Used as "Iron Bird" simulator.
McDonnell Douglas F-18A	161520	NASA 847	21 Sept. 1989	On-going.	

[3] As of the beginning of 2001.

McDonnell Douglas F-18A	161949	NASA 848	28 Dec. 1989	06 Nov. 1990	Returned to NAS Cecil Field, Florida Transferred to VFA-203.
McDonnell Douglas F-18A	161519	NASA 843	07 Feb. 1991	On-going	Chase/support. Second F-18 designated NASA 843 (previously 817).
McDonnell Douglas F-18A	161703	NASA 850	24 Feb.1993	On-going	Chase/support.
McDonnell Douglas F-18A	161705	NASA 851	25 Feb. 1993	On-going	Chase/support.
McDonnell Douglas F-18A	161744	NASA 853	04 Mar. 1999	On-going	Advanced Aeroelastic Wing (AAW) project.
McDonnell Douglas TF-18	161355	NASA 846	01 Mar. 1991	On-going	Chase/support. Later redesignated an F-18B.
McDonnell Douglas TF-18	161217	NASA 852	07 Sept. 1994	On-going	Chase/support. Later redesignated an F-18B.
McDonnell Douglas TF-18	160781	NASA 845	24 July 1986	On-going	Systems Research Aircraft (SRA). Later redesignated an F-18B.
Northrop/McDonnell Douglas *YF-23	87-0800	—	01 Dec. 1993	02 May 1995	Never flew for NASA. Heat facility planned to use for strain gage studies. To AFFTC Museum. Departed to USAF Museum, 31 March 2000.
Northrop/McDonnell Douglas *YF-23	87-0801	—	01 Dec. 1993	Aug. 1995	Used for spare parts. Transferred to Western Museum of Flight, Hawthorne, CA.
Lockheed SR-71A	61-7980	NASA 844	15 Feb. 1990	—	In flyable storage at NASA DFRC. Last flight 27 Sept. 1999 (as of April 2001).
Lockheed SR-71A	61-7971	NASA 832	19 Mar. 1990 14 Oct. 1999	12 Mar. 1995 —	Returned to the AF and then back to DFRC. In flyable storage at NASA DFRC. Second aircraft numbered 832.
Lockheed SR-71A	61-7967	—	14 Oct. 1999	—	In flyable storage at NASA DFRC.
Lockheed SR-71B	61-7956	NASA 831	25 Jul. 1991	—	In flyable storage at NASA DFRC.

Miscellaneous Models Flown on Experimental Research Flights

AIRCRAFT	SERIAL NUMBER	NACA/NASA CODE	RECEIVED DATE	TRANSFER DATE	REMARKS
Beechcraft UC-45	44-47110	NACA 105	May 1947	—	Transferred from Langley.
Beechcraft Beech 99 PD-280	N1031S	—	Dec. 1974	—	
Beechcraft T-34C	160945	N819NA	June 1996	On-going	Support of DarkStar research program and others.
Beechcraft Beech 200 Super KingAir	N7NA	NASA 7	Sept. 1996	On-going.	Flown for many years as a "shuttle" support aircraft from JPL before transferring to Dryden.
Beechcraft Beech 200 Super King Air	N701NA	N801NA	03 Oct. 1997	On-going	Research flights.
Lockheed T-33A	49-0939	—	26 Nov. 1957	20 Sept. 1961	To salvage at Davis-Monthan AFB, AZ.
Lockheed T-33A	51-4299	—	25 Apr. 1961	31 May 1961	To salvage at Davis-Monthan AFB, AZ.
Lockheed T-33A	51-6692		2 June 1961	26 December 1962	Crashed at Norton AFB.
Lockheed T-33A	55-4351	NASA 815	09 Jan. 1963	10 Sept. 1973	To Redding, CA. First Flight 08 Mar. 1963.
*Lockheed NT-33A[4]	51-4120	---	Nov. 1981	Dec. 1981	Control laws, AFTI/F-16 calibration. Calspan aircraft often flown by a non-NASA pilot. Had earlier flown at Edwards to simulate the low-lift-over-drag reentry characteristics of the X-15, in the spring and summer of 1965 to simulate the lifting bodies, and again in 1979 and 1980. Returned in March 1982.
Lockheed YO-3A	69-18010	NASA 818	27 June 1997	—	Engine noise & exhaust studies. Helicopter acoustic studies. 1978 doing acoustic tests at Edwards for Ames. In flyable storage at NASA DFRC.
Lockheed ER-2	80-1063	N806NA	03 Nov. 1997	On-going	High-altitude science experiments.
Lockheed ER-2	80-1098	N809NA	06 Nov. 1997	On-going	High-altitude science experiments.
Cessna T-37A	54-2737	—			Ames aircraft. Damaged in mishap with ETF-51D 15 Apr. 1959.
Cessna T-37B	56-3480		22 Mar. 1962	08 Feb. 1965	Transferred to EAFB.
Cessna T-37B	60-0084	NASA 807	24 June 1974	08 Nov. 1982	First Flight 23 Aug. 1974. Vortex probe, RPRV. Fatal crash 08 Nov. 1982.
Cessna L-19A	50-1675	—	Feb. 1962	—	

[4] The ordering of these miscellaneous aircraft is problematical. Generally, the ordering is chronological, with all airplanes by a given manufacturer (and of a given type) kept together regardless of date. However, it seemed logical to keep all the aircraft leased for General Aviation studies together, regardless of manufacturer, and it also seemed logical to list the two XV-15s by number even though the #2 aircraft flew at the center before the #1 XV-15 did.

Cessna TO-1A (L-19A)	51-12220	—	29 Nov. 1962	—	To evaluate pilot visibility using monocular telescopes.
Cessna TO-1A (L-19G)	144128	—	14 Nov. 1963	24 Apr. 1964	
Cessna U-3A	57-5921	—	13 Oct. 1969	—	
*Cessna U206C Stationair					
*Cessna U206C Stationair	N3927G	—	Oct. 1991	Dec. 1996	Mothership for the Spacewedge and Wedge.
*Cessna Caravan 208B	N1030N	—	09 July 1996	1997	Used for remotely operated sensor systems (ROSS)/tether observation system (TOS) studies.
*Boeing Stearman	N69056	—	May 1962	—	On loan, towed the Paresev.
Rockwell Model 680F Aero Commander	N6285X	—	07 Feb. 1963	—	On loan until NASA FRC received the one ordered. Replaced with N6297X.
Rockwell Model 680F Aero Commander	N6297X	N801NA	14 May 1963	14 Mar. 1979	To Customs Air Branch, San Diego, CA.
Beechcraft Beech C33 Debonair	N430T	—	24 Aug. 1964	12 Jan. 1965	Leased for General Aviation studies.
Beechcraft Beech S35 Bonanza	N5849K	—	12 Nov. 1965	06 Dec. 1965	Leased for General Aviation studies.
Piper PA-23 Apache	N4383P	—	01 June 1965	30 Sept. 1965	Leased for General Aviation studies.
Piper	N7845Y	—	Feb. 1966	May 1966	Leased for General Aviation studies.
Cessna 210	N910V	—	04 Mar. 1965	13 July 1965	Leased for General Aviation studies.
Cessna 310	8199M16-1	—	05 Dec. 1964	29 Mar. 1965	Leased for General Aviation studies.
Wing D-1 Derringer	N7597V	—	Apr. 1968	1968	Leased for General Aviation studies.
Wing D-1 Derringer	N644W	—	Apr. 1968	1968	Leased for General Aviation studies.
American Aviation AA-1 Yankee	N5646L	—	23 Apr. 1969	26 May 1969	Leased for General Aviation studies.
Hawker Siddeley XV-6A/P.1127	64-18264/ XS690	—	Aug. 1966	—	On display at U.S. Army Aviation Museum, Ft. Rucker, AL.
Piper PA-30-160B Twin Commanche	N8351Y	N808NA	05 June 1967	12 May 1995	Used for RPRV tests then as general mission support aircraft. Delivered to Kings River College, Fresno, CA.
Northrop T-38A	65-10353	N821NA	28 Sept. 1972	12 Oct. 1994	Support. First Flight 5 Oct. 1972. Transferred to Johnson Space Center.
Northrop YA-9A	71-1367	NASA 823	05 Apr. 1974	20 Sept. 1980	Evaluating lift damper and unique speedbrakes for alleviating trailing wake vortices. Transferred to Ames, then Castle AFB, then Museum EAFB.
Northrop YA-9A	71-1368	—	10 Apr. 1974	20 Sept. 1980	Transferred to Ames, then March AFB.
*Northrop AT-38B	62-3715	—	17 June 1991	—	Never flew. Used for spare parts. Returned to Air Force.
*Northrop AT-38B	62-3742	—			Never flew. Used for spare parts. Returned to Air Force.

Northrop T-38A	65-10357	N822NA	03 Oct. 1985	08 Nov. 1990	Transferred from ARC to DFRC. Used for support. Returned to Ames.
*Fairchild Republic YA-10		—	09 Oct. 1974	11 Oct. 1974	Loads Lab. For spin chute tests. Returned to the Air Force.
Rockwell International Orbiter *Enterprise*	OV-101	---	Jan. 1977	1982	Last free flight 26 Oct. 1977. Never flew in space. Approach and landing simulations. Eventually transferred to the National Air and Space Museum.
Bell XV-15 #1	N702NA	—	06 Mar. 1981	May 1981	Tilt-rotor research. Departed to Ames, crashed 20 Aug. 1992.
Bell XV-15 #2	N703NA	—	13 Aug. 1980	30 Oct. 1980	Tilt-rotor research. Returned to Ames. In 1994 assigned back to Bell.
Mitsubishi MU-2B	N253MA	—	21 May 1982	09 Sept. 1982	To Mojave Airport.
Gates LearJet 25	N616NA	NASA 616	23 June 1982	—	MSBLS system verification. Lewis aircraft.
Gates LearJet 24B	N705NA	N805NA	13 Feb. 1998	On-going	Icing research aircraft at Langley. Transferred to Dryden 21 Nov. 1997.
Sikorski S.72 RSRA	72-001	N740NA	20 Dec. 1983	03 Oct. 1984	Returned to Ames. Rotor System Research Aircraft.
Sikorski S.72 RSRA/ X-Wing	72-002	N741NA	25 Sept.1986	09 Oct. 1991	For assembly and tests.
*McDonnell Douglas YAV-8B	158394	—	11 Apr. 1984 to ARC	—	Never flew here. Simulator work only. Engine runs at Edwards AFB.
*Scaled Composites Model 281 Proteus	N281PR	—			Flown 1999 at Mojave Airport. Carries two pilots. NASA DFRC Life Support Group supported flights with pressure suits.

Bombers, Motherships, Transports, Tankers

AIRCRAFT	SERIAL NUMBER	NACA/NASA CODE	ARRIVED DATE	DEPARTED DATE	REMARKS
Boeing JTB-29A	45-21800	—	7 Oct. 1946	02 July 1959	Operated by the Air Force until 1955. Mothership for X-1. Departed for Davis-Monthan AFB, AZ.
Boeing P2B-1S	84029	NACA 137	Aug. 1951	05 Aug. 1959	Arrived at Edwards AFB 10 Aug. 1950. D-558-2 Mothership. "Fertile Myrtle." Departed for Davis-Monthan AFB, AZ.

Boeing JB-47A	49-1900	NACA 150	22 July 1952	28 Feb. 1958	Loads, handling-qualities, dynamic stability, landing studies. Aeroelasticity studies at Ames in 1957. Departed for Davis-Monthan AFB, AZ.
Boeing JC-47D	43-48273	—	26 June 1960	29 Aug. 1960	To Davis-Monthan AFB.
Douglas R4D-6	50831	—	07 July 1952	05 Jan. 1956	
Douglas R4D-5	17136	NASA 817	05 Jan. 1956	22 Aug. 1979	Redesignated C-47H in 1962. Towed the M2-F1. Given to Mississippi State Police.
*Boeing KC-135A	55-3124	—	06 Mar. 1958	06 June 1958	On 90 day loan from the Air Force. Returned to AFFTC 1997 for runway calibrations.
Boeing NKC-135A	55-3129	NASA 837	09 Nov. 1977	22 May 1981	Winglet study. To Wright-Patterson AFB, OH.
Boeing EB-50A	46-006	—	---	09 Nov. 1951	Mothership. Destroyed in explosion of X-1-3 (46-064) during defueling.
Boeing EB-50A	46-011	—	---	12 May 1953	Mothership. Damaged in in-flight explosion X-2-2.
*Boeing NB-52A	52-0003	—	---		Arrived at Edwards AFB 14 Nov. 1958. Mothership for X-15 and lifting bodies. Retired by the AF in 1969. On display at Pima Air Museum, Tucson, AZ.
Boeing NB-52B	52-0008	—	Loaned to NASA 26 Apr. 1976	On-going	Arrived at Edwards AFB 08 June 1959. Mothership for HiMAT, DAST, X-15, lifting bodies, X-38, Pegasus.
Boeing B-52G	59-2586	—	19 Sept. 1990	22 Apr. 1994	Never flown by Dryden. Intended as a replacement for the NB-52B. Went to Davis-Monthan AFB.
*Douglas B-26B	N9417H	—	29 Jan. 1963	15 Feb. 1963	Variable stability aircraft flown by Cornell Aeronautical Laboratory.
Lockheed C-140A JetStar	N814NA`	NASA 814	08 May 1963	—	GPAS, laminar flow studies, prop fans, aircraft flying qualities, MSBLS testing. Last flight 25 Oct. 1987. Storage.
Martin B-57B	52-1576	NASA 809	14 Oct. 1969	06 May 1987	Atmospheric conditions; clear-air turbulence. Not continuously at the FRC/Dryden. Retired from flight status and transferred to AFFTC Museum.
*Martin WB-57F	63-13501	N925NA	01 May 1973	—	Returned to NASA Johnson Space Center.
*North American XB-70 #2	62-0207	—	—	—	Arrived Edwards AFB 17 Jul 1965. Destroyed in mid-air collision with F-104 (813) 08 June 1966.
North American XB-70 #1	62-0001	—	28 Mar. 1967	04 Feb. 1969	Arrived at Edwards AFB 21 Sept. 1964. 23 NASA flights. Transferred to the USAF Museum, WPAFB.
*Convair CV-990	–	NASA 711	28 Aug. 1968	—	ARC aircraft; flown 28 times at the FRC by Dryden and Ames pilots; last flight 19 Dec. 1968.

Convair CV-990	N5617NA	NASA 810	06 Mar. 1989	20 Sept. 1996	Shuttle landing gear study. Landing system research aircraft. In storage at Mojave Airport.
Boeing 727			1973	1974	Wake vortex strudies. Had smoke generators.
Boeing 747-123	N905NA	NASA 905	18 July 1974	On-going	Wake vortex studies. Shuttle carrier aircraft.
Boeing 747-SR-46	N911NA	NASA 911	07 July 1999	On-going	Shuttle carrier aircraft.
Boeing 747-121	N7470	—	1979	—	Wake vortex studies.
*Lockheed L-1011	N1011	—	12 July 1977	Aug. 1980	Wake vortex study and adaptive performance. Contracted from Lockheed.
*Lockheed L-1011	N140SC	—	—	—	Flown by Orbital Sciences for Pegasus and X-34.
Convair *NC-131H	53-7793	—	Sept. 1979	Oct. 1979	Cornell Aeronautical Laboratory TIFS. Orbiter landing evaluation.
Rockwell International *B-1A	USAF 74-0160	—	08 Apr. 1981	22 Apr. 1981	Fitz Fulton made two flights for handling qualities evaluation. Data compared to XB-70 and YF-12 data.
Boeing 720B	N23	NASA 833	21 Mar. 1983	08 Dec. 1984	Destroyed in remotely-piloted intentional crash. CID program.
*McDonnell Douglas MD-11	N90178	—	1995	—	29 Aug. 1995. Made first-ever safe landing using engine power only. Propulsion-controlled aircraft (PCA).
Tupolev *Tu-144LL	RA-77114	—	29 Nov. 1996 Sept. 1998	11 Feb. 1998 Apr. 1999	19 flights in first series, 7 in second set. Russian built supersonic transport. Flights made in Russia.
Douglas DC-8-72	N817NA	NASA 817	27 June 1997	On-going	Airborne laboratory.
Lockheed NC-130B	58-0712	N707NA	30 June 1997	19 Nov. 1999	To National Center for Atmospheric Research (NCAR).

Remotely Piloted Research Vehicles (RPRVs)

NASA FRC "Mother"	—	—	1962	On-going	Radio-controlled model flown to launch other models.
NASA FRC Hyper III	—	—	1969	1979	Launched from helicopter. One flight on 12 Dec. 1969.
McDonnell Douglas F-15 RPRV/SRV - #1, #2, and #3 nose config.	—	3/8 scale model	04 Dec. 1972	—	Three vehicles launched from B-52. F-15 RPRVs until nose reconfigurations when they became SRVs. #1 crashed Oct. 1974. Last flight 15 July 1981, drop # 53. In storage at DFRC.

NASA FRC Mini-Sniffer, I, II, III.	—	—	Mar. 1974	1977	I, 12 flights; II, 21 flights; III, 1 flight on 11-23-76.
Teledyne Ryan BQM-34F Firebee II	72-1564	—	Nov. 1975	09 Mar. 1979	DAST program. BQM-34F Firebee II transferred to WR-ALC, Robins AFB, GA, 10 Sept. 1987.
Teledyne Ryan DAST 1 ARW-1	72-1557	—	13 Sept. 1979	—	Crashed 12 June 1980.
Teledyne Ryan DAST 2 ARW-1R	72-1558		29 Oct. 1982	—	Crashed 01 June 1983.
NASA Oblique Wing			1976	—	Total of three flights: 06 Aug., 16 Sept., 20 Oct. 1976. Data used for AD-1 aircraft.
Free Wing RPRV		NASA 15	Apr. 1977		Model. Demonstrate torsion-free wing with free stabilizer.
Rockwell HiMAT #1	870	NASA 839	27 July 1979	27 Aug. 1982	Total 14 flights. Highly Maneuverable Aircraft Technology (HiMAT). RPRV.
Rockwell HiMAT #2	871	NASA 840	24 July 1981	12 Jan. 1983	Total 12 flights. Highly Maneuverable Aircraft Technology (HiMAT). RPRV.
RAE Tornado RPRV	ADV-B	—	31 July 1981	08 Dec. 1983	Sept. 1981, 4 drops. Oct. 1983, 2 drops.
RAE Tornado RPRV	ADV-C	—	11 May 1981	23 Jan. 1982	Sept. 1981, 1 drop.
RAE Tornado RPRV	ADV-D	—	31 July 1981	08 Dec. 1983	Sept. 1981, 4 drops. Oct. 1983, 3 drops.
RAE Tornado RPRV	IDS-I	—	31 July 1981	08 Dec. 1983	Sept. 1981, 1 drop. Sept. 1983, 1 drop.
RAE HIRM–1	Hirmon	—	21 Sept. 1983	Dec. 1986	Oct. 1983, 5 drops. Nov. 1986, 10 drops.
RAE HIRM-2	Hermes	—	21 Sept. 1983	Dec. 1986	Oct. 1983, 5 drops. Nov. 1986, 2 drops.
General Atomics *GNAT 750			Oct. 1994		Flew at El Mirage, CA.
AeroVironment, Inc. Pathfinder. ERAST	—	—	13 Sept. 1995	—	Solar powered. Converted to Pathfinder Plus. Flown 1995-1997. Flew at DFRC in 1993 in a DoD program.
AeroVironment, Inc. Pathfinder Plus	—	—	—	—	Solar powered. Flown on 06 Aug. 1998 over Kaui, Hawaii, to a record altitude of 80,201 feet.
AeroVironment, Inc. Centurion			1998	—	Solar powered; low altitude tests, three flights on 10 Nov., 19 Nov., 03 Dec. 1998. Converted to Helios Prototype.
AeroVironment, Inc. Helios Prototype			July 1999	14 Dec. 1999	Last flight at NASA DFRC, 08 Dec. 1999. Used on-board batteries to power its 14 electric motors.
Lockheed Martin *RQ-3A DarkStar.Tier 3 minus	695	—	06 Sept. 1995	—	2 flights 29 Mar. 1996 and 22 Apr. 1996. Crashed on take-off on second flight.
Lockheed Martin *RQ-3A DarkStar	696	—	16 Oct. 1997	18 Feb. 1999	Last flight 09 Jan. 1999. Program cancelled 28 Jan. 1999. To Lockheed Martin Skunk Works, Palmdale, CA.

Accurate Automation Corp. *LoFLYTE	YH-X	—	06 Aug. 1996	—	At DFRC to fly at low subsonic speeds to explore take-off and landing control issues. Attempted flights on 20 Nov. & 21 Nov. 1996. Stored in AFFTC hangar.
NASA DFRC Utility RPRV				01 Oct. 1998	Crash on Rosamond Dry Lake. Used as a mothership to launch an X-33 model.
Rans S-12 Airaile			1992	—	Nickname “Ye Better Duck.” NASA DFRC storage. This was a 700 lb. model. Two drops of the Spacewedge #2.
Spacewedge-1			23 Apr. 1992	—	Launched from Tehachapi hills.
Inert Spacewedge-2			10 June 1992	—	Phase I, 36 flights on Wedges 1 & 2. Phase II, 45 flights on Wedge 2.
Wedge–3			14 June 1995	—	Phase III, 34 flights on Wedge 3 (Army).
Wedge-4					Never flew. Was a backup vehicle.
Aurora Flight Sciences Perseus A ERAST	AU-002		18 Dec. 1992	1994	
Aurora Flight Sciences Perseus A ERAST	AU-003		1993	1994	Autopilot gyro failed 22 Nov. 1994. Descended on flight-termination-system parachute.
Aurora Sciences Perseus B ERAST	AU-004		30 Apr. 1998 first flight	—	Crashed 01 Oct. 1999.
Aurora Flight Sciences Theseus RPV.	AU-007		24 May 1996	—	Structural failure in flight 12 Nov. 1996.
Lockheed D-21B	513	DA013	02 June 1994	On-going	Storage.
Lockheed D-21B	525	DA010	01 June 1994	On-going	Loaned to Blackbird Airpark, Palmdale, CA.
Lockheed D-21B	529	DA015	01 June 1994	On-going	Storage.
Lockheed D-21B	537	DA007	02 June 1994	On-going	Storage.
General Atomic Aeronautical Sys. Inc., Altus I, ERAST			1996	1997	Equipped with a single stage turbocharger. (1997).
General Atomic Aero. Sys. Inc., Altus II, ERAST			1997	1998	Equipped with a two-stage turbocharger. (1998). Became Altus II. Reached an altitude of 55,000 feet at Kauai, Hawaii, in 1999.
Scaled Composites X-38	V131	—	04 June 1997	On-going	Reconfigured to X-38 V131R, which arrived at Dryden 11 July 2000.
Scaled Composites X-38	V132	—	09 Sept. 1998	9 May 2000	To German air show 09 May 2000.
*McDonnell Douglas Boeing X-36-1	—	—	02 July 1996	—	ARC program; DFRC host and support – in storage. First flight 17 May 1997.

*McDonnell Douglas Boeing X-36-2	—	—	30 Nov. 1996	—	ARC program; DFRC host and support – wrapped and stored at NASA DFRC. 12 Nov. 1999 last flight
Orbital Science X-34-A1	—	—	24 Feb. 1999	—	29 Jun. 1999 first captive flight. Project cancelled March 2001.

Helicopters, Gliders
and Ultralights

Bell 47G-3B1	N822NA	NASA 822	04 Nov. 1973	21 June 1985	Departed to Napa Fire Department.
*Bell UH-1H	69-15231	—	1975		
*Bell UH-1H	69-15491	—	1999		
*Gossamer Albatross II			20 Mar. 1979		Six week program. Human powered.
*Gossamer Penguin			July 1980		Solar-powered. Flew 07 Aug. 1980
Eiri-Avion in Finland. PIK-20E	N202NA	NASA 803	11 Aug. 1981	—	Low Reynolds number. Airflow over lifting surfaces. Inactive status at NASA DFRC, Oct. 1991.
Morgan Aircraft. Vulmer Jensen Ultralight	LRV-2	---	1981	1983	Low Reynolds Number Vehicle. First flight 20 Oct. 1982.
Schweizer SGS 1-36	---	NASA 810	28 Dec. 1982	—	High angle of attack research. Total of 20 flights. Storage at NASA DFRC.
*MIT Daedalus 87			23 Dec. 1986	—	Human powered. Crashed 07 Feb. 1988.
*MIT Daedalus 88			23 Feb. 1988	11 Mar. 1988	Human powered.
*Michelob Light Eagle (MLE)			1987	1988	Human powered.

Acronyms

ACTIVE	Advanced Controls Technology for Integrated Vehicles
AF	Air Force
AFB	Air Force Base
AFTI	Advanced Fighter Technology Integration
AMAS	Automated Maneuvering and Attack System
ARI	Aileron/rudder interconnect
CAWAP	Cranked-Arrow Wing Aerodynamics Project
CID	Controlled Impact Demonstration.
DoD	Department of Defense
EAFB	Edwards AFB
ERAST	Environmental Research Aircraft and Sensor Technology
GPAS	General Purpose Airborne Simulator
HARV	High Angle-of-attack Research Vehicle
JPL	Jet Propulsion Laboratory
JSC	NASA Johnson Space Center
MSBLS	Microwave Scanning Beam Landing System
NACA	National Advisory Committee for Aeronautics
NAS	Naval Air Station
NASA	National Aeronautics and Space Administration
NAWC	Naval Air Warfare Center (Weapons Division, China Lake)
RAE	(British) Royal Aircraft Establishment
RPRV	Remotely Piloted Research Vehicle
RSRA	Rotor System Research Aircraft
TIFS	Total In-Flight Simulator
VFA	Designation of a Navy fighter attack squadron
WPAFB	Wright-Patterson Air Force Base
WR-ALC	Warner Robbins Air Logistics Center

Sources and note: Compiled by Betty Love and Peter Merlin from flight logs in the Dryden Pilots Office; NASA Ames-Dryden Aircraft Inventories of 9 April 1984 and 6 May 1986; Inventory of NASA FRC [sic] Aircraft of 1980; NASA Aircraft Inventory Report 20 Oct. 1978; various other aircraft maintenance records, 1972-1993; Roy Bryant's FRC Vehicle ResumÈ of 14 July 1967; George Sitterle's daily logs 1979-1993; Dryden Research Aircraft Operations Division NASA Aircraft Inventory of 28 February 1991; and Peter Merlin's daily log, 16 June 1997-29 March 2001. These sources often contradict one another, and there were many questions regarding which vehicles to include and which did not qualify as research aircraft. This appendix represents the best information available.

Appendix F
X-1 Program Flight Chronology, 1946-58

This chronology covers all the flights of X-1 series aircraft built and flown. The NACA operated the X-1 #2, X-1A, X-1B, and X-1E (the rebuilt X-1#2). In the interest of completeness, and because of the close NACA-Air Force-Bell relationship in the entire program, flights of the other aircraft are also listed. The X-1 series aircraft were air-launched from modified Boeing B-29 or B-50 Superfortress bombers.

XS-1 #1 (X-1-1), Serial 46-062, Flights

Bell Contractor Flights, Pinecastle AAF, Florida, 1946

Date	Remarks
25 Jan.	Bell flight 1, Jack Woolams, pilot familiarization; 1st glide flight.
5 Feb.	Bell flight 2, Woolams; glide flight.
5 Feb.	Bell flight 3, Woolams; glide flight.
8 Feb.	Bell flight 4, Woolams. Gear retracted, left wing damaged; glide flight.
19 Feb.	Bell flight 5, Woolams. Nosewheel retracted on landing runout. Landing-gear door damaged; glide flight.
25 Feb.	Bell flight 6, Woolams. Static directional stability investigation; glide flight.
25 Feb.	Bell flight 7, Woolams. Longitudinal and directional stability investigation; glide flight.
26 Feb.	Bell flight 8, Woolams. Dynamic stability check; glide flight.
26 Feb.	Bell flight 9, Woolams. Rate of roll investigation; glide flight.
6 Mar.	Bell flight 10, Woolams. Static longitudinal stability investigation; glide flight.

Bell at Muroc Dry Lake, California, 1947

Date	Remarks
10 Apr.	Bell flight 11, Chalmers Goodlin. Glide flight and stall check; 1st flight of 8-percent wing/6-percent tail.
11 Apr.	Bell flight 12, Goodlin. Nosewheel damaged. First powered flight of XS-1 #1 aircraft.
29 Apr.	Bell flight 13, Goodlin. Handling qualities check
30 Apr.	Bell flight 14, Goodlin. Same as flight 13.
5 May	Bell flight 15, Goodlin. Same as flight 13.
15 May	Bell flight 16, Goodlin. Buffet-boundary investigation. Aileron-damper malfunction
19 May	Bell flight 17, Goodlin. Buffet-boundary investigation.
21 May	Bell flight 18, Goodlin. Same as flight 17.
5 June	Bell flight 19, Goodlin. Demonstration flight for Aviation Writers Association.

Air Force Flights, 1947

Date	Remarks
6 Aug.	AF glide flight 1, Capt. Charles E. Yeager. Pilot familiarization; glide flight.
7 Aug.	AF glide flight 2, Yeager. Same as flight 1.
8 Aug.	AF glide flight 3, Yeager. Same as flight 1.
29 Aug.	AF powered flight 1, Yeager. Mach 0.85.
4 Sept.	AF powered flight 2, Yeager. About Mach 0.89. Telemeter failure required repeat of this flight.
8 Sept.	AF powered flight 3, Yeager. Repeat of flight 2.
10 Sept.	AF powered flight 4, Yeager. Mach 0.91. Stability and control investigation.
12 Sept.	AF powered flight 5, Yeager. Mach 0.92. Check of elevator and stabilizer effectiveness; buffet investigation.
3 Oct.	AF powered flight 6, Yeager. Same as flight 5.

8 Oct.	AF powered flight 7, Yeager. Airspeed calibration. Plane attained Mach 0.945.
10 Oct.	AF powered flight 8, Yeager. Stability and control investigation. Plane attained Mach 0.997.
14 Oct.	AF powered flight 9, Yeager. World's first supersonic flight by a piloted aircraft. XS-1 #1 attained Mach 1.06 at 43,000 ft., approximately 700 mph.
27 Oct.	AF powered flight 10, Yeager. Electric power failure. No rocket.
28 Oct.	AF powered flight 11, Yeager. Telemeter failure.
29 Oct.	AF powered flight 12, Yeager. Repeat of flight 11.
31 Oct.	AF powered flight 13, Yeager.
3 Nov.	AF powered flight 14, Yeager.
4 Nov.	AF powered flight 15, Yeager.
6 Nov.	AF powered flight 16, Yeager. Mach 1.35 at 48,600 ft. Maximum altitude 51,434 ft.

Air Force Flights, 1948

16 Jan.	AF powered flight 17, Yeager. Airspeed calibration. Mach 1.048.
22 Jan.	AF powered flight 18, Yeager. Pressure distribution survey. Mach 1.2.
30 Jan.	AF powered flight 19, Yeager. Same as flight 18. Mach 1.1. No data worked up.
24 Feb.	AF powered flight 20, Capt. James T. Fitzgerald, Jr. Engine fire after launch forced jettisoning of propellants; completed as a glide flight.
11 Mar.	AF powered flight 21, Yeager. Attained Mach 1.25 in dive.
26 Mar.	AF powered flight 22, Yeager. Attained Mach 1.45 at 40,130 ft. (957 mph) during dive. Fastest flight ever made in original XS-1 aircraft.
31 Mar.	AF powered flight 23, Yeager. Engine shutdown after launch. Propellants jettisoned, completed as glide flight.
6 Apr.	AF powered flight 24, Fitzgerald. Pilot-check flight. Mach 1.2, during 4-cylinder run at 41,000 ft.
7 Apr.	AF flight 25, Maj. Gustav E. Lundquist. Glide flight only.
7 Apr.	AF flight 26, Fitzgerald. Familiarization flight.
9 Apr.	AF flight 27, Lundquist. Powered pilot-check flight.
16 Apr.	AF flight 28, Lundquist. Pressure distribution survey. Only cylinders 2 and 4 ignited.
26 Apr.	AF flight 29, Fitzgerald. Aborted because of inconsistent rocket operation. Reached Mach 0.9.
29 Apr.	AF flight 30, Lundquist. Pressure distribution survey. Attained Mach 1.186.
4 May	AF flight 31, Fitzgerald. Same as flight 30. Mach 1.22.
21 May	AF flight 32, Lundquist. Stability and control and buffeting investigation. Mach 0.92. No data worked up.
25 May	AF flight 33, Fitzgerald. Buffet investigation, wing and tail loads. Mach 1.08.
26 May	AF flight 34, Yeager. Same as flight 33. Mach 1.094. Reached altitude of 63,917.
3 June	AF flight 35, Lundquist. Left main gear door opened in flight. Nosewheel collapsed on landing.
1 Dec.	AF flight 36, Yeager. Handling qualities and wing and tail loads at Mach 1.
13 Dec.	AF flight 37 Yeager. Same as flight 36.
23 Dec.	AF flight 38, Yeager. Wing and tail loads during supersonic flight at high altitudes. Mach 1.09.

Air Force Flights, 1949

5 Jan.	AF flight 39, Yeager. Rocket takeoff from the ground.
11 Mar.	AF flight 40, Capt. Jack Ridley, pilot. Familiarization flight. Mach 1.014. Max altitude, 32,636 ft. Small engine fire due to loose igniter.
16 Mar.	AF flight 41, Col. Albert Boyd, pilot. Familiarization flight. In-flight engine fire and shutdown.
21 Mar.	AF flight 42, Maj. Frank Everest, Familiarization flight. Mach 1.067. Max altitude 39,750 ft.

25 Mar.	AF flight 43, Everest. Check of pressure suit for altitude operation. Mach 1.227. Max alt. 50,131 ft. Rocket fire and automatic engine shutdown.
14 Apr.	AF flight 44, Ridley. Accelerated stall check at transonic speeds. Mach 1.136. Max alt. 39,306 ft.
19 Apr.	AF flight 45, Everest, Altitude attempt. Only 2 cylinders fired.
2 May	AF flight 46, Yeager. Partial engine malfunction, faulty engine ignition plug. Still reached Mach 1.427.
5 May	AF flight 47, Everest. Engine chamber exploded, jamming rudder. Everest landed safely.
25 Jul.	AF flight 48, Everest. Altitude attempt. Attained 66,846 ft. altitude.
8 Aug.	AF flight 49, Everest. Altitude attempt. Attained 71,902 ft. altitude.
25 Aug.	AF flight 50, Everest. First use of partial pressure suit to save life of pilot during flight at high altitude. X-1 #1 lost cockpit pressurization about 69,000 ft. Everest made safe emergency descent.
6 Oct.	AF flight 51, Lt. Col. Patrick Fleming, pilot. Pilot familiarization.
26 Oct.	AF flight 52, Maj. Richard L. Johnson, pilot. Pilot familiarization.
29 Nov.	AF flight 53, Everest. High-altitude wing-and-tail loads investigation. Reached 71,040 ft.
2 Dec.	AF flight 54, Everest. Same as flight 53, but only reached 67,900 ft.

Air Force Flights, 1950

21 Feb.	AF flight 55, Everest. Wing-and-tail-loads investigation.
26 Apr.	AF flight 56, Yeager. Lateral stability and control investigation.
5 May	AF flight 57, Ridley. Buffeting, wing and tail loads.
10 May	AF flight 58, Ridley. Same as flight 57.
12 May	AF flight 59, Yeager. Last flight of X-1 #1. Flight made for camera footage for motion picture *Jet Pilot*. Aircraft subsequently retired and presented to the Smithsonian Institution.

XS-1 #2 (X-1-2), Serial 46-063, Flights
Bell Contractor Flights, 1946

11 Oct.	Bell flight 1, Chalmers Goodlin, pilot. Glide flight, pilot familiarization.
14 Oct.	Bell flight 2, Goodlin. Glide flight. Handling characteristics.
17 Oct.	Bell flight 3, Goodlin. Glide flight, stall check.
2 Dec.	Bell flight 4, Goodlin. Glide flight, check of fuel-jettison system.

XS-1 #2 (X-1-2), Serial 46-063, Flights
Air Force Flights, 1946-47

9 Dec. 1946	AF flight 1, Goodlin. First XS-1 powered flight. Mach 0.79 at 35,000 ft. Minor engine fire.
20 Dec. 1946	AF flight 2, Goodlin. Familiarization powered flight.
8 Jan. 1947	AF flight 3, Goodlin. Buffet boundary investigation. Mach 0.80 at 35,000 ft.
17 Jan.	AF flight 4, Goodlin. Buffet boundary investigation. Full-power climb. Plane reached Mach 0.82.
22 Jan.	AF flight 5, Goodlin. Buffet boundary investigation and full-power climb. Telemetry failure.
23 Jan.	AF flight 6, Goodlin. Buffet boundary investigation and full-power climb.
30 Jan.	AF flight 7, Goodlin. Accelerated stalls. Partial power due to faulty engine igniters. Mach 0.75.
31 Jan.	AF flight 8, Goodlin. Buffet boundary investigation. Mach 0.7.
5 Feb.	AF flight 9, Goodlin. Machmeter calibration.
7 Feb.	AF flight 10, Goodlin. Buffet boundary investigation.
19 Feb.	AF flight 11, Goodlin. Accelerated stalls.

21 Feb.	AF flight 12, Goodlin. Flight aborted after drop because of low engine-chamber pressure.
22 May	AF flight 13, Alvin M. Johnston. Pilot familiarization flight. Mach 0.72, 8 g pullout.
29 May	AF flight 14, Goodlin. Airspeed calibration flight to Mach 0.72. End of Bell contractor program.
25 Sept.	AF flight 15, Airworthiness demonstration. Capt. Charles E. Yeager. Number 4 cylinder burned out.

XS-1 #2 (X-1-2), Serial 46-063, Flights
NACA Flights, 1947-48

21 Oct. 1947	NACA glide-familiarization flight for NACA pilot Herbert H. Hoover. Stall check. Nosewheel collapsed on landing.
16 Dec.	NACA powered flight 1, Hoover. Familiarization. Mach 0.765. Max altitude 20,000 ft.
17 Dec.	NACA flight 2, Hoover. Same as flight 1. Mach 0.8.
6 Jan. 1948	NACA flight 3, Hoover. Turns and pull-ups to buffet. Mach 0.85. Max altitude 30,025.
8 Jan.	NACA flight 4, Hoover. Turns and pull-ups to buffet. Mach 0.88. Max altitude 31,371.
9 Jan.	NACA flight 5, Howard C. Lilly. Pilot familiarization.
15 Jan.	NACA flight 6, Lilly. Turns and pull-ups to buffet. Sideslips. Mach 0.9. Max altitude 30,400.
21 Jan.	NACA flight 7, Hoover. Stabilizer effectiveness investigation. Mach 0.877 at 29,000 ft. Max altitude 34,600.
23 Jan.	NACA flight 8, Hoover. Attempted high-speed run aborted at Mach 0.89, drop in chamber pressure.
27 Jan.	NACA flight 9, Hoover. High-speed run to Mach 0.946 at 38,000 ft. Cylinders 2 and 3 failed to fire.
4 Mar.	NACA flight 10, Hoover. High-speed run to Mach 1.029 at 40,000 ft. First NACA supersonic flight. First civilian supersonic flight.
10 Mar.	NACA flight 11, Hoover. Mach 1.055. Nosewheel failed to extend for landing. Minor damage.
22 Mar.	NACA flight 12, Hoover. Stability and loads investigation. Mach 1.172.
30 Mar.	NACA flight 13, Hoover. Same as flight 12 Mach 1.045.
31 Mar.	NACA flight 14, Lilly. Same as flight 12. Plane attained Mach 1.127.
5 Apr.	NACA flight 15, Lilly. Engine failed to ignite. Propellants jettisoned, completed as glide flight.
9 Apr.	NACA flight 16, Lilly. Same as flight 12. Mach 0.986.
16 Apr.	NACA flight 17, Lilly. Same as flight 12. Plane's nosewheel collapsed on landing. Moderate damage.
1 Nov.	NACA flight 18, Hoover. Stability and control. Mach 0.948. Number 4 cylinder failed to fire.
15 Nov.	NACA flight 19, Hoover. Same as flight 18. Also pressure-distribution survey. Mach 0.981.
23 Nov.	NACA flight 20, Robert A. Champine. Pilot familiarization. Check on handling qualities and pressure distribution. Mach .839
29 Nov.	NACA flight 21, Champine. Check on handling qualities and pressure distribution. Mach 1.070.
30 Nov.	NACA flight 22, Champine. Same as flight 21. Mach 0.999
2 Dec.	NACA flight 23, Champine. Same as flight 21. Mach 1.12.

XS-1 #2 (X-1-2), Serial 46-063, Flights
NACA Flights, 1949-50

6 May 1949	NACA flight 24, Champine. Check on airplane instrumentation. Mach 0.985 at 40,000 ft.
13 May	NACA flight 25, Champine. Spanwise pressure distribution, stability and control. No records.
27 May	NACA flight 26, Champine. Same as flight 25. Mach 1.002. Stabilizer found more effective than the elevator during pull-ups at Mach 0.91. (indicated airspeed)
16 June	NACA flight 27, Champine. Same as flight 25. Rolls and pull-ups around Mach 1.078.
23 June	NACA flight 28, Champine. Same as flight 25. Rolls, pull-ups, check of stabilizer effectiveness. Mach 0.860.
11 Jul.	NACA flight 29, Champine. Same as flight 25. Rolls, pull-ups, check of stabilizer effectiveness. Mach 1.02. Number 2 cylinder failed to fire.

19 Jul.	NACA flight 30, Champine. Same as flight 25. Rolls, pull-ups, check of stabilizer effectiveness. Mach 1.022. Number 2 cylinder failed to fire.
27 Jul.	NACA flight 31, Champine. Same as flight 25. Rolls, pull-ups, check of stabilizer effectiveness. Mach 0.974
4 Aug.	NACA flight 32, Champine. Same as flight 25. Sideslips, rolls, check of stabilizer effectiveness. Mach 1.123.
23 Sept.	NACA flight 33, John H. Griffith. Pilot familiarization. Mach 0.998.
30 Nov.	NACA flight 34, Griffith. Same as flight 33. Mach 1.067.
12 May 1950	
16 May	NACA flight 36, Griffith. Same as flight 25. Push-downs and pull-ups. No data worked up.
26 May	NACA flight 37, Griffith. Same as flight 25. Push-downs, pull-ups, rolls. Mach 1.189. Nosewheel collapsed on landing.
9 Aug.	NACA flight 38, Griffith. For pressure distribution and stability and control data. Check of stabilizer effectiveness. Mach 0.825.
11 Aug.	NACA flight 39, Griffith. Same as flight 38. Mach 1.014.
21 Aug.	NACA flight 40, Griffith. Same as flight 38. Also drag investigation. Pull-ups. Mach 0.998.
4 Oct.	NACA flight 41, Griffith. Same as flight 40. Mach 0.927.

XS-1 #2 (X-1-2), Serial 46-063, Flights
NACA Flights, 1951

6 Apr.	NACA flight 42, Capt. Charles E. Yeager. Flight for RKO film *Jet Pilot*. Slight engine fire but no damage.
20 Apr.	NACA flight 43, A. Scott Crossfield. Pilot familiarization. Reached Mach 1.113.
27 Apr.	NACA flight 44, Crossfield. Plane and instrument check. Mach 1.337.
15 May	NACA flight 45, Crossfield. Wing loads and aileron effectiveness. Aileron rolls at Mach 1.000.
12 Jul.	NACA flight 46, Crossfield. Same as flight 45. Aileron rolls at Mach 1.140.
20 Jul.	NACA flight 47, Crossfield. Same as flight 45. Abrupt rudder fixed aileron rolls left and right, from Mach 0.70 to Mach 0.88. Max speed on this flight of was Mach 0.997.
31 Jul.	NACA flight 48, Crossfield. Same as flight 45. Mach 0.989.
3 Aug.	NACA flight 49, Crossfield. Same as flight 45. Mach 0.985.
8 Aug.	NACA flight 50, Crossfield. Same as flight 45. Elevator and stabilizer pull-ups. Mach 0.987.
10 Aug.	NACA flight 51, Crossfield. Same as flight 45. Elevator and stabilizer pull-ups, clean stalls. Mach 0.975.
27 Aug.	NACA flight 52, Joseph A. Walker. Pilot familiarization. Reached Mach 1.219 at 44,000 ft during four cylinder run. Max altitude 47,008 ft.
5 Sept.	NACA flight 53, Crossfield. Fuselage pressure distribution survey. Number 1 cylinder failed to fire. Stabilizer pull-ups at Mach 1.097.
23 Oct.	NACA flight 54, Walker. Vortex-generator investigation. Engine cut out after two ignition attempts; propellants jettisoned and flight completed as glide flight. Flap actuator failed, so landing made flaps-up. Plane subsequently grounded after a non-data flight 55 on 2 Nov. because of possibility of fatigue failure of nitrogen spheres. Later rebuilt as the Mach 2+ X-1E.

X-1 #3 (X-1-3), Serial 46-064, Flights

20 Jul. 1951	Bell flight 1, Joseph Cannon, pilot. Glide flight for familiarization. Nosewheel collapse on landing.
9 Nov.	Bell flight 2, Cannon. Captive flight with B-50 for propellant jettison test. X-1-3 destroyed in postflight explosion and fire on ground. B-50 launch plane also lost and Cannon injured.

X-1A, Serial 48-1384, Bell Contractor Flights

14 Feb. 1953	Bell flight 1, Jean Ziegler, pilot. Familiarization. Fuel jettison test. Glide flight only.
20 Feb.	Bell flight 2, Ziegler. Planned as powered flight, but completed as glide flight following propellant-system difficulties.
21 Feb.	Bell flight 3, Ziegler. First powered flight. False fire warning.
26 Mar.	Bell flight 4, Ziegler. Plane demonstrated successful 4-cylinder engine operation.
10 Apr.	Bell flight 5, Ziegler. Pilot noted low-frequency elevator buzz at Mach 0.93, did not proceed above this speed, pending buzz investigation.
25 Apr.	Bell flight 6, Ziegler. Buzz again noted at Mach 0.93. Turbopump overspeeding caused pilot to terminate power and jettison remaining fuel.

X-1A, Serial 48-1384, Air Force Flights
(After USAF took over remaining Bell program on X-1A and initiated its own flight program.)

21 Nov. 1953	Flight 7, Maj. Charles E. Yeager. First Air Force flight. Reached Mach 1.15. Familiarization purposes.
2 Dec.	Flight 8, Yeager. Mach 1.5.
8 Dec.	Flight 9, Yeager. First high-Mach flight attempt by X-1A. Mach 1.9 attained at 60,000 ft during slight climb.
12 Dec.	Flight 10, Yeager. Plane attained Mach 2.44 (1,612 mph) but met violent instability above Mach 2.3. Tumbled at 50,000 ft, wound up in subsonic inverted spin. Yeager recovered to upright spin, then into normal flight at 25,000 ft.

NOTE: Fourteen Air Force flight attempts for high altitudes were made in the spring and summer of 1954. Of these, only four flights were successful. The rest were aborted for various malfunctions, including ruptured canopy seal, failure of gear doors to close fully, turbine overspeed, faulty ignition operation. Of the four successful flights, one was Maj. Arthur Murray's checkout flight. The rest were successful high-altitude tries by Murray. The successful altitude flights were:

28 May 1954	Flight 16, Murray. X-1A attained 87,094 ft, unofficial world altitude record for piloted aircraft.
4 June	Flight 17, Murray. X-1A reached 89,750 ft. Encountered same instability Yeager had, but at Mach 1.97. Murray recovered after tumbling 20,000 ft down to 66,000 ft.
26 Aug.	Flight 24, Murray. Murray attained 90,440 ft. Air Force then turned X-1A over to the NACA.

X-1A, Serial 48-1384, NACA Flights

20 July 1955	NACA flight 1, Joseph A. Walker. Familiarization. Walker attained Mach 1.45 at 45,000 ft. Noted severe aileron buzz at Mach 0.90 to 0.92.
8 Aug.	Planned as NACA flight 2. Shortly before launch from B-29, X-1A suffered low order explosion, later traced to detonation of Ulmer leather gaskets. Walker exited into B-29 bomb bay. Extent of damage prohibited landing crippled X-1A, NACA B-29 launch crew jettisoned it into desert. It exploded and burned on impact.

X-1B, Serial 48-1385, Air Force Flights

24 Sept. 1954	Air Force flight 1, Lt. Col. Jack Ridley, pilot. Glide flight, because of turbopump over-speeding.
6 Oct.	Air Force flight 2, Ridley. Glide flight, aborted power flight because of evidence of high lox tank pressure.
8 Oct.	Air Force flight 3, Maj. Arthur "Kit" Murray. First powered flight.
13 Oct.	Air Force flight 4, Maj. Robert Stephens.

19 Oct.	Air Force flight 5, Maj. Stuart R. Childs.
26 Oct.	Air Force flight 6, Col. Horace B. Hanes.
4 Nov.	Air Force flight 7, Capt. Richard B. Harer.
26 Nov.	Air Force flight 8, Brig. Gen. J. Stanley Holtoner (Commander, Air Force Flight Test Center).
30 Nov.	Air Force flight 9, Lt. Col. Frank K. Everest.
2 Dec.	Air Force flight 10, Everest. Mach 2.3 (approx. 1,520 mph) at 65,000 ft.

X-1B, Serial 48-1385, NACA Flights

Note: John B. McKay pilot on flights 1-13, Neil A. Armstrong pilot on flights 14-17

15 Aug. 1956	NACA flight 1. Pilot check; nose landing gear failed on landing, minor damage.
29 Aug.	NACA flight 2. Cabin pressure regulator malfunction caused inner canopy to crack; only low-speed, low-altitude maneuvers made.
7 Sept.	NACA flight 3. Speed run to 57,676 ft and Mach 1.782. Limited heating data gathered.
18 Sept.	NACA flight 4. Glide flight, due to erratic engine start.
28 Sept.	NACA flight 5. Three-chamber engine run to 60,676 ft and Mach 1.721 to obtain heating data.
3 Jan. 1957	NACA flight 6. Mach 1.936 aerodynamic heating investigation (end of heating program).
22 May	NACA flight 7. Control pulses at Mach 1.45 at 60,000 ft. Flight for instrumentation check. Reached Mach 1.572 and an altitude of 62,866 ft.
7 June	NACA flight 8. Supersonic maneuvers to Mach 1.3 at ca. 60,000 ft. to determine the dynamic and static stability and control characteristics.
24 June	NACA flight 9. Supersonic maneuvers to Mach 1.49 at 60,322 ft to determine the dynamic and static stability and control characteristics.
10 Jul.	NACA flight 10. Aborted after launch, indication of open landing-gear door. Propellants jettisoned, completed as a glide flight.
19 Jul.	NACA flight 11. Mach 1.77 at 61,858 ft. Control pulses, sideslips, and a 2-g wind-up turn.
29 Jul.	NACA flight 12. Enlarged wing tips installed to simulate those to be used with reaction controls. Mach 1.485 at 60,000 ft.
8 Aug.	NACA flight 13. Stability and control investigation. Ca. Mach 1.5 at 60,000 ft, accelerated maneuvers, control pulses, and pull-ups. Film lost; no precise data.
15 Aug.	NACA flight 14. Pilot check for Armstrong. Nose landing gear failed on landing, minor damage.
27 Nov.	NACA flight 15. First reaction-control flight.
16 Jan. 1958	NACA flight 16. Low altitude, low Mach reaction-control investigation.
23 Jan.	NACA flight 17. Reaction-control investigation. Mach 1.5 at 55,000 ft. Last NACA flight.

X-1D, Serial 48-1386, Flights

24 Jul. 1951	Bell flight 1, Jean Ziegler, pilot. Glide flight for familiarization. Nose landing gear broken on landing. Following repairs, plane turned over to the Air Force.
22 Aug. 1951	AF flight 1, Lt. Col. Frank K. Everest. Launch aborted, but X-1D suffered low order explosion during pressurization for fuel jettison. Plane jettisoned from B-50. X-1D exploded on impact with desert. Everest managed to get into B-50 bomb bay before drop. B-50 not damaged, no personal injuries.

X-1E, Serial 46-063, NACA Flights

Note: Joseph Walker pilot for flights 1-21, John McKay pilot for flights 22-26

3 Dec. 1955	Captive flight.
12 Dec.	NACA flight 1. Glide flight for pilot check-out and low speed evaluation.
15 Dec.	NACA flight 2. First powered flight. Engine ran at excessive pressure, four overspeeds of turbopump and two automatic shutdowns. Power terminated by pilot.
3 Apr. 1956	NACA flight 3. Mach 0.835 at 29,636 ft. Damping characteristics good; number 1 cylinder failed to fire.
30 Apr.	NACA flight 4. Turbopump did not start; no engine operation.
11 May	NACA flight 5. Wind-up turns to $C_{L_{max}}$ from Mach 0.69 to 0.854; also control pulses.
8 June	NACA flight 6. Mach 1.578 at 50,503 ft (Approx. 1,020 mph). Longitudinal and lateral trim changes in transonic region found annoying to pilot.
18 June	NACA flight 7. Mach 1.723 at 59,749 ft (approx. 1,150 mph). Damaged on landing.
26 Jul.	NACA flight 8. Subsonic because cylinders 3 and 4 would not fire.
31 Aug.	NACA flight 9. Mach 1.988 at 60,683 ft (approx. 1,340 mph). Sideslips, pulses, rolls.
14 Sept.	NACA flight 10. Mach 2.1 at 62,566 ft (approx. 1,385 mph). Stabilizer, rudder, and aileron pulses.
20 Sept.	NACA flight 11. Brief engine power only; flight aborted, unspecified engine malfunction.
3 Oct.	NACA flight 12. Only 60-sec rocket operation; intermittent pump operation. Flight aborted, turbopump and engine replaced.
20 Nov.	NACA flight 13. No engine operation, ignition failure and lack of manifold pressure.
25 Apr. 1957	NACA flight 14. Mach 1.761 at 66,590 ft. (approx. 1,130 mph). Aileron and rudder pulses.
15 May	NACA flight 15. Mach 2.016 at 73,458 ft. (approx. 1,325 mph). Aileron pulses and rolls, sideslips, and wind-up turns. Plane severely damaged upon landing.
19 Sept.	NACA flight 16. Planned Mach number not attained, loss of power during pushover from climb.
8 Oct.	NACA flight 17. Mach 2.22 (approx. 1,480 mph).
14 May 1958	NACA flight 18. First flight with ventral fins; longitudinal and lateral stability and control maneuvers. Engine airstart made at 70,000 ft.
10 June	NACA flight 19. Flight aborted after only 1 cylinder of engine fired. Plane damaged on landing.
10 Sept.	NACA flight 20. Stability and control investigation with ventral fins.
17 Sept.	NACA flight 21. Stability and control with ventral fins and a new stabilizer bell crank permitting greater stabilizer travel.
19 Sept.	NACA flight 22. Checkout flight for John McKay.
30 Sept.	NACA flight 23. Checkout flight for McKay, also check of low-speed stability and control.
16 Oct.	NASA flight 24. First flight with elevated chamber pressure; cut short because overcast obscured pilot's view of lakebed.
28 Oct.	NASA flight 25. Elevated chamber pressure; good stability and control data gathered.
6 Nov.	NASA flight 26. Elevated chamber pressure; low altitude and low Mach investigation of U-Deta fuel. Last NASA flight.

Sources: Richard P. Hallion, *Supersonic Flight: Breaking the Sound Barrier and Beyond - The Story of the Bell X-1 and Douglas D-558* (New York: Macmillan Company in association with Smithsonian Institution, 1972), pp. 209-20; data from Computer Office (containing the women who computed the flight data) in NASA Dryden Historical Reference Collection.

Appendix G

Douglas D-558 Program Flight Chonology, 1946-1958

This chronology covers flights by the three Douglas D-558-1 Skystreaks and the three Douglas D-558-2 Skyrockets.

The D-558-1 Skystreak was a turbojet-powered aircraft that took off from the ground under its own power. It featured a straight wing and tail section.

The D-558-2 Skyrocket was powered both by a turbojet engine and a liquid-propellant rocket engine. It also took off from the ground initially. In 1950, however, the D-558-2 #2 (BuAer no. 37974) was modified for all-rocket air-launch from a P2B-1S (Navy version of the B-29) mothership, enhancing greatly its safety and performance potential. Another D-558-2 #3 (BuAer no. 37975) was also modified for air-launch, but it retained both its turbojet and rocket engine. The D-558-2 #1 (BuAer no. 37973) was likewise later modified for all-rocket operation but completed only one flight before termination of the entire Skyrocket program.

D-558-1 #1, BuAer No. 37970 Flight Highlights

This aircraft completed 101 flights during its Douglas contractor program. Douglas delivered it to the NACA on 21 April 1949, but the NACA never flew it, relegating it to spares support for D-558-1 #3.

Date	Remarks
14 Apr. 1947	Douglas flight 1, Eugene F. May, pilot. For familiarization. Partial power loss forced immediate landing after takeoff.
17 Jul.	Douglas flight 14, May. Beginning of performance investigations at high Mach numbers. Mach 0.81.
20 Aug.	Douglas flight 25, Comdr. Turner F. Caldwell, Jr., USN. Set new world airspeed record of 640.663 mph.
29 Sept. 1948	Douglas flight (86?), May. Plane exceeded Mach 1 during 35-degree dive, only time a Skystreak attained Mach 1.
3 Nov. 1949	Douglas flight 101. Obtained low airspeed calibration with XF3D-1 in a clean configuration and with the gear and flaps down. Last flight for this airplane.

D-558-1 #2, BuAer No. 37971, Flights

Howard C. Lilly piloted these missions. The 27 previous flights made by Douglas, Navy, and Marine pilots beginning with flight 1 by Eugene May on 15 Aug. 1947)

25 Nov. 1947	NACA flight 1. Pilot familiarization; instrumentation malfunction.
26 Nov.	NACA flight 2. Landing gear would not lock up.
16 Feb. 1948	NACA flight 3. Attempted airspeed calibration; instrumentation malfunction.
31 Mar.	NACA flight 4. Landing gear door would not lock.
	NACA flight 5. Landing gear door would not lock.
1 Apr.	NACA flight 6. Landing gear door would not lock.
7 Apr.	NACA flight 7. Landing gear door would not lock.
8 Apr.	NACA flight 8. Attempted airspeed calibration; radar beacon failure.
	NACA flight 9. Airspeed calibration, 30,000 ft.
9 Apr.	NACA flight 10. Airspeed calibration, 30,000 ft.
12 Apr.	NACA flight 11. Airspeed calibration, tower fly-by.
	NACA flight 12. Airspeed calibration, 30,000 ft.
14 Apr.	NACA flight 13. Smoke in cockpit after takeoff necessitated landing. Smoke due to burning 400-cycle inverter in nose compartment; inverter replaced.
20 Apr.	NACA flight 14. Sideslips at 10,000 ft from Mach 0.50 through 0.85, for static directional stability.
23 Apr.	NACA flight 15. Sideslips at 30,000 ft from Mach 0.50 through 0.85, for static directional stability.
28 Apr.	NACA flight 16. Right landing gear would not retract.
29 Apr.	NACA flight 17. Two speed runs; Mach 0.70 at 41,000 ft, Mach 0.88 at 36,000 ft. Left and right rudder kicks at 10,000 ft.
3 May	NACA flight 18. Landing gear would not retract. NACA flight 19. Crash after takeoff due to compressor disintegration; Lilly killed.

D-558-1 #3, BuAer No. 37972, Flights 1949-1950
(Four flights made in early 1948 by Douglas pilots and Howard Lilly, at least one flight in early 1949 by Eugene May)

22 Apr. 1949	NACA flight 1, Robert A. Champine, pilot. For pilot familiarization.
28 Apr.	NACA flight 2, Champine. Pilot check; dive to Mach 0.86.
12 Aug.	NACA flight 3, Champine. Handling qualities (rudder kicks, aileron roll, sideslips); dive to Mach 0.875.
18 Aug.	NACA flight 4, Champine. Handling qualities; dive to Mach 0.88.
19 Aug.	NACA flight 5, John H. Griffith, pilot check, handling qualities; trim run to Mach 0.84.
23 Aug.	NACA flight 6, Griffith. Airspeed calibration using tower passes.
24 Aug.	NACA flight 7, Champine. Handling qualities; dive to Mach 0.87.
31 Aug.	NACA flight 8, Champine. Aileron effectiveness investigation; 22 rolls made, 4 at Mach 0.86.
28 Sept.	NACA flight 9, Griffith. Aileron effectiveness investigations; 16 rolls made, 4 above Mach 0.875.
28 Oct.	NACA flight 10, Griffith. Beginning of pressure- distribution survey.
21 Nov.	NACA flight 11, Griffith. Pressure-distribution investigation.
23 Nov.	NACA flight 12, Champine. Pressure-distribution investigation and climb measurements.
26 Jan. 1950	NACA flight 13, Champine. Calibration of airspeed instrumentation and pressure distribution.
7 Feb.	NACA flight 14, Champine. Aborted, engine malfunction.
5 Apr.	NACA flight 15, Griffith. Pressure-distribution investigation. Mach 0.95 attained.
11 Apr.	NACA flight 16, Griffith. Pressure-distribution investigation. Mach 0.98 attained.
5 May	NACA flight 17, Griffith. Vortex generator- investigation to determine effectiveness in reducing/ delaying wing flow separation. Mach 0.97 attained.
11 May	NACA flight 18, Griffith. Vortex generator investigation. Mach 0.87 attained.
18 May	NACA flight 19, Griffith. Vortex generator investigation. Mach 0.98 attained.
31 May	NACA flight 20, Griffith. Vortex generator investigation. Mach 0.98 attained.
8 June	NACA flight 21, Griffith. Vortex generator investigation. Mach 0.98 attained.
13 June	NACA flight 22, Griffith. Vortex generator investigation. Mach 0.98-1.0. Conclusion of pressure-distribution investigation.
26 Oct.	NACA flight 23, Griffith. Instrument check flight in preparation for the buffeting, tail loads, and longitudinal stability investigation.
29 Nov.	NACA flight 24, A. Scott Crossfield. Beginning of buffeting, tail-loads, and longitudinal-stability program.
12 Dec.	NACA flight 25, Crossfield. Buffeting, tail-loads, longitudinal-stability investigation.
18 Dec.	NACA flight 26, Crossfield. Buffeting, tail-loads, longitudinal-stability investigation.
20 Dec.	NACA flight 27, Crossfield. Buffeting, tail-loads, longitudinal-stability investigation.
26 Dec.	NACA flight 28, Crossfield. Buffeting, tail-loads, longitudinal-stability investigation.

D-558-1 #3, BuAer No. 37972, Flights 1951

5 Jan. 1951	
23 Jan.	NACA flight 30, Crossfield. Buffeting, tail-loads, longitudinal-stability investigation. Aborted, fuel leak.
25 Jan.	NACA flight 31, Crossfield. Buffeting, tail-loads, longitudinal-stability investigation.
8 Feb.	NACA flight 32, Crossfield. Airspeed calibration, 5 tower passes.
13 Feb.	NACA flight 33, Walter P. Jones. Pilot check; some buffeting, tail-loads, and longitudinal-stability data taken.
20 Feb.	NACA flight 34, Jones. Aborted after Jones suffered anoxia due to a faulty O_2 regulator. Some buffeting, tail-loads, longitudinal-stability data.
2 May	NACA flight 35, Jones. Buffeting, tail-loads, longitudinal-stability investigation. Mach 0.85.
1 June	NACA flight 36, Crossfield. Buffeting, tail-loads, longitudinal-stability investigation. Mach 0.84.
13 June	NACA flight 37, Crossfield. Buffeting, tail-loads, longitudinal-stability investigation. Mach 0.86.
21 June	NACA flight 38, Crossfield. Buffeting, tail-loads, longitudinal-stability investigation. Mach 0.835.
28 June	NACA flight 39, Jones. Buffeting, tail-loads, longitudinal stability investigation. Mach 0.85.
29 June	NACA flight 40, Joseph A. Walker, Pilot check. Mach 0.82.
3 Jul.	NACA flight 41, Walker. Buffeting, tail loads, pilot checkout.
17 Jul.	NACA flight 42, Walker. Buffeting, tail loads, longitudinal stability. (Cut short for low fuel—flight made without tip tanks.)
20 Jul.	NACA flight 43, Walker. Buffeting, tail loads, longitudinal stability.
26 Jul.	NACA flight 44, Walker. Buffeting, tail loads, longitudinal stability. Mach 0.83; cut short, cloud formation preventing tests at altitudes above 15,000 ft.
30 Jul.	NACA flight 45, Walker. Buffeting, tail loads, longitudinal stability. Mach 0.85.
2 Aug.	NACA flight 46, Walker. Buffeting, tail loads, longitudinal stability. Mach 0.84.
7 Aug.	NACA flight 47, Jones. Buffeting, tail loads, longitudinal stability. Mach 0.86.
10 Aug.	NACA flight 48, Walker. Flight shortened due to a fuel leak.
20 Aug.	NACA flight 49, Walker. Buffeting, tail loads, longitudinal stability. Mach 0.875 (0.90 true).
22 Aug.	NACA flight 50, Walker. Flight cut short, hydraulic failure.
30 Aug.	NACA flight 51, Walker. Buffeting, tail-loads, longitudinal-stability investigation. Mach 0.86.
6 Sept.	NACA flight 52, Walker. Buffeting, tail loads, longitudinal stability. Mach 0.84.
14 Sept.	NACA flight 53, Walker. Buffeting, tail loads. Mach 0.84.
18 Oct.	NACA flight 54, Walker. Lateral stability and landing programs. Mach 0.86.
19 Oct.	NACA flight 55, Stanley P. Butchart.. Pilot check.
9 Nov.	NACA flight 56, Butchart. Buffeting, tail loads, lateral stability, and landing programs.

D-558-1 #3, BuAer No. 37972, Flights 1952-1953

27 June 1952	NACA flight 57, Crossfield. Beginning of lateral stability and control (aileron effectiveness) investigation.
2 Jul.	NACA flight 58, Crossfield. Lateral stability and control. Mach 0.81.
17 Jul.	NACA flight 59, Butchart. Lateral stability and control.
22 Jul.	NACA flight 60, Butchart. Lateral stability and control.
31 Jul.	NACA flight 61, Butchart. Lateral stability and control.
6 Aug.	NACA flight 62, Butchart. Lateral stability and control.
12 Aug.	NACA flight 63, Butchart. Lateral stability and control. Completion of lateral stability (aileron effectiveness) program.
29 Jan. 1953	NACA flight 64, Butchart. Dynamic stability investigation.
6 Feb.	NACA flight 65, Butchart. Dynamic stability investigation.
11 Feb.	NACA flight 66, Butchart. Dynamic stability investigation.
17 Feb.	NACA flight 67, Butchart. Dynamic stability investigation.
20 Feb.	NACA flight 68, Butchart. Dynamic stability investigation.
27 Mar.	NACA flight 69, John B. McKay. Pilot check.
1 Apr.	NACA flight 70, McKay. Flight for dynamic stability fill-in data.
2 Apr.	NACA flight 71, McKay. Flight for dynamic stability fill-in data.
7 May	NACA flight 72, McKay. Beginning investigation of tip tanks' effect on Skystreak's buffet characteristics. Aborted, leak in tip tank.
12 May	NACA flight 73, McKay. Tip tank-buffet investigation. No records taken.
13 May	NACA flight 74, McKay. Tip tank-buffet investigation.
20 May	NACA flight 75, McKay. Buffeting.
2 June	NACA flight 76, McKay. Tip tank-buffet investigation.
3 June	NACA flight 77, McKay. Tip tank-buffet investigation.
10 June	NACA flight 78, Crossfield. Low-speed stability-and-control-in-coordinated-turns investigation. Last research flight flown by Skystreak.

D-558-2 #1, BuAer No. 37973, Flights

This aircraft completed 122 flights during its Douglas contractor program. The first flight was on 4 Feb. 1948, by John F. Martin. After initial flight testing and addition of its rocket engine, Douglas began performance investigation in the aircraft on 25 Oct. 1949 on the 57th flight by William Bridgeman. Douglas delivered the craft to the NACA on 31 Aug. 1951. The NACA sent it to Douglas in 1954 for all-rocket air-launch modification, for external stores tests at supersonic speeds. The aircraft returned to Edwards on 15 Nov. 1955. NACA research pilot John McKay completed a familiarization flight on 17 Sept. 1956, but the NACA subsequently canceled the remaining planned program.

D-558-2 #2, BuAer No. 37974, Flights

NACA Jet-Powered Flights, Robert A. Champine and John H. Griffith, pilots

24 May 1949	NACA flight 1, Champine. Pilot and instrument check, general handling qualities. Mach 0.74.
1 June	NACA flight 2, Champine. Longitudinal and lateral stability and control, wing loading, bending, and twisting. Mach 0.85.
13 June	NACA flight 3, Champine. Longitudinal and lateral stability and control, wing and tail loads.
21 Jul.	NACA flight 4, Champine. Unsuccessful airspeed calibration, airspeed-altitude recorder failure.
27 Jul.	NACA flight 5, Champine. Successful air-speed calibration, using tower passes.
3 Aug.	NACA flight 6, Champine. Lateral control investigation.
8 Aug.	NACA flight 7, Champine. Longitudinal stability and control; inadvertent pitch-up to 6 g during a 4 g turn at Mach 0.60.
24 Aug.	NACA flight 8, Champine. Longitudinal stability and lateral control investigation during maneuvering flight. Mach 0.855.
30 Aug.	NACA flight 9, Champine. Aborted after takeoff, fluctuations in engine RPM and oil pressure.
12 Sept.	NACA flight 10, Griffith. Longitudinal and lateral stability and control. Only partial completion of mission; one JATO bottle failed to drop.
13 Sept.	NACA flight 11, Griffith. Longitudinal and lateral stability and control. High engine temperatures.
10 Oct.	NACA flight 12, Champine. Longitudinal and lateral stability and control, stall characteristics.
14 Oct.	NACA flight 13, Griffith. Same as flight 12.
1 Nov.	NACA flight 14. Inadvertent pitch-up and snap-roll, later stall, pitch-up followed by a spin.
21 Nov.	NACA flight 15, Champine. Lateral control and static lateral and directional stability investigation (aileron rolls). Mach 0.855.
22 Nov.	NACA flight 16, Griffith. Same as flight 15.
23 Nov.	NACA flight 17, Griffith. Same as flight 15.
7 Dec.	NACA flight 18, Champine. Same as flight 15.
30 Dec.	NACA flight 19, Griffith. Stall investigation with tufts.
6 Jan. 1950	NACA flight 20, Griffith. Same as flight 19.
6 Jan.	NACA flight 21, Griffith. Same as flight 19.

Douglas Air-Launch Rocket Flight

William B. Bridgeman, pilot

Note: D-558-2 #2 (37974) arrived at Edwards from Douglas via B-29 (P2B-1S) launch aircraft on 8 Nov. 1950. D-558-2 #2 was turned over to the NACA on 31 Aug. 1951.

26 Jan. 1951	Douglas flight 24. Air launch at 32,000 ft, climb to 41,000 ft, level run to Mach 1.28. Dutch-roll oscillation, loss of elevator effectiveness noted.
5 Apr.	Douglas flight 25. Drop at 34,000 ft, maximum Mach of about 1.36 at 46,500 ft. Severe lateral oscillation forced Bridgeman to shut off engine prematurely. Rudder lock subsequently installed to control rapid rudder oscillation.
18 May	Douglas flight 26. Launch at 34,540 ft., maximum Mach of 1.7 at 61,650 ft. Rudder locked at all speeds above Mach 1.0.
11 June	Douglas flight 27. Mach 1.79 at 58,600 ft. Low lateral dynamic stability, also a lightly damped longitudinal oscillation noted at burnout.
23 June	Douglas flight 28. Mach 1.85 at 62,000 ft. Violent oscillations in all 3 axes necessitated engine shutdown. Wing rolling ± 80 deg. (1.5 radians per sec).
7 Aug.	Douglas flight 29. Mach 1.88 at 68,600 ft. Dynamic lateral instability not as severe on this flight, for Bridgeman did not push over to as low an angle of attack as on previous flights.
15 Aug.	Douglas flight 30. Altitude flight to 77,800 ft. Unofficial world's altitude record.

NACA Air-Launch Rocket Flights, 1951-1952

Aircraft delivered to NACA HSFRS 31 Aug. 1951

28 Sept.	NACA flight 1, A. Scott Crossfield. Pilot check, Mach 1.2, rough engine operation. Exploratory transonic flight.
12 Oct.	NACA flight 2, Crossfield. Stick impulses and rudder kicks, Mach 1.28.
13 Nov.	NACA flight 3, Crossfield. Mach 1.11. Longitudinal and lateral stability and control, loads data, and aileron effectiveness.
16 Nov.	NACA flight 4, Crossfield. Same as flight 3. Maximum Mach ca. 1.65 at 60,000 ft.
13 June 1952	NACA flight 5, Crossfield. Lateral stability and control, vertical tail loads. Mach 1.36.
18 June	NACA flight 6, Crossfield. Stability and control, loads in low supersonic flight. Mach 1.05.
26 June	NACA flight 7, Crossfield. Same as flight 6. Mach 1.35.
10 Jul.	NACA flight 8, Crossfield. Longitudinal stability and tail loads. Mach 1.68 at 55,000 ft.
15 Jul.	NACA flight 9, Crossfield. Intended to investigate high lift at maximum Mach No. Engine malfunction limited speed to Mach 1.05.
23 Jul.	NACA flight 10, Crossfield. High lift investigation at maximum Mach. Mach ca. 1.51.
13 Aug.	NACA flight 11, Crossfield. Aborted after launch, lox prime valve remained open.
10 Oct.	NACA flight 12, Crossfield. Longitudinal stability at supersonic speeds. Mach 1.65. Pitch-up due to trim variation.
23 Oct.	NACA flight 13, Crossfield. Severe instability in a wind-up turn at Mach 0.9. Maximum Mach 1.10.

NACA Air-Launch Rocket Flights, 1953

26 Mar 1953	NACA flight 14, Crossfield. Turn and sideslips at supersonic speeds. Mild pitch-up at Mach 1.4.
2 Apr.	NACA flight 15, Crossfield. Lateral stability and handling qualities investigation. Beginning of series of flights to evaluate lateral stability at various angles of attack above Mach 1.
3 Apr.	NACA flight 16, Crossfield. Lateral stability investigation.
21 Apr.	NACA flight 17, Crossfield. Lateral stability investigation.
9 June	NACA flight 18, Crossfield. Supersonic lateral stability investigation. Mach 1.7.
18 June	NACA flight 19, Crossfield. Aborted after drop; rockets malfunctioned, no data.
5 Aug.	NACA flight 20, Crossfield. Lateral stability investigation. Mach 1.878.
14 Aug.	NACA flight 21, Lt. Col. Marion Carl, USMC. Unsuccessful altitude attempt.
18 Aug.	NACA flight 22, Carl. Unsuccessful altitude attempt.
21 Aug.	NACA flight 23, Carl. Successful altitude flight to 83,235 ft.
31 Aug.	NACA flight 24, Carl. Maximum Mach flight attempt. Mach 1.5. Violent lateral and longitudinal motions.
2 Sept.	NACA flight 25, Carl. Maximum Mach flight attempt, to Mach 1.728 at 46,000 ft.
17 Sept.	NACA flight 26, Crossfield. 1st flight with nozzle extensions. Mach 1.85 at 74,000 ft.
25 Sept.	NACA flight 27, Crossfield. Lateral directional and longitudinal stability and control. Mach 1.8 at ca. 55,000 ft. Temporarily uncontrollable in roll.
7 Oct.	NACA flight 28, Crossfield. Effects of angle of attack on lateral dynamic behavior with nozzle extensions.
9 Oct.	NACA flight 29, Crossfield. To obtain data on effect of rocket-nozzle extensions on rudder-hinge-moment parameter.
14 Oct.	NACA flight 30, Crossfield. Dynamic lateral behavior and longitudinal control. Attained ca. Mach 1.96.
29 Oct.	NACA flight 31, Crossfield. Lateral stability, longitudinal control. No. 2 chamber failed to ignite, engine shut down prematurely. Subsonic flight only.
4 Nov.	NACA flight 32, Crossfield. Aerodynamic loads and longitudinal control research flight.
13 Nov.	NACA flight 33, Crossfield. Lateral and longitudinal stability and control, loads research.
20 Nov.	NACA flight 34, Crossfield. First Mach 2.0 flight. Plane attained Mach 2.005 in slight dive at 62,000 ft.
11 Dec.	NACA flight 35, Crossfield. Aborted, rocket system malfunction.
16 Dec.	NACA flight 36, Crossfield. For rudder- hinge-moment data with rocket-nozzle extensions.

NACA Air-Launch Rocket Flights, 1954

9 Jul. 1954	NACA flight 37, Crossfield. Dynamic lateral stability and structural temperature.
14 Jul.	NACA flight 38, Crossfield. Dynamic stability and control, wing and tail loads.
21 Jul.	NACA flight 39, Crossfield. Stability and control, loads, and pressure distributions.
23 Jul.	NACA flight 40, Crossfield. Static and dynamic stability and control, loads, and pressure distribution. Mach 1.7 at 60,000 ft.
6 Aug.	NACA flight 41, Crossfield. Dynamic stability-and-control data.
13 Aug.	NACA flight 42, Crossfield. Same as flight 40. Pitch-up encountered at Mach 1.08, plane pitched to 5.8g with heavy buffeting.
20 Aug.	NACA flight 43, Crossfield. Structural loads, pressure distribution, and dynamic stability.
17 Sept.	NACA flight 44, Crossfield. Pressure distributions and structural loads.
22 Sept.	NACA flight 45, Crossfield. Dynamic stability.
4 Oct.	NACA flight 46, Crossfield. To obtain dynamic lateral stability data at speeds up to Mach 1.5.
27 Oct.	NACA flight 47, Crossfield. To obtain dynamic stability data up to Mach 1.8. Engine shut down, pump overspeed during climb.

NACA Air-Launch Rocket Flights, 1955

18 Mar 1955	NACA flight 48, Crossfield. For pressure- distribution and buffeting data.
29 Apr.	NACA flight 49, Joseph A. Walker. Pilot familiarization and lateral-stability data.
5 May	NACA flight 50, Lt. Col. Frank K. Everest, Jr., USAF. Pilot familiarization in preparation for X-2 program. Mach 1.46 at 68,000 ft.
6 May	NACA flight 51, Walker. For lateral stability-and-control data at low supersonic speeds.
12 May	NACA flight 52, Crossfield. For wing and horizontal stabilizer pressure- distribution data to Mach 1.75.
19 May	NACA flight 53, Crossfield. To gather lateral stability and structural loads data to Mach 1.6; aborted when fire warning indicator came on. Subsequently, nozzle extensions removed from plane.
8 June	NACA flight 54, Crossfield. Lateral stability and aerodynamic loads data to Mach 1.67 at 60,000 ft.
21 June	NACA flight 55, Crossfield. Static and dynamic stability investigation to Mach 1.4. End of pressure-distribution program. Recording manometers removed from aircraft.
1 Jul.	NACA flight 56, Crossfield. Supersonic dynamic stability and structural loads investigation.
27 Jul.	NACA flight 57, Crossfield. Same as flight 56.
3 Aug.	NACA flight 58, Crossfield. Same as flight 56.
12 Aug.	NACA flight 59, Crossfield. Same as flight 56.
24 Aug.	NACA flight 60, Crossfield. Same as flight 56.
2 Sept.	NACA flight 61, Crossfield. Dynamic stability investigation. Beginning of vertical tail-loads research program. One rocket cylinder failed to ignite, so plane limited to Mach1.25 at 45,000 ft.
16 Sept.	NACA flight 62, John B. McKay. Pilot familiarization, but some data on stability and control and tail loads taken. McKay had to use emergency hydraulic system to lower landing gear on this flight.
4 Nov.	NACA flight 63, Walker. Dynamic stability and structural loads investigation. Mach 1.34. Following this flight, nozzle extensions were again fitted to the LR-8 engine.
10 Nov.	NACA flight 64, McKay. Structural heating survey. Stable Mach number and altitude not held for sufficient time to stabilize temperatures.
5 Dec.	NACA flight 65, McKay. Same as flight 64, Mach 1.2.

NACA Air-Launch Rocket Flights, 1956

24 Jan. 1956	NACA flight 66, McKay. Same as flight 64, Mach 1.25. Structural heating investigation program canceled after this flight because temperatures did not stabilize.
22 Mar.	NACA flight 67, McKay. Plane jettisoned in inflight emergency from P2B (runaway prop on #4 engine). McKay jettisoned propellants and made safe landing on lakebed. P2B required extensive repairs.
17 Aug.	NACA flight 68, McKay. Vertical tail-loads investigation to Mach 1.1.
25 Sept.	NACA flight 69, McKay. Same as flight 68. Completion of vertical tail-loads research program.
9 Oct.	NACA flight 70, McKay. Static and dynamic stability investigation to approximately Mach 1.5.
19 Oct.	NACA flight 71, McKay. Same as flight 70.
1 Nov.	NACA flight 72, McKay. Same as flight 70.
7 Nov.	NACA flight 73, McKay. Same as flight 70.
14 Dec.	NACA flight 74, McKay. For dynamic stability data at Mach 1.4, and to obtain overall sound-pressure levels in aft fuselage at subsonic and supersonic speeds.
20 Dec.	NACA flight 75, McKay. Same as flight 74. This was last NACA research flight on D-558-2 #2.

D-558-2 #3, BuAer No. 37975, Flights, 1950-1951

15 Douglas flights completed before aircraft modified to air-launch configuration. First flight, 8 Jan. 1949. First rocket-powered flight 25 Feb. 1949. Last flight, 9 Sept. 1949. Eugene F. May, pilot

8 Sept. 1950	Douglas flight 16; Bridgeman pilot, 1st airdrop. Flight aborted after launch, airspeed system malfunction.
20 Sept.	Douglas flight 17, Bridgeman, 2d airdrop. To obtain airspeed calibration data.
29 Sept.	Douglas flight 18, Bridgeman, 3d airdrop. Same purpose as flight 17.
6 Oct.	Douglas flight 19, Bridgeman. Airspeed Calibration.
17 Nov.	Douglas flight 20, Bridgeman. Airspeed calibration, air-launch demonstration.
27 Nov.	Douglas flight 21, Bridgeman. Airspeed calibration, air-launch demonstration. Turbojet engine malfunction, premature rocket shutdown.
15 Dec.	Plane delivered to NACA HSFRS, designated NACA 145.
22 Dec.	NACA flight 1, A. Scott Crossfield. Pilot and instrument check, jet engine only.
27 Dec.	NACA flight 2, Crossfield. Same as flight 1.
27 Mar. 1951	NACA flight 3, Crossfield. Slat-loads investigation, jet only. Stalls, turns, rolls, to Mach 0.7.
20 Apr.	NACA flight 4, Crossfield. Stability, tail loads, and landing problems.
17 May	NACA flight 5, Crossfield. First NACA rocket-jet flight. Jet engine shut off, flame instability. Mach 0.86 maximum. Frosting of windshield.
17 Jul.	NACA flight 6, Crossfield. Jet only, rocket failed to fire, valve failure. Mach 0.84 maximum.
20 Jul.	NACA flight 7, Walter P. Jones, Pilot check, jet only. Mach 0.73.
9 Aug.	NACA flight 8, Crossfield. Rolls and accelerated turns to Mach 1.14. Jet and rocket.
14 Aug.	NACA flight 9. Brig. Gen. Albert Boyd, USAF. Pilot check. Jet and rocket.
22 Aug.	NACA flight 10, Jones. Jet and rocket, lateral and longitudinal stability investigation. Aileron rolls, elevator pulses to Mach 1.10.
18 Sept.	NACA flight 11, Jones. Jet only, rocket failure. Longitudinal, lateral stability investigation with accelerated pitching maneuver in landing configuration. Pitch-up followed by spin and normal recovery.
26 Sept.	NACA flight 12, Jones. Longitudinal, lateral, and directional maneuvers. Jet and rocket flight to Mach 0.93. Rolls, sideslips, elevator pulses, accelerated turns.
18 Oct.	NACA flight 13, Jones. Beginning of pitch-up investigation. Evaluation of outboard wing fences at Mach 0.73. Fences markedly aid recovery.
9 Nov.	NACA flight 14, Jones. Same as flight 13. Mach 0.95. Fences subsequently removed.
19 June 1952	NACA flight 15, Crossfield. Jet only. Pitch-up investigation with slats locked open. Mach 0.7.
3 Jul.	NACA flight 16, Jones. Same as flight 15. Mach 0.96.
31 Jul.	NACA flight 17, Crossfield. Jet and rocket. Slat investigation, aborted in climb, faulty cabin heating. Some low-speed data.
8 Aug.	NACA flight 18, Crossfield. Jet and rocket. Same as flight 15. Mach 0.96. Inboard wing fences subsequently removed. Plane now in clean, no-fence configuration.
14 Aug.	NACA flight 19, Crossfield. Slats still locked open. Flight to check effect of removing wing fences. Removal indicates inboard fences had little effect on aircraft behavior. Following flight, slats moved and locked in half open position.
8 Oct.	NACA flight 20, Crossfield. Jet and rocket. Evaluation of effect of slats half open on pitch-up. Plane pitched to 36°. Mach 0.97. Slats subsequently restored to free-floating condition.
22 Oct.	NACA flight 21, Crossfield. Jet and rocket. Plane in no-fence configuration. Longitudinal and lateral stability and control investigation. Pitch-ups encountered during turns. Chord extensions subsequently installed on outer wing panels.

D-558-2 #3, BuAer No. 37975, Flights, 1953-1954

27 Feb. 1953	NACA flight 22, Crossfield. Jet only. First flight with chord extensions. Mach 0.7. Wing-up turns and 1 g stalls. Maneuvers terminated when decay in longitudinal or lateral stability became apparent.
8 Apr.	NACA flight 23, Crossfield. Jet only; rocket failed to fire, frozen valve. Wind-up turns, aileron rolls, sideslips, 1 g stalls.
10 Apr.	NACA flight 24, Crossfield. Jet and rocket. Same as flight 23. Pitch-up not alleviated by chord extensions, so extensions removed after flight and slats reinstalled on wings.
15 June	NACA flight 25, Crossfield. Jet and rocket. Slats locked open, "soft" bungee on control stick. Accelerated longitudinal stability maneuvers.
25 June	NACA flight 26, Crossfield. Jet only. Slats locked open, stiff bungee. Combination of bungee and slats out apparently improved controllability.
26 June	NACA flight 27, Stanley P. Butchart. Pilot checkout. Stiff bungee installed. Jet only.
24 Jul.	NACA flight 28, Crossfield. Jet and rocket. Plane in basic configuration. Transonic lateral and directional stability and control. Mach 1.03.
28 Jul.	NACA flight 29. Lt. Col. Marion Carl USMC. Pilot check out in D-558-2 #3 before flying all-rocket D-558-2 #2. Jet power only. No data.
30 Jul.	NACA flight 30, Carl. Jet and rocket. Same as flight 29.
9 Sept.	NACA flight 31, Crossfield. Lateral and directional stability investigation to Mach 1.04.
14 Sept.	NACA flight 32, Crossfield. Same as flight 31.
22 Sept.	NACA flight 33, Crossfield. Same as flight 31. Because of a malfunction, only 2 rocket chambers fired.
10 Dec.	NACA flight 34, Crossfield. Transonic longitudinal stability investigation. Turns, stalls.
22 Dec.	NACA flight 35, Crossfield. Same as flight 34. Jet only, rocket did not ignite. Plane subsequently modified for external-stores program.
6 May 1954	NACA flight 36, Joseph A. Walker. Pilot checkout, plane in basic configuration. Jet only.
12 May	NACA flight 37, Walker. Same as flight 36. Jet and rocket. Mach 0.97.
2 June	NACA flight 38, Crossfield. First flight with external-stores pylons. Jet only. Evaluation of handling qualities to Mach 0.72.
16 June	NACA flight 39, Crossfield. First flight with external stores (1000-lb bomb shapes). Jet only. No apparent adverse effects except higher drag. Mach 0.72.
8 Jul.	NACA flight 40, Crossfield. Jet and rocket. Stores decreased transonic performance and increased buffet. Mach 1.05. Stores shapes later removed as being too small.
19 Jul.	NACA flight 41, Crossfield. Jet and rocket. Plane in clean configuration. Transonic directional and longitudinal stability and control. Mach 1.04. Sideslips, elevator and rudder pulses.
23 Jul.	NACA flight 42, Crossfield. Jet and rocket. Transonic lateral stability and control investigation. Rolls to Mach 1.04.
28 Jul.	NACA flight 43, Crossfield. Jet and rocket. Same as flight 42. Mach 1.04.
9 Aug.	NACA flight 44, Crossfield. Jet and rocket. Dynamic stability investigation to Mach 1.04. Elevator, aileron, and rudder pulses/kicks.
11 Aug.	NACA flight 45, Crossfield. Jet and rocket. Dynamic stability investigation to Mach 0.94. Elevator, aileron, and rudder pulses/kicks.
18 Aug.	NACA flight 46, Crossfield. Jet and rocket. Dynamic stability investigation to Mach 0.92. Elevator, aileron, and rudder pulses/kicks.
30 Aug.	NACA flight 47, Crossfield. Jet and rocket. Slats unlocked, flight for longitudinal stability and control and buffet characteristics of aircraft in this configuration.
8 Oct.	NACA flight 48, Crossfield. Resumption of stores investigation. Handling qualities with 150 gal tanks. Jet only. Mach 0.73. More buffeting than a clean airplane.
21 Oct.	NACA flight 49, Crossfield. Jet and rocket. Otherwise, same as flight 48. No adverse effects, but pilot noted drag rise and heavier buffet in longitudinal maneuvers. As a result of strain-gauge-loads measurements, stores program again temporarily suspended while Douglas checked strength factor of pylon and wing.
23 Dec.	NACA flight 50, Lt. Col. Frank K. Everest, Jr., USAF. Pilot checkout, jet-and-rocket flight in clean configuration in preparation for Bell X-2 program.
29 Dec.	NACA flight 51, John B. McKay. Pilot check in clean configuration, jet only.

D-558-2 #3, BuAer No. 37975, Flights, 1955-1956

27 Apr. 1955	NACA flight 52, McKay. Jet and rocket. Underwing pylons installed. Sideslips, rolls, elevator and rudder pulses. For handling qualities, wing and pylon loads, and buffet data. Mach 1.0.
23 May	NACA flight 53, McKay. Jet and rocket, 150-gal stores attached. Same maneuvers as flight 52. Buffet levels higher with stores than with pylons only.
3 June	NACA flight 54, McKay. Jet and rocket. Pylons only. Same maneuvers as flight 52. Mach 1.0.
10 June	NACA flight 55, McKay. Jet and rocket. Same as flight 54.
17 June	NACA flight 56, McKay. Jet and rocket. 150-gal stores attached. Same maneuvers as flight 52.
24 June	NACA flight 57, McKay. Jet and rocket. Same as flight 56.
28 June	NACA flight 58, McKay. Jet and rocket. Same as flight 56.
30 Aug.	NACA flight 59, McKay. Jet and rocket. Same as flight 56 but with 180-inch stores. Plane slightly damaged on landing when tail cone touched lake first.
2 Nov.	NACA flight 60, Butchart. Jet and rocket. Same purpose as flight 59.
8 Nov.	NACA flight 61, McKay. Jet and rocket. Same as flight 59.
17 Nov.	NACA Flight 62, McKay. Jet and rocket. Same as flight 59.
8 Dec.	NACA flight 63, McKay. Jet and rocket. Same as flight 59. Concluded stores- investigation program. Plane returned to clean configuration.
1 Feb. 1956	NACA flight 64, McKay. Jet and rocket. To obtain wing-loads data for comparison with external-stores data previously acquired. Lateral, directional, and longitudinal maneuvers. Mach 1.0.
3 Feb.	NACA flight 65, McKay. Jet and rocket. Rocket engine pump overspeed prevented acquisition of data at Mach 0.9. Flight for same purpose as flight 64, so one more flight scheduled to complete research program.
28 Aug.	NACA flight 66, McKay. Jet and rocket.Same as flight 64. Mach 0.96. Completed research program on this aircraft.

Sources: Hallion, *Supersonic Flight*, pp. 221-34; pilot reports and transcripts of radio communications among pilots, chase, and control room; project engineers' progress reports; records of the women computers, all in NASA Dryden Historical Reference Collection.

Appendix H
X-2 Program Flight Chronology, 1954-1956

While NACA pilots never flew the X-2 research aircraft, the High-Speed Flight Station supported the X-2 program with advice and data analysis. During the program, the X-2 spent many weeks in the NACA Calibration Hangar (Building 4801) for instrumentation installation. Also, the NACA expected to receive the aircraft following its Air Force flight-test program.

Two X-2 aircraft were built, the X-2 #1 (46-674) and the X-2 #2 (46-675). The X-2 #2 was lost in an inflight explosion near the Bell plant at Wheatfield, New York during captive flight trials on 12 May 1953. Two crewmen were killed. The X-2 #1 arrived at Edwards AFB for testing in the summer of 1954. The following chronology is for the X-2 #1, 46-674, from 1954 through the crash of this aircraft in 1956. It was air-launched from a modified Boeing B-50 bomber.

Bell X-2 #1 (46-674)

Date	Pilot	Remarks
13 July 1954	Maj. F. K. Everest	Captive flight at Wheatfield, New York
5 Aug	Everest	Captive flight at Edwards AFB, California
5 Aug.	Everest	1st glide flight. Damaged on landing.
5 Feb. 1955	Everest	Captive flight, propellant check
21 Feb.	Everest	Captive flight, propellant check
1 Mar.	Everest	Captive flight, propellant and pressure check
8 Mar.	Everest	2nd glide flight. Propellant system check. Minor damage on landing.
5 Apr.	Everest	Planned glide flight aborted. Completed as captive flight.
6 Apr.	Everest	3rd glide flight. Damaged on landing. Following flight, plane returned to Bell plant for landing gear system modifications and installation of its rocket engine.
25 Oct.	Everest	Planned powered flight aborted. Completed as 4th glide flight.
17 Nov.	Everest	Planned powered flight aborted. Completed as captive flight.
18 Nov.	Everest	1st powered flight. Mach 0.992 at 36,750 ft. Slight fire damage from engine bay fire.
10 Dec.	Everest	Planned powered flight aborted. Completed as captive flight.
10 Dec.	Everest	Planned powered flight aborted. Completed as captive flight.
18 Dec.	Everest	Planned powered flight aborted. Completed as captive flight.
24 Mar. 1956	Everest	Flight 1-56 (2nd powered flight), Mach 0.985 at 47,185 ft.
25 Apr.	Everest	Flight 2-56 (3rd powered flight), Mach 1.381 at 52,087 ft.
1 May	Everest	Flight 3-56 (4th powered flight), Mach 1.684 at 55,716 ft.
11 May	Everest	Flight 4-56 (5th powered flight), Mach 1.887 at 59,149 ft.
22 May	Everest	Flight 5-56 (6th powered flight), Mach 2.531 at 60,824 ft.
25 May	Capt. I. C. Kincheloe	Flight 6-56 (7th powered flight), pilot checkout, Mach 1.097 at 46,159 ft.
11 Jul.	Everest	Flight 7-56 (8th powered flight), premature engine shutdown. Mach 1.577 at 67,780 ft.
23 Jul.	Everest	Flight 8-56 (9th powered flight), Mach 2.87 at 70,348 ft.
3 Aug.	Kincheloe	Flight 9-56 (10th powered flight), Mach 2.567, 87,413 ft.
8 Aug.	Kincheloe	Flight 10-56 (11th powered flight), premature engine shutdown. Mach 1.725 at 78,610 ft.
7 Sept.	Kincheloe	Flight 11-56 (12th powered flight), Mach 2.442 at 119,774 ft. Reached 126,200 ft.
13 Sep.	Kincheloe	Planned powered flight aborted. Completed as captive flight.
14 Sep.	Kincheloe	Planned powered flight aborted. Completed as captive flight.
17 Sep.	Kincheloe	Planned powered flight aborted. Completed as captive flight.
27 Sep.	Capt. M. Apt	Flight 12-56 (13th powered flight), Mach 3.2 at 72,224 ft. Subsequent loss of control from inertial coupling led to the destruction of the aircraft and the death of the pilot.

Source: X-2 flight progress reports, 1954-1956, as corrected by flight log of Roxannah Yancey from the Computing Service office, held in the NASA Dryden Historical Reference Collection. Information on captive flights from Henry Matthews, *The Saga of Bell X-2, First of the Spaceships: The Untold Story* (Beirut, Lebanon: HPM Publications, 1999), p. 125.

Appendix I
X-3 Program Flight Chronology, 1954-1956

The X-3 (serial number 49-2892) arrived at Edwards AFB on 11 September 1952. Bill Bridgeman conducted the first phase of testing for the Douglas Aircraft Company. Following completion of contractor testing (1953), the NACA received the sole Douglas X-3 (49-2892) for testing in 1953, with Air Force pilots flying the initial flights. All subsequent NACA flights were flown by High-Speed Flight Station research pilot Joseph A. Walker. On 28 December 1956, the X-3 was delivered to the USAF Museum in Dayton, Ohio, for permanent display.

Pilots included:

William "Bill" Bridgeman, Douglas Aircraft Co.	26 flights
Lt. Col. Frank "Pete" Everest, USAF	3 flights
Maj. Charles E. "Chuck" Yeager, USAF	3 flights
Joseph A. Walker, NACA	20 flights

X-3 Flights

Contractor (Douglas) Flights (Bill Bridgeman)

Flight	Date	Remarks
	15 Oct. 1952	Inadvertant takeoff during high-speed taxi test. Flew one mile.
1	20 Oct.	Official first flight. Aircraft check out. Mach 0.796. Alt. 13,560 ft.
2	31 Oct.	Mach 0.512. Alt. 7,282 ft. Low-speed stability and control.
3	30 Apr. 1953	Mach 0.910. Alt. 16,692 ft.
4	5 Jun.	Mach 0.874. Alt. 16,252 ft.
5	11 Jun.	Mach 0.979. Alt. 26,073 ft.
6	25 Jun.	First supersonic run: Mach 1.034. Alt. 33,847 ft.
7	15 Jul.	Mach 1.110. Alt. 36,814 ft.
8	22 Jul.	Mach 1.196. Alt. 36,723 ft.
9	28 Jul.	Fastest flight: Mach 1.208. Alt. 37,264 ft.
10	6 Aug.	Mach 1.166. Alt. 33,548 ft.
11	13 Aug.	Mach 1.065. Alt. 28,570 ft.
12	18 Aug.	Mach 1.132. Alt. 27,850 ft.
13	20 Aug.	Mach 1.153. Alt. 29,173 ft.
14	26 Aug	Mach 1.179. Alt. 28,441 ft.
15	28 Aug.	Mach 1.142. Alt. 28,111 ft.
16	2 Sept.	Mach 1.153. Alt. 27,418 ft.
17	4 Sept.	Mach 0.941. Alt. 25,450 ft.
18	11 Sept.	Mach 0.910. Alt. 16,555 ft.
19	17 Sept.	Mach 1.143. Alt. 34,714 ft.
20	18 Sept.	Mach 0.896. Alt. 2,200 ft.
21	24 Sept.	Mach 1.167. Alt. 25,333 ft.
22	1 Oct.	Mach 0.876. Alt. 20,637 ft.
23	21 Oct.	Mach 1.180. Alt. 18,738 ft.
24	21 Nov.	No data worked up.
25	1 Dec.	No data worked up. Last contractor flt.

X-3 Flights

USAF Flights (Lt. Col. "Pete" Everest and Maj. Charles "Chuck" Yeager)[1]

Flight	Date	Remarks
1	23 Dec. 1953	Pilot check out. Everest. Mach 1.091. Alt. 35,697 ft.
2	29 Dec	Mach 0.977. Alt. 26,866 ft. Yeager, check out.
3	2 Jul. 1954	Mach 0.912. Alt. 41,318 ft (highest flight). Everest.
4	15 Jul.	Mach 1.057. Pilot check out. Yeager.
5	27 Jul.	Mach 1.075. Alt. 27,706 ft. Everest.
6	29 Jul.	Yeager. No data. Last USAF flight.

X-3 Flights

NACA Flights (Joseph A. Walker)

Flight	Date	Remarks
1	23 Aug 1954	Pilot check out. Mach 1.073.
2	3 Sept.	Static longitudinal stability and control, wing and tail loads, and pressure.
3	9 Sept.	Static longitudnal stability and control, wing and tail loads, pressure distribution. Mach 1.171. Alt. 37,767 ft.
4	9 Sept.	Static longitudinal stability and control, wing and tail loads, pressure distribution. Mach 1.159. Alt. 34,328 ft.
5	16 Sept.	Static longitudinal stability and control, wing and tail loads, pressure distribution. Mach 0.830. Alt. 30,183 ft.
6	19 Oct.	Static longitudinal stability and control, wing and tail loads, pressure distribution. Mach 1.155. Alt. 36,655 ft.
7	21 Oct.	Static longitudinal stability and control, wing and tail loads, pressure distribution. Mach 1.100. Alt. 34,899 ft.
8	21 Oct.	Static longitudinal stability and control, wing and tail loads, pressure distribution. Mach 0.992. Alt. 32,562 ft.
9	21 Oct.	Lateral directional stability and control. Mach 1.126. Alt. 30,712 ft.
10	27 Oct.	Lateral directional stability and control. Mach 1.154. Alt. 32,356 ft. Aircraft experienced violent roll coupling and was grounded for inspection.
11	20 Sept. 1955	Static longitudinal stability and control, wing and tail loads, pressure distribution. Mach 1.143. Alt. 36,755 ft.
12	22 Sept.	Static longitudnal stability and control, wing and tail loads, pressure distribution. Mach 1.047. Alt. 33,012 ft.
13	6 Oct.	Directional stability and control, vertical tail loads. Mach 1.115. Alt. 34,647 ft.
14	12 Oct.	Directional stability and control, vertical tail loads. Drag chute inadvertently deployed in flt. Mach 0.986. Alt. 33,684 ft.
15	20 Oct.	Directional stability and control, vertical tail loads. Mach 1.060. Alt. 33,782 ft.
16	21 Oct.	Directional stability and control, vertical tail loads. Mach 0.901. Alt. 25,798 ft.
17	25 Oct.	Directional stability and control, vertical tail loads. Mach 1.207. Alt. 30,780 ft.
18	13 Dec.	Control system evaluation. One engine damaged from ingestion of pressure probe; aircraft grounded for repairs. Mach 1.069. Alt. 30,853 ft.
19	6 Apr. 1956	Pressure distribution measurements. Fire damaged test instrumentation in nose compartment. Mach 1.131. Alt. 29,401 ft.
20	23 May	Lateral control investigation. Last flight. Mach 1.127. Alt. 29,573 ft. Aircraft retired.

Sources: X-3 flight progress reports, 1954-1956; records of the women computers; flight log compiled by Peter Merlin. Last two sources in the NASA Dryden Historical Reference Collection.

[1] The NACA counted these as NACA flights even though the pilots were from the Air Force.

Appendix J
X-4 Program Flight Chronology, 1950-1953

The NACA High-Speed Flight Research Station operated the Northrop X-4 #2 research airplane (46-677) from 1950 through 1953. The NACA also had the X-4 #1 (46-676) but used it only for spares support, although Northrop test pilot Charles Tucker had flown it at Muroc in ten flights from 15 December 1948 to 26 January 1950. Tucker also flew 46-677 in a series of 20 contractor flights from 7 June 1949 to 17 February 1950. In cooperation with the NACA, the Air Force Air Materiel Command then ran a brief program on the second craft (46-677) during the summer of 1950 before delivering it to the NACA. Since NACA instrumentation was carried and data was collected on these flights, they were also logged as NACA test missions.

X-4 Flights, 1950

Flight	Date	Pilot	Remarks
1	18 Aug. 1950	C.E. Yeager	AF flight, pilot check.
2	22 Aug	F.K. Everest	AF flight, pilot check.
3	22 Aug.	Everest	Aborted, landing gear malfunction.
4	30 Aug.	Everest	Handling qualities.
5	31 Aug.	Everest	Same as flight 4.
6	13 Sept	Yeager	Longitudinal and latitudinal stability and control.
7	15 Sept	Yeager	Aborted, faulty canopy lock.
8	18 Sept.	Yeager	Longitudinal and latitudinal dynamic stability and control.
9	20 Sept.	Yeager	Same as flight 8.
10	21 Sept.	Yeager	Same as flight 8.
11	22 Sept.	Yeager	Same as flight 8.
12	26 Sept.	Yeager	
13	26 Sept	A. Boyd	AF flight, pilot check-out.
14	27 Sept.	J.S. Nash	Same as flight 13; airspeed calibration.
15	28 Sept.	John Griffith	First NACA pilot checkflight.
16	28 Sept.	Fred Ascani	AF pilot check-out.
17	29 Sept.	Arthur Murray	AF pilot check-out.
18	29 Sept.	Jack Ridley	AF pilot check-out.
19	7 Nov.	Griffith	Longitudinal, latitudinal, directional stability, and control.
20	17 Nov.	Griffith	Same as flight 19.
21	6 Dec.	R.L. Johnson	AF flight, pilot check.
22	6 Dec.	A.S. Crossfield	NACA pilot check; aborted.
23	15 Dec	Crossfield	Aborted; instrument malfunction.
24	28 Dec.	Crossfield	Longitudinal, latitudinal, directional stability and control.

X-4 Flights, 1951-1952

Flight	Date	Pilot	Remarks
25	4 Jan. 1951	Crossfield	Same as flight 24.
26	17 Jan	Crossfield	Same as flight 24.
27	19 Jan.	Crossfield	Aborted, landing gear malfunction.
28	24 Jan.	Crossfield	Aborted, instrument malfunction.
29	27 Jan.	Crossfield	Longitudinal, latitudinal directional stability, and control.
30	19 Feb.	Crossfield	Same as flight 29.
31	19 Mar.	Crossfield	Same as flight 29.
32	27 Mar.	Crossfield	Same as flight 29.
33	28 Mar.	Walter P. Jones	NACA pilot check-out.
34	12 Apr.	Crossfield	Same as flight 29.
35	13 Apr.	Crossfield	Same as flight 29.
36	17 Apr.	Crossfield	Same as flight 29.
37	20 Apr.	Jones	Same as flight 29.
38	26 Apr.	Crossfield	Same as flight 29.
39	27 Apr.	Jones	Same as flight 29.
40	3 May	Jones	Same as flight 29.
41	9 May	Crossfield	Same as flight 29.
42	16 May	Crossfield	Same as flight 29.
43	18 May	Crossfield	Same as flight 29.
44	29 May	Crossfield	Same as flight 29.
45	20 Aug.	Jones	First flight with thick trailing edge on speed brakes.
46	2 Oct.	Crossfield	Stability and control with thick trailing edge.
47	5 Oct.	Crossfield	Lift-to-drag variation using various speedbrake settings.
48	9 Oct.	Crossfield	Landings at various lift-to-drag ratios.
49	11 Oct.	Crossfield	Same as flight 48.
50	12 Oct.	Jones	Same as flight 48.
51	17 Oct.	Jones	Constant speed-drag ratios.
52	18 Oct.	Joseph Walker	NACA pilot check; handling qualities.
53	19 Oct.	Walker	Maneuvers and speed runs.
54	24 Oct	Jones	General stability and control.
55	6 Mar. 1952	Jones	Determining lift-to-drag at various speedbrake settings.
56	13 Mar.	Jones	Directional trim change investigation.
57	17 Mar.	Jones	Same as flight 55.
58	21 Mar.	Jones	Lift-to-drag variation studies.
59	25 Mar.	Jones	Dynamic stability investigation.
60	26 Mar.	Jones	Same as flight 59.
61	27 Mar.	S.P. Butchart	NACA pilot check
62	19 May	Jones	Check flight with thickened trailing edge on elevons.
63	6 Aug.	Crossfield	Stability and control with thickened elevons.
64	16 Sept	Crossfield	Aborted, engine malfunction.
65	16 Sept	Crossfield	Aborted, instrumentation malfunction.
66	22 Sept	Crossfield	Stability and control with thickened elevons.
67	25 Sept.	Crossfield	Same as flight 66.

X-4 Flights, 1953

Flight	Date	Pilot	Remarks
68	27 Mar. 1953	Crossfield	Stability and control with thickened elevons.
69	29 Apr.	Crossfield	Airspeed calibration with thickened elevons.
70	30 Apr	Crossfield	High-lift stability and control.
71	29 May	Crossfield	Dynamic stability without thickened elevons.
72	1 Jul.	Butchart	Dynamic stability without thickened elevons.
73	3 Jul.	Butchart	Same as flight 70.
74	4 Aug.	George Cooper	NACA pilot check for Ames pilot.
75	11 Aug	John McKay	NACA pilot check.
76	13 Aug.	McKay	Dynamic stability.
77	19 Aug.	McKay	Dynamic stability.
78	21 Aug.	McKay	Dynamic stability.
79	24 Aug.	Crossfield	Dynamic stability.
80	26 Aug.	McKay	Dynamic stability.
81	29 Sept.	McKay	Dynamic stability.

Sources: NACA X-4 flight reports, 1950-1953, held at the National Archives and Records Administration's Federal Records Center at Laguna Niguel, CA; flight log prepared by Peter Merlin from the flight reports; X-4 chronolgy prepared by Robert Mulac of Langley Research Center.

Appendix K
X-5 Program Flight Chronology, 1952-1955

The NACA High-Speed Flight Station operated the Bell X-5 #1 (50-1838) from 1952 to late 1955. Following the conclusion of the contractor's program in October 1951, the airplane was grounded for installation of an NACA instrument package. In December 1951, the Air Force completed a brief evaluation program involving six flights. Because data were taken, these were considered part of the overall NACA effort and were logged as joint AF/NACA flights. The first all-NACA flight was flight 7. The second X-5 (50-1839) was operated only by Bell and the Air Force and was lost in a spin accident in 1953.

Bell X-5 #1 (50-1838) Flights

Contractor (Bell) Flights (Jean ìSkipî Ziegler) and USAF Flight (Brig. Gen. Albert Boyd)

Flight	Date	Pilot	Remarks
1	20 Jun. 1951	Jean Ziegler	First flight. Aircraft check out. Mach 0.560. Alt. 15,974 ft.
2	25 Jun.	Ziegler	No data.
3	27 Jun	Ziegler	Mach 0.638. Alt. 16,200 ft.
4	28 Jun	Ziegler	Mach 0.696. Alt. 41,028 ft
5	16 Jul.	Ziegler	Mach 0.551. Alt. 25,218 ft.
6	17 Jul.	Ziegler	Mach 0.548. Alt. 20,487 ft.
7	18 Jul.	Ziegler	No data.
8	26 Jul.	Ziegler	Mach 0.838. Alt. 35,649 ft.
9	27 Jul.	Ziegler	First wing sweep test (partial sweep only). Mach 0.562. Alt. 20,995 ft.
10	30 Jul.	Ziegler	Mach 0.847. Alt. 27,694 ft.
11	31 Jul.	Ziegler	Mach 0.821. Alt. 22,538 ft.
12	2 Aug.	Ziegler	Mach 0.508. Alt. 20,960 ft
13	3 Aug.	Ziegler	Mach 0.704. Alt. 21,174 ft.
14	16 Aug.	Ziegler	Mach 0.828. Alt. 16,518 ft.
15	20 Aug.	Ziegler	Mach 0.867. Alt. 30,953 ft.
16	21 Aug.	Ziegler	Mach 0.854. Alt. 22,589 ft.
17	22 Aug.	Ziegler	Mach 0.914. Alt. 45,384 ft.
18	23 Aug.	Ziegler	Mach 0.953. Alt. 40,434 ft.
19	27 Aug.	Boyd	USAF check out flight. No data.
20	8 Oct.	Ziegler	Completion of Phase I (contractor) testing.

Bell X-5 #1 (50-1838) Flights

USAF/NACA Flights (Lt. Col. Frank K. "Pete" Everest)

Flight	Date	Pilot	Remarks
1	20 Dec. 1951	"Pete" Everest	Pilot check out. No data.
2	20 Dec.	Everest	Mach 0.844. Alt. 22,380 ft.
3	21 Dec	Everest	Mach 0.909. Alt. 30,752 ft.
4	27 Dec.	Everest	Mach 0.603. Alt. 12,808 ft.
5	7 Jan. 1952	Everest	Mach 0.847. Alt. 31,317 ft.
6	8 Jan.	Everest	Mach 0.858. Alt. 21,841 ft.

Bell X-5 #1 (50-1838) Flights

NACA Flights, 1952

Flight	Date	Pilot	Remarks
7	9 Jan.	J.A. Walker	Pilot check out. Mach 0.630. Alt. 31,756 ft.
8	14 Jan.	Walker	Static and dynamic longitudinal and latitudinal stability and control. Mach 0.946. Alt. 39,695 ft.
9	21 Jan.	Walker	Same as flight 8. Mach 0.963.
10	23 Jan.	Walker	Airspeed calibration.
11	25 Jan.	Walker	Static and dynamic longitudinal and latitudinal stability and control.
12	1 Feb.	Walker	Same as flight 11. Mach 0.971. Alt. 45,363 ft.
13	5 Feb.	Walker	Same as flight 11.
14	12 Feb.	Walker	Same as flight 11.
15	12 Feb.	J. C. Meyer	Air Force pilot check flight.
16	4 Mar.	Walker	Lateral stability at 60-degree sweep.
17	13 Mar.	Walker	Flight aborted early.
18	17 Mar.	Walker	Poor voltage; no data.
19	17 Mar.	Walker	Latitudinal and longitudinal stability and control at 60-degree sweep.
20	19 Mar.	Walker	Airspeed calibration.
21	19 Mar.	Walker	Static and dynamic longitudinal and latitudinal stability and control.
22	20 Mar.	Walker	Gust loads investigation 20- and 60-degree sweep. Sent to Langley.
23	27 Mar.	Walker	Static longitudinal stability at 45-degree wing sweep.
24	1 Apr.	Walker	Airspeed calibration.
25	2 Apr	A. S. Crossfield	Pilot check.
26	29 Apr.	W. P. Jones	Pilot check.
27	2 May	Jones	Flight discontinued; no data.
28	6 May	Jones	Static and dynamic lateral and longitudinal stability and control.
29	7 May	Jones	Same as flight 27.
30	8 May	Crossfield	Same as flight 27.
31	16 May	Crossfield	Same as flight 27.
32	27 May	Jones	Same as flight 27. Mach 0.977.
33	28 May	Jones	Same as flight 27.
34	28 May	Jones	Aborted; gear door opened in flight.
35	21 June	Jones	Static and dynamic longitudinal and lateral stability and control.
36	25 June	Walker	Same as flight 34.
37	26 June	Walker	Same as flight 34.
38	2 Jul.	Walker	Same as flight 34.
39	10 Jul.	Walker	Same as flight 34.
40	12 Jul.	Walker	Gust loads investigation. No data.
41	16 Jul.	J. Reeder	Pilot check for Langley pilot.
42	17 Jul.	Walker	Static and dynamic longitudinal and lateral stability and control.
43	22 Jul.	Walker	Same as flight 41.
44	25 Jul.	Walker	Static and dynamic longitudinal stability and control. Mach 0.986.
45	1 Aug.	Walker	Same as flight 43. Mach 1.006
46	7 Aug.	Walker	Static longitudinal and lateral stability and control. Mach 1.006.
47	23 Sept.	Walker	Lateral control, longitudinal stability.
48	25 Sept.	Walker	Static longitudinal control. Mach 1.032.
49	26 Sept.	Crossfield	Static longitudinal stability and control. No data.
50	21 Oct.	Walker	Inadvertent spin. Mach 1.025.
51	5 Dec.	Walker	Longitudinal stability and control. Mach 1.082.
52	10 Dec.	Walker	Lateral stability and control.
53	11 Dec.	S. P. Butchart	Pilot check.
54	15 Dec.	Butchart	Pilot check.
55	18 Dec.	Walker	Photographic flight. No data.
56	18 Dec.	Butchart	Aborted, inoperable stabilizer motor.
57	22 Dec.	Walker	Static and dynamic longitudinal and lateral stability and control.

Bell X-5 #1 (50-1838) Flights
NACA Flights, 1953-1954

Flight	Date	Pilot	Remarks
58	8 Jan. 1953	Walker	Vertical tail loads in maneuvers.
59	12 Jan.	Crossfield	Stalls and maneuvers at 20-degree wing sweep.
60	22 Jan.	Crossfield	Vertical tail loads in rolling pullouts. No data.
61	27 Jan.	Walker	Drag study behind F-80 jet. No data.
62	29 Jan.	Walker	Drag study behind B-29 bomber.
63	29 Jan.	A. Murray	AF flight for comparison with AF X-5. Mach 1.017. Alt. 44,572 ft.
64	6 Feb.	Walker	Drag study behind B-29. No data.
65	13 Feb.	Walker	Gust loads investigation. No data.
66	20 Feb.	Walker	Gust loads investigation. No data.
67	24 Feb.	Walker	Effect of wing translation on trim.
68	25 Feb.	Walker	Same as flight 66.
69	25 Feb.	Walker	Gust loads investigation.
70	27 Feb.	Walker	Longitudinal stability and control during wing transition. Mach 1.051.
71	26 Mar.	Walker	Emergency landing, gear failure. Mach 0.919
72	23 Apr.	Walker	Longitudinal stability and control.
73	29 Apr.	Walker	Same as flight 71.
74	30 Apr	Walker	Same as flight 71.
75	1 May	Walker	Same as flight 71.
76	13 May	Walker	Strain gauge response to temperature.
77	4 June	Walker	Buffet-induced tail loads.
78	3 Jul	Walker	Wing and horizontal tail loads. No data.
79	21 Jul.	Walker	Dynamic lateral stability.
80	27 Jul.	Walker	Aborted, cabin pressurization malfunction.
81	28 Jul.	Walker	Buffet-induced tail loads.
82	20 Aug.	Walker	Buffet-induced tail loads.
83	25 Aug.	Walker	Longitudinal stability and control; wing and tail loads. No data.
84	27 Aug.	Walker	Same as flight 82. Mach 0.958.
85	28 Aug.	Crossfield	Airspeed calibration of NACA B-47A.
86	4 Sept.	Crossfield	Pacer for NACA B-47 aircraft. Mach 0.83.
87	12 Nov.	Walker	Lateral stability and control.
88	16 Nov.	Walker	Wing twisting and bending tail loads. Mach 1.021. Alt. 43,571 ft.

Bell X-5 #1 (50-1838) Flights
NACA Flights, 1955-1956

Flight	Date	Pilot	Remarks
89	14 Jan. 1954	Walker	Tail loads.
90	21 Jan	Walker	Longitudinal stability and control.
91	26 Jan.	Walker	Aborted, landing gear door malfunction.
92	29 Jan.	Walker	Lateral stability and control.
93	2 Feb.	Crossfield	Gust loads at various wing sweeps.
94	4 Feb.	Walker	Vertical tail loads.
95	8 Feb.	Walker	Longitudinal stability and control.
96	9 Feb.	Walker	Same as flight 94.
97	10 Feb.	Walker	Lateral stability and control at 45 and 59 degrees, longitudinal stability and control at 59 degrees. Mach 0.965.
98	16 Feb.	Crossfield	Longitudinal stability and control, wing and high tail loads.
99	23 Feb.	Walker	Dynamic pressure effects on buffet.
100	23 Feb.	Walker	Same as flight 98.
101	16 Mar	Walker	Same as flight 98.
102	8 Apr.	Walker	Same as flight 98.
103	9 Apr	Walker	Same as flight 98.
104	13 Apr.	Walker	Same as flight 98. Mach 1.0.
105	15 Apr.	Walker	Same as flight 98.
106	20 Apr.	Walker	Same as flight 98.
107	21 Apr.	Walker	Same as flight 98.
108	23 Apr.	Walker	Same as flight 98.
109	7 June	Butchart	Vertical tail loads. Mach 0.869.
110	11 June	Crossfield	Vertical tail loads. Mach 0.926.
111	14 Dec.	Butchart	Instrumentation check. No data.
112	15 Dec.	John B. McKay	Pilot check.
113	20 Dec.	McKay	Pilot check.
114	21 Dec.	McKay	Pilot check.
115	21 Dec.	McKay	Pilot check.
116	27 Jan. 1955	Butchart	Instrumentation check. Mach 0.922
117	28 Jan.	Butchart	Longitudinal stability and control.
118	3 Feb.	Butchart	Longitudinal stability and control.
119	4 Feb.	Butchart	Longitudinal stability and control.
120	21 Feb.	Butchart	Longitudinal stability and control. Mach 1.011. Alt. 45,267 ft.
121	24 Feb	Butchart	Longitudinal and lateral stability and control.
122	8 Mar.	McKay	Longitudinal stability and control.
123	21 Mar.	McKay	Longitudinal stability and control.
124	23 Mar.	McKay	Longitudinal stability and control.
125	23 Mar.	Butchart	Lateral stability and control.
126	1 Apr.	McKay	Lateral stability and control.
127	6 Apr.	Butchart	Lateral stability and control.
128	6 Apr.	McKay	Lateral stability and control.
129	7 Apr	McKay	Lateral stability and control.
130	8 Apr	McKay	Lateral stability and control.
131	19 Oct.	McKay	Lateral control
132	20 Oct.	McKay	Lateral control
133	24 Oct.	McKay	Lateral control.
134	25 Oct.	Neil A. Armstrong	Pilot check. Landing gear door separated in flight. Plane subsequently was retired.

Sources: NACA X-5 flight reports, 1952-1955 at the Federal Records Center, Laguna Niguel, CA; records of the women computers, NASA Dryden Historical Reference Collection; X-5 chronology prepared by Robert Mulac of Langley Research Center; X-5 chronology prepared by Peter Merlin of the Dryden History Office.

Appendix L
XF-92A Program Flight Chronology, 1953

NACA flight-tested the Convair XF-92A (46-682) during 1953. This program followed earlier flight testing of the aircraft by Convair and the Air Force in 1948-1953. The Air Force flights ran from 13 October 1949 to 6 February 1953. The maximum recorded speed was Mach 1.01 on 2 February 1953. The maximum recorded altitude was 41,443 feet on 8 August 1952. Project pilot for the NACA research was High-Speed Flight Research Station research pilot A. Scott Crossfield.

XF-92A Flights

Flight	Date	Remarks
1	9 Apr. 1953	Pilot check; static longitudinal stability investigation.
2	16 Apr.	Static and dynamic stability and control.
3	21 Apr.	Longitudinal stability and control. Max. alt. 41,364 ft. Highest NACA flight.
4	27 May	Longitudinal stability and control.
5	3 June	Lateral and directional stability and control.
6	5 June	Longitudinal stability and control.
7	9 June	Longitudinal stability and control.
8	11 June	Longitudinal stability and control.
9	16 June	Longitudinal stability and control.
10	19 June	Longitudinal stability and control.
11	24 June	Longitudinal stability and control.
12	24 June	Longitudinal stability and control. Mach 0.962, fastest NACA flight.
13	26 June	Low-speed stability and control.
14	3 Jul.	First flight with wing fences.
15	3 Jul.	Second fence flight.
16	22 Jul.	Modified fence design; fences buckled in flight.
17	17 Aug.	Engine malfunctioned, aborted flight.
18	20 Aug.	Longitudinal stability and control with modified fence design.
19	20 Aug.	Same as flight 18.
20	30 Sept.	Low-speed lateral and directional control with fences.
21	30 Sept.	Same as flight 20.
22	2 Oct.	Same as flight 20.
23	5 Oct.	Same as flight 20.
24	14 Oct.	Low-speed lateral and directional control without fences.
25	14 Oct.	Same as flight 24. Nose landing gear collapsed during landing rollout. Plane was retired.

Sources: XF-92A flight progress reports, 1953; women computers' records in the Dryden Historical Reference Collection.

Appendix M
X-15 Program Flight Chronology, 1959-1968

This chronology covers all flights by the three X-15 series aircraft (the X-15 #1, 56-6670; X-15 #2, 56-6671; and X-15 #3, 56-6672) from 1959 through the program's conclusion in 1968. The X-15s were air launched from modified Boeing B-52 Stratofortress bombers.

The overall program number is at the far left. The flight number includes: aircraft number-flight number-B-52 carry number. Thus, 2-10-21 is the 10th flight of the 2nd aircraft and the 21st time a B-52 had carried that particular X-15 aloft, whether or not it was launched. (Thus, flight 3 in the program was 2-2-6 and flight 4 was 2-3-9, meaning that the number 2 aircraft had two aborted missions between flights 3 and 4, which were flights 2 and 3 of X-15 #2.) Altitude is given in feet and meters, above mean sea level; speed is shown in Mach number, miles per hour, and kilometers per hour.

X-15 Flights

Program No.	Date	Pilot	Flight No.	Maximum Speed (Mach-mph-kph)	Max Altitude (feet-meters)	Remarks
1	8 June 1959	Crossfield	1-1-5	0.79-522-840	37,550-11,445	Planned glide flight.
2	17 Sept.	Crossfield	2-1-3	2.11-1,393-2,241	52,300-15.941	First powered flight.
3	17 Oct.	Crossfield	2-2-6	2.15-1,419-2,283	61,781-18,831	
4	5 Nov.	Crossfield	2-3-9	1.00-660-1,062	45,463-13,857	Engine fire; fuselage structural failure on landing.
5	23 Jan. 1960	Crossfield	1-2-7	2.53-1,669-2,685	66,844-20,374	
6	11 Feb.	Crossfield	2-4-11	2.22-1,466-2,359	88,116-26,858	
7	17 Feb.	Crossfield	2-5-12	1.57-1,036-1,667	52,640-16,045	
8	17 Mar.	Crossfield	2-6-13	2.15-1,419-2,283	52,640-16,045	
9	25 Mar.	Walker	1-3-8	2.00-1,320-2,124	48,630-14,822	First govt. flight.
10	29 Mar	Crossfield	2-7-15	1.96-1,293-2,080	49,982-15,235	
11	31 Mar.	Crossfield	2-8-16	2.03-1,340-2,156	51,356-15,653	
12	13 Apr.	White	1-4-9	1.90-1,254-2,018	48,000-14,630	
13	19 Apr.	Walker	1-5-10	2.56-1,689-2,718	59,496-18,134	
14	6 May	White	1-6-11	2.20-1,452-2,336	60,938-18,574	

15	12 May	Walker	1-7-12	3.19-2,111-3,397	77,882-23,738	
16	19 May	White	1-8-13	2.31-1,590-2,558	108,997-33,222	
17	26 May	Crossfield	2-9-18	2.20-1,452-2,336	51,282-15,631	
18	4 Aug	Walker	1-9-17	3.31-2,196-3,533	78,112-23,809	
19	12 Aug.	White	1-10-19	2.52-1,772-2,851	136,500-41,605	
20	19 Aug.	Walker	1-11-21	3.13-1,986-3,195	75,982-23,159	
21	10 Sept.	White	1-12-23	3.23-2,182-3,510	79,864-24,343	
22	23 Sept.	Petersen	1-13-25	1.68-1,108-1,783	53,043-16,168	
23	20 Oct.	Petersen	1-14-27	1.94-1,280-2,059	53,800-16,398	
24	28 Oct.	McKay	1-15-28	2.02-1,333-2,145	50,700-15,453	
25	4 Nov.	Rushworth	1-16-29	1.95-1,287-2,071	48,900-14,905	
26	15 Nov	Crossfield	2-10-21	2.97-1,960-3,154	81,200-24,750	First flight with XLR-99 engine.
27	17 Nov.	Rushworth	1-17-30	1.90-1,254-2,018	54,750-16,688	
28	22 Nov.	Crossfield	2-11-22	2.51-1,656-2,665	61,900-18,867	First restart with XLR-99 and demonstration of throttle ability.
29	30 Nov.	Armstrong	1-18-31	1.75-1,155-1,858	48,840-14,886	
30	6 Dec.	Crossfield	2-12-23	1,881-3,027	53,374-16,268	
31	9 Dec	Armstrong	1-19-32	1.80-1,188-1,911	50,095-15,269	First ball-nose flight.
32	1 Feb. 1961	McKay	1-20-35	1.88-1,211-1,949	49,780-15,170	
33	7 Feb.	White	1-21-36	3.50-2,275-3,660	78,150-23,820	Last XLR-11 flight.
34	7 Mar.	White	2-13-26	4.43-2,905-4,674	77,450-23,610	First govt. XLR-99 flight. First Mach 4 flight, any airplane.
35	30 Mar.	Walker	2-14-28	3.95-2,760-4,441	169,600-51,700	
36	21 Apr.	White	2-15-29	4.62-3,074-4,946	105,000-32,000	
37	25 May	Walker	2-16-31	4.95-3,307-5,321	107,500-32,850	
38	23 June	White	2-17-33	5.27-3,603-5,797	107,700-32,830	First Mach 5 flight for any aircraft.
39	10 Aug.	Petersen	1-22-37	4.11-2,735-4,401	78,200-23,830	

40	12 Sept.	Walker	2-18-34	5.21-3,618-5,821	114,300-34,840	
41	28 Sept.	Petersen	2-19-35	5.30-3,600-5,792	101,800-31,030	
42	4 Oct.	Rushworth	1-23-39	4.30-2,830-4,553	78,000-23,770	First flight with lower ventral off.
43	11 Oct.	White	2-20-36	5.21-3,647-5,868	217,000-66,150	First flight above 200,000 ft. Outer panel of left wind-shield cracked
44	17 Oct.	Walker	1-24-40	5.74-3,900-6,275	108,600-33,100	Aero-heating data.
45	9 Nov	White	2-21-37	6.04-4,093-6,586	101,600-30,950	Design speed achieved. First Mach 6 flight for any acrft.
46	20 Dec.	Armstrong	3-1-2	3.76-2,502-4,026	81,000-24,700	First flight for X-15 #3.
47	10 Jan. 1962	Petersen	1-25-44	0.97-645-1,038	44,750-13,640	Emergency landing on Mud Lake after engine failed to light.
48	17 Jan.	Armstrong	3-2-3	5.51-3,765-6,058	133,500-40,690	MH-96 evaluation.
49	5 Apr.	Armstrong	3-3-7	4.12-2,850-4,586	180,000-54,860	
50	19 Apr.	Walker	1-26-46	5.69-3,866-6,220	154,000-46,940	
51	20 Apr.	Armstrong	3-4-8	5.31-3,789-6,097	207,500-63,250	Longest X-15 flt. at 12 min., 28.7 sec.
52	30 Apr.	Walker	1-27-48	4.94-3,489-5,614	246,700-75,190	Certified altitude record.
53	8 May	Rushworth	2-22-40	5.34-3,524-5,670	70,400-21,460	Dynamic pressure over 2,000 psi.
54	22 May	Rushworth	1-28-49	5.03-3,450-5,551	100,400-30,600	
55	1 June	White	2-23-43	5.42-3,675-5,913	132,600-40,420	
56	7 June	Walker	1-29-50	5.39-3,672-5,908	103,600-31,580	
57	12 June	White	3-5-9	5.02-3,517-5,659	184,600-56,270	
58	21 June	White	3-6-10	5.08-3,641-5,858	246,700-75,190	
59	27 June	Walker	1-30-51	5.92-4,104-6,603	123,700-37,700	
60	29 June	McKay	2-24-44	4.95-3,280-5,278	83,200-25,360	
61	16 Jul.	Walker	1-31-52	5.37-3,674-5,911	107,200-32,670	
62	17 Jul.	White	3-7-14	5.45-3,832-6,166	314,750-95,940	FAI world altitude record. First flt. above 300,000 ft.
63	19 Jul.	McKay	2-25-45	5.18-3,474-5,590	85,250-25,680	
64	26 Jul.	Armstrong	1-32-53	5.74-3,989-6,418	98,900-30,150	

65	2 Aug.	Walker	3-8-16	5.07-3,438-5,532	144,500-44,040	
66	8 Aug.	Rushworth	2-26-46	4.40-2,943-4,735	90,877-27,700	
67	14 Aug.	Walker	3-9-18	5.25-3,747-6,029	193,600-59,010	
68	20 Aug.	Rushworth	2-27-47	5.24-3,534-5,686	88,900-27,000	
69	29 Aug.	Rushworth	2-28-48	5.12-3,447-5,546	97,200-29,630	
70	28 Sept.	McKay	2-29-50	4.22-2,765-4,450	68,200-20,790	This and all following flights without lower ventral.
71	4 Oct.	Rushworth	3-10-19	5.17-3,493-5,620	112,200-34,200	
72	9 Oct	McKay	2-30-51	5.46-3,716-5,979	130,200-39,700	
73	23 Oct.	Rushworth	3-11-20	5.47-3,764-6,056	134,500-41,000	
74	9 Nov.	McKay	2-31-52	1.49-1,019-1,640	53,950-16,450	Emergency landing at Mud Lake.
75	14 Dec	White	3-12-22	5.65-3,742-6,021	141,400-43,100	
76	20 Dec.	Walker	3-13-23	5.73-3,793-6,103	160,400-48,900	
77	17 Jan. 1963	Walker	3-14-24	5.47-3,677-5,917	271,700-82,810	First civilian flight above 80 km (50 mi).
78	11 Apr.	Rushworth	1-33-54	4.25-2,864-4,608	74,400-22,680	
79	18 Apr.	Walker	3-15-25	5.51-3,770-6,066	92,500-28,190	Nose gear scoop door opened at Mach 3.4.
80	25 Apr.	McKay	1-34-55	5.32-3,654-5,879	105,500-32,160	
81	2 May	Walker	3-16-26	4.73-3,488-5,612	209,400-63,820	
82	14 May	Rushworth	3-17-28	5.20-3,600-5,792	95,600-29,140	
83	15 May	McKay	1-35-56	5.57-3,856-6,204	124,200-37,860	Nose gear scoop door opened at Mach 5.2.
84	29 May	Walker	3-18-29	5.52-3,858-6,208	92,000-28,040	Inner panel of left windshield cracked.
85	18 June	Rushworth	3-19-30	4.97-3,539-5,694	223,700-68,180	
86	25 June	Walker	1-36-57	5.51-3,911-6,293	111,800-34,080	
87	27 June	Rushworth	3-20-31	4.89-3,425-5,511	285,000-86,870	
88	9 Jul.	Walker	1-37-59	5.07-3,631-5,842	226,400-69,010	
89	18 Jul.	Rushworth	1-38-61	5.63-3,925-6,315	104,800-31,940	
90	19 Jul.	Walker	3-21-32	5.50-3,710-5,969	347,800-106,010	

91	22 Aug.	Walker	3-22-36	5.58-3,794-6,105	354,200-107,960	Unofficial world altitude record. Highest X-15 flt.
92	7 Oct.	Engle	1-39-63	4.21-2,834-4,560	77,800-23,710	
93	29 Oct.	Thompson	1-40-64	4.10-2,712-4,364	74,400-22,600	
94	7 Nov.	Rushworth	3-23-39	4.40-2,925-4,706	82,300-25,080	
95	14 Nov.	Engle	1-41-65	4.75-3,286-5,287	90,800-27,680	
96	27 Nov.	Thompson	3-24-41	4.94-3,310-5,326	89,800-27,371	
97	5 Dec.	Rushworth	1-42-67	6.06-4,018-6,465	101,000-30,785	Highest Mach No. for unmodified X-15.
98	8 Jan. 1964	Engle	1-43-69	5.32-3,616-5,818	139,900-42,642	
99	16 Jan.	Thompson	3-25-42	4.92-3,242-5,216	71,000-21,641	
100	28 Jan.	Rushworth	1-44-70	5.34-3,618-5,821	107,400-32,736	
101	19 Feb.	Thompson	3-26-43	5.29-3,519-5,662	78,600-23,957	
102	13 Mar.	McKay	3-27-44	5.11-3,392-5,458	76,000-23,165	
103	27 Mar.	Rushworth	1-45-72	5.63-3,827-6,158	101,500-30,937	
104	8 Apr.	Engle	1-46-73	5.01-3,468-5,580	175,000-53,340	
105	29 Apr.	Rushworth	1-47-74	5.72-3,906-6,285	101,600-30,968	Right inner windshield cracked.
106	12 May	McKay	3-28-47	4.66-3,084-4,962	72,800-22,189	
107	19 May	Engle	1-48-75	5.02-3,494-5,262	195,800-59,680	
108	21 May	Thompson	3-29-48	2.90-1,865-3,001	64,200-19,568	Premature engine shutdown at 41 sec.
109	25 June	Rushworth	2-32-55	4.59-3,104-4,994	83,300-25,390	First flight of X-15A-2.
110	30 June	McKay	1-49-77	4.96-3,334-5,364	99,600-30,358	
111	8 Jul.	Engle	3-30-50	5.05-3,520-5,664	170,400-51,938	
112	29 Jul.	Engle	3-31-52	5.38-3,623-5,250	78,000-23,774	
113	12 Aug.	Thompson	3-32-53	5.24-3,535-5,688	81,200-24,750	
114	14 Aug.	Rushworth	2-33-56	5.23-3,590-5,776	103,300-31,486	Nose gear extended above Mach 4.2; tires failed on landing.
115	26 Aug.	McKay	3-33-54	5.65-3,863-6,216	91,000-27,737	

116	3 Sept.	Thompson	3-34-55	5.35-3,615-5,817	78,600-23,957	
117	28 Sept.	Engle	3-35-57	5.59-3,888-6,256	97,000-29,566	
118	29 Sept.	Rushworth	2-34-57	5.20-3,542-5,699	97,800-29,809	Nose-gear scoop door opened at above Mach 4.5.
119	15 Oct.	McKay	1-50-79	4.56-3,048-4,904	84,900-25,878	
120	30 Oct.	Thompson	3-36-59	4.66-3,113-5,009	84,600-25,786	
121	30 Nov	McKay	2-35-60	4.66-3,089-4,970	87,200-26,579	
122	9 Dec.	Thompson	3-37-60	5.42-3,723-5,990	92,400-28,164	
123	10 Dec.	Engle	1-51-81	5.35-3,675-5,913	113,200-34,503	
124	22 Dec.	Rushworth	3-38-61	5.55-3,593-5,781	81,200-24,750	
125	13 Jan. 1965	Thompson	3-39-62	5.48-3,712-5,973	99,400-30,297	
126	2 Feb.	Engle	3-40-63	5.71-3,885-6,253	98,200-29,931	
127	17 Feb.	Rushworth	2-36-63	5.27-3,539-5,696	95,100-28,986	
128	26 Feb.	McKay	1-52-85	5.40-3,750-6,034	153,600-46,817	
129	26 Mar.	Rushworth	1-53-86	5.17-3,580-5,760	101,900-31,059	
130	23 Apr.	Engle	3-41-64	5.48-3,580-5,760	79,700-24,293	
131	28 Apr.	McKay	2-37-64	4.80-3,273-5,266	92,600-28,224	
132	18 May	McKay	2-38-66	5.17-3,541-5,697	102,100-31,120	
133	25 May	Thompson	1-54-88	4.87-3,418-5,100	179,800-54,803	
134	28 May	Engle	3-42-65	5.17-3,754-6,040	209,600-63,886	
135	16 June	Engle	3-43-66	4.69-3,404-5,477	244,700-74,585	
136	17 June	Thompson	1-55-89	5.14-3,541-5,697	108,500-33,071	
137	22 June	McKay	2-39-70	5.64-3,938-6,336	155,900-47,518	
138	29 June	Engle	3-44-67	4.94-3,432-5,522	280,600-85,527	
139	8 Jul.	McKay	2-40-72	5.19-3,659-5,887	212,600-64,800	

140	20 Jul.	Rushworth	3-45-69	5.40-3,760-6,050	105,400-32,126	
141	3 Aug.	Rushworth	2-41-73	5.16-3,602-5,796	208,700-63,612	
142	6 Aug.	Thompson	1-56-93	5.15-3,534-5,686	103,200-31,455	
143	10 Aug.	Engle	3-46-70	5.20-3,550-5,712	271,000-82,601	
144	25 Aug.	Thompson	1-57-96	5.11-3,604-5,799	214,100-65,258	
145	26 Aug.	Rushworth	3-47-71	4.79-3,372-5,426	239,600-73,030	
146	2 Sept.	McKay	2-42-74	5.16-3,570-5,744	239,800-73,091	
147	9 Sept.	Rushworth	1-58-97	5.25-3,534-5,686	97,200-29,627	
148	14 Sept.	McKay	3-48-72	5.03-3,519-5,662	239,000-72,847	
149	22 Sept.	Rushworth	1-59-98	5.18-3,550-5,712	100,300-30,571	
150	28 Sept.	McKay	3-49-73	5.33-3,732-6,005	295,600-90,099	
151	30 Sept.	Knight	1-60-99	4.06-2,718-4,373	76,600-23,350	
152	12 Oct.	Knight	3-50-74	4.62-3,108-5,001	94,400-28,770	Auxiliary power unit shutdown 1.5 sec. After launch.
153	14 Oct.	Engle	1-61-101	5.08-3,554-5,718	266,500-81,230	
154	27 Oct.	McKay	3-51-75	5.06-3,519-5,662	236,900-72,210	
155	3 Nov.	Rushworth	2-43-75	2.31-1,500-2,414	70,600-21,520	First flight with external tanks (empty).
156	4 Nov.	Dana	1-62-103	4.22-2,765-4,450	80,200-24,440	
157	6 May 1966	McKay	1-63-104	2.21-1,434-2,307	68,400-20,850	Premature engine shutdown at 32 sec.
158	18 May	Rushworth	2-44-79	5.43-3,689-5,936	99,000-30,170	
159	1 Jul.	Rushworth	2-45-81	1.54-1,023-1,646	44,800-13,720	First heavy tank flight; engine shutdown at 32 sec.
160	12 Jul.	Knight	1-64-107	5.34-3,652-5,876	130,000-39,620	
161	18 Jul.	Dana	3-52-78	4.71-3,217-5,176	96,100-29,290	
162	21 Jul.	Knight	2-46-83	5.12-3,568-5,741	192,300-58,610	
163	28 Jul.	McKay	1-65-108	5.19-3,702-5,957	241,800-73,700	
164	3 Aug.	Knight	2-47-84	5.03-3,440-5,535	249,000-75,890	

165	4 Aug.	Dana	3-53-79	5.34-3,693-6,376	132,700-40,450	
166	11 Aug.	McKay	1-66-111	5.21-3,590-5,776	251,000-76,500	
167	12 Aug.	Knight	2-48-85	5.02-3,472-5,586	231,100-70,440	
168	19 Aug	Dana	3-54-80	5.20-3,607-5,804	178,000-54,250	
169	25 Aug.	McKay	1-67-112	5.11-3,543-5,701	257,500-78,490	
170	30 Aug.	Knight	.2-49-86	5.21-3,543-5,701	100,200-30,540	
171	8 Sept.	McKay	1-68-113	2.44-1,602-2,578	73,200-22,310	Premature engine shutdown at 38 sec.
172	14 Sept.	Dana	3-55-82	5.12-3,586-5,770	254,200-22,980	
173	6 Oct.	Adams	1-69-116	3.00-2,900-4,666	75,400-22,980	
174	1 Nov.	Dana	3-56-83	5.46-3,750-6,034	306,900-93,540	
175	18 Nov.	Knight	2-50-89	6.33-4,250-6,838	98,900-30,140	Unofficial world speed record.
176	29 Nov	Adams	3-57-86	4.65-3,120-5,020	92,000-28,040	
177	22 Mar. 1967	Adams	1-70-119	5.59-3,822-6,150	133,100-40,570	
178	26 Apr.	Dana	3-58-87	1.80-1,163-1,871	53,400-16,280	Premature engine shutdown.
179	28 Apr.	Adams	1-71-121	5.44-3,720-5,985	167,000-50,900	
180	8 May	Knight	2-51-92	4.75-3,193-5,138	97,600-29,750	
181	17 May	Dana	3-59-89	4.80-3,177-5,112	71,100-21,670	
182	15 June	Adams	1-72-125	5.12-3,606-5,802	229,300-69,890	
183	22 June	Dana	3-60-90	5.34-3,611-5,810	82,200-25,050	
184	29 June	Knight	1-73-126	4.17-2,870-4,618	173,000-52,730	Total power failure; landed at Mud Lake, Nev.
185	20 July	Dana	3-61-91	5.44-3,693-5,942	84,300-25,700	
186	21 Aug.	Knight	2-52-96	4.94-3,368-5,419	91,000-27,740	First flight w/full ablative coating.
187	25 Aug.	Adams	3-62-92	4.63-3,115-5,012	84,400-25,720	
188	3 Oct.	Knight	2-53-97	6.70-4,520-7,273	102,100-31,120	Unofficial world speed record, (full ablative, tanks, dummy ramjet, mechanical eyelid).
189	4 Oct.	Dana	3-63-94	5.53-3,897-7,270	251,100-76,530	

190	17 Oct.	Knight	3-64-95	5.53-3,856-6,204	280,500-85,500	
191	15 Nov.	Adams	3-65-97	5.20-3,570-5,744	266,000-81,080	Fatal accident, aircraft destroyed.
192	1 Mar. 1968	Dana	1-74-130	4.36-2,878-4,631	104,500-31,850	
193	4 Apr.	Dana	1-75-133	5.27-3,610-5,808	187,500-57,150	
194	26 Apr	Knight	1-76-134	5.05-3,545-5,704	209,600-63,886	
195	12 Jun.	Dana	1-77-136	5.15-3,563-5,733	220,100-67,090	
196	16 Jul.	Knight	1-78-138	4.79-3,382-5,442	221,500-67,510	
197	21 Aug.	Dana	1-79-139	5.01-3,443-5,540	267,500-81,530	
198	13 Sept.	Knight	1-80-140	5.37-3,723-5,990	254,100-77,450	
199	24 Oct.	Dana	1-81-141	5.38-3,716-5,979	255,000-77,720	Last flight.

Sources: FRC release on X-15, 1969; flight logs compiled by Roy Bryant and Betty J. Love; records in DFRC Flight Operations.

Appendix N

Lifying Body Program Flight Chronology, 1966-1975

This chronology covers flight operations of the M2-F2 and M2-F3, HL-10, and X-24A and X-24B lifting bodies from 1966 to 1976. It does not include earlier flight tests of the plywood M2-F1, as records for this craft's glide flights are incomplete. The M2-F1 was often towed above the lakebed behind a modified Pontiac automobile or towed and released behind a Douglas C-47 aircarft. There were approximately 400 car tows and 77 air tows for the M2-F1 beginning on 1 March 1963, with the first air tow occuring on 16 August 1963, and the last one on 16 August 1966. The other lifting bodies were air-launched from a modified Boeing B-52 Stratofortress. In the table, the flight number stands for: vehicle-free flight number-B-52 carry number for that vehicle. Vehicle abbreviations are: M = M2-F2. M3=M2-F3. H = HL-10. X = X-24A. B=X-24B.

Piloted Lifitng Body Flight Log

No.	Date	Flight No.	Pilot	Max Altitude (ft-m)	Max Speed (Mach-mph-kph)	Flight Time (sec)	Remarks
1	12 Jul. 1966	M-1-8	Thompson	45,000-13,720	0.646-452-727	217	First flight.
2	19 Jul.	M-2-9	Thompson	45,000-13,720	0.598-394-634	245	
3	12 Aug.	M-3-10	Thompson	45,000-13,720	0.619-408-656	278	
4	24 Aug.	M-4-11	Thompson	45,000-13,720	0.676-446-718	241	
5	2 Sept.	M-5-12	Thompson	45,000-13,720	0.707-466-750	226	Thompson's last lifting-body (L/B) flight.
6	16 Sept.	M-6-13	Peterson	45,000-13,720	0.705-466-750	210	360-degree approach.
7	20 Sept.	M-7-14	Sorlie	45,000-13,720	0.635-421-677	211	
8	22 Sept.	M-8-15	Peterson	45,000-13,720	0.661-436-702	233	
9	28 Sept.	M-9-16	Sorlie	45,000-13,720	0.672-443-713	225	
10	5 Oct.	M-10-17	Sorlie	45,000-13,720	0.615-430-692	234	Sorlie's last L/B flight.
11	12 Oct.	M-11-18	Gentry	45,000-13,720	0.662-436-702	227	
12	26 Oct.	M-12-19	Gentry	45,000-13,720	0.605-399-642	261	
13	14 Nov.	M-13-20	Gentry	45,000-13,720	0.681-445-716	230	
14	21 Nov.	M-14-21	Gentry	45,000-13,720	0.695-457-735	235	
15	22 Dec.	H-1-3	Peterson	45,000-13,720	0.693-457-735	187	First HL-10 flight.
16	2 May 1967	M-15-23	Gentry	45,000-13,720	0.623-411-661	231	
17	10 May	M-16-24	Peterson	45,000-13,720	0.612-403-648	223	Landing accident. Peterson's last L/B flight.

18	15 Mar. 1968	H-2-5	Gentry	45,000-13,720	0.609-425-684	243	First flight of modified HL-10.
19	3 Apr.	H-3-6	Gentry	45,000-13,720	0.690-455-732	242	
20	25 Apr.	H-4-8	Gentry	45,000-13,720	0.697-459-739	258	
21	3 May	H-5-9	Gentry	45,000-13,720	0.688-455-732	245	
22	16 May	H-6-10	Gentry	45,000-13,720	0.678-447-719	265	
23	28 May	H-7-11	Manke	45,000-13,720	0.657-434-698	245	
24	11 Jun.	H-8-12	Manke	45,000-13,720	0.635-433-697	246	
25	21 Jun.	H-9-13	Gentry	45,000-13,720	0.637-423-681	271	
26	24 Sept.	H-10-17	Gentry	45,000-13,720	0.682-449-722	245	XLR-11 engine installed.
27	3 Oct.	H-11-18	Manke	45,000-13,720	0.714-471-758	243	
28	23 Oct.	H-12-20	Gentry	39,700-12,100	0.666-449-722	189	First powered flight of HL-10. Premature shutdown.
29	13 Nov.	H-13-21	Manke	42,650-13,000	0.840-524-843	385	Only 2 chambers lit, 186.1-sec powered flight.
30	9 Dec.	H-14-24	Gentry	47,420-14,450	0.870-542-872	394	
31	17 Apr. 1969	X-1-2	Gentry	45,000-13,720	0.718-474-763	217	First flight of X-24A. Glide.
32	17 Apr.	H-15-27	Manke	52,740-16,070	0.994-605-973	400	Only 3 chambers lit.
33	25 Apr.	H-16-28	Dana	45,000-13,720	0.701-462-743	252	Glide.
34	8 May	X-2-3	Gentry	45,000-13,720	0.693-457-735	253	
35	9 May	H-17-29	Manke	53,300-16,250	1.127-744-1,197	410	First supersonic L/B flight.
36	20 May	H-18-30	Dana	49,100-14,970	0.904-596-959	414	
37	28 May	H-19-31	Manke	62,200-18,960	1.236-815-1,311	398	Only 2 chambers lit.
38	6 June	H-20-32	Hoag	45,000-13,720	0.665-452-727	231	Glide.
39	19 June	H-21-33	Manke	64,100-19,540	1.398-922-1,483	378	Only 2 chambers lit.
40	23 June	H-22-34	Dana	63,800-19,450	1.271-839-1,350	373	Only 2 chambers lit.
41	6 Aug.	H-23-35	Manke	76,100-23,190	1.540-1020-1,641	372	First 4-chambered flight.
42	21 Aug.	X-3-5	Gentry	40,000-12,190	0.718-473-761	270	
43	3 Sept.	H-24-37	Dana	77,960-23,760	1.446-958-1,541	414	All 4 chambers lit.
44	9 Sept.	X-4-7	Gentry	40,000-12,190	0.594-402-647	232	
45	18 Sept.	H-25-39	Manke	79,190-24,140	1.256-833-1,340	426	
46	24 Sept.	X-5-8	Gentry	40,000-12,190	0.596-396-637	257	
47	30 Sept.	H-26-40	Hoag	53,750-16,380	0.924-609-780	436	Only 2 chambers lit.

48	22 Oct.	X-6-10	Manke	40,000-12,190	0.587-387-623	238	
49	27 Oct.	H-27-41	Dana	60,610-18,470	1.577-1,041-1,675	417	
50	3 Nov.	H-28-42	Hoag	64,120-19,540	1.396-921-1,482	439	
51	13 Nov.	X-7-11	Gentry	45,000-13,720	0.646-427-687	270	
52	17 Nov.	H-29-43	Dana	64,590-19,690	1.594-1,052-1,693	408	
53	21 Nov.	H-30-44	Hoag	79,280-24,160	1.432-952-1,532	378	
54	25 Nov.	X-8-12	Gentry	45,000-13,720	0.685-454-730	266	
55	12 Dec.	H-31-46	Dana	79,960-24,370	1.310-871-1,401	428	
56	19 Jan. 1970	H-32-47	Hoag	86,660-26,410	1.351-869-1,398	410	
57	26 Jan.	H-33-48	Dana	87,684-26,730	1.351-897-1,443	411	
58	18 Feb.	H-34-49	Hoag	67,310-20,520	1.861-1,228-1,976	380	Fastest L/B flight.
59	24 Feb.	X-9-14	Gentry	47,000-14,326	0.771-509-819	258	
60	27 Feb.	H-35-51	Dana	90,303-27,524	1.314-870-1,400	416	Highest L/B flight.
61	19 Mar.	X-10-15	Gentry	44,400-13,533	0.865-571-919	424	First powered X-24 flight.
62	2 Apr.	X-11-16	Manke	58,700-17,892	0.866-571-919	435	
63	22 Apr.	X-12-17	Gentry	57,700-17,587	0.925-610-981	408	
64	14 May	X-13-18	Manke	44,600-13,594	0.748-494-795	513	Only 2 chambers lit.
65	2 June	M-17-26	Dana	45,000-13,716	0.688-469-755	218	First M2-F3 flight.
66	11 June	H-36-52	Hoag	45,000-13,716	0.744-503-809	202	Powered landing study using hydrogen-peroxide rockets instead of the XLR-11.
67	17 June	X-14-19	Manke	61,000-18,593	0.990-653-1051	432	
68	17 Jul.	H-37-53	Hoag	45,000-13,716	0.733-499-803	252	Last L/B flight for Hoag, HL-10; powered landing study.
69	21 Jul.	M-18-27	Dana	45,000-13,716	0.660-440-708	228	
70	28 Jul.	X-15-20	Gentry	58,100-17,678	0.938-619-996	388	
71	11 Aug.	X-16-21	Manke	63,900-19,477	0.986-651-1,047	413	
72	26 Aug.	X-17-22	Gentry	41,500-12,649	0.694-458-737	479	Only 2 chambers lit.
73	14 Oct.	X-18-23	Manke	67,900-20,696	1.186-784-1,261	411	First supersonic X-24 flight.
74	27 Oct	X-19-24	Manke	71,400-21,763	1.357-899-1,446	417	Highest X-24 flight.
75	2 Nov.	M-19-28	Dana	45,000-13,716	0.630-429-690	236	

76	20 Nov.	X-20-25	Gentry	67,600-20,604	1.370-905-1,456	432	
77	25 Nov.	M-20-29	Dana	51,900-15,819	0.809-534-859	377	First M2-F3 powered flight.
78	21 Jan. 1971	X-21-26	Manke	57,900-15,819	1.030-679-1,093	462	
79	4 Feb.	X-22-27	Powell	45,000-13,716	0.659-435-700	235	
80	9 Feb.	M-21-30	Gentry	45,000-13,716	0.707-469-755	241	Gentry's last L/B flight.
81	18 Feb.	X-23-28	Manke	67,400-20,544	1.511-998-1,606	447	
82	26 Feb.	M-22-31	Dana	45,000-13,716	0.773-510-821	821	Only two chambers lit.
83	8 Mar.	X-24-29	Powell	56,900-17,343	1.002-661-1,064	437	
84	29 Mar.	X-25-30	Manke	70,500-21,488	1.600-1,036-1,667	446	Fastest X-24 flight
85	12 May	X-26-32	Powell	70,900-21,610	1.389-918-1,477	423	Delayed engine light.
86	25 May	X-27-33	Manke	65,300-19,903	1.191-786-1,265	548	Only 3 chambers lit.
87	4 June	X-28-34	Manke	54,400-16,581	0.817-539-867	517	Only 2 chambers lit. Final X-24A flight.
88	23 Jul.	M-23-34	Dana	60,500-18,440	0.930-614-788	353	
89	9 Aug.	M-24-35	Dana	62,000-18,898	0.974-643-1,035	415	
90	25 Aug.	M-25-37	Dana	67,300-20,513	1.095-723-1,163	390	First supersonic M2-F3 flight
91	24 Sept.	M-26-38	Dana	42,000-12,802	0.728-480-772	210	Engine malfunction, Rosamond landing.
92	15 Nov.	M-27-39	Dana	45,000-13,716	0.739-487-784	215	Glide flt to check out new jettison location.
93	1 Dec.	M-28-40	Dana	70,800-21,580	1.274-843-1,356	391	
94	16 Dec.	M-29-41	Dana	46,800-14,265	0.811-535-861	451	
95	25 Jul. 1972	M-30-45	Dana	60,900-18,562	0.989-652-1,049	420	1st command-augmentation-system flight.
96	11 Aug.	M-31-46	Dana	67,200-20,480	1.101-726-1,168	375	
97	24 Aug.	M-32-47	Dana	66,700-20,330	1.266-835-1,344	376	
98	12 Sept.	M-33-48	Dana	46,000-14,020	0.880-581-935	387	Engine malfunction.
99	27 Sept.	M-34-49	Dana	66,700-20,330	1.340-885-1,424	366.5	
100	5 Oct.	M-35-50	Dana	66,300-20,210	1.370-904-1,455	376	
101	19 Oct.	M-36-51	Manke	47,100-14,360	0.905-597-961	359	
102	1 Nov.	M-37-52	Manke	71,300-21,730	1.213-803-1,292	378	
103	9 Nov.	M-38-53	Powell	46,800-14,260	0.906-597-961	364	
104	21 Nov.	M-39-54	Manke	66,700-20,330	1.435-947-1,524	377	Planned Rosamond landing.
105	29 Nov.	M-40-55	Powell	67,500-20,570	1.348-890-1,432	357	

106	6 Dec.	M-41-56	Powell	68,300-20,820	1.191-786-1,265	332	
107	13 Dec.	M-42-57	Dana	66,700-20,330	1.613-1,064-1,712	383	Fastest M-2 flight. Used L/D rockets.
108	20 Dec.	M-43-58	Manke	71,500-21,790	1.294-856-1,377	390	Last M2-F3 flight, also highest.
109	1 Aug. 1973	B-1-3	Manke	40,000-12,190	0.640-422-679	252	First X-24B (glide) flight.
110	17 Aug.	B-2-4	Manke	45,000-13,720	0.650-429-690	267	
111	31 Aug.	B-3-5	Manke	5,000-13,720	0.716-472-759	277	
112	18 Sept.	B-4-6	Manke	45,000-13,720	0.687-450-724	271	
113	4 Oct.	B-5-9	Love	45,000-13,720	0.704-465-748	279	
114	15 Nov.	. B-6-13	Manke	52,764-16,080	0.930-614-988	404	First powered X-24B flight.
115	12 Dec.	. B-7-14	Manke	62,604-19,080	0.987-645-1,038	434	
116	15 Feb. 1974	B-8-15	Love	45,000-13,720	0.696-459-738	307	
117	5 May	B-9-16	Manke	60,334-18,390	1.086-708-1,139	437	First supersonic X-24B flight.
118	30 Apr.	B-10-21	Love	52,040-15,860	0.876-578-930	419	
119	24 May	B-11-22	Manke	55,979-17,060	1.140-753-1,212	448	
120	14 June	B-12-23	Love	65,400-19,934	1.228-810-1,303	405	
121	28 June	B-13-24	Manke	68,150-20,770	1.391-920-1,480	427	
122	8 Aug.	B-14-25	Love	73,380-22,370	1.541-1,022-1,644	395	
123	29 Aug.	B-15-26	Manke	72,440-22,080	1.098-727-1,170	467	
124	25 Oct.	B-16-27	Love	72,150-21,990	1.752-1,164-1,873	417	Fastest flight for X-24B.
125	15 Nov.	B-17-28	Manke	72,060-21,960	1.615-1,070-1,722	481	
126	17 Dec.	B-18-29	Love	68,780-20,960	1.585-1,036-1,667	420	
127	14 Jan. 1975	B-19-30	Manke	72,787-22,180	1.748-1,157-1,862	477	
128	20 Mar.	B-20-32	Love	70,373-21,450	1.443-955-1,537	409	
129	18 Apr.	B-21-33	Manke	57,900-17,650	1.204-795-1,279	450	
130	6 May	B-22-34	Love	73,400-22,370	1.444-958-1,541	448	
131	22 May	B-23-35	Manke	74,100-22,580	1.633-1,084-1,744	461	Highest X-24B flight.
132	6 June	B-24-36	Love	72,100-21,980	1.677-1,110-1,786	474	
133	25 June	B-25-38	Manke	58,000-17,680	1.343-887-1,427	426	
134	15 Jul.	B-26-39	Love	69,480-21,180	1.585-1,049-1,688	415	
135	5 Aug.	B-27-40	Manke	57,000-17,373	1.190-785-1,264	420	1st runway landing, Manke's last flight.

136	20 Aug.	B-28-41	Love	71,100-21,671	1.548-1,025-1,650	420	Runway landing, Love's last flight.
137	9 Sept.	B-29-42	Dana	69,700-21,244	1.481-980-1,577	435	
138	23 Sept.	B-30-43	Dana	56,800-17,313	1.157-764-1,229	438	Last rocket-powered flight, Dana's last flight.
139	9 Oct.	B-31-44	Enevoldson	45,300-13,807	0.705-450-724	251	
140	21 Oct.	B-32-45	Scobee	45,000-13,720	0.696-459-739	255	
141	3 Nov.	B-33-46	McMurtry	45,300-13,807	0.702-463-745	248	
142	12 Nov.	B-34-47	Enevoldson	45,000-13,720	0.702-463-745	241	
143	19 Nov.	B-35-48	Scobee	45,000-13,720	0.700-462-743	249	
144	26 Nov.	B-36-49	McMurtry	44,800-13,655	0.713-471-757	245	Last lifting-body flight.

Sources: DFRC fact sheet, March 1976 as corrected by Jack Kolf's Heavy Weight Lifting Body Flight Log in Robert G. Hoey, "Testing Lifting Bodies at Edwards," Book 1 of "Air Force/NASA Lifting Body Legacy History Project" by PAT Projects, Inc., Sept. 1994, pp. 167-172, also available on the Internet at http://www.dfrc.nasa.gov/History/Publications/LiftingBodies/ap_c_2.html.

Appendix O
XB-70 Program Flight Chronology, 1967-1969

The FRC operated the XB-70A #1 (62-0001) aircraft from 1967 through early 1969. This aircraft was the sole survivor of two prototypes. The second aircraft (62-0207) was destroyed in a mid-air collision on 8 June 1966. By this time, the two aircraft had accumulated a combined total of 95 flights, 49 by #1, and 46 by #2.

A joint Air Force-NASA program began in November 1966 and lasted through January 1967. The all-NASA program (with Air Force support) began with the 107th flight of the XB-70A series, on 25 April 1967. Because the XB-70A #1 formally began its NASA research career at that point, that flight has been chosen to head this chronology.

XB-70A Flights

Flight	Date	Pilot/Copilot	Remarks[1]
107 (1-61)	25 Apr. 1967	Cotton/Fulton	Flight aborted after crew entry door opened and landing gear malfunctioned.
108 (1-62)	12 May	Fulton/Cotton	Low-speed handling qualities; airspeed calibration.
109 (1-63)	2 June	Cotton/Van Shepard	Mach 1.43; handling qualities and airspeed calibration.
110 (1-64)	22 June	Fulton/Donald Mallick	Pilot checkout; acceleration to Mach 1.83.
111 (1-65)	10 Aug.	Cotton/Col. E. Sturmthal	Pilot checkout; Mach 0.92. Pressurization malfunction
112 (1-66)	24 Aug.	Fulton/Mallick	Mach 2.24 at 57,700 ft.
113 (1-67)	8 Sept.	Cotton/Sturmthal	Mach 2.30; inlet studies.
114 (1-68)	11 Oct.	Fulton/Mallick	Mach 2.43 at 58,000 ft.
115 (1-69)	2 Nov.	Cotton/Sturmthal	Mach 2.55; inlet studies, longitudinal handling qualities.
116 (1-70)	12 Jan. 1968	Fulton/Mallick	Mach 2.55 at 67,000 ft.; stability and control.
117 (1-71)	13 Feb.	Mallick/Cotton	Mach 1.18; handling qualities.
118 (1-72)	28 Feb.	Fulton/Sturmthal	Landing gear malfunction.
119 (1-73)	21 Mar.	Cotton/Fulton	Gear-down, low-speed studies.
120 (1-74)	11 June 1968	Mallick/Fulton	Landing gear malfunction.
121 (1-75)	28 June	Sturmthal/Cotton	Mach 1.23, structural dynamics.
122 (1-76)	19 Jul.	Mallick/Fulton	Mach 1.62, structural dynamics.
123 (1-77)	16 Aug.	Fulton/Sturmthal	Mach 2.47, structural dynamics. Inlet unstart, loss of #6 engine.
124 (1-78)	10 Sept.	Mallick/Fulton	Mach 2.5 at 62,800 ft.
125 (1-79)	18 Oct.	Sturmthal/Fulton	Mach 2.18, structural dynamics.
126 (1-80)	1 Nov.	Sturmthal/Fulton	Mach 1.62, structural dynamics, stability and control.
127 (1-81)	3 Dec.	Fulton/Mallick	Mach 1.64, same as flight 126.
128 (1-82)	17 Dec.	Fulton/Sturmthal	Mach 2.53, same as flight 126.
129 (1-83)	4 Feb. 1969	Fulton/Sturmthal	Subsonic, ferry flight to USAF Museum, Wright-Patterson AFB, Dayton, Ohio.

[1] These remarks are necessarily brief and do not include all of the flight research conducted on each flight.

XB-70A Program Summary

Total Flights	
XB-70A #1	83
XB-70A #2	46

Total Flying Time (both aircraft): 252 hours, 48 minutes

Total time Mach 1 to Mach 1.9: 55 hours, 50 minutes.

Total time Mach 2 to Mach 2.9: 49 hours, 32 minutes.

Total time above Mach 3: 1 hour, 48 minutes

Source: XB-70A program flight log, prepared by Betty J. Love.

Appendix P
XB-70 Program Flight Chronology, 1967-1969

The NASA YF-12 program flew three aircraft, YF-12A #60-6935 (935), YF-12A #60-6936 (936), and "YF-12C" #60-6937 (937). The "YF-12C" was, in fact, a then-secret SR-71A whose true serial number was 61-7951. The 937 serial number actually belonged to an also secret A-12 and was assigned to hide the fact that NASA had an SR-71. YF-12A 936 completed 62 flights, primarily by Air Force flight test crews, before being lost in an inflight fire on 24 June 1971. The Air Force crew ejected safely.

NASA's program on the YF-12A and YF-12C aircraft (hereafter so-called to coincide with usage at the time) lasted from 1969 through 1978. The first NASA flight occurred on 5 March 1970, when test pilot Fitzhugh L. Fulton piloted YF-12A 936 on a checkout flight. He followed this with flights on 9 March and 11 March. NASA's first flight in YF-12A 935 was flown by Donald Mallick on 1 April 1970. The first NASA flight in YF-12C 937 was by Fitzhugh L. Fulton on 24 May 1972.

The FRC flight crews for the YF-12 were pilots Fitzhugh Fulton and Donald Mallick, and flight-test engineers Victor Horton and Ray Young. Before retirement, aircraft 935 was used to check out other center pilots on familiarization flights.

YF-12A 935 Flights

Flight	Date	Pilot/Test Engineer	Remarks
1	11 Dec. 1969	Col. Joseph Rogers/Maj. Gary Heidlebaugh	USAF test, functional check flight (FCF)
2	17 Dec.	Maj. William Campbell/Maj. Sam Ursini	USAF test
3	6 Jan. 1970	Rogers/Ursini	USAF test
4	14 Jan	Campbell/Heidlebaugh	USAF test
5	19 Jan.	Col. Hugh Slater/Heidlebaugh	USAF test
6	21 Jan.	Slater/Ursini	USAF test
7	27 Jan.	Slater/Heidlebaugh	USAF test.
8	11 Feb.	Campbell/Ursini	USAF test, ventral fin damaged in sideslip
9	26 Mar.	Campbell/Victor Horton	FCF, first flight w ith NASA engineer.
10	1 Apr.	Donald Mallick/Ursini	FCF, first flight of 935 with NASA pilot
11	8 Apr.	Mallick/Ursini	Stability and deflection data
12	14 Apr.	Fitzhugh Fulton/Horton	Stability, deflection points, phugoids
13	17 Apr.	Fulton/Horton	SST flight-control design data
14	28 Apr.	Mallick/W. Ray Young	Stability, deflection points, phugoids
15	1 May	Fulton/Horton	Stability, deflection points, phugoids
16	7 May	Mallick/Young	SST flight-control design data
17	15 May	Fulton/Horton	Stablity and control (S&C), phugoids
18	22 May	Mallick/Young	S&C, phugoids
19	27 May	Fulton/Horton	S&C, phugoids, check of wing camera.
20	2 June	Mallick/Young	Propulsion tests, performance, camera FCF
21	11 June	Fulton/Horton	Performance, SST handling qualities
22	16 June	Fulton/Young	SST handling qualities, phugoids, deflection, served as radar target for YF-12A 936
23	22 March 1971	Mallick/Horton	FCF with ventral fin on. Fin removed following flight to assess performance of plane without ventral fin.
24	7 Apr.	Fulton/Horton	Ventral fin off.
25	16 Apr.	Fulton/Horton	S&C, pilot induced oscillation (PIO) data
26	29 Apr.	Mallick/Young	S&C, two unstarts
27	5 May	Fulton/Horton	S&C, loads, deflection points, PIO
28	23 June	Mallick/Young	1st flight with ventral back on, refueling, deflections
29	9 Jul.	Fulton/Horton	Loads, wing camera data, SST handling qualities
30	13 Jul.	Mallick/Young	Air Force intercepts, loads, deflections, temperature data
31	20 Jul.	Mallick/Horton	Aborted early due to SAS malfunction
32	27 Jul	Mallick/Horton	Air Force intercepts, loads, deflection, temperature data
33	3 Aug.	Lt. Col. R. Jack Layton/Young	Same as flight 32, three unstarts
34	10 Aug.	Fulton/Young	Air Force intercepts, loads, deflection, temperature data
35	17 Aug.	Fulton/Horton	Air Force intercepts, loads, deflection, temperature data

36	22 Oct.	Mallick/Horton	Low-speed FCF, PIO tracking investigation
37	22 Oct.	Fulton/Horton	Aborted early due to hydraulic systems failure
38	29 Oct.	Mallick/Horton	Loads, level accelerations
39	29 Oct.	Fulton/Young	Level acceleration/deceleration, loads
40	2 Nov.	Mallick/Horton	Loads data, temperature profile, climb
41	9 Nov.	Fulton/Young	Loads data, temperature profile, handling qualities
42	16 Nov.	Mallick/Young	Same as flight 40, plus airframe/propulsion interaction
43	23 Nov.	Fulton/Horton	Loads, simulated single-engine approach
44	30 Nov.	Mallick/Young	Same as flight 42
45	7 Dec.	Fulton/Horton	Bypass, angle-of-attack (AOA) lag, diverted to Palmdale
46	7 Dec.	Fulton/Horton	Ferry from Palmdale to Edwards
47	14 Dec.	Mallick/Young	Air Force intercepts, loads, sideslips, handling qualities
48	21 Dec.	Fulton/Horton	Handling qualities, loads, stability, airframe/propulsion interaction, low-speed phugoids
49	11 Jan. 1972	Mallick/Young	Loads, handling qualities
50	18 Jan.	Fulton/Horton	Bypass tests, AOA lag, airframe/propulsion interaction
51	26 Jan.	Fulton/Young	Loads data, level cruise (Mach 3.2)
52	26 Jan.	Fulton/Horton	Loads, airframe/propulsion interaction
53	23 Feb.	Mallick/Young	Refueled over El Paso, Texas for maximum time at Mach 3.2 cruise speed. Aircraft grounded for more than a year for studies in FRC heat loads laboratory.
54	12 Jul. 1973	Fulton/Horton	Low-speed FCF, first flight since loads laboratory tests
55	26 Jul.	Mallick/Young	FCF, autopilot check, stabilized cruise points
56	3 Aug.	Fulton/Horton	High-speed FCF, boundary-layer, sideslip data
57	23 Aug.	Fulton/Young	Boundary-layer, aft-facing step (AFS), sideslip data
58	6 Sept.	Mallick/Horton	Boundary-layer and AFS experiments
59	13 Sept.	Fulton/Horton	Boundary-layer, AFS, phugoids, pitch pulses
60	11 Oct.	Mallick/Larry Barnett	Boundary-layer, AFS, emergency gear extension check
61	11 Oct.	Mallick/Young	Autopilot baseline checks, boundary-layer
62	23 Oct.	Fulton/Horton	Steady-state side-slips, boundary-layer, AFS
63	7 Nov.	Mallick/Young	Boundary-layer, AFS, phugoids
64	16 Nov.	Fulton/Horton	Boundary-layer, AFS, autopilot baseline data
65	3 Dec.	Mallick/Young	Boundary-layer, AFS, autopilot baseline data
66	13 Dec.	Fulton/Horton	Boundary-layer, AFS
67	11 Jan. 1974	Mallick/Young	Boundary-layer, AFS
68	25 Jan.	Fulton/Horton	Boundary-layer, AFS
69	25 Jan.	Mallick/Young	Boundary-layer, AFS
70	4 Mar.	Fulton/Horton	Boundary-layer, AFS, boat-tail drag, autopilot test
71	8 Mar.	Mallick/Young	Boundary-layer, AFS, boat-tail drag
72	15 Mar.	Fulton/Horton	Boundary-layer, AFS, boat-tail drag
73	21 Mar.	Mallick/Young	Boundary-layer, AFS, boat-tail drag, altitude hold
74	28 Mar.	Fulton/Horton	Boundary-layer, AFS, boat-tail drag, turns, autopilot
75	18 Apr.	Mallick/Young	Aborted early due to engine oil-pressure malfunction
76	2 May	Fulton/Horton	Boundary-layer, AFS, boat-tail drag
77	9 May	Mallick/Young	Boundary-layer, AFS, boat-tail drag
78	16 May	Fulton/Young	Boundary-layer, AFS, boat-tail drag, handling, autopilot
79	23 May	Mallick/Young	Aborted early due to generator malfunction
80	30 May	Mallick/Young	Boundary-layer, AFS, boat-tail drag
81	6 Jun.	Fulton/Horton	Boundary-layer, AFS, boat-tail drag
82	11 Sept.	Fulton/Horton	Low-speed FCF, AFS, boat-tail drag, ventral flow field survey rake
83	17 Sept.	Fulton/Horton	AFS, flow field survey rake, boat-tail drag
84	3 Oct.	Fulton/Horton	AFS, flow field survey rake, boat-tail drag
85	18 Oct.	Fulton/Horton	FCF of Type K engines, AFS, flow field survey rake, boat-tail drag, rudder doublet, water dump

86	25 Oct.	Fulton/Young	Simulated Coldwall track/profile, same as flight 83
87	1 Nov.	Mallick/Young	Same as flight 86
88	7 Feb. 1975	Fulton/Horton	FCF with Coldwall, aborted early
89	14 Feb.	Mallick/Young	Coldwall FCF aborted, subsonic S&C data
90	27 Feb.	Mallick/Young	Rudder doublets, plane shed ventral at Mach 0.9; landed safely, grounded for repairs. No ventral until Flight 97.
91	11 Jul.	Fulton/Horton	Coldwall structural and stability data
92	24 Jul.	Mallick/Young	Coldwall structural demonstration with insulation blanket installed, ventral flow field data
93	7 Aug.	Fulton/Horton	Coldwall insulation removal test; insulation ingested in both inlets, causing unstarts
94	21 Aug.	Fulton/Horton	Flow field survey with Coldwall removed
95	28 Aug.	Mallick/Young	Flow field survey and stability with Coldwall removed
96	5 Sept.	Fulton/Horton	Stability and control investigation.
97	16 Jan. 1976	Mallick/Young	First flight with Lockalloy ventral fin
98	27 Jan.	Fulton/Horton	Touch-and-go landings for high-speed taxi data
99	5 Feb.	Mallick/Young	Same as flight 98, runway roughness tests
100	12 Feb.	Fulton/Horton	Lockalloy ventral fin envelope expansion
101	4 Mar.	Fulton/Young	Lockalloy ventral envelope expansion, low L/D landing
102	23 Mar.	Fulton/Horton	Lockalloy ventral fin envelope expansion
103	2 Apr.	Fulton/Young	Lockalloy ventral fin envelope expansion
104	12 Apr.	Fulton/Young	Ventral tufts, landing gear data
105	13 May	Mallick/Young	Lockalloy ventral envelope expansion, landing gear data
106	20 May	Fulton/Horton	Lockalloy ventral envelope expansion, landing gear data
107	15 Jul.	Mallick/Young	S&C with Coldwall, ventral fin, and camera pods
108	22 Jul.	Fulton/Horton	S&C, Coldwall alignment
109	10 Aug.	Mallick/Young	Coldwall profile, Hotwall data
110	31 Aug.	Fulton/Horton	Same as flight 109 plus landing flare data
111	13 Sept.	Fulton/Horton	Coldwall skin-friction balance-cooling studies
112	28 Sept.	Fulton/Horton	Coldwall profile, Hotwall data, skin friction
113	21 Oct.	Fulton/Horton	Premature loss of Coldwall insulation, no data
114	10 Nov.	Fulton/Horton	Gust vane calibration, handling qualities, flow visulaization, autopilot altitude hold tests
115	9 Dec.	Mallick/Young	Coldwall profile, Hotwall data, fuselage tuft photos
116	3 Mar. 1977	Fulton/Horton	Premature loss of Coldwall insulation, no data
117	2 June	Fulton/Horton	Vortex flow visualization photos, premature loss of Coldwall insulation
118	23 June	Fulton/Horton	Vortex flow visualization, good Coldwall data.
119	21 July	Fulton/Horton	Bad unstarts due to ingestion of Coldwall insulation, no Colwall data
120	30 Sept.	Fulton/Horton	Coldwall data, U-2 intercept, fly-by of DFRC
121	13 Oct.	Fulton/Horton	Final Coldwall flight, good results
122	18 Nov.	Mallick/Young	Boundary-layer, handling-qualities
123	1 Dec.	John Manke/Horton	Pilot familiarization
124	1 Dec.	William Dana/Horton	Pilot familiarization.
125	9 Dec.	Gary Krier/Young	Pilot familiarization.
126	13 Dec.	Einar Enevoldson/Young	Pilot familiarization.
127	14 Dec.	Tom McMurtry/Horton	Pilot familiarization.
128	28 Feb. 1978	Mallick/Young	Dual-mode landing gear baseline stiffness tests
129	28 Feb.	Mallick/Young	Dual-mode landing gear baseline stiffness tests
130	7 Mar.	Fulton/Horton	Large volume landing gear stiffness tests
131	7 Mar.	Fulton/Horton	Large volume landing gear stiffness tests
132	15 Mar.	Mallick/Young	Mixed volume landing gear stiffness tests
133	15 Mar.	Mallick/Young	Mixed volume landing gear stiffness tests
134	23 Mar.	Fulton/Horton	Space Shuttle approach simulations

135	31 Mar.	Enevoldson/Young	Space Shuttle approach simulations
136	31 Mar.	Dana/Horton	Space Shuttle approach simulations
137	22 Nov.	Fulton/Horton	Shaker-vane study
138	1 Dec.	Fulton/Horton	Shaker-vane study
139	24 Jan. 1979	Mallick/Young	Shaker-vane study
140	16 Feb.	Fulton/Horton	Shaker-vane study.
141	8 Mar	Mallick/Young	Shaker-vane study
142	15 Mar	Fulton/Horton	Shaker-vane study.
143	29 Mar	Stephen Ishmael/Horton	Pilot familiarization
144	29 Mar	Michael Swann/Young	Pilot familiarization
145	31 Oct.	Fulton/Horton	Last NASA flight of 935
146	7 Nov.	Col. James Sullivan/ Col. Richard Uppstrom	Ferry to USAF Museum, Wright-Patterson AFB, Ohio.

YF-12A 936 FLIGHTS

Flight	Date	Pilot/Test Engineer	Remarks
1	3 Mar. 1970	Slater/Heidelbaugh	USAF test, functional check flight (FCF)
2	5 Mar	Fulton/Horton	Pilot checkout #1
3	9 Mar	Fulton/Horton	Pilot checkout #2
4	11 Mar.	Fulton/Horton	Pilot checkout #3
5	24 Mar.	Slater/Heidelbaugh	Aborted
6	31 Mar	Slater/Heidelbaugh	Intercepts and turn performance, FCF
7	10 Apr.	Mallick/Ursini	Final checkout (#3) flight for Mallick, air data system calibration, radar tracking & intercepts
8	16 Apr.	Slater/Ursini	Supersonic, subsonic controlled intercepts
9	21 Apr.	Slater/Heidelbaugh	Radar intercepts
10	24 Apr.	Slater/Heidelbaugh	Simulated supersonic target intercepts
11	30 Apr.	Slater/Ursini	Mission Control data link check
12	7 May	Slater/Heidelbaugh	Air data and pitch trim calibration, fuel consumption and CG check, simulated supersonic ID intercept
13	17 May	Rogers/Heidelbaugh	Armed Forces Day Airshow fly-by at Edwards AFB
14	26 May	Campbell/Ursini	Supersonic cruise altitude control
15	5 June	Slater/Heidelbaugh	Air data system calibration
16	11 June	Slater/Ursini	Supersonic intercept using YF-12A 935 as target, air data system calibration
17	16 June	Slater/Heidelbaugh	Supersonic intercept, Air Force documentary photos
18	29 Aug.	Campbell/Heidelbaugh	FCF, inlet instrumentation check
19	5 Sept.	Layton/Heidelbaugh	Pilot checkout #1
20	6 Sept.	Layton/Heidelbaugh	Pilot checkout #2
21	6 Sept.	Layton/Heidelbaugh	Pilot checkout #3
22	11 Sept.	Layton/Heidelbaugh	Pilot checkout #4.
23	12 Sept.	Layton/Heidelbaugh	Final pilot checkout #5
24	15 Sept.	Layton/Heidelbaugh	F-106 autonomous intercept, B-57 controlled intercept
25	17 Sept.	Layton/Heidelbaugh	USAF test
26	6 Oct.	Layton/Heidelbaugh	High-speed altitude control, level decelration
27	9 Oct.	Layton/Heidelbaugh	Aborted
28	16 Oct.	Layton/Heidelbaugh	Air data system calibration
29	23 Oct.	Layton/Heidelbaugh	F-106 and B-57 controlled intercepts
30	27 Oct.	Layton/Heidelbaugh	Radar intercepts, turn performance
31	29 Oct.	Layton/Heidelbaugh	F-106 supersonic target ID and radar intercepts, lag stair step profile
32	3 Nov.	Mallick/Heidelbaugh	Pilot recurrency, phugoids
33	3 Nov.	Layton/Heidelbaugh	U-2 and F-4 controlled radar intercepts, F-106 autonomous radar intercepts, phugoids

34	17 Nov.	Layton/Maj. Wm. Curtis	FCO check-out, U-2 and F-106 radar intercepts
35	20 Nov.	Layton/Curtis	F-106 radar intercepts, aborted early due to oil problem
36	25 Nov.	Campbell/Heidelbaugh	F-106 radar intercepts, sideslip, level deceleration
37	1 Dec.	Layton/Heidelbaugh	Radar intercepts, aborted due to hydraulic system failure
38	17 Dec.	Layton/Curtis	Aborted due to liquid nitrogen system failure
39	19 Jan. 1971	Layton/Curtis	Pulse code modulation (PCM) telemetry failed
40	22 Jan.	Mallick/Heidelbaugh	Pilot recurrency, B-57 radar intercept, sideslip and roller coaster maneuver, low engine oil pressure
41	29 Jan.	Layton/Curtis	Aborted due to low engine oil pressure
42	2 Feb.	Layton/Heidelbaugh	B-57 ECM target, F-106 conversion target, sideslip, engine malfunction (oil pressure fluctuation)
43	10 Feb.	Layton/Heidelbaugh	FCF, sideslip, phugoid, inlet tests
44	18 Feb.	Layton/Curtis	Conversion targets, inlet tests
45	24 Feb.	Layton/Curtis	PCM system noise checks, intercepts, bypass, modified roller coaster maneuver
46	2 Mar.	Layton/Curtis	B-57 intercepts, roller coaster maneuver, inlet tests
47	5 Mar.	Mallick/Heidelbaugh	Pilot proficiency check, B-57 intercepts, inlet tests
48	9 Mar.	Layton/Heidelbaugh	Intercepts, handling qualities, aborted due to low engine oil pressure.
49	18 Mar.	Layton/Curtis	Intercepts, phugoids
50	23 Mar.	Layton/Heidelbaugh	Intercepts, handling qualities
51	25 Mar.	Layton/Curtis	Intercepts, handling qualities, phugoid
52	30 Mar.	Layton/Heidelbaugh	Intercepts, handling qualities
53	13 Apr.	Layton/Curtis	USAF test
54	20 Apr.	Layton/Curtis	C-131 radar target (propeller signatures), intercepts
55	27 Apr.	Layton/Heidelbaugh	Speed stability trim point
56	6 May	Layton/Curtis	Speed stability trim point
57	13 May	Layton/Heidelbaugh	Transonic acceleration and deceleration data, roller coaster maneuvers
58	4 June	Layton/Curtis	Lakebed takeoff, transonic accelerations, inlet cruise data calibration, auto inlet parameters
59	10 June	Layton/Heidelbaugh	Throttle advance/inlet noise correlation
60	15 June	Layton/Curtis	Handling qualities, intervalometer data
61	22 June	Layton/Heidelbaugh	Steady-state inlet data, handling qualities
62	24 June	Layton/Curtis	Handling qualities. Fire in right engine resulted in crash. Crew ejected safely.

"YF-12C" 937 Flights

Flight	Date	Pilot/Test Engineer	Remarks
1	16 Jul. 1971	Maj. Mervin Evenson/Maj. Charles McNeer	Delivery to FRC. Aircraft grounded for installation of NASA instrumentation.
2	24 May 1972	Fulton/Horton	FCF, S&C, first NASA flight of 937
3	6 June	Mallick/Young	Airspeed calibration
4	14 June	Fulton/Horton	Airspeed calibration, propulsion
5	21 June	Mallick/Young	Propulsion performance baseline data, S&C
6	18 Jul.	Fulton/Horton	Performance, airspeed calibration, engine check
7	26 Jul.	Mallick/Young	Propulsion baseline data, S&C, phugoids
8	1 Aug.	Fulton/Horton	Propulsion, served as target for Navy F-14 intercept
9	15 Aug.	Mallick/Young	Airspeed lag calibration, S&C
10	22 Aug.	Fulton/Young	Propulsion baseline data, S&C, phugoids
11	29 Aug.	Mallick/Young	Performance, airspeed calibration, phugoids
12	15 Nov.	Fulton/Horton	RPM trim, intentional unstart, bypass tests
13	22 Nov.	Mallick/Young	Propulsion studies
14	5 Dec.	Fulton/Horton	Propulsion studies

15	12 Dec.	Mallick/Young	Propulsion studies
16	11 Jan. 1973	Fulton/Horton	Propulsion studies
17	18 Jan.	Mallick/Young	Performance tests
18	24 Jan.	Fulton/Horton	Performance tests.
19	1 Feb.	Fulton/Young	Propulsion studies
20	8 Feb.	Mallick/Horton	Propulsion, handling qualities
21	15 Feb.	Fulton/Young	Performance tests
22	22 Feb.	Mallick/Horton	Performance tests
23	22 Mar.	Fullton/Horton	FCF, propulsion studies
24	5 Apr.	Mallick/Horton	Support U-2 Target Radiation Intensity Measurement (TRIM) test, propulsion studies, radar target for F-14
25	12 Apr.	Fulton/Young	Propulsion studies
26	20 Apr.	Mallick/Young	Propulsion studies, low L/D approach
27	26 Apr.	Fullton/Horton	Propulsion studies, low L/D approach
28	2 May	Mallick/Young	Stuck inlet spike caused high fuel consumption and emergency landing at NAS Fallon, Nevada
29	3 May	Mallick/Young	Subsonic ferry flight from NAS Fallon to Edwards AFB
30	10 May	Fulton/Horton	Propulsion studies
31	17 May	Mallick/Young	Propulsion studies.
32	31 May	Fulton/Horton	S&C, handling qualities, phugoids
33	8 June	Mallick/Young	Performance, low L/D approach
34	11 Jul. 1974	Fulton/Horton	FCF, data acquisition system checkout
35	26 Jul.	Mallick/Young	Performance tests
36	13 Sept.	Mallick/Young	Propulsion studies
37	25 Sept.	Mallick/Young	Propulsion studies
38	7 Nov.	Fulton/Horton	Performance tests
39	19 Dec.	Mallick/Young	Low-speed FCF, inlet transducer lag checks
40	19 Dec.	Mallick/Young	High-speed FCF, performance modeling data
41	17 Jan. 1975	Fulton/Horton	Performance modeling data, Turbine Inlet Gas Temperature (TIGT) system check
42	24 Jan.	Mallick/Horton	Performance modeling data, TIGT system check
43	24 Apr.	Fulton/Horton	Performance modeling data, TIGT system check
44	5 June	Mallick/Young	TIGT system check
45	12 June	Fulton/Horton	TIGT system tests, inlet spike tip data
46	20 June	Mallick/Young	Performance modeling data, TIGT system check
47	26 June	Fulton/Horton	TIGT system tests
48	3 Jul.	Mallick/Young	Propulsion studies
49	7 Aug.	Mallick/Young	Chase for YF-12A 935, inlet spike tip tests
50	14 Aug.	Mallick/Young	Propulsion studies
51	11 Sept.	Mallick/Young	Propulsion studies
52	24 Sept.	Fulton/Horton	TIGT tests, S&C, co-op controls, altitude hold
53	16 Oct.	Mallick/Young	Propulsion studies, cockpit and suit cooling malfunction
54	30 Oct.	Fulton/Horton	Engine compressor stalls, co-op control interactions, autopilot tests
55	16 Sept. 1976	Mallick/Young	Low-speed FCF, boattail drag
56	30 Sept.	Mallick/Young	High-speed FCF, boattail drag, autothrottle data
57	21 Oct.	Mallick/Young	Chase for YF-12A 935, boattail drag
58	9 Nov.	Mallick/Young	Propulsion, handling qualities
59	19 Nov.	Mallick/Young	Inlet tests, boattail drag, handling qualities
60	2 Dec.	Mallick/Young	Inlet tests, boattail drag, handling qualities
61	3 Mar. 1977	Mallick/Young	Chase for YF-12A 935, inlet tests, boattail drag
62	18 Mar.	Fulton/Horton	Propulsion studies, autothrottle tests
63	24 Mar.	Fulton/Young	Propulsion, autothrottle tests, gust vane experiment
64	1 Apr.	Mallick/Horton	Propulsion, sine/step function generator data, gust vane
65	12 May	Fulton/Horton	Autothrottle tests

66	19 May	Mallick/Young	Propulsion studies, autothrottle, gust vane
67	26 May	Fulton/Horton	Propulsion studies, autothrottle, gust vane
68	2 June	Mallick/Young	Chase for YF-12A 935, propulsion studies
69	15 June	Mallick/Young	Propulsion studies, autothrottle tests
70	16 June	Fulton/Horton	Airframe/propulsion interactions, autothrottle tests, handling qualities
71	23 June	Mallick/Young	Chase for YF-12A 935, boattail drag
72	14 Jul.	Fulton/Horton	Propulsion, autothrottle, airframe/propulsion interactions
73	21 Jul.	Mallick/Young	Chase for YF-12A 935, severe unstarts
74	8 Sept.	Mallick/Young	Propulsion, airframe/propulsion interactions, sine/step generator, autothrottle
75	16 Sept.	Fulton/Horton	Airframe/propulsion interactions, propulsion, autothrottle tests
76	22 Sept.	Mallick/Young	Airframe/propulsion interactions, propulsion
77	30 Sept.	Mallick/Young	Chase for YF-12A 935, U-2 intercept, fly-by of DFRC
78	13 Oct.	Mallick/Young	Chase for YF-12A 935, airframe/propulsion interactions, propulsion studies
79	26 May 1978	Mallick/Young	Functional check of Cooperative Airframe/Propulsion Control System (CAPCS) digital computer system
80	16 June	Fulton/Horton	CAPCS envelope expansion.
81	17 Jul.	Mallick/Young	Aborted due to air data transducer failure
82	3 Aug.	Fulton/Horton	CAPCS test
83	18 Aug.	Mallick/Young	CAPCS test
84	31 Aug.	Fulton/Horton	CAPCS test,
85	7 Sept.	Mallick/Young	CAPCS test, co-op control test
86	13 Sept.	Fulton/Horton	CAPCS test, co-op control test
87	25 Sept.	Fulton/Horton	CAPCS test, co-op control test.
88	28 Sept.	Mallick/Young	CAPCS test, early due to bypass door failure
89	27 Oct.	Col. James Sullivan/Maj. William Frazier	Transfer to USAF, ferry flight to Palmdale
90	22 Dec.	Lt. Col. Calvin Jewett/Frazier	Last flight of aircraft. Stored at Palmdale.

YF-12 Program Summary

Total Flights in Program

YF-12A 935:	146
YF-12A 936:	62
YF-12C 937:	89

Total Flying Time (935 plus 937)
Approximately 450 flight hours.

Total Flying Time at or above Mach 3
Approximately 37 flight hours.

Sources: YF-12 program flight requests and reports; information supplied by flight crews; information from the files of Richard Klein, Gene Matranga, Ming Tang, and Paul Reukauf, DFRC. DFRC Daily Flight Logs as researched by Peter Merlin.

Appendix Q

Space Shuttle Orbiter Approach and Landing Test Program Flight Chronology, 1977

During 1977, in conjunction with the NASA Johnson Space Center, DFRC undertook verification testing of the Shuttle's approach and landing behavior.

Tests air-launched the Space Shuttle orbiter OV-101, *Enterprise*, from a modified Boeing 747-100 jet transport. The approach and landing test (ALT) program was intended to certify the low-speed airworthiness of the Shuttle orbiter, as well as its pilot-guided and automatic approach and landing capabilities. For this reason, the *Enterprise* differed in a number of respects from a Shuttle ready for orbital spaceflight. During the initial testing, the *Enterprise* was fitted with a drag-reducing and airflow-smoothing tailcone. Subsequently, two glide flights were made without the tailcone and with the Shuttle's engine installation simulated so as to acquire data more closely approximating the Shuttle's configuration when returning from orbit.

The ALT program had three phases: a captive-inert phase with the *Enterprise* unpiloted and with its systems inert; a captive-active phase, with the *Enterprise* piloted and all systems operational; and, finally, a free-flight test phase, with the *Enterprise* actually launched in flight from the back of the 747 aircraft, becoming, in effect, the world's largest glider.

Space Shuttle Orbitor OV-101, Enterprise

Captive-Inert Tests

Flight	Date	Crew	Duration	Remarks
1	18 Feb. 1977	–	2 hrs 10 min	250 knots at 16,000 ft
2	22 Feb.	–	3 hrs 15 min	288 knots at 22,000 ft
3	25 Feb.	–	2 hrs 30 min	282 knots at 26,000 ft
4	28 Feb.	–	2 hrs 11 min	283 knots at 28,000 ft
5	2 Mar.	–	1 hr 40 min	278 knots at 30,100 ft

During this test phase, the crew on each flight of the *Enterprise* was a team of astronauts–either Fred Haise and Charles Gordon Fullerton or Joseph Engle and Richard Truly. Haise was a former FRC research pilot; Engle had flown the X-15 at the center as well; and Fullerton later became a Dryden research pilot. Truly, of course, became the NASA Administrator from 14 May 1989 to 31 March 1992.

Captive-Active Tests

Flight	Date	Crew	Duration	Remarks
1A	18 June 1977	Haise/Fullerton	55 min 45 sec	180 knots at 15,630 ft
1	28 June	Engle/Truly	1 hr 3 min	270 knots at 24,190 ft
3	26 July	Haise/Fullerton	59 min 53 sec	272 knots at 30,250 ft
4				Cancelled because it was unnecessary.

Free-Flight Tests

Flight	Date	Crew	Duration (min-sec)	Tail Cone	Launch Altitude	Launch Speed	Landing Speed
1	12 Aug. 1977	H/F	5-22	On	(26,000 ft)	500 kph (270 knots)	343 kph (185 knots)
2	13 Sept.	E/T	5-31	On	(26,000 ft)	556 kph (300 knots)	359 kph (194 knots)
3	23 Sept.	H/F	5-35	On	(27,300 ft)	463 kph (250 knots)	354 kph (191 knots)
4	12 Oct.	E/T	2-35	Off	(22,400 ft)	445 kph (240 knots)	369 kph (199 knots)
5	26 Oct.	H/F	2-06	Off	(19,000 ft)	454 kph (245 knots)	350 kph (189 knots)

Crew: H/F = Haise/Fullerton. E/T = Engle/Truly.
Launch altitude: Altitude at separation from 747.

Sources: NASA, *Space Shuttle Orbiter Approach and Landing Test: Final Evaluation Report* (Washington, DC: NASA JSC-13864, Feb. 1978); Linda Neuman Ezell, *NASA Historical Data Book*, Vol. III: *Programs and Projects 1969-1978* (Washington, DC: NASA SP-4012, 1988), p. 118; NASA Shuttle Carrier Aircraft Test Team, *Space Shuttle Orbiter Approach & Landing Test: Captive Inert Flight Test Program Summary* (Edwards, CA: NASA DFRC SOD 40.1, Jun. 1977); NASA Approach and Landing Test Evaluation Team, *Space Shuttle Orbiter Approach and Landing Test Evaluation Report: Captive-Active Flight Test Summary* (Houston, TX: JSC-13045, Sept. 1977).

Appendix R
Accident Statistics, 1954-1999

Calendar Year	Number of Accidents Accidents	Flight Hours	Total Flights
1954	1	346	379
1955	1	521	521
1956	1	469	461
1957	0	411	463
1958	0	508	448
1959	1	532	520
1960	0	685	695
1961	0	672	650
1962	3	1,016	926
1963	0	1,806	1,472
1964	0	1,747	N/A
1965	0	1,833	N/A
1966	1	1,493	1,258
1967	2	1,731	1,430
1968	0	1,963	1,496
1969	0	1,887	1,688
1970	0	1,736	1,596
1971	0	1,777	1,613
1972	0	1,868	1,571
1973	0	1,675	1,459
1974	0	1,585	1,327
1975	0	1,679	1,474
1976	0	1,476	1,231
1977	0	1,666	1,441
1978	0	1,991	1,693
1979	0	1,546	1,339
1980	1 (remotely piloted)	1,596	1,442
1981	0	1,544	1,440
1982	1	1,624	1,468
1983	1 (remotely piloted)	1,546	1,401
1984	0	1,347	1,159
1985	0	1,617	1,375
1986	0	1,639	1,504
1987	0	1,496	1,395
1988	1	1,331	1,222
1989	0	1,166	1,028
1990	0	1,287	1,066
1991	0	1,298	1,121
1992	0	1,123	1,135
1993	0	1,331	1,218

1994	1 (remotely piloted)	1,217	1,155
1995	1	1,063	842
1996	1 (remotely piloted)	1,213	838
1997	0	1,159	836
1998	0	1,998	1,052
1999	0	2,108	1,070

Accident Rate

Period	Rate per 100,000 Flight Hours	Rate per 100,000 Flights
1996-1999	15.4	26.3
1991-1995	33.2	36.6
1986-1990	14.5	16.1
1981-1985	28.9	29.2
1976-1980	12.1	14.0
1971-1975	0	0
1966-1970	34.1	40.2
1961-1965	42.4	N/A[a]
1956-1960	76.8	77.3
1954-1955	230.7	223

a Not available; data lacking for total number of flights in 1964 and 1965.

HSFS/FRC/DFRC Flight Accidents

Calendar Year	Aircraft	Remarks
1948	Douglas D-558-1	Engine failure on takeoff Howard Lilly killed.
1953	Convair XF-92A	Landing gear failure on rollout after landing. Aircraft retired.
1954	North American F-100A	Collision with hangar after emergency landing. Minor damage to aircraft.
1955	Bell X-1A	Inflight explosion before launch. Plane jettisoned, destroyed.
1956	Bell X-1E	Landing accident; nose-gear collapse. Plane damaged.
1959	North American F-107A	Takeoff accident. Plane damaged beyond repair.
1962	North American X-15 #2	Inflight powerplant failure, followed by tailskid collapse on landing. Pilot Jack McKay seriously injured; plane virtually destroyed but rebuilt as X-15A-2.
1962	Lockheed T-33A	Pilot undershot landing at Norton AFB; plane damaged beyond repair.
1962	Lockheed F-104A	Inflight asymmetric flap deployment caused uncontrollable rolling. Pilot Milton Thompson ejected safely.
1966	Lockheed F-104N	Mid-air collision with XB-70A #2. Pilot Joseph Walker killed, as was XB-70A copilot. Both aircraft lost.
1967	Northrop M2-F2	Landing accident. Pilot Bruce Peterson seriously injured; plane rebuilt as M2-F3.
1967	North American X-15 #3	Break-up during reentry; pilot Michael Adams killed.

1980	DAST (Drones for Aero-dynamic and Structural Testing; 72-1557	Right wing failed in flight. Aircraft destroyed. Remotely piloted.
1982	Cessna T-37B (60-0084)	Crashed in a spin, killing pilot Richard Gray. Not a research flight.
1983	DAST (72-1558)	Crashed when recovery parachute deployed immediately after launch, destroying the remotely piloted vehicle.
1988	McDonnell Douglas F-18A	Crashed due to a flight-control malfunction; non-fatal; aircraft destroyed.
1994	Aurora Flight Sciences Perseus A	Remotely piloted aircraft destroyed when it crashed due to a gyro malfunction.
1995	X-31 (164-584)	Crashed due to icing on the Kiel probe airspeed boom, destroying the aircraft. The pilot, Karl Heinz-Lang parachuted to the ground, injuring his back.
1995	Aurora Flight Sciences Corporation Theseus	Remotely piloted aircraft destroyed as a result of a structural failure.

Sources: DFRC Safety Office files for data before 1976; flight hours and total flights compiled from DFRC Flight Operations Office files by Judy Duffield; accidents since 1967 taken from NASA DFRC Aircraft Accidents/Incidents, compiled by Peter W. Merlin.

Appendix S

Boeing 720B Controlled Impact Demonstration Flight Chronology

In 1984, Dryden remotely piloted a Boeing 720 (N833NA) to a landing on iron posts on Rogers Dry Lake to test an anti-misting kerosene for its ability to suppress a fire. In preparation, Dryden pilots flew a series of rehearsal flights to prepare for the final mission.

Flight	Date	Ground Pilot/Aircrew	Remarks
1	7 Mar. 1984	E. Schneider, G.P. F. Fulton/ T. McMurtry/ R. Young/J. Cooper	Check-out flight.
2	15 Mar.	E.Schneider,G.P. F.Fulton/ T. McMurtry/ R.Young	Auto-throttle and autopilot tests.
3	3 May	Schneider, G.P. Fulton/McMurtry/Young/ V. Horton/D.Dennis	Uplink/downlink tests, auto-throttle and autopilot tests, theta vs. flap angle calibration.
4	9 May	Fulton, G.P. McMurtry/Schneider/ Young/Horton	Ground effects data, airspeed system calibration, handling qualities.
5	13 Jul.	Fulton, G.P. McMurtry/Schneider/ Horton/Young/ Dennis	Fuel degrader tests, handling qualities during CID flight profile.
6	8 Aug.	Fulton, G.P. McMurtry/D.Mallick/Young/ Horton/Dennis	Ground steering, ground effects during approach airspeed system calibration, fuel degrader tests.
7	17 Aug.	Fulton, G.P. McMurtry/Schneider/ Horton/Young	Practice CID approaches, qualify all degraders on Jet-A fuel, remotely controlled takeoff.
8	28 Aug.	Fulton, G.P. McMurtry/Schneider/ Young/Horton	Qualify fuel degraders on Jet-A fuel, practice CID approaches, remotely controlled takeoff to qualify degrader #3 on AMK fuel.
9	18 Sept.	Fulton, G.P. McMurtry/Schneider/ Horton/Young	Successful flight.
10	2 Oct.	Fulton, G.P. McMurtry/Schneider/ Young/Horton	Successful flight
11	25 Oct.	Fulton, G.P. McMurtry/Schneider/ Horton/Young	Successful dress-rehearsal flight
12	5 Nov.	Fulton, G.P. McMurtry/Schneider/ Horton/Young	
13	15 Nov.	Fulton, pilot McMurtry/Schneider/ Horton/M. Knutson.	Piloted systems- readiness checkout.
14	26 Nov.	Fulton, G.P. McMurtry/Schneider/ Horton/Young	
15	1 Dec.	Fulton, G.P.	CID profile. Controlled impact on lakebed. Aircraft destroyed.

Note: G.P.=ground pilot.

Sources: Flight reports for flights 2-8, *X-Press*es for 1984. Flight log compiled by Peter W. Merlin in July 1998.

Appendix T
NF-111A (63-9778) Flight Chronology

NASA obtained the 13th service test F-111A from the U.S. Air Force through a loan agreement, signed on 3 February 1972. It was initially used for the joint Air Force/NASA Transonic Aircraft Technology (TACT) program. Modified with supercritical wings, it was redesignated NF-111A. It was later flown by NASA and Air Force crewmembers for the Natural Laminar Flow (NLF), Advanced Fighter Technology Integration (AFTI), and Mission Adaptive Wing (MAW) research programs. Following completion of the MAW program in December 1988, the aircraft was placed in storage at NASA DFRC. It was transferred to the Air Force Flight Test Center Museum on 29 June 1990.

NF-111A Flight Log

Flight No.	Date	Crew	Remarks
1	18 Feb. 1972	Einar K. Enevoldson/Maj. Stuart R. Boyd	NASA acceptance flight
2	19 Jul.	Maj. Boyd/Enevoldson	Functional check flight (FCF) of standard F-111A
3	25 Jul.	Maj. Boyd/Enevoldson	Thrust, performance, airspeed calibration
4	28 Jul.	Maj. Boyd/Enevoldson	Airspeed calibration, Tracking
5	2 Aug.	Maj. Boyd/Enevoldson	Stability and control (S&C), performance, airspeed calibration
6	2 Aug.	Maj. Boyd/Enevoldson	S&C, performance, airspeed calibration
7	4 Aug.	Enevoldson/Maj. Boyd	S&C, specific excess power, buffet
8	16 Aug.	Enevoldson/Maj. Boyd	S&C, specific excess power, buffet
9	16 Aug.	Maj. Boyd/Enevoldson	S&C, specific excess power, performance
10	18 Aug.	Enevoldson/Maj. Boyd	S&C, specific excess power, tracking
11	18 Aug.	Maj. Boyd/Enevoldson	S&C, specific excess power, buffet
12	23 Aug.	Enevoldson/Maj. Boyd	Performance, airspeed calibration
13	23 Aug.	Maj. Boyd/Enevoldson	Performance, specific excess power
14	25 Aug.	Enevoldson/Maj. Boyd	Performance, airspeed calibration
15	25 Aug.	Maj. Boyd/Enevoldson	Performance, specific excess power
16	11 Oct.	Maj. Boyd/Enevoldson	High "G" stability and control
17	13 Oct.	Enevoldson/Maj. Boyd	High "G" stability and control, buffet
18	17 Oct.	Maj. Boyd/Enevoldson	Performance, airspeed calibration, tracking
19	24 Oct.	Enevoldson/Maj. Boyd	Performance, tracking
20	25 Oct.	Maj. Boyd/Enevoldson	Performance
21	26 Oct.	Enevoldson/Maj. Boyd	Performance
22	26 Oct.	Maj. Boyd/Enevoldson	Performance

23	6 Nov.	Enevoldson/Maj. Boyd	Performance
24	9 Nov.	Maj. Boyd/Enevoldson	Performance, tracking
25	1 Nov. 1973	Enevoldson/Maj. Boyd	FCF with supercritical wing installed. A/C designated NF-111A
26	23 Jan. 1974	Maj. Boyd/Enevoldson	Performance, S&C, loads, flutter
27	28 Jan.	Enevoldson/Maj. Boyd	Performance, S&C, loads, flutter
28	21 Feb.	Maj. Boyd/Enevoldson	Performance, S&C, loads, flutter
29	12 Mar.	Enevoldson/Maj. Boyd	Performance, S&C, flutter, airspeed calibration
30	20 Mar.	Maj. Boyd/Enevoldson	Flutter, first supersonic TACT flight
31	26 Mar.	Enevoldson/Capt. Stanley E. Boyd	S&C, flutter, pilot check out
32	26 Mar.	Enevoldson/Gary E. Krier	Performance, flutter, pilot check out
33	3 Apr.	Maj. Boyd/Thomas C. McMurtry	Performance, pilot check out
34	11 Apr.	Enevoldson/Maj. Boyd	Performance, flutter
35	15 May	Maj. Boyd/Enevoldson	S&C, flutter, loads
36	12 Jun.	Maj. Boyd/Enevoldson	Performance (supersonic)
37	20 Jun.	Enevoldson/Maj. Boyd	Base drag, Strip-A-Tube reference
38	9 Sep.	Capt. Boyd/Enevoldson	Base drag, Strip-A-Tube reference
39	16 Sep.	Capt. Boyd/Enevoldson	Performance (supersonic)
40	19 Sep.	Enevoldson/Capt. Boyd	S&C, specific excess power, buffet, tracking
41	10 Oct.	Capt. Boyd/Enevoldson	Aborted early due to cowl malfunction
42	17 Oct.	Capt. Boyd/Enevoldson	Base drag, Strip-A-Tube reference
43	22 Oct.	Capt. Boyd/Krier	Performance
44	31 Oct.	Enevoldson/McMurtry	S&C, specific excess power, buffet, tracking
45	19 Nov.	Capt. Boyd/Enevoldson	Base drag, Strip-A-Tube (Configuration 2)
46	26 Nov.	Enevoldson/Capt. Boyd	Base drag, Strip-A-Tube (Configuration 2)
47	27 Nov.	Capt. Boyd/Enevoldson	Base drag, Strip-A-Tube (Configuration 3)
48	27 Nov.	Enevoldson/Capt. Boyd	Base drag, Strip-A-Tube (Configuration 3)
49	5 Dec.	Capt. Boyd/Enevoldson	Base drag, Strip-A-Tube (Configuration 4)
50	5 Dec.	Enevoldson/Capt. Boyd	Base drag, Strip-A-Tube (Configuration 4)
51	11 Dec.	Enevoldson/Victor W. Horton	S&C
52	12 Dec.	Enevoldson/Horton	S&C, tracking
53	13 Dec.	Capt. Boyd/Horton	S&C
54	18 Dec.	Capt. Boyd/Enevoldson	Strip-A-Tube reference
55	20 Dec.	Capt. Boyd/Fitzhugh L. Fulton	S&C
56	7 Jan. 1975	Enevoldson/Capt. Boyd	S&C, specific excess power, buffet
57	7 Jan.	Capt. Boyd/Enevoldson	Strip-A-Tube reference, S&C

58	15 Jan.	Enevoldson/Capt. Boyd	Performance
59	15 Jan.	Capt. Boyd/Enevoldson	High "G" stability and control
60	21 Jan.	Enevoldson/Weneth D. Painter	Ferry flight to McClellan AFB, California for maintenance
61	27 Mar.	Enevoldson/Capt. Boyd	Ferry flight to Edwards AFB from McClellan AFB
62	20 Jun.	Capt. Boyd/Fulton	S&C, high-frequency buffet, performance
63	27 Jun.	Capt. Boyd/Fulton	Wing static pressure data
64	27 Jun.	Capt. Boyd/Fulton	Wing static pressure data
65	9 Jul.	Enevoldson/Fulton	Tracking
66	16 Jul.	Enevoldson/Fulton	S&C, high "G" derivatives
67	16 Jul.	Enevoldson/Fulton	S&C, high "G" derivatives
68	18 Jul.	Capt. Boyd/Enevoldson	Wing pressure correlation
69	22 Jul.	Capt. Boyd/Fulton	Wing pressure correlation
70	25 Jul.	Enevoldson/Capt. Boyd	Performance (fixed nozzle correlation)
71	30 Jul.	Capt. Boyd/Enevoldson	Performance
72	6 Aug.	Capt. Boyd/Fulton	High-frequency pressure data
73	13 Aug.	Capt. Boyd/Capt. Ronald J. Grabe	High-frequency pressure data
74	13 Aug.	Capt. Boyd/Horton	High-frequency pressure data
75	3 Sep.	Enevoldson/Capt. Boyd	Performance
76	15 Sep.	Capt. Boyd/Enevoldson	Performance (supersonic)
77	17 Sep.	Enevoldson/Lt. Col. Edward D. McDowell Jr.	Performance
78	22 Sep.	Capt. Boyd/Grabe	Performance
79	1 Oct.	Enevoldson/Ronald Gerdes	Performance
80	3 Oct.	Capt. Boyd/Grabe	Performance
81	8 Oct.	Grabe/Enevoldson	Performance
82	10 Oct.	Fulton/Enevoldson	Performance
83	7 Nov.	Enevoldson/Fulton	Performance, buffet, energy meter evaluation
84	21 Nov.	Enevoldson/Grabe	Performance
85	26 Nov.	Enevoldson/Grabe	Performance, tracking
86	14 Jan. 1976	Enevoldson/Capt. Boyd	Performance and flutter with external stores (twelve 500-lb. M.117 dummy bombs on two wing pylons)
87	19 Jan.	Capt. Boyd/Enevoldson	Same as flight 86
88	27 Jan.	Enevoldson/Grabe	Same as flight 86
89	29 Jan.	Grabe/Enevoldson	Same as flight 86
90	3 Feb.	Capt. Boyd/Krier	Same as flight 86
91	5 Mar.	Enevoldson/Maj. Francis R. "Dick" Scobee	Boundary layer rake (trailing edge)
92	12 Mar.	Scobee/Capt. Boyd	Boundary layer rake (trailing edge)

93	17 Mar.	Capt. Boyd/Scobee	Boundary layer rake (trailing edge)
94	22 Mar.	Scobee/Enevoldson	Dynamic pressure distribution, boundary layer rake
95	20 Aug.	Enevoldson/Horton	Dynamic pressure distribution
96	17 Nov.	Scobee/Enevoldson	Trailing edge boundary layer survey
97	19 Nov.	Scobee/Enevoldson	Dynamic performance (engine mapping)
98	30 Nov.	Scobee/Donald L. Mallick	Trailing edge boundary layer survey
99	26 Jan. 1977	Enevoldson/Scobee	Trailing edge boundary layer survey
100	1 Feb.	Enevoldson/Scobee	Trailing edge boundary layer survey
101	4 Feb.	Scobee/Enevoldson	Joint U.S./U.K. wing buffet research
102	8 Feb.	Scobee/Enevoldson	Dynamic performance
103	10 Feb.	Enevoldson/Scobee	Dynamic performance
104	11 Feb.	Enevoldson/Scobee	Performance (rate effect)
105	23 Jun.	Scobee/Enevoldson	Functional check flight, overwing fairing pressure investigation
106	24 Jun.	Enevoldson/Scobee	Overwing fairing pressure
107	30 Jun.	Scobee/Stephen D. Ishmael	Overwing fairing pressure
108	7 Jul.	Enevoldson/Ishmael	Overwing fairing pressure, dry hook-up with KC-135 tanker
109	4 Aug.	Ishmael/Enevoldson	Overwing fairing pressure
110	9 Aug.	Scobee/Lt. Col. Richard Cooper	Overwing fairing pressure, tracking
111	18 Aug.	Enevoldson/Scobee	Overwing fairing pressure, aerial refueling
112	19 Aug.	Scobee/Enevoldson	Overwing fairing pressure, aerial refueling
113	2 Sep.	Enevoldson/Scobee	Overwing fairing pressure, aerial refueling
114	4 Oct.	Enevoldson/Capt. Vincent T. Baker	Joint U.S./U.K. wing buffet research
115	4 Oct.	Enevoldson/Baker	Joint U.S./U.K. wing buffet research
116	7 Oct.	Enevoldson/Birk	Trailing edge boundary layer survey
117	7 Oct.	Enevoldson/Birk	Trailing edge boundary layer survey
118	1 Nov.	Enevoldson/Scobee	Trailing edge boundary layer survey
119	22 Nov.	Scobee/Maj. Paul D. Tackabury	Joint U.S./U.K. wing buffet research
120	22 Nov.	Scobee/Capt. Peter Tait	Joint U.S./U.K. wing buffet research
121	2 Dec.	Enevoldson/Scobee	Joint U.S./U.K. wing buffet research
122	6 Dec.	Scobee/Enevoldson	Joint U.S./U.K. wing buffet research
123	6 Dec.	Enevoldson/Scobee	Joint U.S./U.K. wing buffet research
124	21 Mar. 1978	Scobee/Enevoldson	FCF following installation of traversing probe on upper surface of wing to measure boundary-layer, NLF pre-test S&C evaluation, low L/D Orbiter landing investigation
125	22 Mar.	Enevoldson/Scobee	Boundary-layer growth, low L/D Orbiter landing investigation
126	27 Jun.	Enevoldson/Michael R. Swann	Boundary-layer growth, joint U.S./U.K. wing buffet research

127	7 Jul.	Enevoldson/Maj. Frank T. Birk	Pre-NLF evaluation of flying qualities with inboard wing spoilers locked out
128	2 Aug.	Enevoldson/Birk	Overwing fairing flow visualization
129	9 Aug.	Enevoldson/Birk	Performance modeling into the partial-power region of operation
130	11 Aug.	Enevoldson/Birk	Boundary-layer growth, control system data for structural purposes
131	31 Aug.	Enevoldson/Swann	Boundary-layer pressure data over upper wing surface
132	7 Sep.	Enevoldson/Swann	Boundary-layer growth
133	14 Sep.	Enevoldson/Swann	Airspeed and thermometer calibration
134	21 Sep.	Birk/Swann	Ferry flight to McClellan AFB for maintenance (pyrotechnics replacement)
135	19 Jan. 1979	Enevoldson/Swann	Ferry flight to Edwards AFB from McClellan AFB
136	16 Mar.	Swann/Enevoldson	No-flaps takeoff and landing demonstration
137	16 Mar.	Swann/McMurtry	Airspeed calibration, pitot-static probe check
138	12 Apr.	Swann/Birk	Boundary-layer growth, no-flaps takeoff and landing
139	15 May	Swann/Capt. Richard A. Schroeder	Fiberglass test samples, heavy-duty tires, airspeed calibration, boundary-layer
140	24 May	Birk/Capt. Francis R. Smith	Flight Deflection Measurement System (FDMS), boundary-layer growth
141	6 Jun.	Swann/Lt. Col Frederick A. Fiedler	Fiberglass test samples, heavy-duty tires, boundary-layer growth
142	21 Jun.	Swann/Birk	FDMS evaluation
143	11 Jul.	Swann/Maj. Robert D. Muldrow	Boundary-layer flight aborted due to flight control system anomaly
144	15 May 1980	Swann/Enevoldson	FCF following installation of Natural Laminar Flow (NLF) wing gloves, handling qualities, flutter clearance
145	15 May	Swann/Enevoldson	Flutter clearance, stability, NLF data, structure verification, drag reduction
146	22 May	Swann/Enevoldson	Flutter clearance, NLF data, drag reduction
147	18 Jun.	Enevoldson/Swann	NLF data, base drag reduction, controllability, lateral PIO tendencies
148	1 Jul.	Swann/Enevoldson	NLF data, base drag reduction
149	3 Jul.	Enevoldson/Swann	NLF data, base drag reduction
150	9 Jul.	Swann/Horton	NLF data, base drag reduction
151	10 Jul.	Swann/Horton	NLF data, base drag reduction, engine noise influence on flow
152	15 Jul.	Swann/Horton	NLF data, base drag reduction
153	17 Jul.	Swann/Horton	NLF data, boundary layer data, base drag reduction
154	18 Jul.	Swann/Horton	NLF data, boundary layer data, axisymmetric base drag
155	21 Jul.	Swann/Horton	NLF data (clean configuration), axisymmetric base drag, transition noise
156	22 Jul.	Swann/Birk	NLF data, boundary layer data, axisymmetric base drag, transition noise
157	23 Jul.	Birk/Swann	NLF data, boundary layer data, axisymmetric base drag, transition noise
158	23 Jul.	Swann/Horton	NLF data, boundary layer data, axisymmetric base drag, transition noise
159	25 Jul.	Swann/Birk	NLF data, boundary layer data, transition study
160	25 Jul.	Swann/Birk	NLF data, boundary layer data, transition study
161	8 Aug.	Enevoldson/Horton	NLF data, boundary layer data, base drag reduction

162	8 Aug.	Enevoldson/Birk	NLF data, boundary layer data
163	24 Feb. 1981	Swann/Ishmael	FCF following installation of Pratt & Whitney TF30-P-9 engines
164	26 Feb.	Swann/Birk	Limited engine baseline data, base drag reduction
165	26 Feb.	Swann/Birk	Engine baseline performance data, axisymmetric base drag reduction, lateral maneuvers to aid in Boeing AFTI/F-111 simulation development
166	3 Mar.	Birk/Enevoldson	Same as Flight 165
167	3 Mar.	Birk/Enevoldson	Same as Flight 165
168	10 Mar.	Swann/Horton	Engine baseline performance data, axisymmetric base drag reduction, airspeed calibration
169	10 Mar.	Birk/Swann	Same as Flight 165
170	12 Mar.	Enevoldson/Birk	Same as Flight 165
171	12 Mar.	Enevoldson/Swann	Flutter, autopilot characteristics and lateral maneuvers for AFTI/F-111 simulation development
172	30 Mar.	Enevoldson/Swann	Engine baseline performance data, flutter, and autopilot characteristics for AFTI/F-111 simulation development
173	18 Oct. 1985	Birk/ Capt. Smith	FCF following modification to AFTI/Mission Adaptive Wing (MAW) configuration
174	22 Nov.	Capt. Smith/Birk	Stability and control, wing sweep and rolling tail handling qualities
175	26 Feb. 1986	Capt. Smith/Capt. Scott Parks	Envelope clearance, pilot evaluation, MAW system operator (MAWSO) familiarity, emergency landing due to hydraulic leak
176	28 Mar.	Birk/Parks	Envelope clearance
177	8 Apr.	Parks/ Rogers E. Smith	Envelope clearance
178	15 Apr.	Birk/Enevoldson	Envelope clearance, pilot familiarity
179	1 May	Enevoldson/Parks	Pilot familiarity, refueling, performance
180	6 May	R. Smith/Birk	Envelope clearance, performance, controllabillity
181	30 May	Enevoldson/Capt. Smith	Performance
182	3 Jun.	Birk/Parks	Performance, stability and control, pressure distribution check
183	3 Jun.	Parks/Capt. Smith	Performance, stability and control
184	1 Jul.	Birk/Maj. Kermit Rufsvold	Envelope clearance, pilot familiarity
185	3 Jul.	Capt. Smith/Rufsvold	Envelope clearance
186	8 Jul.	Birk/Enevoldson	Envelope clearance
187	10 Jul.	Rufsvold/Capt. Smith	Envelope clearance, first MAW supersonic flight
188	28 Aug.	Parks/Rufsvold	Performance, mountain fly-ats, FDMS evaluation
189	4 Sep.	Rufsvold/Capt. Smith	Envelope clearance, FDMS evaluation
190	16 Sep.	Enevoldson/Rufsvold	Envelope clearance, FDMS evaluation
191	30 Sep.	Parks/Rufsvold	Performance, pressure distribution, FDMS evaluation

192	7 Oct.	Enevoldson/Rufsvold	Performance, pressure distribution, maneuver camber control (MCC), cruise camber control (CCC), maneuver load control (MLC), angle of attack (AOA) clearance
193	14 Oct.	Capt. Smith/Parks	Performance, angle of attack clearance
194	14 Oct.	Parks/Capt. Smith	Performance
195	21 Oct.	Rufsvold/Parks	Performance, pressure distribution, loads clearance
196	23 Oct.	Enevoldson/Rufsvold	Unsteady aerodynamics, loads clearance
197	4 Nov.	Rufsvold/Capt. Smith	Performance
198	14 Nov.	Capt. Smith/Rufsvold	Loads clearance, pressure distribution, unsteady aerodynamics
199	6 Aug. 1987	R. Smith/Capt. David W. Eidsaune	FCF following installation of new computers
200	11 Aug.	Eidsaune/Capt. Smith	Loads clearance, pressure distribution, radar acceleration
201	28 Aug.	Smith/James W. Smolka	Pilot familiarity, pacer, loads research, flutter clearance
202	3 Sep.	Eidsaune/Capt. Smith	First automatic mode flight (MCC,CCC)
203	5 Oct.	Capt. Smith/Smolka	Flutter and loads clearance, auto investigation in manual
204	7 Oct.	Smolka/Eidsaune	Flutter and loads clearance, buffet, handling qualities (MCC)
205	23 Oct.	Smolka/Capt. Smith	CCC mode investigation, aborted due to electronic cooling system malfunction
206	28 Oct.	Smolka/Maj. Ronald E. Johnston	CCC and MCC mode investigation, flutter and loads clearance, pilot proficiency
207	3 Nov.	Eidsaune/Johnston	MCC mode handling qualities, performance, and flight control evaluation
208	8 Apr. 1988	Capt. Smith/Johnston	Clearance and check out of MLC and maneuver enhancement/gust alleviation (MEGA), combination of MCC/MLC/MEGA
209	26 Apr.	Johnston/Capt. Smith	Envelope clearance and check out of MLC, MEGA, and combination of automodes
210	24 May	Fullerton/Capt. Smith	Functional check for MEGA and auto combination, manual test points (loads and pressure distribution)
211	3 Jun.	Fullerton/Capt. Timothy R. Seeley	MEGA mode pitch feedback, loads, pressure distribution, wing deflection
212	7 Jun.	Seeley/C. Gordon Fullerton	Loads and buffet
213	17 Jun.	Fullerton/Smolka	Pressure distribution, automode functional checks, loads and pressure distribution (manual)
214	21 Jun.	Fullerton/Seeley	Pressure distribution, loads, buffet
215	24 Jun.	Seeley/Fullerton	Performance, handling qualities, flutter
216	28 Jun.	Smolka/Fullerton	Pressure distribution, loads
217	29 Jun.	Smolka/Seeley	Tracking tasks for MCC, MLC, MCC/MLC, and manual flight control modes
218	10 Aug.	Seeley/Fullerton	MEGA check out and handling qualities, supersonic evaluation of MCC and manual flight control modes
219	19 Oct.	Seeley/Fullerton	MEGA mode evaluation, manual loads, compressor and fan stalls
220	25 Oct.	Fullerton/Seeley	Performance (manual, MCC, MLC, MCC/MLC), MEGA handling qualities

221	26 Oct.	Seeley/Fullerton	Buffet, loads, performance, pressure distribution, compressor stalls
222	4 Nov.	Fullerton/Johnston	Performance (manual and MCC), buffet, flutter, loads, pressure distribution
223	8 Nov.	Johnston/Seeley	Performance (manual, MCC, MCC/MLC), buffet and loads (MCC, MCC/MLC), pressure distribution
224	22 Nov	Fullerton/Seeley	Performance (manual and MCC), buffet, aeroservoelastic clearance in MEGA
225	29 Nov.	Johnston/Fullerton	MEGA mode evaluation, manual research
226	1 Dec.	Fullerton/Capt. Smith	Performance, MEGA light turbulence evaluation, buffet, pressure distribution
227	13 Dec.	Seeley/Fullerton	CCC mode (new software), operational remotely augmented vehicle (RAV) evaluation, performance
228	14 Dec.	Fullerton/Smolka	Operational RAV evaluation, performance in MCC and manual, manual CCC mode evaluation
229	20 Dec.	Johnston/Seeley	Performance (manual and MCC), boundary layer rake, step inputs (MLC/MCC/MEGA)
230	22 Dec.	Seeley/Fullerton	Operational remotely augmented vehicle (RAV) evaluation, calibrated step inputs in automode
231	22 Dec.	Fullerton/Seeley	Operational RAV evaluation, performance, MEGA evaluation, flow visualization. Last flight of NF-111A

Sources: NF-111A program chronology by Peter W. Merlin, DFRC Flight Operations Office Daily Logs, Monthly Aerospace Projects Update memoranda, NF-111A flight reports, Test/Aircraft Initial Schedule sheets

Appendix U

NKC-135A (55-3129) Flight Chronology, 1978-1981

An NKC-135A model arrived at DFRC at the end of 1977 and was instrumented for a research effort to determine the usefulness of winglets in reducing drag and thus enhancing fuel efficiency. Flights with the airplane began in April 1978 before the winglets were installed. Starting in July 1979 and ending in January 1981, the flight research program tested several different winglet configurations (all fabricated by Boeing), subjecting them to envelope expansion, flutter tests, and performance measurements. The research pilots flew the KC-135 at Mach 0.70, 0.75, and 0.80, at a nominal altitude of 36,000 feet. In the remarks below, the winglet configurations refer to the cant angle of the winglet and its incidence. For example, 15/-2 indicates a 15-degree cant angle and a minus 2-degree incidence.

Flight	Date	Crews	Remarks
1	11 Apr. 1978	Gordon Fullerton/Fitz Fulton/ Tom McMurtry	Crew checkout baseline configuration.
2	11 Apr.	Fullerton/Fulton/McMurtry	Same as Flight 1.
3	12 Apr.	Fullerton/Don Mallick/Fulton	Same as Flight 1.
4	12 Apr.	Fullerton/Mallick/Fulton	Same as Flight 1.
5	14 Apr.	McMurtry/Fulton/Mallick	Same as Flight 1.
6	20 Apr.	Fulton/Mallick/McMurtry	Same as Flight 1.
7	24 Apr.	McMurtry/Fulton	Ferry to Wright-Patterson AFB, Ohio.
8	28 Apr.	Mallick/McMurtry	Ferry to Edwards AFB, California.
9	21 Sept.	McMurtry/Capt.PeterTait/Capt.Young	Airspeed calibration and flutter, baseline configuration.
10	25 Sept.	Fulton/McMurtry/Victor Horton	Ferry to Tinker AFB, Oklahoma for wing fatigue inspection and repairs
11	22 Dec.	McMurtry	Functional check flight.
12	17 Jan.1979	McMurtry/Fulton.	Ferry to Edwards AFB, California.
13	14 Mar.	McMurtry/Young/Capt.David Sprinkel	Airspeed calibration and flutter, baseline configuration.
14	23 Apr.	McMurtry/Tait	Flutter and instrumentation checkout. Baseline configuration.
15	26 Apr.	McMurtry/Tait	Instrumentation checkout. Baseline configuration.
16	30 Apr.	McMurtry/Tait	Same as Flight 15. Aircraft down for installation of winglets.

Winglets Flights

17	24 Jul.	McMurtry/Tait	Airspeed calibration and flutter, 15/-2 configuration. First winglets flight.
18	1 Aug.	McMurtry/Tait	Flutter, 15/-2 configuration.
19	2 Aug.	McMurtry/Tait	Same as Flight 17.
20	10 Aug.	McMurtry/Mallick/Tait	Same as Flight 18.
21	24 Aug.	McMurtry/Tait	Performance, 15/-2 configuration.
22	19 Sept.	McMurtry/Tait	Same as Flight 21.
23	21 Sept.	McMurtry/Tait	Airspeed calibration and performance, 15/-2 configuration.
24	26 Oct.	McMurtry/Tait	Flutter, 15/-4 configuration.
25	2 Nov.	McMurtry/Tait	Performance, 15/-4 configuration.
26	9 Nov.	McMurtry/Tait	Flutter, modified baseline configuration.

27	16 Nov.	McMurtry/Tait	Performance, modified baseline configuration.
28	28 Nov.	McMurtry/Fulton	Flutter, 0/-4 configuration.
29	13 Dec.	McMurtry/Tait	Flutter and performance, 0/-4 configuration.
30	16 Jan. 1980	McMurtry/ Maj.Royce Grones	Performance, 0/-4 configuration. Aborted early due to excessive air turbulence and fuel flow meter failure.
31	17 Jan.	McMurtry/Grones	Performance, 0/-4 configuration. Aborted early due to a severe fuel leak in the No. 4 fuel tank.
32	31 Jan.	McMurtry/Tait	Performance, 0/-4 configuration. Inflight fuel leak check. Aircraft down for fuel tank and wing spar repairs.
33	15 Jul.	McMurtry/Grones	Functional Check Flight, 0/-4 configuration.
34	22 Jul.	Fulton/Grones	Flutter, 0/-4 configuration.
35	29 Jul.	McMurtry/Grones	Performance, 0/-4 configuration.
36	1 Aug.	McMurtry/Grones	Performance, 0/-4 configuration. Aborted early due to excessive air turbulence.
37	4 Aug.	McMurtry/Grones	Same as Flight 36.
38	8 Aug.	McMurtry/Grones	Performance, 0/-4 configuration.
39	14 Aug.	McMurtry/Grones	Same as Flight 38.
40	21 Aug.	Fulton/Grones	Performance, modified baseline configuration. Aborted early due to excessive air turbulence.
41	25 Aug.	McMurtry/Grones	Performance, modified baseline configuration.
42	28 Aug.	Fulton/Grones	Same as Flight 41.
43	5 Sept.	McMurtry/Grones	Same as Flight 41.
44	9 Sept.	McMurtry/Grones	Same as Flight 41.
45	11 Sept.	McMurtry/Grones	Same as Flight 41.
46	17 Sept.	McMurtry/Grones	Performance, 15/-4 configuration.
47	23 Sept.	McMurtry/Grones	Same as Flight 46.
48	25 Sept.	McMurtry/Grones	Same as Flight 46. Aborted early due to excessive air turbulence.
49	3 Oct.	McMurtry/Grones	Same as Flight 46.
50	15 Oct.	McMurtry/Grones	Same as Flight 46.
51	17 Dec.	McMurtry/Grones	Same as Flight 46.
52	19 Dec.	McMurtry/Grones	Same as Flight 46.
53	23 Dec.	McMurtry/Grones	Same as Flight 46.
54	24 Dec.	McMurtry/Grones	Same as Flight 46. Aborted early due to a computer malfunction.
55	8 Jan. 1981	Fulton/Grones	Same as Flight 46. Final flight of research program. Winglets removed.
56	27 Mar.	McMurtry/Grones	Functional check flight, baseline configuration.
57	22 May	McMurtry/Grones	Ferry flight to Wright-Patterson AFB, Ohio

Sources: Based on flight logs compiled from a variety of sources by Peter Merlin and formatted by Betty Love and Jay Levine.

Appendix V

F-18 #840 HARV Flight Chronology, 1987-1996

The NASA Dryden Flight Research Center used an F-18 Hornet fighter aircraft as its High Angle-of-Attack (Alpha) Research Vehicle (HARV) in a three-phased flight research program lasting from April 1987 until September 1996. The aircraft completed 385 research flights and demonstrated stabilized flight at angles of attack between 65 and 70 degrees using thrust vectoring vanes, a research flight control system, and (eventually) forebody strakes (hinged structures on the forward side of the fuselage to provide control by interacting with vortices, generated at high angles of attack, to create side forces). This combination of technologies provided carefree handling of a fighter aircraft in a part of the flight regime that was otherwise very dangerous. Flight research with the HARV increased our understanding of flight at high angles of attack, enabling manufacturers of U.S. fighter aircraft to design airplanes that will fly safely in portions of the flight envelope that pilots previously had to avoid.

Flight	Date	Pilot	Flight hours	Mach No.	Altitude (feet)	G	ALPHA[1] (DEG)	REMARKS
1	2 Apr. 1987	Einar Enevoldson	0.5					Functional check flight ; RTB[2] for landing gear problem.
2	6 Apr.	Ed Schneider	1.5					Functional check flight.
3	9 Apr.	Enevoldson	1.2					Accel/decel; test pts. @ AOA = 25°, 30°, 35°, 40°; PGME flow vis. @25° AOA/ 20 KFT.
4	4 June	Schneider	1.4					Flutter clearance; PGME flow vis. @ 25° AOA/20 KFT.
5	8 June	Bill Dana	1.1					Flutter clearance; PGME flow vis. @ 30° AOA/20 KFT.
6	10 June	Dana	0.9					Flutter clearance; Lost tail accel. @ 0.95 M/15 KFT.
7	18 June	Dana	0.7					Flutter clearance

[1] Preceding four columns show maximum flight conditions for each flight—maximum Mach number, altitude, acceleration, and angle of attack, respectively.

[2] See Acronyms and Abbreviations guide at the end of the appendix for such terms.

8	18 June	Dana	0.5					Flutter clearance: test pts. @ AOA = 30°, 35°, 40°.
9	23 June	Schneider	0.5					Flutter clearance; PGME flow vis.–malfunction
10	25 June	Schneider	0.6					Lear Jet photo flight. Flutter clearance; PGME flow vis. @ AOA=30°/20 KFT.
11	26 June	Schneider	0.7					Flutter clearance; PID 25° AOA test pt.
12	7 Jul.	Schneider	0.6					Fly to El Toro MCAS for INS boresight.
13	8 Jul.	Schneider	0.4					Return from El Toro MCAS.
14	10 Jul.	Schneider	0.8					Flutter clearance; PGME flow vis. 35° AOA/20 KFT; test pt. @ AOA=25°, 30°, 35°.
15	20 Aug.	Enevoldson	1.0					Airspeed calibration.
16	21 Aug.	Enevoldson	1.2					Natural flow vis; PGME flow vis. @ 25° AOA/20 KFT; test. pt. @ AOA=20°, 25°, 30°.
17	27 Aug.	Schneider	1.0					Chase wake encounter. AS & AOA calib. Natural flow vis.; PGME flow vis. @ 20° AOA/20 KFT; test. pt. @ AOA=20°, 25°, 32.5°, 15°.
18	8 Oct.	Schneider	1.0					Temp. survey; airspeed calib; sim val. @0.8M/35 KFT; test pt. @ 32.5° AOA.
19	29 Oct.	Schneider	1.0					Airspeed calibration. RTB–lighting & rain, good natural flow vis. PID @ AOA=15°, 25°, 30°, 32°, 35°, 40°.
20	30 Oct.	Schneider	1.0					Airspeed calibration. PID @ 35°, 40°, 45° AOA; PGME vis. @ 32° AOA/20 KFT.
21	3 Nov.	Schneider	1.3					Angle-of-attack calibration. AS accel/decel.; PID @ 35° & 40° AOA/ 20 KFT; PGME vis. @32° AOA/20 KFT.
22	5 Nov.	Schneider	1.1					AS & AOA calibration. Sim val. @ 0.6M/35 KFT; PID @ 36°, 40°, 45° AOA; PGME vis. @ 40° AOA.
23	6 Nov.	Schneider	1.1					AS & AOA calibration. PID & PGME vis.; sim val. @ 20° AOA.

24	10 Nov.	Schneider	1.2					Angle-of-attack calibration. RTB–control sys. Q miscompare. Sim val. PID. test pts. @ AOA =10°; 36°, 45°. PGME flow vis.
25	1 Dec.	Schneider	1.5					RTB–control sys. miscompare. HQ & S/C M @ 30 KFT; HI AOA test pts @ 28°, 34°, 38°, 45°, 50°.
26	4 Dec.	Schneider	1.3					HQ/S&C M; Lat. Agility @ 20°, 25°, 30°, 35° AOA; HI AOA test pts. 28°, 34°.
27	14 Jan. 1988	Schneider	1.2					First flight w/modified seat–side restraint. HQ @ 25°, 30° AOA; lat. agility @ 25°, 30° & 35°; HI AOA test pts. & PGME flow vis. @ 50° AOA.
28	15 Jan.	Schneider	1.2					WUT 10°-30° AOA; PID @ 28°, 34°, 38° AOA; lat. agility @ 25°, 30°, 35° AOA; PGME flow vis.
29	3 Feb.	Schneider	1.3					First smoker flight (use of smoke for flow vis). Test pts. @ 20° AOA; HQ @ 20°, 25°, 30° AOA.
30	5 Feb.	Dana	1.3					Air data calibration. Smoker vis.; lat. agility.
31	11 Feb.	Schneider	1.3					PGME flow vis.; HQ test pts. @ 20°, 25°, 30° AOA.
32	1 Mar.	Dana	1.1					Swivel probe AS calib. @ 35 KFT & 30 KFT.; PID @ 28°, 32.5° & 50° AOA.
33	3 Mar.	Dana	1.6					Lear Jet photo flight. Smoker flow vis.; sim val.; lat. agility.
34	8 Mar.	Schneider	1.1					Swivel probe calib. @ 15 KFT, 20 KFT & 30 KFT; sim val. @ 10 KFT, 30 KFT.
35	8 Mar.	Dana	1.1					Swivel probe calib. @ 25 KFT; PID agility; PGME flow vis. @ 50° AOA.
36	10 Mar.	Schneider	1.2					PID; HQ-WUT & rolls; PGME flow vis. @ 34° AOA.
37	10 Mar.	Dana	1.2					HQ 0.9 M/45 KFT AOA=20°, 25°, 30°.
38	11 Mar.	Schneider	1.3					HQ-loaded roll @ 0.9M, 0.8M, 0.7M, 0.6 M, 0.4 M/AOA up to 35°.
39	15 Mar.	Dana	1.0					Ferry flight to Tucson for painting.

40	16 Apr.	Dana	0.8					Return flight from Tucson.
41	7 June	Dana	1.0					Smoker flow vis. FADS calib.
42	9 June	Schneider	1.0					FADS calib. Agility 1.2 M/35 KFT; PGME flow vis. AOA 10°/20° KFT.
43	9 June	Dana	0.7					Smoker flow vis.; agility to 0.5 M/20 KFT to 40° AOA.
44	21 June	Schneider	0.7					Instr. air data problem. Smoker flow vis.
45	23 June	Dana	1.1					FADS calibration. PGME flow vis.; Pitch agility to 20° AOA; uplink drop outs.
46	23 June	Schneider	0.8					Smoker flow vis.; 25°, 30°, 35° AOA/20 KFT; pitch agility 25°, 30°, 35° AOA/20 KFT.
47	30 June	Dana	1.1					Smoker flow vis. Pitch agility 0.6 M/20 KFT to 30° AOA.
48	30 June	Schneider	0.8					PID; agility–max turn rate. HQ track M.
49	1 Jul.	Schneider	0.9					HQ–track M; agility; LEX PGME flow vis./20 KFT, 30° AOA; sim val.
50	7 Jul.	Dana	0.9					Smoker flow vis. PID. Agility.
51	7 Jul.	Dana	0.6					Agility. Sim val.; PGME flow vis. 48° AOA/30 KFT.
52	11 Aug.	Schneider	0.7					Sim val.; RTB–AV air hot light; tail hook unlatched.
53	23 Aug.	Dana	0.5					Smoker flow vis.; 35° AOA/25 KFT; RTB-tank 4 fuel transfer.
54	30 Aug.	Schneider	0.9					Smoker flow vis. 26° AOA; 45° AOA stab. pt. AOA sweeps.
55	8 Sept.	Schneider	0.9					LEX smoker flow vis; WR/30 KFT-40°, 45° AOA; sim. val.; agility–Ps.
56	13 Sept.	Dana	0.8					LEX smoker flow vis. 40°, 45° AOA; HQ target track; RTB–tail hook.
57	15 Sept.	Schneider	1.0					LEX Smoker flow vis.; sim val.
58	15 Sept.	Dana	0.7					PID; LEX PGME flow vis. Agility.

59	16 Sept.	Schneider	0.9					Sim val. HQ–formation & trk.; agility; PGME flow vis. 34° AOA.
60	20 Sept.	Dana	1.3					RTB–air data gauge malfunction. HQ–formation & trk.; PGME f/bdy & LEX 34° AOA/20 KFT.
61	22 Sept.	Schneider	0.9					LEX smoker flow vis. 20° AOA/30 KFT; agility; PID; sim val. Wing rock (WR) 38° AOA.
62	22 Sept.	Dana	0.8					Agility; PID; sim val. WR 38° AOA; PGME F/B & LEX flow vis. 48° AOA.
63	23 Sept.	Schneider	0.7					Sim val.; agility; PGME F/B & LEX @ 48° AOA/30 KFT.
64	27 Sept.	Dana	0.9					FADS data check pts.; agility test pts.; PGME @ 34° AOA/20 KFT.
65	28 Sept.	Schneider	1.1					HQ–formation tracking PID; agility test pts. ; PGME 26° AOA/20 KFT.
66	5 Oct.	Schneider	1.1					Smoke flow vis. 30°, 38°, 48° AOA/28 KFT; agility; FADS data pts.
67	7 Oct.	Dana	1.1					Steve Ishmael, F-18 chase. LE flap malfunction/lost chase plane/pilot OK. Smoke flow vis. 34° AOA/23 KFT. Smoke flow vis. 45° AOA/26 KFT.
68	16 Mar. 1989	Schneider	0.7					Functional check flt. (FCF). RTB due to AV air hot light (indicating the temperature of the air to cool the AV).
69	23 Mar.	Dana	0.8					FCF; smoke flow vis. 30° AOA/25 KFT, 38° AOA/28 KFT, 48° AOA/28 KFT; PID 0.4 M/30 KFT.
70	30 Mar.	Schneider	0.7					Smoke flow vis. 15° AOA/20 KFT, 30° AOA/23KFT, 26° AOA/23 KFT, PID 35KFT/AOA 10°, 32°, 48°.
71	31 Mar.	Dana	1.3					Smoke flow vis. 15°-30° AOA/20 KFT, 56° AOA/28 KFT; HQ freq. SWP/10° AOA/30 KFT.

72	6 Apr.	Schneider	1.1					Smoke flow vis. 15°-28° AOA/20 KFT, 56°AOA/28 KFT; 30°-50° AOA/28 KFT; HQ freq. SWP 10° AOA/30 KFT; PID 12°, 30°, 45° AOA/30KFT; Aero WR-1 g decel. to 50° AOA/30 KFT; 30°, 35° AOA/25 KFT.
73	7 Apr.	Schneider	1.3					Lear Jet photo flight. Stab. pts. @ 30° & 40° AOA/22 KFT & 28 KFT.
74	11 Apr.	Dana	1.3					Lear Jet photo flight. Stab. pts. @ 25° AOA/20 KFT, 15°-35°AOA/20 KFT, 25°-50°AOA/22 KFT; Aero WR 20°, 25° AOA/20 KFT.
75	13 Apr.	Schneider	1.2					Langley sim.val. 0.9M/10 KFT & 30 KFT; LEX fence smoker flow vis. 15°, 20°, 25° AOA/20 KFT; HQ AOA & theta freq. sweep.
76	14 Apr.	Dana	1.3					Lear Jet photo flight. LEX fence–30°, 35° AOA/20 KFT, 20°-15°-35°-15°AOA SWP/20 KFT; stab. pt. @ 17.5°, 22.5°, 27.5°, 32.5° AOA.
77	18 Apr.	Schneider	1.3					Lear Jet photo flight. LEX fence-OFF; Stab. pt. @ 17.5°, 22.5°, 27.5°, 32.5° AOA; smoke flow vis. @ 20°, 25°, 30° AOA/23 KFT; Beta calibration pts.
78	19 Apr.	Dana	1.1					Smoke flow vis. 30°AOA/23 KFT; 15°-35°-15° AOA SWP/23 KFT; Agility-Pitch cap. Roll to 90°.
79	20 Apr.	Schneider	1.0					Smoke flow vis. 40° AOA/28 KFT; 19° AOA/WLSS/20 KFT; 15°-35°-15° AOA SWP/20 KFT; PID-30° AOA/22 KFT Agility-pitch cap, roll & pitch cap.
80	26 Apr.	Dana	0.9					PID AOA =26°, 32°, 38°, 44°, 50°, 24°, 30°, 36°, 28°, 34°, 40°, 46°, 32°, 38°, 30°, 36°, 20°, 14°. Flow vis., PGME 30° AOA/21 KFT.

81	26 Apr.	Schneider	1.1					LEX fence off. Large gust encounter at TD. WR test pt. @ 10°, 15°, 20° AOA/40 KFT; WR @ 0.6M/40 KFT/25°, 30° AOA; HQ freq. SWP @ 30° AOA/35 KFT.
82	27 Apr.	Dana	0.8					PID @ AOA=42°, 48°, 26°, 32°, 38°, 44°; agility turns; trim @ -15 theta/200 KCAS/3.0 g/20 KFT; PGME flow vis. @ 30° AOA/21 KFT.
83	28 Apr.	Schneider	0.9					WR 0.8M/40 KFT/15° & 20° AOA; PGME flow vis. 47° AOA/31 KFT.
84	12 Jul.	Schneider	0.6					Functional check flight. RTB–AV air hot light.
85	18 Jul.	Dana	1.0					Functional check flight. Wind-up turn (WUT) press. pts. @ 30° AOA/42-35 KFT; 20° AOA/42-35 KFT.
86	19 Jul.	Schneider	1.1					WUT press.pt. @ 15° AOA/42-35 KFT; WR/20 KFT/AOA=36°, 38°, 40°, 42°, 45°; press. calib. @ 10°, 15°, 20°, 30° AOA/22-18 KFT.
87	21 Jul.	Dana	1.0					AOA SWP 15°-35°, 35°-50°/20 KFT & 40 KFT; PID 30° AOA/20 KFT; press. calib. 23°, 26°, 30° AOA/20 KFT; DS press. pts. @ 10° AOA/42-35 KFT.
88	21 Jul.	Dana	0.5					Press. calib. @ 28°, 30°, 34° AOA/20 KFT; RTB-AV hot light.
89	27 Jul.	Dana	0.9					Press. calib. 19°, 30° AOA/22 KFT; WUT press. @ 10° AOA/42-35 KFT; WR 45° AOA/45-35 KFT; press. calib. @ 26°, 34° AOA/40 KFT.
90	3 Aug.	Schneider	0.3					RTB–crimped pressure line.
91	3 Aug.	Schneider	1.1					Press. calib. 48°, 50° AOA/25 KFT; 50° AOA/40 KFT; WUT press. pt. @ 30° AOA/42-35 KFT; WR pt. 40° AOA/45-20 KFT.

92	4 Aug.	Dana	0.9					Press. calib. 52°, 54° AOA/25 KFT; dynamic lift–max. pitch rate to capture 50° AOA from 5°, 15°, 20° AOA/25 KFT/0.4 M/0.5 M/0.7 M; WR @ 45° AOA/45–18 KFT.
93	31 Aug.	Dana	1.2					Functional check flight. Press. calib. 10°, 15°, 19°, 20°, 23°, 26°, 28°, 30° AOA/22 KFT; press. calib. 10°, 15° AOA/42 KFT.
94	1 Sept.	Dana	0.7					RTB–mission computer degrade. Flashing warning light in gear handle. WUT press. pt. @ 20°, 26°, 30°, 34°, 52° AOA/42-38 KFT; press. calib @ 48° AOA/27 KFT.
95	1 Sept.	Dana	1.0					Press. calib. @ 48° AOA/27 KFT; eng. press. effects 30° AOA/20 KFT; Press. calib. @ 19°, 20°, 30°, 34° AOA/22 KFT; press. calib. @ 52° AOA/30 KFT; WUT press. Pt. @ 52°, 30° AOA/42-38 KFT.
	9 Sept.	Dana						Spin chute hi-speed taxi test.
96	15 Sept.	Schneider	1.1					Press. calib. @ 10°, 15°, 20°, 30° AOA/22 KFT; WR @ 36°, 38°, 40°, 42°, 45 AOA/30 KFT; WUT press. pt. @ 30° AOA/42-35 KFT.
97	15 Sept.	Dana	1.3					WUT press. pt. @ 10°. 15°. 20°, 30° AOA/42-35 KFT; WR @ 40°, 45° AOA/45-35 KFT.
98	15 Sept.	Schneider	1.0					Press. calib. @ 19°, 23°, 26° AOA/22 KFT; Press. calib. @ 48°, 50°, 52°, AOA/25 KFT; press. calib. @ 26°, 50°, 52° AOA/40 KFT.
99	18 Sept.	Dana	1.0					Press. calib. 19°, 23°, 26°, 34° AOA/20 KFT; press. calib. @ 34°, 54° AOA/40 KFT; press. calib. @ 52°, 54° AOA/25 KFT.

100	18 Sept.	Schneider	1.2					100th flight since coming to Dryden. Agility-35° AOA rolls, pitch-up captures; engine PID 15°-45° AOA, 25°, 35° AOA; dynamic lift, 20 KFT/0.5M; agility 0.4M/25 KFT/35° AOA.
101	26 Sept.	Dana	0.6					RTB–accidental canopy ejection due to –1g PO M with throttle cycle & unguarded switch. Pilot OK, safe plane recovery & landing. Smoke flow vis @ 45°, 50° AOA/28 KFT; agility pitch captures, 50° AOA; PO –1g to 0° AOA. After this flight, the aircraft was equipped with thrust-vector-control vanes.
102	16 Jan. 1991	Schneider	1.1	0.700	35,156	2.55	32	1st TVCS equipped aircraft flt; RFCS not operational. FSC; loads exp. to 2.5g; propulsion env. expansion (EXP) 35 KFT/200 KCAS.
103	18 Jan.	Dana	1.0	0.740	34,218	2.38	44	Dana's first flight of the TVCS-equipped aircraft. Flt. controls EXP at 1g; PID from 5° to 15° AOA; elevator effectiveness PO 20° to 35° AOA; airdata calibrations.
104	12 Jul.	Schneider	1.0	0.720	35,128	1.87	26	1st flight after galled vane bellcrank bushings were repaired. Functional engines checks; 701E 1g controls/20° AOA, RFCS 1g controls at 5° AOA, OBES SWP at 5° AOA.
105	19 Jul.	Dana	1.0	0.670	34,690	1.66	30	RFCS OBES SWP at 170 Qbar; RFCS PO-PU to 2.5g; 701E controls clearance at 30° AOA, SFO.
106	19 Jul.	Schneider	1.0	0.710	31,300	3.56	43	FCS 1g Controls EXP to 20° AOA, 701E 1g controls EXP to 35° AOA, 701E Loads EXP to 4.5g.
107	1 Aug.	Dana	0.8	0.710	35,200	4.30	29	Flight aborted due to an inboard right vane hydraulic ambiguity failure. Loads EXP in 701E to 4.5g; ASE @ 60 Qbar @34.5 KFT.

108	8 Aug.	Schneider	0.9	0.740	32,560	5.30	41	RFCS flt. env. EXP to 4.5g; smooth flt. & 3.5g in M flt.; alpha EXP to 20° w/control system; 701E env. EXP to 5.3g load factor.
109	9 Aug.	Dana	1.0	0.710	34,670	4.10	24	RFCS lds. flt. env. EXP to 3.5g in M flt. ASE env. EXP to 20° AOA in OBES operation.
110	13 Aug.	Dana	0.7	0.730	33,805	4.22	44	701E only flt. EXP lds. env. to 4.1g; controls env. to 30° AOA in M flt.
111	13 Aug.	Schneider	0.9	0.740	34,960	3.30	35	701E clearance to 35° AOA in M flt. at 0.6 M, 30 KFT.
112	15 Aug.	Dana	1.0	0.780	33,900	3.20	56	701E control system cleared to 55° AOA by flying M in 1g smooth flt.
113	20 Aug.	Dana	0.7	0.880	32,820	5.48	54	Right-hand wing pitot boom bent back at 513 Qbar. 701E only flt. for Mach no. EXP to 0.90; 1g abrupt flt. @ 40° & 55° AOA; lds. EXP to 4.1g & 3.5g RPO.
114	27 Aug.	Schneider	0.8	0.830	35,421	2.98	28	Lds. EXP for max vane defl. @ 0.7 M; functional check small vane OBES SWP.
115	29 Aug.	Dana	0.7	0.830	33,126	4.50	46	RFCS lds. EXP to 5.2g smooth & 3.5g RPO. Controls accel RFCS EXP to 20° AOA.
116	5 Sept.	Schneider	1.0	0.770	34,077	3.35	36	RFCS controls EXP to 20° AOA abrupt & 30° AOA smooth, & ASE EXP to 30° AOA.
117	5 Sept.	Dana	0.8	0.840	36,623	5.40	35	RFCS lds. EXP to 5.2g & -2.0g symetric & 4.1g asymmetric. Max A/B env. pitch & roll reversals for lds. definition.
118	5 Sept.	Schneider	0.8	0.840	34,489	3.18	43	RFCS controls EXP to 35° AOA, 1g smooth; RFCS ASE EXP to 30° AOA.
119	10 Sept.	Dana	0.8	0.830	34,729	3.10	49	RFCS controls EXP to 45° AOA, 1g smooth & ASE EXP to 40° AOA.

120	10 Sept.	Schneider	1.0	0.850	34,631	3.43	55	Pictures, flightline photographs from Lear Jet. RFCS 1g smooth controls EXP to 45° AOA; RFCS M.
121	11 Sept.	Schneider	0.8	0.860	34,757	2.84	54	RFCS 1g abrupt EXP to 45° AOA; smooth accel to 25° AOA. 701E & RFCS WLSS to max control input at 20 KFT & 30° AOA.
122	26 Nov.	Dana	0.2	0.640	8,126	3.15	11	Right swivel probe stuck full down. Acoustic data gathered @ 7,300 ft/0.40M. In-flight abort thereafter.
123	26 Nov.	Dana	0.8	0.851	32,340	3.15	55	In-flight abort due to C/H 2 lateral stick miscompare. (C/H 1 is high; 10 is low.) 4 acoustic runs; 701E cleared for 40°, 45°, 50° AOA with 1g/abrupt M. SFO after an RTB call for FCS fault.
124	27 Nov.	Schneider	0.8	0.850	40,670	3.43	90	1st yaw-rate expansion flight. 701E cleared to 70° AOA & the yaw rate env. opened to 50°/sec.
125	5 Dec.	Schneider	0.9	0.897	41,690	2.99	90	Yaw rate EXP @ 24 & 26% MAC; Airdata accel./decel @ 30 KFT.
126	5 Dec.	Schneider	0.8	0.832	35,020	3.50	53	1g smooth RFCS controls & ASE clearance to 50° AOA.
127	11 Dec.	Schneider	0.8	0.841	34,870	3.12	64	RFCS EXP to 55° & 60° AOA @ 1g/smooth flight.
128	11 Dec.	Schneider	0.6	0.832	35,195	3.34	70	Flight aborted due to LH horizontal stab. CPT instrumentation drifting. Trim effect on high alpha; 1g smooth 60° AOA RFCS EXP; ASE EXP to 60° AOA.
129	12 Dec.	Schneider	0.8	0.869	36,060	2.92	70	RFCS 1g smooth EXP thr. 65°/part of 70° AOA; 1g abrupt EXP to 55° AOA.
130	17 Dec.	Schneider	0.7	0.868	35,082	3.35	70	RFCS cleared for 1g flt to 70° AOA/ smooth & abrupt maneuvers.

131	17 Dec.	Schneider	0.6	0.849	39,017	3.32	90	Flt. aborted, engine EGT instrumentation problems. RFCS 1g M env. EXP to 50° AOA with a PO, aileron roll, heading captures.
132	19 Dec.	Schneider	0.8	0.865	38,450	3.22	80	PID doublets for model eval. RFCS M EXP to 60° AOA.
133	19 Dec.	Schneider	0.8	0.855	41,771	3.07	82	Dutch roll maneuver at 30 KFT/0.40 M was terminated due to vane temp. limits. Airdata calib. PO-PUs were accomplished at 40 KFT & 30 KFT.
134	10 Jan. 1992	Schneider	0.4	0.793	17,890	3.27	11	In-flight abort soon after T/O; instrumentation no-go problems in control room.
135	22 Jan.	Schneider	0.7	0.879	34,146	3.01	72	360° rolls at 25°-45° AOA; roll captures at 60° AOA.
136	29 Jan.	Schneider	0.7	0.790	33,648	2.99	72	Right-hand wing boom static pressure inoperative. 1g smooth RFCS controls EXP at 60° & 65° AOA with rolling captures and PO.
137	31 Jan.	Schneider	0.8	0.825	34,622	3.71	74	Propul. EXP M to 70° AOA & 13° beta; 0-60-60-0 heading thru at 65° AOA.
138	4 Feb.	Schneider	0.9	0.888	32,538	4.00	88	20 hertz oscillation observed in numerous data parameters at low & medium alphas. Source unknown. Airdata calibrations from 30 KFT to 20 KFT.
139	4 Feb.	Schneider	0.8	0.819	40,965	3.39	90	Airdata calib. @ 20 KFT; RFCS cleared to 65° AOA, M flt.; yaw-rate env cleared to 70°/sec. with idle engines.
140	22 Jul.	Schneider	0.8	0.872	43,693	3.65	101	1st flt. after 5 1/2 months down time. FCF; 1 yaw rate EXP pt. completed to 50°/sec. to the right.
141	23 Jul.	Schneider	0.9	0.891	44,956	3.02	90	Yaw rate EXP to left/right completed.
142	23 Jul.	Schneider	0.7	0.868	44,606	3.09	0	3 yaw rate EXP M; SFO.

143	5 Aug.	Schneider	0.8	0.844	33,822	3.29	64	MSDR changed before flt., causing 1 hr. delay of T/O. Velocity vector rolls, downmodes; press. data from 15°-25° AOA.
144	5 Aug.	Schneider	0.8	0.795	26,712	3.32	66	Low alt. press. data from 30°-50° AOA.
145	11 Aug.	Schneider	0.7	0.840	27,210	3.20	72	WR data from transition strips on forward fuselage.
146	11 Aug.	Schneider	0.5	0.799	27,584	2.96	51	INS locked up. Flt. RTB. WR data without transition strips.
147	11 Aug.	Schneider	0.3	--	---	--	--	Flight aborted right after T/O due to loss of rate instrumentation data.
148	18 Aug.	Schneider	0.8	0.835	33,860	3.28	72	LEX & Forebody press. at 55°-70° AOA/20 KFT & 70° AOA/30 KFT; WR data 50° AOA.
149	18 Aug.	Schneider	0.6	0.833	34,465	3.16	70	Forebody & LEX press. at 50°-65° AOA/30 KFT.
150	18 Aug.	Schneider	0.8	0.788	34,695	3.18	71	Forebody & LEX press. data at 15°-65° AOA/30 KFT.
151	20 Aug.	Schneider	0.8	0.706	24,638	3.04	54	4 tower flyby pts. & press. at 20°-50° AOA.
152	20 Aug.	Schneider	0.7	0.692	25,903	3.04	67	Press. data @ 15°-65° AOA; WR pts. @ 35° & 45° AOA.
153	18 Sept.	Schneider	0.8	0.782	25,704	3.08	47	Break turn BFM; PID-OIM 701E pitch maneuver.
154	29 Sept.	Schneider	0.3	---	---	---	--	Flt. aborted due to chase fuel venting problems.
155	29 Sept.	Schneider	0.8	0.879	35,047	4.33	62	2nd FPS-16 radar not operational. RTTM at 30 KFT & PID-SSI latte/dir. data; single ship 701E & RFCS BFM data during split S & WUT w/reversals.
156	29 Sept.	Schneider	0.6	0.927	37,100	2.96	66	RTTM accels & stab. pts. in MAX power at 35 KFT; longitudinal step series for PID-OIM.
157	1 Oct.	Schneider	0.8	0.887	35,050	5.05	46	Problems with the guidance system. Split-S BFM M with target; WUT x/reversal, not a satisfactory M; RTTM data at 35 KFT, mil power; PID-SSI; PID-OIM.
158	1 Oct.	Schneider	0.8	0.790	42,314	3.24	61	Problems with the guidance system. Stab. AOA RTTM @ 35KFT, max power; PID-OIM.

159	1 Oct.	Schneider	0.9	0.785	38,010	3.08	48	Stab. AOA RTTM at 35 KFT, max power; BFM horiz. scissors M.
160	5 Oct.	Schneider	1.0	0.700	26,799	3.09	48	PID-OIM in lat./dir. axis at 20° & 30° AOA & BFM pt. & acquire M.
161	5 Oct.	Schneider	0.8	0.760	31,746	3.59	52	PID-OIM in lat./dir. axis at 20° & 30° AOA; BFM vertical looping scissors & post stall reversal M.
162	22 Oct.	Schneider	0.8	0.735	30,170	3.44	65	Post stall reversals & offensive spirals; RTTM cruise pts.; various engine settings.
163	23 Oct.	Schneider	0.8	0.751	27,800	4.11	160	BFM , 1st Classic Herbst M @ 200 knts/ 55° AOA; ACM; pitch & yaw vane SSI M for PID.
164	23 Oct.	Schneider	0.3	0.725	24,990	3.24	53	One ACM followed by RTB because of weather.
165	28 Oct.	Schneider	0.8	0.848	39,619	3.14	55	PID-OIM at 30° AOA & lat./dir. PID-SSI at 10°-50° AOA using OBES; ACM 90° pass lvl. M; control law eval. with alpha captures from 0.60 M.
166	28 Oct.	Schneider	1.0	0.839	42,385	4.09	66	Longitudinal PID-SSI at 10°-60° AOA using OBES; control law eval. with alpha CAP from 0.60 M/20° AOA; loaded rolls from 0.6 M.
167	28 Oct.	Schneider	0.6	0.740	32,029	3.40	64	Long. PID-SSI at 25°, 40°, 60° AOA; lat/dir. PID-SSI at 60° AOA using OBES; LEX & F/B press. at 20°-55° AOA @ various degrees of beta.
168	10 Nov.	Schneider	0.9	0.903	36,758	3.22	65	Control law eval. for vs. 26 completed; PID-OIM pilot input M & RTTM accel.
169	13 Nov.	Schneider	0.7	0.856	44,874	3.05	70	Airdata info. 20 & 40 KFT during P0-PUs & stabilized pts. Climb data.
170	13 Nov.	Schneider	0.7	0.881	33,504	3.20	73	During climbs, power settings compared with chase aircraft. Airdata calib. during climbs to 20 & 30 KFT; PO-PU @ 30 KFT/0.4 M & 0.3 M; Full pedal beta SWP @ 35 KFT/30°- 60° AOA; decels @ 30, 25 KFT.

171	18 Nov.	Schneider	0.8	0.772	27,551	4.09	64	First flight of RFCS vs. 27. Alpha cap @ 20° AOA/0.6 M/25 KFT; 360° roll capture from 15° AOA.
172	18 Nov.	Schneider	0.7	0.739	26,839	4.46	68	RFCS vs. 27 checkout 360° LRC @ 0.6 M; PO using OBES to vary pitch authority.
173	18 Nov.	Lt. Dave Prater, USN	0.8	0.790	36,193	3.26	62	First Navy evaluation flight; RFCS vs. 26 reloaded for flt. BFM/701E & RFCS; +/-45° & +/-60° roll captures in 701E & RFCS @10°-60° AOA .
174	20 Nov.	Schneider	0.8	0.745	26,097	4.09	47	vs. 27 checkout to qualify FCS for NAVY evaluation. RFCS vs.27 checkout/LRC @ 0.4 M/15°, 25°, 35° AOA/0.6 M @ 25° AOA; 1g alpha captures to 15°-20°-38° AOA/0.6 M.
175	20 Nov.	Prater	0.9	0.740	33,965	4.19	55	Completion of Navy thrust vectoring evaluation. ACM of flat scissors & J-turns; stab. PO with 701E, RFCS, & reduced nose pitch rate.
176	24 Nov.	Schneider	0.8	0.755	27,223	3.17	48	Control law eval. of vs. 27/1g roll captures to 40° AOA; LRC/0.6 M; G cap @ 0.6 M to 3g.
177	24 Nov.	Schneider	0.8	0.757	23,386	4.10	46	Control law eval. of vs. 27 with g & psi cap @0.6 M & 4g; tracking runs @ 0.6 M & 0.45 M; ACM runs with break turn & split S maneuvers.
178	25 Nov.	Schneider	0.8	0.687	25,195	3.89	65	Eval. of vs. 27 software with various BFM.
179	25 Nov.	Schneider	0.7	0.702	26,262	3.76	65	2 ACM runs; 4 "Davidson" tracking runs by fine tracking at 30° AOA.
180	2 Dec.	Schneider	0.8	0.903	40,839	4.62	103	Max A/B level accel. 0.5–0.9M @ 35 KFT; stab. turn @ 0.9 M; 2 high-yaw-rate M @ 60–75 degrees/sec.; WUT & throttle transients @ 20 KFT to look for engine problems.
181	10 Dec.	Schneider	0.8	0.895	43,462	2.91	71	Press. data @ 40 KFT during level decels. & mod.-rate PU; RFCS & 701E for control.

182	16 Dec.	Schneider	0.5	0.744	25,688	3.08	61	Data @ 20 KFT for RTTM research; stab. max. A/B pts. @ 60°, 50°, 40°, 30° AOA.
183	16 Dec.	Schneider	0.8	0.746	29,493	3.02	55	First agility & HANG flts. Agility @ 25° AOA /701E, 701E+ rudder & RFCS & @ 30° AOA/701E, 701E+rudder; 180° roll thrus, left & right; HANG nose up alpha captures (cap), 20°-35° & 20°-50° ranges.
184	18 Dec.	Schneider	0.8	0.811	33,844	3.08	73	Roll & heading thrus @ 0-180°/left & right; in RFCS, data @ 30°-65° AOA; nose-up alpha cap. @ 10°-50° & 10°-35° AOA; throttle transients to investigate engine anomaly.
185	18 Dec.	Schneider	0.8	0.815	32,182	3.51	69	Agility roll thrus @ 35° AOA/701E, 701E + rudder & RFCS; HANG nose-up alpha cap @ 20°-65° AOA.
186	18 Dec.	Schneider	0.6	0.774	28,876	2.90	67	Flt. terminated due to time-of-day constraints. HANG nose-up alpha cap @ 10°-65° & 10°-35° AOA RFCS/reduced pitch; agility 0-90 roll cap @ 25° AOA, 701E & aileron, aileron & rudder; throttle transient to observe engine roll-back prob.
187	21 Dec.	Schneider	0.8	0.882	29,578	3.03	67	Agility roll-thru & roll cap. @ 25°, 30°AOA; HANG nose-up pitch cap. RFCS & pitch rate limiting.
188	21 Dec.	Schneider	0.6	0.900	28,988	4.62	41	Roll cap @ 30° AOA, RFCS; max A/B, RTTM @ 20 KFT.
189	8 Jan. 1993	Schneider	1.0	0.742	30,993	3.08	60	HANG PO @ idle power @ 30°, 40°, 50°, 55° AOA/various pitch-rate reductions; agility roll cap in 701E + rudder @ 30° AOA.
190	20 Jan.	Schneider	0.8	0.750	28,415	4.08	105	Transients RFCS & 701E; WUT/vane loads vs. 28.1; PO/HANG eval. @ 30° & 50° AOA; roll cap. for agility.

191	20 Jan.	Schneider	0.9	0.875	29,856	3.20	69	PO @ 30° & 40° AOA; zoom-climb from 70 knts for HANG research; WUTs/accel. @ 20 KFT for RTTM; agility roll cap @ 35° AOA.
192	20 Jan.	Schneider	0.4	0.684	28,795	3.20	81	Flight cut short due to time of day. Zoom-climb recoveries @ 60 knts; PO @ 40°, 50° AOA; HANG research.
193	22 Jan.	Schneider	0.8	0.743	29,764	3.19	66	Aborted-uncommanded engine rollback 3 times before engine shutdown. Full-stick PO @ 50°, 60° AOA; one heading cap @ 50° AOA.
194	14 Jan. 1994	Schneider	0.5	0.757	27,931	3.42	65	1st flt. in a year–addition of inlet & extended aero instr. Ca. 1/2 of FCF & 1 rake verification pt.; rake strain gauge instrumentation failed. RTB-no-fly parameter.
195	26 Jan.	Jim Smolka	0.9	0.937	25,354	3.81	17	Ctd. FCF; 2 rake verification pts.; then rake strain gauge instrumentation failed.
196	28 Jan.	Schneider	0.6	0.842	40,824	3.34	89	Flameout, same as flt. 193; left engine shut off; single- engine landing w/o incident. FCF; rake verification; 50°/sec yaw rate & 15° AOA press. data; RTB w/left engine shut down.
197	9 Feb.	Smolka	0.8	0.643	28,267	3.02	53	Left engine ECU changed prior to flt. 3 SFO approaches. Vel. vector rolls 15°-60° AOA; inlet data @ 0.3M/10° AOA; inlet rake ref. calib. pt. @ 30° AOA; press. data @ 20° AOA for aero discipline.
198	9 Feb.	Schneider	0.8	0.647	25,646	3.18	99	Aero data from 25°-50° AOA, 35° & 45° AOA; rake ref. data @ 0.6M/20 KFT & 30° AOA stab. pt.
199	9 Feb.	Smolka	0.8	0.672	28,639	3.37	70	30° AOA roll cap. @ 90° & 180°; vel. vector rolls @ 60°, 65° AOA; pressure data @ 55° AOA.

200	11 Feb.	Schneider	0.7	0.830	34,737	2.95	73	200th flight since coming to Dryden. Press. & tail buffet @ 55°, 60°, 65°, 70° AOA.
201	18 Feb.	Smolka	0.7	0.780	32,584	3.10	68	Aero pressure data at high AOA; pilot familiarity velocity-vector roll; a propulsion rake ref. calib. pt.
202	23 Feb.	Schneider	0.9	0.655	28,203	3.08	72	HANG of PO-PU @ 50° AOA; agility heading cap. to 50° AOA; roll cap. to 25° AOA.
203	23 Feb.	Schneider	0.9	0.762	29,329	3.31	90	HANG PO from 30°, 40°, 50° AOA; HANG PO-PU to 40° AOA.
204	23 Feb.	Schneider	0.8	0.690	29,142	2.83	65	HANG alpha cap. to 50° AOA; agility alpha cap. 65° AOA/0.6 M.
205	25 Feb.	Schneider	0.8	0.659	34,689	3.08	61	Inlet rake press. calib. with 0.30M/ 0.40M stab.pt.; stab. 1g & full-pedal sideslips @ 15°-60° AOA.
206	2 Mar.	Smolka	0.8	0.843	40,729	4.30	94	Smolka's 1st high yaw-rate maneuver in HARV. 1 aero tail buffet press. pt @ 70° AOA; vel. vector roll; high yaw rate & 2 BFM.
207	3 Mar.	Schneider	0.8	0.690	30,801	3.13	69	4 rolling pushs with basic elevator power (1 idle power & 30° AOA, 3 @ max. power & 30°, 40°, 50° AOA.
208	3 Mar.	Smolka	0.9	0.874	39,868	4.21	99	Pilot fam. with 1 BFM & 2 high-yaw-rate; max pitch rate reversals with P0-PU thru 40° AOA; 0.6 M @ 30° AOA; PO-PU maneuvers @ 35° AOA.
209	4 Mar.	Smolka	0.9	0.923	40,048	3.29	99	Max pitch rate reversal PO-PUs to 65° AOA; high yaw rate Ms to 80 deg/sec; inlet rake ref. calib. pt. @ 30° AOA.
210	15 Mar.	Schneider	1.1	--	---	---	--	Successful photo mission.
211	16 Mar.	Smolka	1.0	0.699	28,556	2.63	73	No air data because of icing on the air data probe. Stab.& dynamic pts @ 10° & 30° AOA; inlet rake airflow calib. @ 0.6 M/20 KFT.
212	16 Mar.	Schneider	0.9	0.793	35,253	3.40	75	Stab. & dynamic M; 10° & 60° AOA.

213	16 Mar.	Smolka	0.9	0.808	38,078	3.06	84	A/B glowouts on final tst. pt. @ 40° AOA/0.4 M/30 KFT. Stab. & dynamic M; 10° to 40° AOA.
214	18 Mar.	Smolka	0.8	0.785	36,922	3.19	75	Stab. & dynamic M; 10° to 60° AOA.
215	18 Mar.	Smolka	0.9	0.710	27,839	3.25	69	Stab. & dynamic M; 10° to 30 ° AOA.
216	18 Mar.	Smolka	0.8	0.831	38,726	3.34	31	Stab. maneuvers; 10° to 30° AOA.
217	22 Mar.	Lt. Chuck Sternberg	0.9	0.808	39,608	4.58	63	1st flt. for Navy eval. for Lt. Chuck Sternberg. Phasing M, 701E & RFCS; throttle transients @ 20 KFT/250 knts; roll cap. 10° to 60° AOA, 701E & RFCS.
218	22 Mar.	Capt. Rick Traven	0.8	0.808	32,142	5.08	68	1st flt. for Capt. Rick Travens, CAF. Mil. pwr. T/O-phasing M @ 20 KFT; roll cap 10° to 60° AOA; WLSS in 701E & left WUT 5° to 30° AOA.
219	23 Mar.	Sternberg	0.9	0.758	25,630	4.46	60	Phasing M; LGA M @ 30°, 45°, 60° AOA; OL reversal @ 30° AOA.
220	23 Mar.	Traven	0.7	0.795	26,460	5.07	59	HARV0008.1.408 loaded into MC2 with lateral acquisition (lat. acq.) bars. Phasing; LGA @ 30° AOA in 701E & RFCS; OL roll in 701E @ 30° AOA; in RFCS @ 45° & 60° AOA.
221	24 Mar.	Sternberg	0.7	0.708	27,163	4.08	74	HARV008.1.408 loaded into MC2 only with lat. acq. bars. Phasing; OL reversal @ 30° AOA with RFCS; defensive tolls @ 30° AOA/701E & 60° AOA/RFCS; J-turns & 1v1.
222	24 Mar.	Traven	0.6	0.730	26,198	4.91	71	HARV0008.1.408 loaded into MC2 only with lat. dir. bars. Phasing; defensive roll M in 701E @ 30° AOA & RFCS @ 30° & 45° AOA; J-turn & 1v1 engagement in RFCS.
223	25 Mar.	Sternberg	0.7	0.786	32,642	3.97	68	HARV0008.1.408 loaded into MC2 only with lat. acq. bars. Phasing; OL roll; LGA; 1v1 engagement; flat scissor M; PO @ mil pwr for inlet pressure work.

224	25 Mar.	Traven	0.4	0.771	33,485	5.31	66	HARV0008.1.408 installed in MC2 only for lat. dir. bars; last WUT terminated-assymetric loads exceeded. Mil T/O & phasing, RFCS on flat scissor; PO from 30° to -5° AOA; WUT to 15° then to 30° AOA; RTB, no damage.
225	1 Apr.	Smolka	0.8	0.731	25,215	3.38	54	LGA @ 30°, 45°, 60° AOA; long.-lat. tracking @ 30° AOA/160 & 200 knts.
226	1 Apr.	Schneider	0.8	0.781	32,425	---	66	Long. PID SSI thru 50° AOA; lat./dir. PID SSI thru 30° AOA; agility heading cap @ 65° AOA.
227	1 Apr.	Schneider	0.7	0.786	27,720	4.52	53	Agility cap @ 35°, 30° AOA/0.6 M.
228	5 Apr.	Smolka	0.8	0.898	44,288	3.13	54	Tracking & acquisition M @ 0.50 M; 1g tail buffet with LEX fence on @ 20°, 30°, 40° AOA & 0.6 M/30° AOA.
229	5 Apr.	Schneider	0.8	0.904	45,288	4.38	45	LEX fences removed for tail buffeting work. Buffet data @ 1g/20°, 30°, 40° AOA & 0.6 M/30° AOA; long.-lat. tracking & LoGA @ 0.6 M/30° AOA.
230	22 Apr.	Smolka	0.9	0.740	30,362	3.28	87	LEX fences reinstalled. Loss of uplink guidance system. Inlet research data @ 0.3 M/3°, 10°, trim AOA; @ 0.4 M/3°, 10°, trim AOA.
231	22 Apr.	Schneider	0.8	0.861	37,762	3.20	69	Inlet research data @ 0.3M/50°, 60° AOA; @ 0.4M/30°, 50° AOA.
232	28 Apr.	Smolka	0.8	0.810	37,237	3.14	73	Inlet research data @ 0.3M/30°, 50°, 60° AOA; @ 0.4 M/40° AOA.
233	28 Apr.	Smolka	0.8	0.897	36,266	3.55	86	Inlet research data @ 0.3 M/60° AOA; @ 0.4 M/40° AOA; planar wave data @ 25 KFT, accels. & decels.
234	4 May	Schneider	0.8	0.899	37,847	3.16	41	Inlet press. calib.; planar wave data @ 25 KFT, mil power; stab. inlet data @ 10° AOA & various betas & 40° AOA.

245	13 May	Smolka	0.7	0.821	37,483	3.23	80	RFCS software vs. 28.1 was reloaded before this flight. Dynamic maneuver @ 40° AOA/0.4 M & 60° AOA/0.3 M; +/- 5 beta flown.
246	17 May	Schneider	0.8	0.931	25,286	3.29	15	Right inlet bleed air door locked down & plugged. Stabilized pts @ 4 different thrust settings & 4 different Mach numbers.
247	17 May	Schneider	0.7	0.855	37,983	3.29	75	Data were difficult to gather due to difficult start conditions. Dynamic inlet press. data @ 60° AOA/0.3 M & -5 beta; stab. data in mil power @ 40° AOA/0.3 M @ +/-5 & +/-10 beta.
248	17 May	Schneider	1.0	0.740	34,401	3.11	55	Optimal input maneuvers for PID performed by pilot inputs in the lat-dir axis at 20°, 30°, 35°, 40° AOA. These were all accomplished in 701E FCS; PID data using OBES at 10°-50° AOA.
249	20 May	Schneider	0.7	0.903	47,196	3.50	39	Fuel panel door was left open during this flight. LEX fence on; tail buffet data @0.6 M & 20°, 25°, 30° AOA during helix M; split S M @ 0.6M & 30°, 35° AOA; formation flight to gather wing tip vortices data, HARV press. data system.
250	25 May	Schneider	0.9	0.813	31,137	3.05	63	Lat-dir. PID data with OBES inputs @ 10°-30° & 50°-60° AOA for SSI; @ 20° & 30° AOA long. pilot input; optimal input data in long.-lat. axis @ 20° & 40° AOA.
251	25 May	Schneider	0.6	0.717	25,981	3.33	62	RTTM data dur. stab. pts. @ 20 KFT & at 30°-60° AOA in mil. & max. power.
252	25 May	Schneider	0.6	0.852	44,988	3.45	41	LEX fence was removed for this flight. Tail buffet data @ 0.6 M & 20°, 25°, 35°, 30° AOA.
253	1 June	Smolka	0.8	0.789	32,383	3.22	65	Data for PID, SSI @ 20°-60° AOA; smooth WWT turn @ 6°-55° AOA.

245	13 May	Smolka	0.7	0.821	37,483	3.23	80	RFCS software vs. 28.1 was reloaded before this flight. Dynamic maneuver @ 40° AOA/0.4 M & 60° AOA/0.3 M; +/- 5 beta flown.
246	17 May	Schneider	0.8	0.931	25,286	3.29	15	Right inlet bleed air door locked down & plugged. Stabilized pts @ 4 different thrust settings & 4 different Mach numbers.
247	17 May	Schneider	0.7	0.855	37,983	3.29	75	Data were difficult to gather due to difficult start conditions. Dynamic inlet press. data @ 60° AOA/0.3 M & -5 beta; stab. data in mil power @ 40° AOA/0.3 M @ +/-5 & +/-10 beta.
248	17 May	Schneider	1.0	0.740	34,401	3.11	55	Optimal input maneuvers for PID performed by pilot inputs in the lat-dir axis at 20°, 30°, 35°, 40° AOA. These were all accomplished in 701E FCS; PID data using OBES at 10°-50° AOA.
249	20 May	Schneider	0.7	0.903	47,196	3.50	39	Fuel panel door was left open during this flight. LEX fence on; tail buffet data @0.6 M & 20°, 25°, 30° AOA during helix M; split S M @ 0.6M & 30°, 35° AOA; formation flight to gather wing tip vortices data, HARV press. data system.
250	25 May	Schneider	0.9	0.813	31,137	3.05	63	Lat-dir. PID data with OBES inputs @ 10°-30° & 50°-60° AOA for SSI; @ 20° & 30° AOA long. pilot input; optimal input data in long.-lat. axis @ 20° & 40° AOA.
251	25 May	Schneider	0.6	0.717	25,981	3.33	62	RTTM data dur. stab. pts. @ 20 KFT & at 30°-60° AOA in mil. & max. power.
252	25 May	Schneider	0.6	0.852	44,988	3.45	41	LEX fence was removed for this flight. Tail buffet data @ 0.6 M & 20°, 25°, 35°, 30° AOA.
253	1 June	Smolka	0.8	0.789	32,383	3.22	65	Data for PID, SSI @ 20°-60° AOA; smooth WWT turn @ 6°-55° AOA.

254	1 June	Smolka	0.9	0.764	35,735	3.27	61	Stab. RTTM data @ 30 KFT & 60°-10° AOA; large amplitude pilot input doublets during WWT @ 30KFT.
255	1 June	Smolka	0.7	0.772	36,841	3.22	62	Stab. pits for RTTM @ 30 KFT from 60°-20° AOA using max power.
256	3 June	Smolka	0.8	0.707	27,043	2.97	58	1st flight of revised NASA 1A control laws. Data of E-D of 1A control laws; doublets from 5°-45° AOA; SHSSs from 5°-25° AOA & roll captures from 5°-30° AOA.
257	3 June	Smolka	0.8	0.748	29,159	3.22	61	Roll cap. from 30°-45° AOA; doublets, 45°-60° AOA; SHSS @ 45° AOA; alpha cap. to 30°, 45° & PO to 10° AOA.
258	3 June	Smolka	0.8	0.707	29,240	3.35	85	Tracking data @ 40°, 30° AOA; AAC from 20°-45° & 45°-10°; SHSS data @ 55° AOA & heading cap. data @ 60° AOA. FS PU-PO from 15° AOA.
259	7 June	Smolka	0.8	0.708	28,696	3.27	72	AAC cap from 20°-60° & 60°-20° AOA; SHSS data @ 65° AOA; FS PU-PO from 20° AOA & PU from 15° AOA; roll cap. @ 25°, 45° AOA.
260	7 June	Smolka	0.8	0.719	29,747	4.12	69	Abrupt AAC to 15° AOA; SHSS data @ 60° AOA & 2g & 3g heading cap.; FS PU-PO from 35°, 40°, 55° AOA.
261	7 June	Smolka	1.0	0.739	27,484	4.35	64	Abrupt AAC to 45° & 60° AOA; LRC @ 15° & 25° AOA; LRC @ 5° AOA/0.6 M; full stick PU-PO 15° AOA/0.6 M.
262	9 June	Smolka	0.8	0.747	31,289	4.28	64	LRC to 15°, 25° AOA/0.6 M & 35° AOA/0.4 M; FS PU-PO from 15° & 30° AOA/0.6 M; 360 heading cap @ 55° AOA.
263	9 June	Smolka	0.7	0.698	25,710	3.68	49	Long.-lat. tracking tasks @ 0.45 M & 0.60 M @ 30° AOA.

264	14 June	Smolka	0.7	0.724	31,404	4.01	69	360 AIL heading cap, left & right @ 60° AOA; tracking runs @ 45° AOA; LoGA @ 30° AOA; smooth WWT for PID.
265	14 June	Smolka	0.5	0.724	25,638	4.03	51	LoGA @ 0.6 M/30° AOA; on a repeat, turning vane overloaded. Turning-vane load limits exceeded due to maneuvering flight, RTB.
266	14 June	Smolka	0.9	0.681	26,045	3.65	55	LoGA @ 30°-60° AOA; LaGA @ 30° AOA.
267	16 June	Smolka	0.8	0.762	32,434	3.37	70	Ail. roll cap @ 60°, 65° AOA; PU from 30° AOA; LaGA @ 60° AOA.
268	21 June	Schneider	0.8	0.683	28,337	3.80	50	Oil flow flt. test technique @ 10 KFT & 30° AOA; AAC to 30° & 45° AOA; aileron roll cap. @ 45° & 25° AOA; long.-lat. Tracking.
269	21 June	Smolka	0.8	0.750	25,472	3.44	74	Long.-lat. tracking @ 60° AOA; LaGA @ 30° & 45° AOA.
270	21 June	Schneider	0.6	0.671	25,702	3.71	65	LoGA & LaGA @45° AOA; Split-S M in 701E & RFCS for control-law eval.
271	23 June	Schneider	0.7	0.695	27,054	2.88	63	Grit strips were placed on both sides of forward nose section. Flow vis. with oils of different viscosities; BFM, Split-S, flat scissors, with 701E & RFCS, & PID @ 60° AOA.
272	23 June	Smolka	0.8	0.676	30,609	3.32	63	Grit strips applied to forward nose area of aircraft. J-turn M with/without RFCS & 60° AOA LaGA; PID/lat.-dir. doublets during WWT M.
273	23 June	Schneider	0.6	0.667	27,193	3.18	56	Grit strips removed for this flight. PID data with pilot inputs in pitch & roll/yaw axis @ 45° & 30° AOA; PU from 30°-55° AOA.
274	28 June	Schneider	0.8	0.853	33,116	3.07	76	Oil flow data @ 30°, 25°, 20° AOA; pilot inputs used for PID data @ 20° & 5° AOA; PID SSI data with OBES @ 70° AOA; airdata accel/decel @ 20 KFT.

275	30 June	Maj. Billie Flynn	0.7	0.869	40,958	3.26	90	Guest pilot, Maj. Billie Flynn, Canadian Air Force. RFCS E-D M & high AOA stable pts; roll cap @ 30° & 60° AOA; high yaw rate expansion to 80°/sec.
276	30 June	Flynn	0.6	0.731	26,898	3.77	72	Final flt., Maj. Billie Flynn. Pitch power with nose-up & nose-down; BFM with slow scissors & J-turn; ACM with flat scissors.
277	30 June	Schneider	0.6	0.632	27,874	3.08	65	Last flight of Phase II. 30° & 60° AOA tracking data; 360° roll cap @ 60° AOA; HANG rolling push M & stab. pts. RTTM @ 40°, 30° AOA.
278	15 Mar. 1995	Smolka	0.6	0.475	11,454	1.27	7	1st flt. Strake Nose on HARV; abort due to gear handle light. Max A/B takeoff; gear swings, dumping of fuel, aircraft made uneventful full-stop landing.
279	16 Mar.	Smolka	1.0	0.749	30,056	2.40	55	1st high AOA flight with strake nosecone installed. FCF; max A/B takeoff; SRM check & airstarts on both engines; 1g AOA env. expanded to 55° AOA.
280	22 Mar.	Smolka	1.0	0.842	37,376	3.00	58	RFCS; exp. pts. in 701E with stab. pts.; alpha cap; FS PO; roll/yaw M.
281	22 Mar.	Smolka	0.8	0.791	31,721	3.42	69	Roll/yaw axis with stick & rudder rolls; FS PU-PO @ 30 KFT; lat.-dir. data @ 30 KFT/0.6 M @ 1g & 20° AOA.
282	24 Mar.	Smolka	0.8	0.883	41,878	3.50	99	Flt. env. expansion with roll/yaw M @ 0.6 M & 40°/sec. yaw-rate M.
283	24 Mar.	Smolka	0.6	0.869	41,845	3.16	97	Yaw rate expansion pts. to 60° & 80°/ sec.; forward cg; WR , 2 stab. pts. @ 42° & 38° AOA.
284	28 Mar.	Schneider	0.9	0.886	41,936	3.45	91	80°/sec. high-yaw-rate M. RTB, high forward fuselage loads. Pilot fam rolls @ 30KFT; stab RFCS pt.; high temp on aft structure; high-yaw-rate @ 40°, 60°, 80°/sec.
285	28 Mar.	Schneider	1.0	0.771	25,942	4.51	73	STEMS research PU-PO, J-turns; heading changes.

286	30 Mar.	Smolka	0.8	0.851	37,289	3.36	59	Aggravated inputs MSRM; WR pts. in CAS & MSRM.
287	30 Mar.	Traven	1.0	0.907	43,501	3.15	95	Aggravated input with MSRM. High-yaw-rate M @ 45, 52, 59 deg./sec.
288	30 Mar.	Capt. C.J. Loria	1.0	0.923	43,646	3.19	90	Flight for Capt. Loria, USMC. Aggravated input with MSRM; high-yaw-rate M @ 54 deg./sec; spiral; 53 deg/sec.
289	31 Mar.	Traven	0.8	0.865	38,749	3.20	62	Aggravated input with MSRM; WR with CAS; WR with MSRM.
290	31 Mar.	Loria	0.9	0.880	41,824	3.05	92	This completed the Navy work. The 2nd MSRM resulted in very good data in falling leaf mode. Cross control aggravated input M with MSRM; yaw M to 60 deg/sec yaw rate; WR with CAS; WR with MSRM.
291	6 Apr.	Schneider	0.8	0.696	28,611	4.81	70	STEMS for pitch rate reserve, high alpha roll cap. & reversal, rolling defense, & nose-up theta cap.
292	6 Apr.	Schneider	0.8	0.692	24,893	4.05	70	STEMS: noseup theta cap; 45° LoGA; 45° LaGA; duel attack M.
293	7 Apr.	Schneider	0.8	0.715	23,895	4.56	49	LaGA @ 45°; tracking high AOA SWP; high AOA tracking & crossing target acquisition.
294	7 Apr.	Schneider	0.8	0.724	29,005	3.67	60	Nose high start J-Turns, trk., slice back, during high AOA SWPS. Phase 1 of the STEMS research completed.
295	11 Jul.	Smolka	0.7	0.794	38,950	3.00	61	1st flt. with ANSER & strakes (S) active. TV, STV, S &701E mode; long. PO from 20°-60° AOA to 10° AOA; trail damped op. to 60° AOA; asymmetry of 10L/47R noted for strakes @ 60° AOA; strake-retract switch used successfully.

296	11 Jul.	Schneider	0.6	0.749	30,272	3.02	65	Lat. stick M, 360° rolls, range 30°-60° AOA; dir. M, left & right full-pedal inputs to 30° AOA, to left @ 45° AOA.
297	13 Jul.	Smolka	0.8	0.702	32,605	3.60	68	Rolls @ 45° & 60° AOA in TV mode; alpha cap. to 45° AOA in gain sets 0, 1, 2; 360° roll cap., gain sets 0, 1; gain set 1 desirable for these Ms.
298	13 Jul.	Schneider	0.7	0.743	28,231	2.88	64	Long. & lat. tracking; pitch down alpha cap. from 60° to 10° AOA; gain set 1 better overall.
299	13 Jul.	Smolka	0.8	0.775	31,601	3.07	49	Split S M, gain set determination; lat-dir ANSER TV exp. at 1g/0.6M/30 KFT.
300	20 Jul.	Schneider	2.0	0.835	37,111	3.48	42	In-flight refueling, first time from KA-6D tanker. Substantial increase to quantity of RFCS data gathered. Left & right 360° rolls using lateral stick, then pedal for 20° & 30° AOA; engine inlet pts, PO to -10° AOA; strake extensions.
301	27 Jul.	Smolka	0.6	0.806	37,414	4.69	60	Tanker problems; no in-flight refueling. Inadvertant bit check initiation while switching out of pressure calib. S extension maneuver 30° position @ 45° AOA; 60° @ 35° AOA; 90° @ 25° AOA; lds check 5 g WUT; photo's @ 25 KFT/0.6M of left & right S extensions.
302	27 Jul.	Smolka	0.6	0.788	36,573	3.05	72	Inadvertant bit check initiation while switching out of press. calib. 30° S position pts. @ 50°, 55° AOA; 60° S position @ 40°, 45° AOA; 90° S position @ 30°, 35° AOA; lds pts @ 90° left & right S positions @ 30° AOA.

303	27 Jul.	Smolka	0.6	0.802	35,990	3.12	67	Hi ambient temps. created problems with instrumentation during T/O & phasing maneuvers; 35 KFT instrumentation normal. 30° S position @ 60°, 65° AOA; 60° position @ 50° AOA; 90° position @ 40°; lds pts for 90° AOA left & right S positions.
304	3 Aug.	Smolka	0.4	0.787	32,436	3.24	23	Control room problems lead to a delayed T/O & RTB to fix loads equations. After T/O, successful phasing maneuvers revealed problems with Alpha true reading and erroneous loads equations.
305	3 Aug.	Smolka	0.6	0.795	37,413	3.13	68	Correct lds. equations installed in control room. T/O & phasing M; S position 30° @ 70° AOA; 60° position @ 60° AOA; 90° position 50° AOA.
306	8 Aug.	Smolka	0.7	0.800	37,275	3.27	67	T/O & phasing M; left & right S extensions @ 50° AOA & left @ 60° AOA for lds exp.; left S extensions 60° & 90° to max 65° AOA.
307	8 Aug.	Smolka	0.6	0.745	29,525	3.35	62	Continued from flt. 306. Lds exp. pts. to 60° AOA @ 1g; lds exp. pts. @ 2g for 20°, 30°, 40° AOA.
308	8 Aug.	Smolka	0.7	0.743	30,121	3.33	43	Stab. @ alt. prior to first pt. to cool instrumentation of engine nozzle parameters. Lds exp. left strake extended to 2g @ 40° AOA; left & right S extensions to 3g @ 20° AOA; right S extension to 3g @ 30° AOA using Split S maneuver.
309	10 Aug.	Schneider	0.9	0.880	30,177	3.58	72	In-flight refueling, KC-135 tanker for HARV and chase acrft. Lds exp. S extended @ 3g to 70° AOA; S doublets @ 1g, 20°, 30°, 40°, 50°, 60°, 70° AOA.
310	15 Aug.	Smolka	0.7	0.761	31,387	3.15	65	Lds exp. S doublets @ 2g @ 20°, 30°, 40° AOA; mode transition @ 20°, 30°, 50° AOA; TV-STV-701E @ 65° AOA.

311	16 Aug.	Schneider	0.6	0.727	30,547	3.01	66	Encountered RFCS downmode to TV during STV rolling PO & in S mode maneuvers. TV-STV-S-TV-S-701E @ 56° AOA; PO 30°, 45°, 60°, AOA to 5° AOA in STV & S modes; rolling PO @ 30°, 45°, 55° AOA in STV mode.
312	24 Aug.	Schneider	1.5	0.786	31,556	3.07	64	In-flight refueling, KC-135. Closed loop exp. in S mode @ 30°, 45°, 60° AOA; 30° AOA doublets, sideslips, alpha cap., & lateral control PID; 360° rolls, all modes, @ 25° AOA.
313	24 Aug.	Schneider	0.4	0.661	26,923	3.06	46	Right strake position light during S mode. 45° AOA doublets & sideslips in STV & S modes.
314	29 Aug.	Smolka	2.8	0.750	31,681	3.13	65	In-flight refueling , KC-135. Smolka's first trip to the KC-135 with the HARV. AAC @ 45° & 60° AOA, TV & STV modes; 360° rolls @ 35° AOA all modes; TV, STV, S modes @ 45°, 50°, 55° AOA; doublets @ 60° AOA STV, S along with lat. PID maneuver S mode; SHSS @ 50°, 55° AOA.
315	30 Aug.	Schneider	2.3	0.768	31,484	3.96	67	In-flight refueling, KC-135, twice; photo & video of both; 360° rolls TV, STV, S modes @ 55° AOA, 65° AOA, TV mode. Left & right full pedal SHSS @ 55°, 60° AOA STV, S modes; alpha cap. TV, STV modes from 20° to 60° AOA; alpha cap @ 0.6 M, 30°, 45°, 60° AOA TV, STV modes.
316	31 Aug.	Schneider	0.8	0.700	32,486	3.58	66	Roll cap @ 0.4 M/35° AOA TV, STV modes; roll cap in TV, STV, S modes @ 25° AOA /0.6 M. Could not do 360° rolls @ 65° AOA in TV, STV, S modes due to excessive spin chute temperatures.

317	31 Aug.	Smolka	0.8	0.751	25,705	4.36	40	Roll thru maneuvers @ 0.6 M /25° AOA in TV, STV, S modes; roll cap @ 0.6 M/35° AOA all modes.
318	31 Aug.	Smolka	0.4	0.637	25,000	2.99	18	Aborted flt. after phasing maneuvers due to loss of primary health parameters.
319	7 Sept.	Schneider	3.0	0.850	31,194	3.32	61	In-flight refueling, KC–10, three times. Long-lat trk @ 30° AOA /0.45 M, all modes; LaGA @ 30° AOA all modes; LoGA @ 30° AOA TV; tail buffet in TV @ 20°, 25°, 30°, 35°, 40°, 50°, 60° AOA.
320	19 Sept.	Smolka	1.5	0.737	25,687	4.16	54	In-flight refueling, KA-6D. Long.-lat trk @ 45° AOA, all modes gain set 1 in TV, S modes; 0.6 M trk. all modes.
321	19 Sept.	Schneider	1.4	0.713	25,565	3.36	62	In-flight refueling, KA-6D. Long.-lat trk @ 45° AOA TV, S modes gain set 0 & 1; LaGA S mode @ 45° AOA.
322	21 Sept.	Smolka	1.8	0.691	25,225	3.53	64	In-flight refueling, KA-6D. LaGA @ 45°, 60° AOA in TV, STV, S modes; LoGA @ 45° AOA in TV mode; Long & Lat trk task @ 60° AOA in TV, STV, S modes.
323	21 Sept.	Schneider	1.4	0.668	25,545	3.73	65	In-flight refueling, KA-6D. LoGA @ 60° AOA TV mode; scissors maneuvers, RFCS off, STV, S modes; J-Turns, RFCS off & STV mode.
324	3 Oct.	Smolka	0.7	0.713	25,506	4.01	42	Long & lat trk @ 30° AOA, TV, STV, S, med. gain set & TV, low gain set.
325	3 Oct.	Schneider	1.6	0.769	25,514	4.22	68	In-flight refueling, KA-6D. LoGA @ 45° AOA TV mode; @ 60° AOA TV, STV, S; LaGA @ 45° AOA TV, STV; @ 60° AOA TV, STV, S; J-Turn S mode. Tracking.

326	5 Oct.	Smolka	0.7	0.704	27,835	3.58	47	KA-6D tanker dummy store broken during refueling rendezvous. Trk. @ 30° AOA TV, STV, S modes, med. gain; TV mode low gain; 45° AOA lat. input PID M.
327	6 Oct.	Smolka	0.8	0.693	24,793	4.18	60	LaGA @ 30° AOA TV, STV, S; LoGA @ 30°, 60° TV; 30°, 45°, 60° AOA in TV using low gain setting.
328	10 Oct.	Schneider	1.7	0.751	32,823	3.50	70	In-flight refueling, KA-6D, twice. Roll caps. & roll thrus @ 25°, 35° AOA; heading caps. & heading thrus @ 50°, 65° AOA; 180° LRC.
329	12 Oct.	Schneider	1.6	0.755	30,942	3.45	65	In-flight refueling, KA-6D, twice; 180° roll thrus @ 25°, 35° AOA; 90° roll caps. @ 35° AOA; 180° LRC & thrus @ 35° AOA; controls PID M @ 5°, 20°, 30°, 40°, 45°, 60° AOA TV & 30°, 45°, 60° S mode.
330	27 Oct.	Smolka	0.7	0.686	28,170	3.11	42	Tail buffet pts. @ 30°, 40° AOA, S deflections 30°, 90° each side during left & right sideslips; PID M @ 30° AOA S mode.
331	27 Oct.	Smolka	0.6	0.726	29,088	3.16	62	Tail buffet pts. @ 50° AOA for S deflections 20/20, 35/5, 5/35, 90° each side, left & right sideslips; 60° AOA tail buffet pts. for S deflections of 90/0, left & right sideslips.
332	31 Oct.	Schneider	0.7	0.687	29,388	3.12	42	LEX fences off. Controls PID M @ 5°, 20° AOA; aero S extension M @ 10° S deflection @ 30°, 40°, 50° AOA @ 25 KFT; 90° deflection @ 20° AOA; 30° AOA pts. 20/0, 0/20, 30/0, 0/30 S positions.
333	31 Oct.	Smolka	0.7	0.708	25,252	2.99	52	LEX fences off. Aero S extension M @ 90° S deflection @ 25°, 30° AOA @ 25 KFT; S deflection 20/0, 0/20 @ 40°, 50° AOA & 0/10 @ 50° AOA.
334	31 Oct.	Schneider	0.7	0.803	37,129	3.23	51	LEX fences off. Aero S extension M @ 90°, 60°, 30° S deflection @ 40° AOA; 20/20 deflection @ 50° AOA/35 KFT.

335	2 Nov.	Schneider	3.4	0.784	39,411	3.12	53	LEX fences off; in-flight refueling, KC-135, 4 times. Aero S extension M @ 30°, 35°, 40°, 50° AOA; PID M & controls PID M @ 30° AOA, TV, S, & TV @ 45° AOA.
336	7 Nov.	Schneider	0.7	0.706	29,947	3.04	52	LEX fences off. Aero S extension M @ 20°, 30°, 40°, 50° AOA, S deflections 90/0, 0/90 @ 20° AOA; 10/0, 0/10, 30/0, 0/30, 0/60, 0/90, 20/0, 0/20 @ 30° AOA; 10/0, 0/10 @ 40° AOA; 90/0, 0/90 @ 50° AOA.
337	7 Nov.	Schneider	0.6	0.773	35,886	3.02	52	LEX fences off. Aero S extension M @ 40°, 45°, 50° AOA, 20/0, 0/20, 30/0, 0/30 @ 50° AOA/35 KFT; 35/5, 0/60 @ 50° AOA; 0/20 @ 45° AOA; 20/0, 30/10, 5/35, 35/5, 20/20 @ 40° AOA/25 KFT.
338	8 Nov.	Schneider	0.9	0.747	34,341	3.03	52	LEX fences off. In-flight refueling with KC-135 aborted for AOA mismatch. Aero S extension M @ 40°, 50° AOA; 25/15, 15/25 @ 50° AOA/35KFT; 25/15, 15/25, 9/90 @ 40° AOA/25KFT.
339	21 Nov.	Schneider	0.8	0.764	35,245	3.04	61	LEX fences Off. Right wing tufted & left fuselage in front of engine inlet. Aero S extension M @ 40°, 50° AOA, 10/30; 60/0, 5/35, 0/20 @ 50° AOA; right sideslips @ 30°, 40° AOA, 0/90 S position; wing tufts photo @ 60°, 50°, 20°, 15°, 5° AOA with zero S deflection.
340	21 Nov.	Smolka	0.6	0.717	29,499	3.12	52	LEX fence off. Right wing tufted for airflow visualization. Aero S extension sideslip M @ 40° AOA S extensions 30/0, 0/30; left & right pedal sideslips @ 50° AOA, S 20/20, 35/5, 5/35; @ 40° AOA 90/0 extension.

341	21 Nov.	Smolka	0.6	0.756	29,235	3.15	61	LEX fences off. Right wing tufted. Aero S extension sideslip M @ 40° AOA, 0/90; left & right pedal sideslips @ 50° AOA, 90/0, 0/90; 55° AOA, 90/0, 0/90.
342	28 Nov.	Schneider	2.9	0.848	31,384	3.11	67	LEX fences off. In-flight refueling, KC-135, four times. Aero S extension sideslip M @ 60° AOA, 90/0, 0/90, 0/0; @ 60° AOA, 20/0, 0/20, 30/0, 0/30, 60/0, 0/60; @ 65° AOA, 90/0, 0/90; PID TV 40°, 45°, 60° AOA; S, TV @ 60° AOA/0.4 M.
343	30 Nov.	Schneider	1.7	0.747	34,393	3.00	66	LEX fences off. In-flight refueling, KC-135, twice. Aero S extension M @ 50°, 60° AOA; Zero S positions @ 40° AOA/35 KFT; 25°, 65° AOA/25 KFT; 50° AOA, 0/90 @ 0.4M; 90/0 @ 0.6M/40° AOA; @ 30° AOA, 90/0, 0/90 @ 0.6 M.
344	1 Dec.	Schneider	0.6	0.684	29,493	4.15	73	Offensive spiral M in STV; ACM side-by-side M in TV, STV, S modes.
345	5 Dec.	Schneider	1.7	0.788	26,123	4.10	74	In-flight refueling, KA-6D. STEMS, high AOA SWP, trk. & acquisitions TV, STV modes; STEMS 18, acquisitions behind tanker before refueling.
346	5 Dec.	Schneider	0.6	0.714	24,697	3.95	71	In-flight refueling aborted from a KA-6D due to an alpha probe strike. STEMS 4, dual attack in STV mode @ 250, 150 knts; nose up gross acquisition @ 250 knts in STV mode.
347	6 Dec.	Schneider	2.2	0.763	30,175	3.63	73	In-flight refueling, KC-135, twice. Nose up theta cap. in TV, STV; 180° heading change, J-Turn, rolling defense, PU, pitch-rate reserve, 45° roll cap, 80° PU-PO & roll reversal.
348	7 Dec.	Schneider	1.2	0.717	29,499	3.12	52	In-flight refueling, KA-6D. STEMS 10, 45° AOA gross acquisition in TV, STV; crossing target in STV; dual attack in STV.

349	7 Dec.	Smolka	0.9	0.736	26,048	3.58	63	In-flight refueling aborted, problems w/KA-6D store. Crossing target, 45° AOA trk., gross acquisition, dual attack, in STV mode.
350	14 Dec.	Smolka	2.6	0.889	30,644	4.41	91	In-flight refueling, KC-135, twice. STEMS Ms.
351	15 Mar. 1996	Smolka	0.6	0.760	35,407	3.21	62	LEX fences off. Aborted 1st T/O, FCS caution due to nose-wheel failed warning. LEX fences off. Aero pts. with S extensions @ 20/0 @ 40°, 50° AOA; 50° AOA/35 KFT with S 0/10; 60° AOA, 30/0; flow vis. pts using 50° AOA, 0/90 S position.
352	19 Mar.	Smolka	0.7	0.663	29,046	3.14	51	LEX fences off. PID @ 5°, 20° AOA; aero pts S position 10/10 @ 40°, 30° AOA; 15/5, 5/15 @ 30°, 40° AOA; 35/25, 25/35, 40/20, 20/40 @ 50° AOA.
353	19 Mar.	Phil Brown, LaRC pilot	0.7	0.688	29,564	3.01	64	Software version 151.1; LEX fences on. AOA cap 60° to 10°; 360° rolls using 701E, TV, S modes @ 35° AOA; 360° rolls @ 55° AOA, TV, S modes; 0.4 M loaded rolls @ 35° AOA, TV, S modes.
354	22 Mar.	Flt. Lt. Dan Griffith, Royal Air Force	0.8	0.728	26,820	3.44	68	Version 151.1; LEX fences on; 360° rolls @ 35° AOA, 701E, TV; 55° AOA in TV, S; 65° AOA in TV; 45° AOA trk task, TV; scissors M & J-turns, TV.
355	22 Mar.	Brown	0.7	0.666	27,129	3.90	57	Version 151.1; LEX fences off; 45° AOA alpha cap in TV mode; 30° AOA long-lat trk task in TV, TV Gain 1, S modes; flow vis 30° AOA with S 90/0, 0/90, TV; Two TV downmodes, S mode 35° AOA/360° roll.

356	28 Mar.	Mark Stucky	0.7	0.723	27,686	3.21	69	Version 151.1; LEX fences off. 35° AOA 360° roll in TV mode, repeated @ 55° AOA; @55° AOA roll, S mode; AOA SWP in TV mode.
357	29 Mar.	Stucky	0.6	0.671	31,119	3.10	78	Version 152.0; LEX fences off. Aero pts @ 50° AOA, TV mode @ S positions 45/15, 15/45, 30/30; 60° AOA pt in TV, 30/30; J-Turn; long-lat PID Ms @ 30°, 45° AOA.
358	29 Mar.	Stucky	0.6	0.794	31,813	3.39	63	Version 152.0; LEX fences off. Pilot fam. Trk. @ 45° AOA, TV, S modes; long.-lat. PID M & aero pts@ 60° AOA, TV mode S @ 35/25, 25/35.
359	4 Apr.	Stucky	0.6	0.671	31,119	3.10	78	Sym S schedule 360° rolls @ 45° AOA, S, STV modes; 50° AOA, S mode; long.-lat trk task @ 30° AOA, S mode.
360	9 Apr.	Stucky	0.7	0.833	40,910	3.35	96	Version 152.0; LEX fences on; 50°/sec yaw rate spin in 2 attempts. Abort, left wing tip AOA probe broken.
361	16 Apr.	Schneider	0.7	0.640	29,828	3.60	52	Version 152.0; LEX fences on; 45° AOA trk task; 35° loaded roll; lat. PID M @ 35°, 45°, 50° AOA.
362	19 Apr.	Schneider	0.7	0.700	28,426	3.09	52	Version 151.1; LEX fences off; 40°, 50° AOA, 0/90 S position flow vis.; attempted S-mode 360° roll but aborted due to TV downmode.
363	19 Apr.	Larry Walker, McDonnell pilot	0.7	0.737	28,175	3.84	75	Version 151.1; LEX fences on; 270° roll cap, 701E, TV mode @ 35° AOA; TV, S @ 55° AOA; TV @ 65°; trk @ 45° AOA, scissors M, J-Turn, STEMS trk, TV mode.
364	23 Apr.	Schneider	0.6	0.758	27,489	3.17	54	Version 151.1; LEX fences off. Flow vis. pts. @ 40°, 50° AOA, S 90/0; 30° AOA, S 0/90.

365	26 Apr.	Smolka	0.8	0.706	27,284	3.32	41	Version 154.0; LEX fences off. ASE pts. @ 5°, 10°, 20° AOA; flow vis. @ 30°, 40° AOA S SWP from 0 to +90 to –90 to 0; TV recover experienced 35° AOA S mode rolls.
366	30 Apr.	Smolka	1.4	0.737	29,072	3.11	64	Version 153.0. LEX fences on; in-flight refueling, KC-135. Long.-lat. PID M version 153.0, @ 5°, 20°, 30°, 40°, 45°, 50°, 60° AOA.
367	30 Apr.	Schneider	1.7	0.794	31,384	2.95	63	Version 153.0. In-flight refueling, KC-135, twice. LEX fences off. Long.-lat. PID M @ 5°, 20°, 30°, 45°, 60° AOA; lat. Ms @ 30°, 45°, 60° AOA, TV, STV, S modes.
368	1 May	Jeff Peer, CALSPAN pilot	0.6	0.678	27,092	3.60	72	Version 153.0; LEX fences on; 360° rolls in TV @ 35°, 55°, 65° AOA; 55° AOA in S mode; 45° AOA trk task, TV, STV; J-Turn, TV mode.
369	3 May	Smolka	0.8	0.684	31,994	3.01	51	Version 154.0; LEX fences off. Flow vis trkng task, 30° AOA, aborted, downmode. Lat. & rudder PID M @ 5°, 20° AOA, TV mode; flow vis pts @ 35°, 50° AOA; ASE pts @ 30°, 40° AOA.
370	3 May	Schneider	0.6	0.688	34,336	3.05	62	Version 154.0; LEX fences on. Lat. PID M @ 30°, 45°, 60° AOA, STV mode; ASE pts. @ 40°, 50°, 60° AOA.
371	8 May	Rogers Smith	0.7	0.856	42,028	3.04	98	Version 154.0; LEX fences on. Spins, 60° alpha cap in TV; 360° rolls @ 35°, 55° AOA, TV, S modes; loaded rolls @ 0.4 M/35° AOA TV, S modes.
372	8 May	Stucky	0.8	0.868	41,608	3.26	107	Version 152.0; LEX fences on. Spins, PID M, TV @ 5°, 45°, 50°, AOA; agility 180° roll under cap @ 35° AOA, STV mode.

373	10 May	Lt. Greg Fenton, USN pilot	0.7	0.701	28,331	3.17	69	Version 152.0; LEX fences on; 360° rolls, TV mode @ 35°, 55°, 65° AOA; 55° AOA, S, STV modes; scissors; 45° AOA trk; J-Turn; rolling scissors in TV mode.
374	10 May	Lt. Bob Roth, USN pilot	0.7	0.685	28,251	3.11	70	Version 152.0; LEX fences on; 360° rolls, TV mode @ 35°, 55°, 65° AOA; 55° AOA in S, STV modes; scissors, 45° AOA trk, J-Turn, TV mode; 45° AOA trk, S mode.
375	14 May	Stucky	0.6	0.724	31,874	3.08	69	Version 152.0; LEX fences off. Aero pts. @ 30° AOA, 0/0 S; 40° AOA SHSS 10/10 S; 60° AOA 40/20, 20/40 S; 45/15, 15/45 S with/without sideslip; 30/30 S sideslip.
376	15 May	Smith	0.7	0.740	28,372	3.74	60	Version 154.0; LEX fences off; 45° AOA alpha cap, TV; 30° AOA trk, TV, S modes; 45° AOA trk, TV; flow vis @ 40°, 50° AOA, S SWP 0 to 90°; 360° roll @ 55° AOA in S mode.
377	15 May	Smith	0.8	0.707	25,449	3.38	70	Version 154.0; LEX fences on; 60° AOA trk, TV; LaGA @ 45° AOA, TV, S, STV modes; J-Turns, TV, STV.
378	17 May	Brown	0.8	0.688	27,253	3.78	79	Version 154.0; LEX fences on; 30° AOA trk, TV from 1,500 ft & 3,000 ft in trail; LaGA @ 45° AOA, TV, S, STV; alpha cap & J-Turns, TV mode.
379	17 May	Brown	0.7	0.695	25,404	3.19	76	Version 154.0; LEX fences on; 45° AOA trk, TV, S modes; LoGA @ 45° AOA, TV; J-Turn in S mode.
380	21 May	Tom McMurtry	0.8	0.903	44,516	3.23	91	Version 151.1; LEX fences on; 360° rolls, TV mode @ 35°, 55°, 65° AOA; 55° roll, S mode; spins, 55° & 70°/sec yaw rates; J-Turn, TV mode.

381	22 May	Gordon Fullerton	0.9	0.887	42,448	3.30	93	Version 151.1; LEX fences on; 360° rolls, 701E, TV modes @ 35° AOA; trk task @ 45° AOA, TV; spins to 50 & 70 deg/sec yaw rates; J-Turns, TV & 360° roll @ 55° AOA in S mode.
382	29 May	Dana Purifoy	0.9	0.872	41,740	3.02	94	Version 151.1; LEX fences on; 360° rolls, TV, S @ 55° AOA; trk task @ 45° AOA, TV; spins to 50 & 70 deg./sec yaw rates; J-Turn, TV mode.
383	29 May	Schneider	0.7	0.686	25,240	3.31	75	Version 151.1; LEX fences on. J-Turns, S, STV modes; 45° AOA trk task, TV, S; attempted 60° AOA trk, S mode. Fly-by over building.
384	4 Sept.	Smolka	1.0					FCF. Research press. data; fuel flow test for ferry flt.
385	6 Sept.	Schneider	0.7					Only research A/C that could fly for DFRC Open House. Fuel flow tests. Good fly-by.
386	10 Sept.	Smolka	1.4					High alpha conference, Kirtland AFB, Albuquerque, NM. First leg of ferry flight to NASA Langley.
387	10 Sept.	Schneider	1.3					McConnell AFB, Wichita, KA.
388	10 Sept.	Schneider	1.3					Scott AFB, St. Louis, MO.
389	10 Sept.	Smolka	0.9					Wright-Patterson AFB, Dayton, OH.
390	15 Sept.	Smolka	1.1					NASA Langely Research Center; Langley AFB, Norfolk, VA.
391	19 Sept.	Smolka	1.3					Wright-Patterson AFB, Dayton, OH. First leg of flight to McDonnell at St. Louis, MO.
392	19 Sept.	Smolka	1.0					McDonnell Aircraft, St. Louis, MO.
393	10 Oct.	Smolka	1.1					McConnell AFB, Wichita, KA.
394	10 Oct.	Smolka	1.0					Cannon AFB, NM.
395	10 Oct.	Smolka	1.3					Luke AFB, AZ.
396	10 Oct.	Smolka	0.9					NASA Dryden Flight Research Center, Edwards AFB, CA.

Source: Flight logs kept by Ronald "Joe" Wilson, NASA Dryden research engineer for flights 1 through 101 and 384-396. His sources were his personal records; F-18 HARV flight cards; flight cards and notes from Jim Cooper, F-18 HARV flight test engineer; comments from Al Bowers, F-18 HARV project engineer; conversations with Ed Schneider and Jim Smolka; and flight records in the Pilots Office. For flights 102 through 383, flight test engineer Karen Richards kept a computerized flight log, which is the source for the information for those flights. Her sources for the log included her knowledge from preparations of flight cards, her participation in the control room during the flights, and telemetered information on maximum flight conditions.

Acronyms and abbreviations:

AAC	Aggressive Alpha (angle-of-attack) Captures
A/B	Afterburner
ACM	Air Combat Maneuvers
Aero	Aeronautical
AIL	Aileron
ANSER	Actuated Nose Strake Enhanced Rolling
AOA	Angle of attack (also known as alpha)
AS	Airspeed
ASE	Aero-Structure Elasticity
AV	Avionics
Beta	Angle of sideslip
BFM	Basic Fighter Maneuvers
CAF	Canadian Air Force
CALIB., calib.	Calibration
Cap.	Captures
CAS	Command Augmentation System
C/H	Cooper/Harper Scale (for rating handling qualities)
CPT	Control position transducer or transmitter
'Davidson'	Name of a person from Langley
ECU	Engine Control Unit
E-D	Engagement-Disengagement
EGT	Exhaust Gas Temperature
Env.	Envelope
EXP	Expanded-Expansion
FADS	Flush Air Data System
FAM	Familiarity
F/B (f/bdy)	Forebody
FCF	Functional Check Flight
FCS	Flight Control System
FS	Full-stick
FSC	Functional Systems Check
G	Gravity (1 g=acceleration equal to 1 times the force of gravity)
HANG	High Angle-of-Attack Nosedown Guidelines
HERBST M	Maneuver involving a high Angle-of-Attack velocity vector roll, named after German aerodynamicist Dr. Wolfgang Herbst
HI, Hi	High
HQ	Handling Qualities
INS	Inertial Navigation System
J-Turn	A rapid 180 deg. reversal
KFT	Thousands of feet
Knts	Knots
LaGA	Lateral Gross Acquisition
LaRC	Langley Research Center
Lat.	Lateral

LE	Leading Edge
LEX	Leading Edge Extension
LH	Left half
LoGA	Longitudinal Gross Acquisition
Long.	Longitudinal
LRC	Loaded Roll Captures
M	Maneuver; Mach number
MAC	Mean aerodynamic chord
MAX	Maximum
MCAS	Marine Corps Air Station
MC2	Mission Computer Two
MSDR	Maintenance System Data Recorder
MSRM	Manual Spin Recover Mode
OBES	On-Board Excitation System
OIM	Optimal Input Maneuver
OL	Offensive loaded
PID	Parameter Identification System—a computerized method that enabled researchers to determine precisely the differences between values predicted from wind tunnels and those actually encountered in flight
PID OIM	Parameter Identification System Optimal Input Maneuver
PID SSI	Parameter Identification System Sequential Single-Surface Input
PGME	Propylene Glycol Monomethyl Ether
PLANAR WAVE	Standing pressure wave in front of the engine inlet
PO-PU	Push-over pull-up
PRESS, press.	Pressure
Ps.	Sustained power
Qbar	Dynamic pressure, measured in pounds per square foot
Ref.	Reference
RFCS	Research Flight Control System
RPO	Rolling pull-out
RTB	Return to Base
RTTM	Real Time Telemetry Data
S	Strakes
S/C	Stability and Control
SFO	Simulated Flame-Out
SHSS	Steady Heading Side-Slip
Sim val.	Simulator validation
SRM	Spin Recovery Mode
SSI	See PID SSI
STEMS	Standard Evaluation Maneuver Set
STV	Strake plus thrust vane
SWP	Sweep
Temp.	Temperature
Theta	Angle between the nose of the airplane and the horizon

Theta freq.	Check by the pilot to see how quickly the airplane responds to commanded changes in pitch (angle of the nose with the sweep horizon)
T/O	Takeoff
TRK, trk.	Tracking
TV	Thrust Vane
TVCS	Thrust Vane (Vector) Control System
USMC	United States Marine Corps
Vis.	Visualization
Vs.	Version (software)
WLSS	Wings Level Side-Slip
WR	Wing Rock
WUT	Wind Up Turn
WWT	Windup-Window Turn
1v1	One aircraft versus one aircraft
10L/47R	10 degrees on the left/ 47 degrees on the right
701E	F-18 flight control computer, standard F-18 configuration

Appendix W
X-29 Flight Chronology, 1984-1992

Two X-29 aircraft, featuring one of the most unusual designs in aviation history, were flown at the NASA Dryden Flight Research Center as technology demonstrators to investigate a host of advanced concepts and technologies. The multi-phased program, conducted from 1984 to 1992, provided an engineering database that is available for the design and development of future aircraft.

X-29A-1, Air Force Serial Number 82-0003, Flights

Flight	Date	Time (hours	Maximum Alt (feet)	Maximum Mach No.	Pilot	Purpose and Comments
1	14 Dec. 1984	1.2	15,000	0.43	Ch. A. Sewell, Grumman	Functional eval. of X-29 systems. Instrumentation validation.
2	4 Feb. 1985	1.3	15,000	0.56	Sewell	Functional Checks, JFS,[1] EPU, gear. Touch-and-go, handling qualities (HQ).
3	22 Feb.	1.4	15,000	0.61	Sewell	HQ, ND, DR, AR, ASE. Simulated Flame-Out (SFO).
4	1 Mar.	1.2	15,000	0.60	Kurt Schroeder, Grumman	S&C, ND, DR, AR, ASE, JFS. Angle-of-attack calibration.
5	2 Apr.	1.4	15,310	0.55	Stephen Ishmael, Dryden	HQ tracking, AR/PA. First NASA flight: pilot familiarization.
6	4 Apr.	1.4	15,516	0.57	Lt Col. Theodore "Ted" Wierzbanowski, USAF	HQ tracking, AR/PA. First Air Force flight, pilot fam.
7	9 Apr.	1.5	20,120	0.60	Rogers Smith, Dryden	HQ tracking, AR/PA. Control-room fam and training.
8	16 Apr.	1.5	20,376	0.62	Sewell	A/B Take-off, Landing, AR & DR. Functional test.

[1] The acronyms are too numerous to spell them all out on first use. See the definitions at the end of the appendix.

9	21 May	1.4	30,188	0.61	Schroeder	ASE & FCS stability clearance. Instrumentation; functional test.
10	29 May	1.2	31,237	0.60	Ishmael, NASA project pilot	ASE & FCS stab clearance, loads. Gyro problems delayed flight.
11	29 May	1.4	30,861	0.60	Wierzbanowski	ASE & FCS stab clearance, loads. EPU functional check, JFS check.
12	6 June	1.3	30,831	0.61	Smith	ASE, FCS stab, loads clearance. Flow visualization, formation flying.
13	6 June	1.1	20,200	0.56	Schroeder	ASE, FCS stab, loads clearance. Flow visualization, formation, throttle transients.
14	11 June	1.2	16,108	0.62	Ishmael	Engine-inlet compatibility, loads maneuver envelopes. Formation, flow visualization.
15	11 June	1.4	30,228	0.58	Sewell	Engine-inlet comp., loads. Formation, flow visualization.
16	13 June	1.2	25,374	0.61	Wierzbanowski	Flow visualization, loads build-up. Formation, HQ.
17	14 Aug.	1.5	30,000	0.70	Smith	Envelope expansion, check of gunsight. First expanded envelope flight.
18	22 Aug.	1.3	30,000	0.75	Schroeder	Envelope expansion, air-to-air tracking. PAC mode evaluation.
19	22 Aug.	1.2	30,000	0.70	Ishmael	Nz expansion, tracking. PAC mode evaluation.
20	1 Nov.	1.5	40,904	0.80	Wierzbanowski	Engine funct., envelope expansion. Flap tab shaker functional.
21	7 Nov.	1.4	40,373	0.79	Ishmael	Envelope expansion. Post-flight JFS failure.
22	19 Nov.	1.0	40,308	0.83	Smith	Env. expansion. Oxygen-system anomaly.
23	20 Nov.	0.8	40,561	0.83	Schroeder	Env. expansion. Roll-rate gyro fail indication.
24	27 Nov.	1.5	40,000	0.90	Smith	MCC functional. Env. expansion.
25	6 Dec.	1.4	41,113	0.94	Wierzbanowski	Envelope expansion. AR mode landing.
26	13 Dec.	1.0	40,000	1.03	Ishmael	Envelope expansion. ITB-2 0.85M.
27	20 Dec.	1.0	40,000	1.07	Schroeder	Envelope expansion. ITB-2 0.85M.
28	20 Dec.	1.0	40,000	1.10	Wierzbanowski	Envelope expansion. ITB-2 O.85M.

29	20 Dec.	1.6	40,000	0.95	Smith	ITB-2, 0.95 M/40 KFT. Buffet wind-up-turn 0.6 M/15 KFT.
30	15 Jan. 1986	0.3	9,800	0.40	Ishmael	Attitude heading reference system. AHRS failure; flight aborted.
31	23 Jan.	1.3	30,000	0.85	Schroeder	Envelope expansion. High AOA investigations.
32	23 Jan.	0.7	40,000	1.20	Wierzbanowski	Envelope expansion. AR landing with crosswind.
33	7 Feb.	1.0	40,000	1.10	Smith	Maneuver envelope expansion. Early termination for potential loss of range time.
34	7 Feb.	1.0	30,000	0.85	Ishmael	Maneuver envelope expansion. MCC clearance 0.6M/30 KFT.
35	19 Feb.	0.7	15,000	0.60	Ishmael	Envelope clearance aborted due to T/M loss. Clearance 0.6M/10 and 15 KFT.
36	21 Feb.	1.3	40,000	1.12	Ishmael	Envelope clearance. MCC Clearance 0.6M/20, 30 KFT & 0.7M/40 KFT.
37	21 Feb.	1.4	30,000	0.73	Ishmael	MCC Clearance 0.7 M/20 KFT. HQ; doublets 0.4 M–0.6, 20 KFT & 30 KFT; MCC maneuvers.
38	27 Feb.	1.2	40,000	1.05	Smith	FCS software verification @ 1.05 M/40 KFT. WUT @ 0.85 M/30 KFT.
39	27 Feb.	1.1	40,000	1.10	Wierzbanowski	Envelope expansion. ITB-2 @ 1.1 M/40 KFT, 0.9 M/30 KFT.
40	10 June	1.2	40,000	0.95	Schroeder	Alpha/beta nose-boom interaction test. ITB-2 @ 0.8 M/40 and 30 KFT.
41	12 June	0.7	20,000	0.55	Ishmael	High-AOA maneuvering. Flt. cut short due to AHRS failure on rotation.
42	12 June	1.2	30,000	0.95	Ishmael	Load factor expanded. High-AOA maneuvering, MCC clearance.
43	12 June	0.8	40,000	1.30	Smith	High-AOA maneuvering. Envelope expansion.
44	12 June	0.7	40,000	1.30	Wierzbanowski	Envelope expansion. Maneuver exp. ITB-1, 2.
45	11 July	1.0	40,000	0.95	Schroeder	MCC clearance. Constant Mach/const AOA WUT.
46	15 July	1.1	40,000	1.20	Wierzbanowski	MCC clearance. Envelope expansion; HQDT.

47	15 July	0.7	40,000	1.40	Smith	Envelope expansion in ACC. MCC maneuvering.
48	18 Jul	0.8	30,000	1.05	Ishmael	Envelope expansion. MCC clearance.
49	18 July	1.0	30,000	1.10	Maj. Harry Walker, USAF	Envelope expansion. HQ evaluation; pilot fam.
50	18 July	0.7	40,000	1.30	Smith	Maneuver exp. ITB-2. MCC expansion.
51	24 July	0.9	40,000	1.20	Schroeder	Envelope expansion, ITBs-1 & 2. Performance WUT @ design point-9M.
52	24 July	0.5	20,000	0.90	Ishmael	Env expansion in ND-ACC, AR-UA.
53	30 July	0.7	30,000	1.20	Walker	Ctd. env. expansion. Maneuver expansion.
54	30 July	0.7	30,000	1.20	Smith	Envelope expansion. ITB-1, -2.
55	30 July	0.7	30,000	1.30	Schroeder	VMAX 1.3M. Maneuver exp.
56	1 Aug.	0.5	30,000	1.30	Ishmael	Envelope expansion. VMAX @ FL300 (1.3M).
57	1 Aug.	0.5	30,000	1.30	Schroeder	Maneuver expansion pt. FL300/1.2M.
58	1 Aug.	0.5	30,000	1.20	Ishmael	MCC clearance (FL 300/1.2 M).
59	8 Aug.	0.6	20,000	1.03	Smith	Envelope expansion. Maneuver expansion (FL 200/0.95 M).
60	8 Aug.	0.5	20,000	1.10	Schroeder	Envelope expansion.
61	8 Aug.	0.5	20,000	1.10	Ishmael	Envelope expansion. Maneuver expansion (FL 200/1.05 M).
62	8 Aug.	0.4	20,000	1.175	Smith	VMAX @ (FL 200/1.175 M).
63	13 Aug.	0.7	30,000	1.20	Schroeder	DR investigation (FL200/0.9M). Performance WUTs @ design point.
64	13 Aug.	0.6	25,000	1.20	Ishmael	ITB-1, 10 KFT/0.6 M. Maneuver expansion (FL 200/1.1 M).
65	13 Aug.	0.6	25,000	1.20	Smith	Wing pressure analysis (FL 250/0.925 M). MCC maneuvering, same condition.
66	27 Aug.	0.9	30,000	0.925	Walker	Envelope and maneuver expansion (15 KFT/0.8 M).
67	27 Aug.	0.6	40,000	0.90	Schroeder	Investigation of Dgr normal (FL 400/0.6 M). AHRS failure on rotation.
68	27 Aug.	0.9	15,000	0.90	Schroeder	Envelope expansion.

69	27 Aug.	0.8	15,000	0.85	Ishmael	Envelope and maneuver expansion (10 KFT/0.7 M/0.8 M).
70	5 Sept.	0.7	15,000	0.97	Smith	Envelope expansion.
71	5 Sept.	0.9	20,000	0.97	Walker	Envelope and maneuver expansion (15 KFT/0.95 M), AR.
72	24 Oct.	0.6	30,000	1.03	Schroeder	Validation of normal mode gain change. Buffet WUT (6,800 FT/300 KIAS).
73	29 Oct.	0.5	15,000	1.10	Ishmael	Envelope expansion.
74	29 Oct.	0.5	30,000	1.10	Smith	Envelope expansion, AR. Buffet WUTs (FL 300/0.6 M to 15.6° AOA).
75	29 Oct.	0.7	34,000	0.90	Walker	Envelope expansion. Buffet WUTs (FL 340/0.9 M, FL 290/0.8 M).
76	29 Oct.	0.5	15,000	1.16	Ishmael	VMAX @ 15,000 FT (1.15 M). Envelope expansion.
77	7 Nov.	0.7	43,000	0.95	Schroeder	Envelope expansion. Buffet WUT (FL 400/0.9 M).
78	7 Nov.	0.7	34,000	0.97	Smith	Envelope expansion. Buffet WUTs (FL 300/0.6 M, FL 340/0.9 M).
79	7 Nov.	0.5	8,000	1.03	Walker	Envelope expansion.
80	7 Nov.	0.4	10,000	1.05	Ishmael	Recheck phase/gain @ 8 KFT/1.03 M. Maneuver expansion.
81	14 Nov.	0.4	10,000	1.12	Schroeder	ASE Clearance (10 KFT/1.1 M). VMAX @ 10 KFT (1.12M).
82	14 Nov.	0.4	10,000	1.10	Smith	Envelope expansion completed with ITB-1/2 @ 10 KFT/1.1 M.
83	14 Nov.	1.0	20,000		Walker	Handling qualities evaluation.
84	14 Nov.	1.0	20,000		Ishmael	Handling qualities evaluation.
85	19 Nov.	0.4	15,000	1.05	Walker	Divergence envelope clearance. (15 KFT/1.05 M) (10 KFT/0.85 M).
86	19 Nov.	0.5	10,000	1.05	Ishmael	Divergence envelope clearance. (10 KFT/0.9 M/0.95 M/1.05 M).
87	19 Nov.	0.5	22,000	1.20	Walker	Divergence envelope clearance. (FL 220/1.2 M).

88	3 Dec.	1.0	20,000		Schroeder	Handling qualities evaluation.
89	3 Dec.	0.6	45,000	1.20	Ishmael	Buffet WUTs, MCC research. Shaker functional check-out.
90	3 Dec.	1.1	20,000		Smith	Handling qualities evaluation.
91	5 Dec.	0.6	41,700	1.33	Walker	MCC, loads expansion. Shaker check.
92	5 Dec.	0.8	45,300	1.02	Smith	Buffet, loads expansion. ASE damping with load factor.
93	5 Dec.	0.5	39,200	1.27	Ishmael	Buffet, loads expansion.
94	5 Dec.	0.8	25,900	1.09	Schroeder	Loads expansion. ASE damping with load factor, speed stability evaluation.
95	10 Dec.	0.5	41,100	1.43	Walker	Flap flutter. Buffet, loads expansion.
96	10 Dec.	0.6	36,200	1.29	Schroeder	Flap flutter. MCC (10 KFT/0.9 M).
97	12 Dec.	0.5	22,000	1.20	Ishmael	Loads expansion (15 KFT). Buffet, MCC, envelope expan (15 KFT).
98	12 Dec.	0.6	18,000	1.10	Smith	Loads expansion (10 KFT). Buffet, MCC, envelope expan (15 KFT).
99	17 Dec.	0.6	47,600	1.45	Walker	Flap flutter, 5 KFT expansion. Repeat MCC points.
100	17 Dec.	0.6	50,200	1.46	Ishmael	Flap flutter, VMAX Tropopause. Envelope expansion (5 KFT).
101	17 Dec.	0.8	20,000	0.90	Smith	Envelope expansion (5 KFT). Speed stability evaluation.
102	23 Dec.	0.8	30,300	1.00	Walker	RAV functional checkout, MCC WUTs. HQDT simulated refueling.
103	23 Dec.	1.1	22,300	0.80	Lt. Cmdr. Ray Craig	Navy evaluation.
104	23 Dec.	1.5	15,500	0.70	Craig	Navy evaluation.
105	19 June 1987	0.8	39,500	1.27	Ishmael	Functional flight. AOA/airspeed calibs. FCS AR mod check.
106	26 June	0.7	31,500	0.996	Walker	AOA/airspeed calibrations. FCS AR mod check, RAV ADI needles.
107	26 June	0.9	46,100	1.39	Ishmael	AOA/airspeed calibrations. FCS AR mod check.

108	26 June	0.8	30,300	1.10	Walker	AOA/airspeed calibrations. Loads.
109	30 June	0.7	31,700	0.985	Smith	AOA/airspeed calibrations. Loads, EPU tests for A/C #2.
110	30 June	0.8	45,150	1.35	Ishmael	AOA/airspeed calibrations.
111	24 July	0.7	46,100	1.01	Schroeder	Aero pressure survey.
112	24 July	0.5	35,970	1.22	Walker	Aero pressure survey.
113	24 July	0.6	25,039	0.91	Smith	Aero pressure survey.
114	29 July	0.4	20,280	1.17	Ishmael	MCC divergence. FCS AR mod check, loads expansion.
115	29 July	0.7	15,080	0.925	Walker	MCC divergence, speed stab. eval. FCS AR mod check, loads expansion.
116	5 Aug.	0.9	45,512	0.94	Smith	MCC divergence.
117	5 Aug.	0.6	27,300	0.95	Ishmael	MCC divergence.
118	5 Aug.	0.6	33,128	0.95	Walker	MCC divergence. Aero pressure survey.
119	7 Aug.	0.5	12,616	1.05	Smith	MCC divergence, aero pressure survey. Loads and fuel flow checks.
120	19 Aug.	0.4	43,677	1.24	Ishmael	Performance.
121	19 Aug.	0.6	50,707	1.21	Walker	Performance.
122	19 Aug.	0.5	37,927	1.21	Smith	Performance.
123	19 Aug.	0.4	33,178	1.23	Ishmael	Performance.
124	9 Sept.	0.7	32,287	0.94	Walker	Performance, loads expansion.
125	9 Sept.	0.6	38,180	1.31	Ishmael	Performance.
126	11 Sept.	0.5	32,432	1.13	Walker	Performance, loads expansion.
127	11 Sept.	1.1	33,596	0.92	Ishmael	Performance, loads expansion.
128	11 Sept.	0.7	25,585	1.21	Walker	Performance, loads expansion.
129	11 Sept.	0.9	43,339	1.01	Ishmael	Gust evaluation.
130	9 Oct.	0.6	30,853	1.33	Walker	Performance.
131	9 Oct.	0.4	31,399	1.33	Ishmael	Performance.
132	9 Oct.	0.6	17,325	1.08	Walker	Performance, loads expansion.
133	9 Oct.	0.4	20,415	1.09	Ishmael	Performance.

134	14 Oct.	0.7	39,505	0.97	Walker	Loads expansion, buffet research.
135	14 Oct.	0.8	44,561	0.96	Ishmael	PID (RAV) research.
136	14 Oct.	0.6	40,174	1.01	Walker	ASE alpha evaluation.
137	16 Oct.	0.6	31,221	0.99	Ishmael	ASE ALPHA evaluation, performance.
138	16 Oct.	0.7	39,305	1.13	Walker	PID (RAV) research, buffet research.
139	16 Oct.	0.4	44,815	1.12	Ishmael	Buffet research.
140	4 Nov.	0.9	38,489	0.98	Smith	Asymmetric loads expansion, buffet research.
141	6 Nov.	0.5	37,656	1.27	Walker	Performance.
142	6 Nov.	0.8	40,752	1.04	Ishmael	Asymmetric loads expansion. K-27 gain handling qualities eval.
143	18 Nov.	0.7	35,929	0.96	Smith	BLK-VIII FCS functional checkout, asymmetric loads, buffet research.
144	18 Nov.	0.8	33,692	0.96	Walker	Asymmetric loads expansion, buffet research.
145	18 Nov.	0.8	35,146	0.95	William Dana, Dryden	Guest pilot handling qualities evaluation, buffet research.
146	2 Dec.	0.7	34,325	0.96	Ishmael	Performance, K-27 gain HQ evaluation. Speed stability evaluation.
147	4 Dec.	0.6	40,461	1.48	Walker	Performance, asymmetric loads expansion.
148	4 Dec.	0.9	40,802	1.06	Rod Womer, Grumman	New pilot HQ evaluation, asymmetric loads expansion, PID (RAV).
149	9 Dec.	0.6	33,245	1.01	Ishmael	Symmetric/asymmetric loads expan.
150	9 Dec.	0.6	49,564	1.33	Walker	RAV sweeps, PID (RAV), buffet research.
151	9 Dec.	0.6	34,873	1.31	Womer	PID (RAV), symmetric loads expan.
152	11 Dec.	0.8	40,822	0.99	Ishmael	Symmetric loads expansion, shaker fairing baseline, buffet research. Engine vibration.
153	11 Dec.	0.6	46,971	1.31	Walker	Symmetric loads expansion PID (RAV).
154	11 Dec.	1.0	30,739	1.21	James W. Smolka, Dryden	Guest pilot HQ evaluation, shaker fairing evaluation. Engine vibration.

155	18 Dec.	0.8	31,594	0.96	Smith	Asymmetric loads expansion PID (AV). FCS sweeps, performance.
156	8 Jan. 1988	0.6	22,437	0.98	Ishmael	Asymmetric loads expansion; GW/CG effects on loads. Engine vibration.
157	13 Jan.	0.8	31,425	0.96	Womer	Military utility evaluation. Engine vibration.
158	13 Jan.	0.7	30,562	0.95	Smith	GW/CG effects on loads, military utility evaluation, asymmetric loads expansion. Engine vibration.
159	13 Jan.	0.5	30,199	1.02	Maj. Alan Hoover, USAF project pilot	New pilot; HQ evaluation; military utility evaluation. Engine vibration.
160	22 Jan.	0.7	31,889	1.03	Walker	Military utility evaluation. Engine vibration.
161	22 Jan.	0.8	32,662	1.01	Hoover	Military utility evaluation. Engine vibration.
162	22 Jan.	0.7	26,937	1.16	Womer	FCS oscillation check, military utility evaluation. Engine vibration.
163	27 Jan.	0.9	37,589	0.94	Smith	Abrupt symm expansion, roll clearance, military utility evaluation. Boost pump check, engine vibration.
164	27 Jan.	0.8	35,391	0.96	Smith	Roll clearance, military utility evaluation, lat/dir PID. Engine vibration.
165	5 Feb.	0.2	26,201	0.96	Ishmael	Aborted due to in-flight TM data dropout.
166	12 Feb.	0.7	30,307	0.94	Hoover	Shaker hydraulic-lines fairings eval., PID (RV), FDMS check-out. MCC loads exp., engine vibration.
167	12 Feb.	0.6	32,924	1.24	Ishmael	Shaker hyd.-lines fairings (removed) evaluation, PID (RAV). Military utility evaluation; MCC loads expansion; engine vibration.
168	12 Feb.	0.7	38,315	1.04	Thomas C. McMurtry, Dryden	Guest pilot, HQ evaluation. Engine vibration.
169	16 Mar.	0.7	31,223	0.97	Smith	Alpha bias roll coupling eval., military utility evaluation. ASE alpha eval, engine vibration.
170	16 Mar.	0.6	42,983	1.18	Hoover	FCS oscillation check, aero pressure survey. Engine vibration.

171	16 Mar.	1.0	26,399	0.91	Edward T. Schneider, Dryden	Guest pilot, HQ evaluation. Engine vibration.
172	23 Mar.	0.7	44,348	1.11	Womer	Aero pressure survey, ASE alpha evaluation. Loads expansion, engine vibration.
173	23 Mar.	0.6	36,789	1.23	Ishmael	Loads expan., aero pressure survey. Engine vibration.
174	23 Mar.	0.6	34,406	1.21	Hoover	Loads expan., aero pressure survey. Engine vibration.
175	30 Mar.	0.7	50,390	1.27	Womer	Loads expansion, MCC divergence (1.20 Mach). Engine vibration.
176	30 Mar.	0.5	48,883	1.26	Ishmael	Loads expansion, MCC divergence (1.20 Mach). Engine vibration.
177	30 Mar.	0.5	33,799	1.20	Womer	Loads expansion, mil utility eval. MCC divergence (1.20 Mach), engine vibration.
178	6 Apr.	0.5	48,398	1.21	Smith	Loads expansion, MCC divergence (1.20 Mach). Military utility, engine vibration.
179	6 Apr.	0.6	46,670	1.27	Hoover	Loads expansion, MCC divergence (1.20 Mach). Military utility, engine vibration.
180	6 Apr.	0.6	26,674	1.22	Ishmael	Loads expansion, mil.utility eval. Engine vibration.
181	13 Apr.	0.7	17,824	0.89	Womer	Loads expansion, mil.utility eval. Engine vibration.
182	15 Apr.	0.5	32,591	1.22	Hoover	Loads expansion, MCC divergence (1.20 Mach), mil. utility eval. Engine vibration.
183	22 Apr.	0.7	26,719	0.87	Ishmael	Loads expansion, mil.utility eval. Engine vibration.
184	22 Apr.	0.5	36,205	1.21	Smith	MCC divergence, (1.20 Mach) military utility evaluation. Engine vibration.
185	22 Apr.	1.0	31,081	1.08	C. Gordon Fullerton, Dryden	Guest pilot HQ evaluation. Engine vibration.
186	22 Apr.	0.5	41,683	1.26	Hoover	MCC divergence (1.20 Mach). Engine vibration.
187	20 May	0.5	33,522	1.24	Ishmael	Block VIII-AC FCF, MCC divergence (1.20 Mach). Speed stability evaluation.

188	20 May	0.5	28,271	1.22	Smith	Block VIII-AC FCF, MCC divergence (1.20 Mach). Speed stability evaluation.
189	20 May	0.4	28,688	1.21	Hoover	Block VIII-AC FCF, MCC divergence (1.20 Mach).
190	20 May	0.3	30,320	1.31	Ishmael	PID (RAV).
191	25 May	0.5	24,808	1.05	Smith	Revised ACC schedule loads eval., PID (RAV).
192	25 May	0.6	46,243	1.29	Ishmael	Revised ACC schedule loads eval., PID (RAV) wing/canard loads interact.
193	25 May	0.4	29,849	1.21	Smith	Revised ACC schedule loads eval., MCC divergence (1.20 Mach). FCS (AR) ITB-1.
194	1 June	0.7	35,551	1.07	Hoover	Wing/canard loads interaction. Performance, minimum landing speeds, buffet, FCS (AR-UA).
195	1 June	0.7	37,530	1.06	Ishmael	Wing/canard loads interaction, buffet, performance. Minimum landing speeds.
196	1 June	0.6	27,610	1.21	Hoover	Wing/canard loads interaction, MCC divergence (1.20 Mach). Loads ACC eval., throttle transients.
197	8 June	0.8	29,788	0.96	Smith	Wing/canard loads interaction, BFF, long stick mod eval. Block VIII-AD K-27 check, throttle transients, min. landing speeds.
198	8 June	0.8	26,078	0.98	Ishmael	Wing/canard loads interaction, long stick mod. eval. Min landing speeds, buffet.
199	8 June	0.8	31,336	0.97	Hoover	Wing/canard loads interaction, long stick mod. eval. Performance, min. landing speeds.
200	8 June	0.6	46,225	1.23	Smith	Wing/canard loads interaction, divergence (1.20 Mach), buffet. HQ simulated refueling with KC-135.
201	6 July	0.5	32,992	1.03	Womer	Revised ACC schedule loads eval., BFF expansion.
202	6 July	0.4	20,354	1.03	Hoover	Revised ACC schedule loads eval., BFF expansion.
203	6 July	0.7	37,102	0.96	Smith	Revised ACC schedule loads eval., wing/canard loads interaction. Min. landing speeds evaluation, buffet.

204	13 July	0.6	22,977	0.96	Womer	Wing/canard loads interaction, revised ACC schedule loads eval.
205	13 July	0.8	31,242	0.94	Hoover	Revised ACC schedule loads eval., military utility evaluation. Wing/canard loads interaction.
206	13 July	0.6	31,598	1.20	Smith	Wing/canard loads interaction, military utility evaluation. Min. landing speeds evaluation.
207	22 July	0.6	21,623	1.01	Ishmael	BFF expansion, military utility eval., min. landing speeds eval.
208	22 July	0.9	23,713	0.82	Lt. Col. Gregory Lewis	Air Force pilot HQ evaluation.
209	22 July	0.8	21,575	1.01	Smith	Revised ACC schedule loads eval., gust response (F-104 comparison). Military utility evaluation.
210	27 July	0.4	15,397	0.57	Womer	Smithsonian movie.
211	27 July	1.1	17,061	0.78	Hoover	Smithsonian movie.
212	27 July	0.9	21,904	0.77	Maj. Erwin B. (Bud) Jenschke	Air Force pilot HQ evaluation.
213	27 July	0.7	20,688	0.68	Smith	Smithsonian movie.
214	6 Oct.	0.6	30,580	0.63	Hoover	FCS/ASE exp. (BLK VIII-AF).
215	12 Oct.	1.0	16,411	0.81	Smith	FCS/ASE exp. (BLK VIII-AF), loads/FQ clearance (BLK VIII-AF)
216	12 Oct.	0.8	16,441	0.92	Womer	Same as flight 215.
217	12 Oct.	1.0	22,518	0.96	Smith	Same as flight 215.
218	18 OCT.	0.6	21,250	0.96	Hoover	Loads clearance (BLK VIII-AF), airshow practice.
219	18 Oct.	0.8	20,223	0.90	Smith	Loads/FQ clearance (BLK VIII-AF).
220	18 Oct.	0.8	20,587	0.96	Hoover	Loads/FQ clearance (BLK VIII-AF), buffet research.
221	20 Oct.	0.7	40,006	0.97	Womer	Buffet research, loads clearance, formation HQ.
222	20 Oct.	0.8	16,982	0.87	Hoover	FCS HQ clearance, airshow practice, gust response (F-104 comparison).

223	3 Nov.	0.8	21,404	0.95	Smith	JFS In-flight Start, Loads Clearance and FCS Evaluation (BLK VIII-AF)
224	3 Nov.	1.0	23,432	0.85	Lt. Col. Jeffrey Riemer.	Air Force pilot, HQ evaluation.
225	3 Nov.	0.9	20,804	0.88	Womer	Loads clearance and FCS evaluation (BLK VIII-AF), gust response (F-104 comparison). Formation HQ evaluation.
226[2]	9 Nov.	0.4	12,235	1.11	Hoover	Loads clearance (BLK VIII-AF), BFF expansion.
227	9 Nov.	0.6	32,457	1.04	Smith	Loads clearance (BLK VIII-AF), BFF expansion, buffet research. Military utility (agility).
228	9 Nov.	0.6	40,518	1.26	Hoover	Wing/canard loads interaction study, buffet research, mil. utility (agility).
229	9 Nov.	0.8	20,295	0.88	Smith	Military utility (agility), A/A, A/G. Formation HQ.
230	15 Nov.	0.8	21,292	0.86	Hoover	Military utility (agility, A/A, A/G w/ ATLAS system and formation HQ).
231	15 Nov.	0.9	19,602	0.81	Smith	Buffet research, military utility, air-to-ground (ATLAS), ground effects research. Formation HQ, ASE research.
232	15 Nov.	0.8	20,570	0.96	Hoover	Buffet research, loads roll clearance, military utility (agility & A/A)
233	18 Nov.	0.5	41,675	1.21	Smith	Wing/canard loads interaction study, supersonic roll expansion, military utility (agility). SFO.
234	18 Nov.	1.0	22,720	0.92	Maj. Dana D. Purifoy	Air Force pilot HQ evaluation.
235	18 Nov.	0.8	24,060	0.89	Hoover	Military utility (agility). SFO.
236	23 Nov.	0.8	20,599	0.87	Ishmael	Military utility (agility, formation HQ, A/G with ATLAS system).
237	23 Nov.	0.9	23,402	0.84	Smith	Military utility (agility and A/G with ATLAS system).

[2] Flight #226, new X-29 airspeed record of 665 KEAS (1.11 Mach/5,032 feet).

238	23 Nov.	0.5	26,892	1.21	Hoover	Military utility (agility), wing/canard loads interaction study, supersonic roll clearance.
239[3]	8 Dec.	0.5	32,527	1.01	Ishmael	Military utility (agility), HQ (A/A tracking, loads-negative g expansion.
240	8 Dec.	0.9	25,611	0.84	Col. John M. Hoffman	Air Force pilot HQ evaluation.
241	8 Dec.	0.8	20,838	0.92	Lt. Gen. David J. McCloud	Air Force pilot HQ evaluation.
242	8 Dec.	0.5	33,163	0.97	Smith	Military utility (agility), HQ (A/A tracking), buffet research, wing/canard interaction study. Aircraft placed in flyable storage.
243	15 June 1990	0.7	33,598	0.95	Ishmael	Function checks for airshow ferry flights: ECS; FCS modes; PTO shaft; RAV; JFS; EPU; engine; aero; loads.
244	15 June	1.2	44,966	1.01	Purifoy	Functional checks for airshow ferry flights: rudder trim; cruise performance; JFS; engine; aero & loads; speed stability.
245	27 June	1.0	44,900	0.94	Smith	Functional checks for airshow ferry flights: loads; cruise performance. FCS research.
246	27 June	1.1	41,200	1.01	Ishmael	Practice for airshow ferry flights. Cruise performance; project appreciation maneuvers.
247	17 July	1.0	44,900	0.93	Purifoy	Functional checks for airshow ferry flights: engine; cruise performance. FCS research.
248	18 July	1.1	41,000	0.90	Ishmael	Dayton Airshow deployment flight, Edwards, CA, to Albuquerque, NM.
249	18 July	1.2	41,000	0.90	Smith	Dayton Airshow deployment flight, Albuquerque, NM, to Tulsa, OK.

[3] Flight #239, new X-29 negative load factor of –1.9g (equivalent) and angle-of-attack of –4.2° were attained at 0.60 Mach/10,000 feet; new X-29 positive load factor of 6.7g (equivalent) was attained at 0.95 Mach/20,600 feet.

250	18 July	1.3	41,000	0.90	Purifoy	Dayton Airshow deployment flight, Tulsa, OK, to Dayton, OH.
251	23 July	1.1	41,000	0.90	Womer	Oshkosh Airshow deployment flight, Dayton, OH, to Oshkosh, WI.
252	5 Aug.	1.3	41,000	0.90	Ishmael	Dayton/Oshkosh Airshow return, Oshkosh, WI, to Wichita, KS.
253	5 Aug.	1.2	41,000	0.90	Smith	Dayton/Oshkosh Airshow return, Wichita, KS, to Albuquerque, NM.
254	5 Aug.	1.2	41,000	0.90	Ishmael	Dayton/Oshkosh Airshow return, Albuquerque, NM, to Edwards, CA.

Total Flight Time: 200.2 hours

X-29A-2, Air Force Serial Number 82-0049, Flights

Flight	Date	Time (hours)	Max Alt. (feet)	Max Mach Number	Max LD FAC (g)	Max Alpha (Deg)	Pilot	Purpose and Comments
1	23 May 1989	0.9	29,100	0.65	4.4	15.0	Ishmael	Pilot qual.; FCS clearance; systems eval & SFO practice.
2	13 Jun.	0.9	30,100	0.97	5.6	18.5	Maj. Hoover	Pilot qual; FCS clearance; systems eval & SFO practice.
3	13 Jun.	1.1	26,000	0.84	5.4	12.0	Ishmael	Loads check; EPU check & high-speed spin chute deployment.
4	23 Jun.	1.0	30,300	0.83	3.1	18.5	Hoover	FCS clearance; low-speed spin chute deployment; engine-throttle transients & SFO practice.
5	23 Jun.	1.1	30,300	0.87	3.0	13.5	Smith	Pilot qual; FCS clearance; engine-throttle transients; speed stability; PAC eval & SFO practice. End Phase 1 functional flights.
6	11 Oct.	0.9	20,300	0.81	5.3	12.2	Womer	Pilot qual; RAV checkout; airdata calib; FCS clearance & SFO practice.

7	11 Oct.	0.9	30,000	0.95	5.9	16.2	Ishmael	RAV checkout; air data & radar calibration; FCS clearance & SFO practice. End Phase 2 functional flts.
8	19 Oct.	1.0	37,900	0.83	3.1	22.7	Hoover	1g ITB-1 expansion to 20° AOA; SFO practice.
9	8 Nov.	1.0	38,900	0.91	3.1	24.2	Smith	1g ITB-1 expansion to 22.5° AOA; 1g ITB-2 expansion to 10° AOA; SFO practice.
10	8 Nov.	1.1	38,200	0.87	3.1	24.7	Womer	Same as flight 9.
11	28 Nov.	0.7	39,000	0.90	2.9	29.7	Ishmael	1g ITB-1 expansion to 27.5° AOA; in-flt. abort due to TM data loss.
12	19 Dec.	0.9	38,400	0.82	3.0	32.0	Hoover	1g ITB-1 expansion to 30° AOA; 1g ITB-2 expansion to 15° AOA.
13	4 Jan. 1990	0.7	39,200	0.93	3.1	39.0	Smith	1g ITB-1 expansion to 35° AOA.
14	4 Jan.	0.9	38,100	0.91	3.0	36.0	Womer	1g ITB-2 expansion to 15°, 19°, 25° AOA.
15	11 Jan.	0.9	38,200	0.90	3.0	43.0	Ishmael	1g ITB-1 expansion to 40° AOA; 1g ITB-2 expansion to 30° AOA. In-flight abort due to jam of chute-jettison handle.
16	25 Jan.	0.9	39,000	0.87	3.5	42.0	Hoover	WUT ITB-1 expansion to 10°, 15°, and 20° AOA.
17	25 Jan.	1.0	37,400	0.90	3.1	42.0	Smith	WUT ITB-2 expansion to 10° AOA; flow visualization.
18	25 Jan.	1.1	38,900	0.81	3.0	31.0	Purifoy	Pilot familiarization.
19	1 Feb.	0.8	41,300	0.90	3.0	50.5	Womer	1g ITB-1 expansion to 45° AOA. In-flight abort due to high winds & blowing sand.
20	8 Feb.	0.9	38,400	0.92	3.8	48.0	Ishmael	1g directional control check at 45° AOA; 1g ITB-2 expansion to 35° AOA; WUT ITB-1 expansion to 25° AOA.
21	8 Feb.	1.1	38,000	0.95	3.1	51.5	Hoover	Flow visualization.
22	8 Feb.	1.1	37,500	0.85	3.1	46.5	Smith	Flow visualization; WUT ITB-2 expansion to 15° AOA.

23.	15 Feb.	1.0	39,600	0.94	3.2	55.5	Smith	1g directional control check at 50° AOA; 1g ITB-1 expansion to 45° AOA; engine expansion.
24	15 Feb.	0.9	38,300	0.92	3.1	53.5	Hoover	1g directional control check at 50° AOA; 1g ITB-2 expansion to 40° AOA; engine expansion.
25	15 Feb.	0.9	38,500	0.84	3.7	40.8	Purifoy	Pilot familiarization; high AOA qualitative evaluation. In-flight abort due to MCR shutdown.
26	9 Mar.	0.7	39,700	0.90	3.5	52.8	Ishmael	1g directional control expansion to 50° AOA; 160 KCAS/WUT ITB-1 expansion to 35° AOA & ITB-2 expansion to 25°AOA; 200 KCAS/WUT ITB-1 expansion to 20° AOA.
27	9 Mar.	0.4	40,600	0.85	3.4	55.5	Womer	1g directional control checks at 50°AOA; 200 KCAS/WUT ITB-1 expansion to 25° AOA & ITB-2 expansion to 15° AOA; MIL engine expansion. In-flt abort due to alpha failure on recovery from 50° AOA.
28	9 Mar.	1.0	41,200	0.90	3.8	31.3	Smith	200 KCAS/WUT ITB-1 expansion to 25°AOA; MIL engine expansion; in-flight variable gain (90% K2:p/ a) test.
29	22 Mar.	0.9	40,100	0.89	3.1	53.0	Purifoy	Pilot familiarization at 45° and 50°AOA; 200 KCAS/WUT ITB-1 expansion to 25° AOA & ITB-2 expansion to 15° AOA.
30	22 Mar.	1.0	40,400	0.88	3.2	51.0	Ishmael	PID/aero doublets at 35°, 45°, and 50°AOA; 1g ITB-3 expansion to 20°, 25°, 30° and 35° AOA; AB engine expansion.

31	29 Mar.	1.0	42,100	0.92	5.6	42.5	Womer	FCS software functional check; 1g ITB-3 expansion to 20°, 25°, 30° and 35° AOA; in-flight variable gain (80% K2: p/ a) test.
32	29 Mar.	0.9	40,100	0.92	3.3	32.5	Smith	In-flight variable gain (80% K2: p/ a) test; 160 KCAS/WUT ITB-1 reexpansion to 35° AOA.
33	29 Mar.	0.9	39,800	0.87	4.0	44.0	Purifoy	In-flight variable gain (80% K2: p/ a) test; 160 KCAS/WUT ITB –1 reexpansion to 25° AOA.
34	11 Apr.	0.8	40,200	0.92	3.6	31.5	Ishmael	160 KCAS/WUT ITB-2 expansion to 25°AOA; 200 KCAS/WUT ITB-1 expansion to 25° AOA.
35	17 Apr.	0.5	39,900	0.90	5.7	40.5	Smith	160 KCAS/WUT ITB-2 expansion to 30°AOA; 160 KCAS/WUT ITB-3 expansionto 15°, 20°, 25°, and 30° AOA. In-flight abort due to left outboard flap failure light during 30° AOA roll.
36	17 Apr.	0.9	42,600	0.90	6.3	41.8	Womer	Variable gain test (80% & 100% K2:p/ a) 1g ITB-1 expansion to 30° AOA, ITB-2 expansion to 30° AOA; AB engine expansion.
37	17 Apr.	0.7	40,300	0.85	3.2	47.0	Purifoy	AB engine expansion; 200 KCAS/split-S ITB-1 expansion to 30° and 35° AOA & ITB-2 expansion to 20° AOA. RTB due to left & right out-board flap failure lights during 30° AOA Split-S maneuver.

38	27 Apr.	0.8	38,600	0.93	3.4	32.1	Smith	200 KCAS/Split-S ITB-1 expansion to 30° AOA & ITB-2 expansion to 20° and 25° AOA & ITB-3 expansion to 15°, 20°, & 25° AOA.
39	27 Apr.	1.0	40,600	0.91	4.5	49.9	Womer	Airspeed and static pressure.
40	27 Apr.	0.8	40,400	0.92	3.4	32.5	Smith	Variable gain (80% K2:p/ a) 160 KCAS/ITB-2 expansion to 30° AOA & 200 KCAS/ITB-2 expansion to 25° AOA; AB engine exp.
41	9 May	0.8	40,400	0.89	5.0	52.6	Ishmael	1g directional control check at 55° AOA; 200 KCAS/Split-S ITB-1 expansion to 35° AOA; loads clearance.
42	9 May	1.0	40,400	0.81	5.4	50.3	Purifoy	1g directional control check at 55° AOA; 200 KCAS/Split-S ITB-1 expansion to 35° AOA, ITB-2 expansion to 30° AOA & ITB-3 expansion to 30° AOA; variable gain (80% 2:p/ a) 200 KCAS/ITB-2 expansion to 30° AOA; loads clearance.
43	9 May	0.8	40,300	0.93	5.0	67.0	Smith	1g directional control check at 55° AOA; loads clearance; loads expansion; AB & MIL engine expansion.
44	30 May	0.8	41,600	0.90	3.0	19.7	Womer	Vertical tail loads; engine expansion; agility. In-flight abort due to failure of the spin chute continuity test.
45	6 Sept.	0.9	38,600	0.91	3.0	40.4	Purifoy	FCS check; AB engine expansion; engine power/trim effects; military utility/agility; RAV sweeps.
46	6 Sept.	1.0	40,300	0.86	4.4	36.2	Womer	Engine power trim effects; military utility/agility; RAV sweeps.

47	6 Sept.	0.8	39,000	0.87	3.3	38.2	Ishmael	AB engine expansion; military utility/agility.
48	13 Sept.	0.9	35,100	0.87	3.1	44.2	Purifoy	AB engine expansion; military utility/ agility; APU ITB-2 expansion to 15° AOA.
49	18 Sept.	1.0	40,000	0.90	3.3	36.2	Womer	Military utility/agility; APU ITB-2 expansion to 20° AOA.
50	25 Sept.	0.7	40,700	0.90	3.7	36.2	Ishmael	Military utility/agility; APU/160 ICAS ITB-2 expansion to 25° AOA.
51	25 Sept.	1.2	40,700	0.91	3.1	37.1	Purifoy	Military utility/agility; APU/160 KCAS ITB-2 expansion to 30° AOA.
52	25 Sept.	1.0	39,100	0.83	3.3	36.4	Smith	Military utility/agility; APU/160 KCAS ITB-2 expansion to 35° AOA.
53	26 Sept.	0.9	40,500	0.93	4.6	47.3	Womer	Military utility/agility; APU/160 KCAS ITB-2 expansion to 35° AOA.
54	26 Sept.	0.7	38,000	0.89	3.8	37.5	Ishmael	Military utility/agility.
55	9 Nov.	0.3	25,900	0.90	4.4	10.0	Smith	Vertical tail loads data; agility maneuvers; MIMO data. In-flight abort due to loss of SOF parameter in MCR.
56	14 Nov.	1.1	40,900	0.82	3.0	37.0	Purifoy	MIMO data; vertical tail loads data.
57	14 Nov.	1.0	40,300	0.88	3.6	37.3	Smith	Vertical tail loads data; agility maneuvers; MIMO data.
58	11 Dec.	1.0	38,900	0.86	4.4	52.2	Ishmael	FCS software check; low altitude expansion: 1g/20° AOA, 160 KCAS/ITB-2/20° AOA, 200 KCAS/ITB-1/20° AOA.
59	13 Dec.	0.8	40,600	0.88	3.7	52.0	Smith	1g directional control check at 45° and 50° AOA; low altitude expansion: 200 and 250 KCAS/ITB-2/20° AOA.
60	13 Dec.	0.7	41,200	0.91	4.7	53.0	Ishmael	1g directional control check at 45° and 50° AOA; low altitude expansion: 275 KCAS/ITB-2/20° AOA.

61	13 Dec.	0.7	38,500	0.91	4.4	20.5	Smith	Low altitude expansion: 300 KCAS/ITB-2/15° AOA; high altitude expansion: 230 KCAS/ITB-1/20° AOA.
62	18 Dec.	0.8	40,000	0.91	4.0	52.4	Womer	1g directional control check at 45°; high altitude expansion: 230 KCAS/ITB-2/20° AOA; 250 KCAS/ITB-1/20° AOA.
63	18 Dec.	0.8	40,400	0.83	4.7	46.3	Purifoy	1g directional control check at 45°; high altitude expansion: 250 KCAS/ITB-2/20° AOA; 275 KCAS/ITB-2/15° AOA; low altitude aero data: 250 & 275 KCAS/10°, 15°, & 20° AOA.
64	18 Dec	0.9	38,800	0.91	4.7	32.2	Smith	ASE check data at 1g/30°; variable gain functional testing at 1g, 160 & 200 KCAS/10°, 15°, 20° AOA; high altitude aero data: 300 KCAS/10°, & 15° AOA.
65	20 Dec.	1.1	40,700	0.90	3.0	48.0	Purifoy	1g directional control check at 45° AOA with variable gain; MIMO data: SFO practice.
66	20 Dec.	1.0	40,200	0.89	3.0	47.5	Smith	1g directional control check at 45° AOA with variable gain; MIMO data: SFO practice.
67	8 Jan. 1991	0.8	40,300	0.94	3.2	52.6	Womer	FCS software check; low altitude expansion: 1g/ITB-2/40° AOA; 160 KCAS/ITB-1/25° AOA.
68	18 Jan.	0.6	26,300	0.77	3.7	45.0	Ishmael	Low altitude expansion: 1g/ITB-2/35° AOA; 1g/ITB-1/45° AOA; 160 KCAS/ITB-2/35° AOA; 200 KCAS/ITB-1/30° AOA.
69	18 Jan.	0.7	25,700	0.70	3.8	44.0	Purifoy	Low altitude expansion: 200 KCAS/ITB-2/30° AOA; 215 KCAS/ITB-1 and ITB-2/30° AOA; 230 KCAS/ITB-1/30° AOA.

70	23 Jan.	0.4	26,100	0.79	4.7	32.5	Smith	Low altitude expansion: 230 KCAS/ITB-2/30° AOA; 215 KCAS/ITB-2/30° AOA; 250 KCAS/ITB-1/25° AOA.
71	23 Jan.	0.7	44,600	0.92	4.3	31.5	Ishmael	High altitude expansion: 215 KCAS/ITB-1/30° AOA; 230 KCAS/ITB-1/30° AOA; 0.75 Mach/ITB-1/30° AOA; low altitude expansion: 250 KCAS/ITB-2/25° AOA.
72	23 Jan.	0.8	25,500	0.76	3.4	39.0	Smith	Variable gain testing at 1g, 160, & 200 KCAS/10°-35° AOA & 1g/10°-30° AOA.
73	23 Jan.	0.7	25,400	0.73	3.6	39.5	Ishmael	Variable gain testing at 1g/35° AOA, 160 KCAS & 200 KCAS/10° -35° AOA; low altitude aero data: 250 KCAS/25° AOA.
74	25 Jan.	0.8	26,700	0.77	3.4	46.5	Smith	Variable gain testing at 1g/10°-45° AOA; agility: 200 KCAS pitch captures.
75	25 Jan.	0.5	23,500	0.62	3.9	39.0	Ishmael	Variable gain testing at 1g/10°-20° AOA; agility: 200 KCAS pitch captures.
76	25 Jan.	0.7	26,600	0.67	3.8	43.5	Smith	Variable gain testing at 1g/20°, 25° AOA; agility: 200 KCAS pitch and roll captures.
77	25 Jan.	0.5	26,500	0.69	3.4	42.0	Ishmael	Variable gain testing at 1g/30°, 35° AOA; agiliity: 200 KCAS pitch captures and roll transients.
78	25 Jan.	0.5	23,200	0.65	3.8	46.0	Smith	Variable gain testing at 160 KCAS/10°, 15° AOA; agility: 200 KCAS roll captures.
79	7 Feb.	0.7	41,500	0.83	3.5	37.0	Purifoy	Engine functional check; WP AFB agility: 180° aileron rolls; military utility: BFM.
80	7 Feb.	0.9	40,100	0.90	3.7	42.0	Ishmael	WP AFB agility: 180° aileron rolls; slow decel.; theta zoom; abrupt pull-up.
81	7 Feb.	0.8	30,000	0.79	4.0	30.0	Smith	Variable gain testing at 1g/160, 200, 250 KCAS and at 10°, 15°, 20°, 25° AOA/250 KCAS; military utility: BFM.

82	8 Feb.	0.5	34,900	0.83	3.9	43.8	Ishmael	Military utility: BFM with baseline & variable FCS gains.
83	20 Feb.	0.6	25,600	0.76	4.2	41.0	Purifoy	Agiliity:200 KCAS roll captures; military utility: loaded decelerations at 250 & 275 KCAS.
84	20 Feb.	0.9	43,300	0.91	4.4	38.5	Smith	High altitude expansion: 0.75 Mach/ITB-1/25° AOA & ITB-2/15° AOA; military utility loaded decels. at 275 & 300 KCAS.
85	21 Feb.	0.1	6,000	0.50	2.1	8.0	Ishmael	In-flight abort due to FCC inlet temperature and ECS anomalies.
86	24 July	0.7	38,600	0.92	3.5	42.0	Ishmael	Functional flight: ECS system eval; FCS mode switching; 1g/15°, 30° AOA aero checks; smoke point: 1g/35° AOA.
87	25 July	0.9	39,200	0.88	3.1	47.0	Purifoy	Functional flight: 1g/40° & 45° AOA aero checks; smoke point: 1g/40° AOA.
88	6 Aug.	0.9	39,500	0.94	4.6	51.7	Purifoy	Functional flight: 1g/50° AOA directional control check; 1g/15° & 40° AOA aero checks; 0.95 M/23 KFT & 0.60 M/15 KFT ITB-2; 1g, 15° & 25° AOA/200 KCAS aileron rolls.
89	6 Aug.	0.6	27,200	0.67	3.0	49.0	Ishmael	Functional flight: smoke point: 1g/25°-50°-25° AOA sweep; forebody pressure: 1g/15° AOA; variable gain checks: 200 KCAS roll captures @ 1g, 15° & 25° AOA/TW47 & TW53. In-flight abort due to loss of MCR telemetry data.

90	7 Aug.	0.5	33,700	0.80	3.4	42.0	Ishmael	Functional flight: forebody pressure: 1g/40° AOA; variable gain checks: 200 KCAS roll captures @ 15° & 25° AOA/TW53; agility: 30° AOA/200 KCAS/TW47 roll captures; BFM with TW47.
91	7 Aug.	0.6	26,600	0.74	4.8	38.5	Purifoy	Agility: 200 KCAS roll captures @ 30° AOA/TW47 & 25° AOA/TW53; BFM with TW47.
92	7 Aug.	0.4	24,000	0.57	3.1	37.5	Ishmael	Agility: 200 KCAS roll captures @ 25° AOA/TW53; BFM with TW53.
93	16 Aug.	0.8	28,200	0.69	3.0	50.0	Smith	Airdata calibration: 1g decel to 50° AOA; forebody pressures & tufts @ 1g/15°, 20°, 25°, 30°, 35°, 40° & 45° AOA.
94	16 Aug.	0.9	28,500	0.65	3.3	51.0	Smith	Forebody pressures & tufts @ 1g/50° AOA; smoke point: 1g/15°-35° AOA sweep; agility: lat. gross acquisition with TW09, TW47, & TW53.
95	16 Aug.	0.7	26,500	0.66	3.2	46.0	Smith	Smoke point: 1g/15°-35°-15° AOA sweep; agility: 200 KCAS roll captures @ 1g, 15° AOA/TW53; BFM, rolling scissors with TW47.
96	21 Aug.	0.8	27,500	0.60	3.5	53.5	Purifoy	Tuft data: 1g decel. to 50° AOA; 1g/15°, 20°, 25°, 30°, 35° AOA; smoke point: 1g/15°-35°-15° AOA sweep; variable gain testing: 200 KCAS roll captures @ 1g, 15°, 25°, 30° AOA/TW47 & TW09.
97	28 Aug.	0.6	2,400	0.63	3.8	12.0	Ishmael	Airdata calibration: tower fly-bys @ 400, 350, 300, 250, 200 KIAS In-flight abort due to loss of MCR telemetry data.

98	28 Aug.	0.8	35,100	0.91	3.0	52.0	Smith	Airdata calibration: 0.6 Mach decels. to 50° AOA @ 27.5, 20, 35 KFT; 0.5-0.9 Mach accel/decel to 20° AOA @ 27.5, 20, 35 KFT; smoke point: 1g/25° AOA; USAF agility: lat. gross acquisition w/TW09.
99	30 Aug.	0.6	28,300	0.68	3.5	51.5	Ishmael	Forebody pressures; 1g/20 KFT @15°, 20°, 25°, 30°, 35°, 40°, 45°, & 50° AOA; smoke point: 1g/45° AOA.
100	30 Aug.	0.6	41,300	0.84	3.6	51.5	Smith	Forebody pressures; 1g/40 KFT @ 15°, 20°, 25°, 30°, & 50° AOA; smoke point: 1g/20° AOA.
101	30 Aug.	0.5	40,500	0.83	3.9	46.5	Ishmael	Forebody pressures: 1g/40 KFT @ 35°, 40°, & 45° AOA; variable gain testing: 200 KCAS roll captures @ 1g, 15° AOA/TW47 & TW53; 25° AOA/TW47; 30° AOA/TW09.
102	4 Sept.	0.5	22,000	0.62	4.1	42.0	Smith	Forebody pressures @ 0.5 Mach: 40 KFT WUT to 15°, 20°, 25°, 30°, 35°, 40° AOA; 20 KFT WUT 15°, 20°, 25° AOA.
103	4 Sept.	0.4	40,300	0.83	4.2	35.5	Ishmael	Forebody pressures @ 0.5 Mach: 20 KFT WUT to 30° AOA; variable gain testing: 200 KCAS roll captures @ 1g, 15° AOA/TW47 & TW53; 25° AOA/TW47; 30° AOA/TW09.
104	4 Sept.	0.7	22,900	0.67	3.7	38.0	Smith	Variable gain testing: 200 KCAS roll captures @ 1g, 25° AOA/TW09, TW47 & TW53; forebody pressures @ 0.5 Mach: 20KFT WUT to 15°, 20°, 25°, 30° AOA; agility: lat. gross acquisition TW47 & TW53.

105	10 Sept.	1.0	40,300	0.86	2.0	51.5	Ishmael	Forebody pressures & tufts: 1g decel to 50° AOA; 1g/5°-30° AOA (5° intervals).
106	10 Sept.	0.8	27,600	0.68	3.3	37.0	Smith	Forebody pressures: 200 KCAS/40 KFT WUT to 15°, 20°, 25°, 30°, 35° AOA; smoke point: 1g/25° AOA.
107	10 Sept.	1.0	40,100	0.83	4.2	53.7	Ishmael	Forebody pressures & tufts: 1g decel to 50° AOA; 1g @ 5°, 10°, 35° AOA; forebody pressures: 200 KCAS/40 KFT WUT to 40° AOA; smoke point: 1g/15°-35°-15° AOA sweep.
108	13 Sept.	0.9	25,500	0.64	3.0	53.0	Smith	1g directional control check at 25 KFT/52.5° AOA; ITB-1 @ 45° AOA/20 KFT; level flight decel to max AOA in max AB; smoke point: 1g/30° AOA; variable gain testing: 200 KCAS roll captures @ 1g/TW47; BFM, rolling scissors with TW47.
109	13 Sept.	0.9	25,500	0.64	4.1	52.7	Ishmael	1g directional control check at 25 KFT/52.5° AOA; ITB-1 @ 45°, 50° AOA/20 KFT; Smoke point: 1g/15°-35°-15° AOA sweep; agility: 200 KCAS APU to 40° AOA.
110	13 Sept.	0.6	25,900	0.77	3.3	51.5	Dana	High AOA guest pilot evaluation.
111	19 Sept.	0.7	40,200	0.85	3.5	48.0	Smith	Forebody pressures: 1g decel to 50° AOA; smoke point: 1g/35° AOA; BFM rolling scissors with TW47, TW53, TW09.
112	19 Sept.	0.8	26,800	0.68	3.8	52.0	Maj John Rickerson, USAF	High AOA guest pilot evaluation by pilot from Nellis Air Force Base.

113	19 Sept.	0.5	29,000	0.72	4.0	52.0	Ishmael	Forebody pressures: 1g decel to 50° AOA; smoke point: 1g/50° AOA; Langley PID; TW47 PID data @ 1g/15°, 30°, 40° AOA.
114	24 Sept.	0.6	27,800	0.70	3.7	50.0	Smith	TW47 PID data @ 1g/50° AOA; Langley PID; Forebody pressures: 200 KCAS/40 KFT WUT to 20° AOA; 0.50 Mach/20° AOA spiral dive from 27 KFT; smoke point: 0.50 Mach/25° AOA spiral dive from 27 KFT.
115	24 Sept.	0.7	26,500	0.70	3.7	50.5	Purifoy	TW09 PID data @ 1g/50° AOA; smoke point: 1g/40° AOA; BFM, rolling scissors with TW47, TW53: SFO.
116	24 Sept.	0.7	26,100	0.71	3.4	50.0	Schneider	High AOA guest pilot evaluation.
117	26 Sept.	0.9	31,000	0.74	3.6	50.5	Ishmael	Smoke point: 1g/15°-35°-15° AOA sweep; forebody pressures: 0.50 Mach/30° AOA spiral dive from 27 KFT; Langley PID; TW35 PID data @ 1g/45° AOA; roll performance eval. in TW47 @ 40° AOA; airdata calibration: tower fly-bys 350, 300, 250, 200 KIAS.
118	26 Sept.	0.9	27,900	0.73	3.2	68.4	Purifoy	Smoke point: 1g/52.5° AOA; TW32 & TW35 PID data @ 1g/45°, 50° AOA; roll perform. Eval. in TW47 & TW53 @ 40°, 45° AOA; air data calibration: tower fly-bys @ 400, 200 KIAS.
119	26 Sept.	0.9	20,800	0.66	3.7	42.0	Ishmael	Canard streamers @ 1g/15°-40° AOA;handling qualities in TW09, TW47 & TW53.

120	30 Sept.	0.9	30,200	0.76	3.6	52.0	Smith	Smoke point: 1g/25°-50°-25° AOA sweep; Langely PID; MIMO/RAV data @ 0.70 Mach/30 KFT; roll performance eval. in TW47 @ 30°, 35°, 40°, & 45° AOA.

Total Flight Time: 96.2 Hrs.

Vortex Flow Control Flights

In 1992 the U.S. Air Force initiated a program to study the use of vortex flow control (VFC) as a means of providing increased aircraft control at high angles of attack when the normal flight control systems are ineffective. The No. 2 aircraft was modified with the installation of two high-pressure nitrogen tanks and control valves with two small nozzle jets located on the forward upper portion of the nose. VFC was more effective than expected in generating yaw (left-to-right) forces, especially at higher angles of attack where the rudder loses effectiveness. VFC was less successful in providing control when sideslip (relative wind pushing on the side of the aircraft) was present, and it did little to decrease any rocking oscillation of the aircraft.

Flight	Date	Pilot	Duration (hours)	Purpose and Comments
121	12 May 1992	Ishmael	0.5	FCF; nose gear remained down during gear cycle.
122	15 May	Maj. Regis Hancock, Air Force Flight Test Center	0.9	PF; pilot's first X-29 flight.
123	27 May	Ishmael	0.4	First VFC flight, medium nozzles.
124	27 May	Hancock	0.9	VFC.
125	29 May	Fullerton	0.9	VFC; pilot's first high AOA flight in X-29.
126	29 May	Hancock	0.7	VFC.
127	3 June	Ishmael	0.3	VFC.
128	3 June	Fullerton	0.7	VFC.
129	3 June	Hancock	0.6	VFC.
130	10 June	Ishmael	0.5	VFC.
131	10 June	Fullerton	0.6	VFC.

132	10 June	Ishmael	0.5	VFC.
133	10 June	Fullerton	0.4	VFC.
134	12 June	Hancock	0.6	VFC.
135	17 June	Ishmael	0.5	VFC.
136	17 June	Fullerton	0.6	VFC.
137	17 June	Hancock	0.8	VFC.
138	17 June	Ishmael	0.5	VFC.
139	24 June	Fullerton	0.6	VFC.
140	24 June	Hancock	0.7	VFC; downmode to AR on takeoff.
141	24 June	Fullerton	0.7	VFC.
142	24 June	Hancock	0.5	VFC.
143	24 June	Fullerton	0.6	VFC.
144	1 July	Ishmael	0.3	VFC; first flight with large nozzles.
145	1 July	Hancock	0.3	VFC.
146	1 July	Fullerton	0.3	VFC.
147	1 July	Hancock	0.3	VFC.
148	1 July	Fullerton	0.4	VFC.
149	1 July	Hancock	0.4	VFC.
150	10 July	Ishmael	0.3	VFC.
151	10 July	Fullerton	0.4	VFC.
152	10 July	Hancock	0.4	VFC.
153	10 July	Ishmael	0.4	VFC.
154	10 July	Fullerton	0.4	VFC.
155	10 July	Hancock	0.5	VFC.
156	15 July	Ishmael	0.4	VFC.
157	15 July	Fullerton	0.5	VFC.
158	15 July	Hancock	0.6	VFC.
159	15 July	Ishmael	0.4	VFC; departure to 68 degrees true AOA.
160	24 July	Fullerton	0.9	VFC; first flight with small nozzles.
161	24 July	Hancock	0.8	VFC.

162	24 July	Ishmael	0.5	VFC.
163	24 July	Fullerton	0.6	VFC.
164	12 Aug.	Fullerton	0.7	VFC; first flight with one modified regulator and medium nozzles installed.
165	12 Aug.	Smith	0.6	GP.
166	12 Aug.	Fullerton	0.7	VFC.
167	12 Aug.	Smith	0.5	GP.
168	19 Aug.	Hancock	0.4	VFC; first flight with both regulators modified and with non-slotted nozzles.
169	19 Aug.	Ishmael	0.4	VFC.
170	26 Aug.	Fullerton	0.4	VFC.
171	26 Aug.	Hancock	0.5	VFC.
172	26 Aug.	Ishmael	0.5	VFC.
173	26 Aug.	Fullerton	0.3	VFC.
174	26 Aug.	Hancock	0.3	VFC.
175	26 Aug.	Ishmael	0.5	VFC.
176	26 Aug.	Fullerton	0.8	VFC; medium nozzles.
177	26 Aug.	Hancock	0.7	VFC.
178	26 Aug.	Fullerton	0.5	VFC.
179	26 Aug.	Hancock	0.6	VFC.
180	26 Aug.	Ishmael	0.7	VFC; control room fly-by.
181	14 Oct.	Hancock	0.2	Practice for Edwards open house.
182	18 Oct.	Hancock	0.1	Edwards Open House.

Sources: Flight logs compiled by the X-29 program supplemented by available flight reports and the flight logs kept in the Pilots Office.

X-29A Test Program Acronyms

AA, A/A	Air-to-air
AB, A/B	Afterburner
A/C	Aircraft
AC	Aerocharacterization, Flow Visualization
ACC	Automatic Camber Control
ADI	Attitude Direction Indicator
AEROCHARAC	Aerodynamic Characteristics
AG, A/G	Air-to-Ground
AHRS	Attitude Heading Reference System
Alpha, AOA	Angle-of-Attack
APU	Abrupt Pull-Up
AR	Analog Reversion Mode
AR/PA	Analog Reversion/Powered Approach
AR/UA	Analog Reversion/Up and Away
ASE	Aeroservoelastic/Aeroservoelasticity
ATLAS	Adaptable Target Lighting Array System
Beta	Angle of sideslip
BFF	Body Freedom Flutter
BFM	Basic Fighter Maneuver
BLK-VIII-AC	Refers to a software version number or "block" number
CAL	Airdata, INS Calibration
"Dgr normal"	Degraded Normal Mode
DR	Digital Reversion Mode
ECS	Environmental Control System
EE	Envelope Expansion
EPU	Emergency Power Unit
FCC	Flight Control Computer
FCF	Functional Check Flight
FCS	Flight Control System
FDMS	Flight Deflection Measurement System
FL	Flight Level
FLT CON	Flight Condition
FQ	Flying Qualities
G	Acceleration equal to the force of gravity at sea level
GP	Guest Pilot
GW/CG	Gross Weight/Center-of-Gravity
HQ	Handling Qualities
HQDT	Handling Qualities During Tracking
INS	Inertial Navigation System
ITB-1, 2, 3	Integrated Test Block-1, 2, or 3
JFS	Jet Fuel Starter
K2: p/da	Roll Rate Gain

K-27	Lateral Stick Gain
KCAS	Calibrated Airspeed, in Knots
KFT	Altitude in 1,000s of feet
KIAS	Indicated Air Speed, in Knots
LAT/DIR	Lateral/Directional
LONG STICK	Longitudinal Stick
MCC	Manual Camber Control
MCR	Mission Control Room
MIL	Millitary (Power Setting)
MIMO	Minimum Input/Minimum Output
MU	Military Utility
ND	Normal Digital Mode
Nz	Acceleration in Z-axis (i.e. "g" or normal acceleration)
PAC	Power Approach Control Mode
PF	Pilot Familiarization
PID	Parameter Identification
PTO	Power Take-Off Shaft
RAV	Remotely Augmented Vehicle
RTB	Return to Base
S&C	Stability and Control
SFO	Simulated Flame-Out
SOF	Safety of Flight, e.g., flight system hydraulic pressure data relating thereto
T/M, TM	Telemetry
TW09, TW47, TW53	Thumb-wheel settings
USAF	United States Air Force
VFC	Vortex Flow Control
VMAX	Maximum Velocity
VT	Vertical Tail Strain Gauge Data
WP AFB	Wright-Patterson Air Force Base
WUT	Wind-Up Turn

Appendix X
X-31 Flight Chronology

Two X-31 Enhanced Fighter Maneuverability (EFM) demonstrators flew at the NASA Dryden Flight Research Center, Edwards, CA, and at Palmdale, CA, to obtain data that may apply to the design of highly maneuverable next-generation fighters. The program ended in June 1995. The X-31 program demonstrated the value of thrust vectoring (directing engine exhaust flow) coupled with advanced flight control systems to provide controlled flight during close-in air combat at very high angles of attack.

X-31 #1 (Bu. No. 164584) Flight Chronology

Flight No.	Date	Pilot	Purpose	Comments
1-001	10-11-90	Ken Dyson, Rockwell		First flight, 38 minutes long, 340 miles per hour, 10,000 feet. Flown from Palmdale.
1-002	10-17-90	Dyson		ADC failures.
1-003	11-03-90	Dyson		Air intake disconnected; Rt. WoW Sw. during landing.
1-004	11-06-90	Dietrich Seeck, MBB		Rt. WoW Sw. during landing.
1-005	11-08-90	Seeck		ADC failure.
1-006	11-10-90	Seeck		ADC failure.
1-007	11-13-90	Dyson		ADC failure.
1-008	11-15-90	Fred Knox, Rockwell		
1-009	11-21-90	Knox		CCDL failure.
1-010	02-14-91	Dyson		
1-011	02-15-91	Seeck		
1-012	02-20-91	Knox		
1-013	02-27-91	Dyson		Flap switch discrete, FCS reset discrete.
1-014	03-12-91	Seeck	R2, R3 modes FQ.	
1-015	03-15-91	Karl Lang, WTD-61		
1-016	03-28-91	Knox	Line flutter.	LFTVS flag set.
1-017	03-29-91	Dyson	Line flutter.	FTI clock failure.
1-018	04-03-91	Dyson	Line flutter.	FTI clock failure.
1-019	04-05-91	Lang		
1-020	04-12-91	Seeck	Loads.	
1-021	04-18-91	Knox	Line flutter	N_y-TB failure, resonance during TEO asym. flutter.
1-022	04-19-91	Dyson	Line flutter.	Same as flight 1-021.

1-023	04-23-91	Lang	Line flutter.	Same as flight 1-021.
1-024	04-24-91	Bob Trom-badore USMC	GPE	
1-025	04-30-91	Karl-Heinz Mai, GAF	GPE	
1-026	05-02-91	Mai	GPE	FID during landing. Pilot did not turn on tape for landing.
1-027	05-03-91	Trom-badore	GPE	CCDL message timeout.
1-028	05-31-91	Seeck	Plume line.	
1-029	06-04-91	Knox	Line flutter 6.2 & 9.7 kft.	RTB because of ECS problem.
1-030	06-06-91	Dyson	Line flutter 12.8 & 15.8 kft.	FID during landing (Spdbrk, Rt. WOW) LTEI code during engine start.
1-031	06-07-91	Lang	Line & R1 mode.	Alpha oscillation in R1.
1-032	06-12-91	Dyson	TV calib. & FQ.	
1-033	06-14-91	Knox	TV calib.	
1-034	06-18-91	Seeck	FQ, R1.	TEF failures during R1.
1-035	06-20-91	Dyson	Line flutter at 4.7 kft., 0.75 M.	Pitch stick looseness prior to flight.
1-036	06-21-91	Knox	Line flutter at 4.7 kft., 0.75 M.	TEF failures during engine start, PFB exit problem.
1-037	06-21-91	Dyson	Line flutter at 4.7 kft., 0.75 M.	
1-038	07-12-91	Knox	Line flutter 4.7 kft., 0.80 M.	B failure during TEO sym. flutter sweep. Landed at Edwards in R2.
1-039	07-14-91	Dyson	Ferry flight.	Edwards to Palmdale.
1-040	07-16-91	Seeck	Flutter 4.7 & 7.2 kft.	
1-041	07-17-91	Lang	800q. Flutter 7.2 & 10 kft.	
1-042	07-18-91	Knox	Flutter 10 kft.	Main generator failed during WUT. Landed at Edwards.
1-043	07-24-91	Dyson	TVV calibration.	Takeoff from Edwards. ECS buzz at 35 kft.
1-044	07-26-91	Seeck	Loads.	
1-045	07-30-91	Knox	Loads.	
1-046	08-14-91	Dyson	Plume tracking.	
1-047	08-14-91	Seeck	Plume tracking	
1-048	08-16-91	Lang	25° AoA.	
1-049	08-16-91	Knox	25° AoA.	
1-050	08-26-91	Knox	30° AoA.	Chute test aborted, no photo plane. Right WoW FID during landing.
1-051	08-27-91	Knox	Chute test.	Chute test successful.

1-052	09-04-91	Lang	C.L. TV, 20 kft.	TV closed-loop 1st time. Generator failed, reset after 2.5 min.; short in wiring.
1-053	10-18-91	Dyson	Loads.	
1-054	10-18-91	Seeck	Loads.	Left WoW during landing. Engine caution due to low L.O. pressure.
1-055	10-24-91	Knox	Loads.	Rudder & canard solenoid BIT failed twice.
1-056	10-24-91	Lang	Loads.	Left WoW during landing; no FTI data for landing.
1-057	10-28-91	Seeck	Loads, R1.	Right WoW during landing. Low L.O. Press 4 times, LFINS once.
1-058	10-31-91	Knox	R1, FQ maneuvers.	Canard solenoid fail during BIT.
1-059	10-31-91	Lang	Maneuvers.	
1-060	11-18-91	Dyson	Loads.	T/O bird strike, engine caution during negative Gs several times.
1-061	11-20-91	Comdr. Al Groves, USN	GPE.	
1-062	11-20-91	Lang	GPE.	
1-063	11-21-91	Groves	GPE.	N_y failure after landing.
1-064	11-25-91	Groves	GPE.	
1-065	11-25-91	Knox	GPE.	TV3 failure during 'both' decel.
1-066	11-26-91	Seeck	Loads.	Engine caution during negative Gs.
1-067	12-06-91	Lang	TV effectiveness.	
1-068	01-20-92	Seeck	Ferry	To Edwards AFB. Fix-point overflow, SW fix.
1-069	07-02-92	Knox	PST, R-modes.	1st A/C's 1st ITO flight, from Dryden.
1-070	07-02-92	Wisneski	PST.	Lt. Col. Jim Wisneski, USAF pilot.
1-071	07-07-92	Seeck.	PST.	
1-072	07-16-92	Groves	Decel., inverted.	
1-073	07-16-92	Lang	PST.	Hyd. pres. failure during engine start.
1-074	07-16-92	Groves	Decel., inverted.	
1-075	09-10-92	Knox	40° PST.	
1-076	09-10-92	Smith	45° PST.	Rogers E. Smith, NASA DFRC pilot.
1-077	09-11-92	Wisneski	50° PST.	
1-078	09-11-92	Groves	55° PST.	INU bus error. Av/air hot lite. Landed in R2 mode.
1-079	09-16-92	Lang	60° PST	TV3 zero-detect failure after landing, beta oscillation at 60°.
1-080	09-16-92	Smith	62° PST	4 Hz oscillation in beta above 60°.
1-081	09-18-92	Groves	65°, 70° PST, 35 kft.	TV3 zero-detect failure after landing.
1-082	09-18-92	Lang	65°, 70° PST, 35 kft.	

1-083	09-18-92	Smith	55°, 70° PST, 35 kft.	TV3 zero-detect failure after landing.
1-084	09-22-92	Seeck	70°, 35 kft. Loads.	
1-085	09-22-92	Knox	70°, 35 kft. Loads.	
1-086	09-22-92	Wisneski	70°, 35 kft. Loads.	Right WoW FID after touchdown.
1-087	11-24-92	Lang	70° AoA, 360° rolls.	Brake failed during landing.
1-088	12-01-92	Groves	Loads.	
1-089	12-01-92	Smith	Loads.	
1-090	12-01-92	Seeck	Loads.	
1-091	12-08-92	Knox	Loads.	
1-092	12-08-92	Seeck	Loads.	
1-093	12-10-92	Lang	Loads.	
1-094	12-15-92	Wisneski	Pitch pulls, 23 kft.	
1-095	12-15-92	Knox	WUTs 30 kft.	
1-096	12-15-92	Lang	WUTs 30 kft.	
1-097	01-12-93	Knox	20 kft. Turns, alter. Cards.	1st flight with forebody strakes & new LEF schedule, 90% cloud cover.
1-098	01-14-93	Lang	35 kft. PST 40° AoA.	
1-099	01-14-93	Smith	35 kft. PST 50° AoA.	
1-100	01-14-93	Knox	35 kft. PST 60° AoA.	
1-101	01-21-93	Lang	35 kft. PST 60° AoA.	
1-102	02-03-93	Wisneski	30 kft. PST 70° AoA.	
1-103	02-03-93	Smith	30 kft. PST 70° AoA.	
1-104	02-03-93	Knox	30 kft. PST 70° AoA.	
1-105	02-09-93	Lang	30 kft. PST 70° AoA.	
1-106	02-09-93	Wisneski	30 kft. PST Mach 0.4.	Elevated entry.
1-107	02-11-93	Smith	30 kft. PST Mach 0.5.	N_Y TB failure during pull-up.
1-108	02-11-93	Knox	30 kft. PST Mach 0.5.	
1-109	02-11-93	Lang	30 kft. PST Mach 0.5.	
1-110	02-25-93	Groves	30 kft. PST Mach 0.5.	1120 FID on 02-23-93. INU still aligning when entering PST.
1-111	02-25-93	Lang	30 kft. PST Mach 0.5.	
1-112	02-25-93	Knox	30 kft. PST Mach 0.5.	
1-113	02-25-93	Groves	30 kft. PST Mach 0.5.	
1-114	03-17-93	Groves	30 kft. PST Mach 0.5.	
1-115	03-17-93	Wisneski	30 kft. PST Mach 0.5.	
1-116	03-23-93	Knox		
1-117	03-23-93	Lang		
1-118	03-23-93	Groves		

1-119	03-23-93	Groves		
1-120	06-29-93	Lang	30 kft., 0.8 M turns.	RTB because of fuel pres. Sensor. 1st use of HUD & over-the-shoulder videos.
1-121	06-29-93	Smith	30 kft., 0.5 Mach.	AIL SA failure during ground test.
1-122	06-29-93	Lang	23 kft., 720° rolls.	
1-123	06-29-93	Smith	30 kft., 0.6 M.	LR1RQ abrupt pull-up.
1-124	07-08-93	Knox	30 kft., 0.6M.	RTB, T/R caution lamp & FTI power problem.
1-125	07-13-93	Knox	28 kft., 225 Kt.	
1-126	07-13-93	Smith	20 kft., 225 Kt.	
1-127	07-13-93	Knox	28 kft., 225 Kt.	
1-128	07-14-93	Smith	16 kft., 185 Kt.	
1-129	07-14-93	Groves	16 kft., 185 Kt.	
1-130	07-14-93	Knox	16 kft., 185 Kt.	
1-131	07-20-93	Quirin Kim, GAF	Pilot checkout.	RTB because of noisy N2S signal at takeoff.
1-132	07-23-93	Lang	15 kft., 185 Kt.	
1-133	07-27-93	Wisneski	15 kft., 185 Kt.	RTB because of high gen. case temp.
1-134	07-27-93	Knox	Grit effect.	Grit removed from radome.
1-135	07-29-93	Kim	Pilot checkout.	TVVs engaged during R3 because FEST negative.
1-136	08-04-93	Smith	15 kft., 225 kts.	Overlay with lower limit on FEST because FEST negative.
1-137	08-04-93	Lang	15 kft., 225 kts.	INU trip.
1-138	08-04-93	Wisneski	15 kft., 225 kts.	
1-139	08-04-93	Knox	15 kft., 225 kts.	INU trip.
1-140	08-17-93	Lang	15 kft., 225 kts.	
1-141	08-17-93	Wisneski	15 kft., 225 kts.	
1-142	08-17-93	Knox	15 kft., 225 kts.	
1-143	08-19-93	Knox	15 kft., 225 kts.	
1-144	08-19-93	Kim	15 kft., 225 kts.	
1-145	08-19-93	Lang	15 kft., 225 kts.	
1-146	08-23-93	Wisneski	15 kft., 225 kts.	
1-147	08-24-93	Knox	15 kft., 225 kts.	
1-148	08-24-93	Groves		RTB for excessive engine mount vibration.
1-149	08-26-93	Groves	Tactical.	
1-150	08-26-93	Smith	Tactical.	
1-151	08-26-93	Kim	Tactical.	
1-152	08-26-93	Lang	Tactical.	
1-153	08-31-93	Wisneski	21 kft., 185 kts.	Backup PST.

1-154	08-31-93	Smith	30 kft., 225 kts.	
1-155	08-31-93	Kim		1st basic fighter maneuver (BFM) tests.
1-156	08-31-93	Lang	21 kft., 225 kts.	FIDs during power-up.
1-157	09-14-93	Knox	15 kft., 225 kts.	
1-158	09-16-93	Groves	BFM.	
1-159	09-16-93	Wisneski	BFM.	
1-160	09-16-93	Smith	BFM.	
1-161	09-16-93	Lang	BFM.	
1-162	09-20-93	Lang	BFM.	Offensive.
1-163	09-20-93	Knox	BFM.	HSLA.
1-164	09-21-93	Groves	BFM.	HSLA, LSLA, defensive.
1-165	09-21-93	Wisneski	BFM.	HSLA, defensive.
1-166	09-28-93	Groves	BFM.	RADM Mixon in F-18.
1-167	09-28-93	Smith	BFM.	
1-168	09-28-93	Kim	BFM.	
1-169	09-30-93	Lang	BFM.	
1-170	09-30-93	Knox	BFM.	
1-171	09-30-93	Wisneski	BFM.	
1-172	09-30-93	Kim	BFM.	
1-173	10-14-93	Kim	BFM.	
1-174	11-05-93	Kim	CIC.	FCC 4 C/B problem during pre-taxi.
1-175	11-05-93	Groves	CIC.	
1-176	11-05-93	Kim	CIC.	
1-177	11-05-93	Groves	CIC.	
1-178	11-09-93	Kim	CIC.	
1-179	11-09-93	Groves	CIC.	
1-180	11-09-93	Kim	CIC.	
1-181	11-09-93	Groves	CIC.	Left WoW on touchdown.
1-182	11-19-93	Smith	Flutter M 0.9, 0.95.	RLE solenoid fail during PFB.
1-183	11-19-93	Knox	Flutter.	Abort because of beta oscillation at M 0.96. Rt. WoW on touchdown.
1-184	11-24-93	Lang	Flutter M 1.05, 0.90, 0.95.	1st supersonic flight. FID during power-up.
1-185	11-24-93	Wisneski	Flutter M 1.10.	
1-186	11-29-93	Knox	Flutter M 1.15.	
1-187	11-29-93	Lang	Flutter M 1.20.	FID during power-up.
1-188	12-01-93	Wisneski	Flutter M 1.25.	
1-189	12-02-93	Knox	Flutter M 1.25.	
1-190	12-02-93	Lang	Flutter M 1.25.	

1-191	12-07-93	Smith	CIC.	
1-192	12-07-93	Wisneski	CIC.	FID during power-up.
1-193	12-09-93	Smith	CIC.	
1-194	12-09-93	Wisneski	CIC.	Rt. WoW FID during landing.
1-195	12-10-93	Smith	CIC.	
1-196	12-10-93	Wisneski	CIC.	Generator overtemp during pretaxi for flight 1-197.
1-197	12-14-93	Smith	CIC.	
1-198	12-14-93	Wisneski	CIC.	
1-199	12-14-93	Wisneski	CIC.	
1-200	02-22-94	Lang	FCF, CIC.	FID during power-up.
1-201	02-23-94	Hess	Familiarization.	Air Force guest pilot Derek Hess.
1-202	02-24-94	Schmidt	Familiarization.	Navy guest pilot LCDR Steve Schmidt.
1-203	02-24-94	Hess	Familiarization.	
1-204	02-24-94	Schmidt	Familiarization.	
1-205	02-24-94	Hess	Familiarization.	
1-206	02-28-94	Schmidt	Familiarization.	
1-207	03-01-94	Hess	CICP.	High generator temp. RTB.
1-208	03-02-94	Hess	CIC.	
1-209	03-03-94	Schmidt	CIC.	
1-210	03-03-94	Schmidt	CIC.	
1-211	03-03-94	Hess	CIC.	
1-212	03-03-94	Schmidt	CIC.	
1-213	03-30-94	Kim	CIC.	
1-214	03-10-94	Smith	Quasi-tailless.	0% QT Demo. Right LEF fail. PFB once.
1-215	03-10-94	Lang	Quasi-tailless.	30% Stab. 20% de-stable.
1-216	03-10-94	Smith	Quasi-tailless.	40% Stab. 20% de-stable.
1-217	03-17-94	Smith	Quasi-tailless.	50, 60 & 70% Stabilized.
1-218	03-28-94	Lang	Env. exp.	30 kft. 265 kt. PST entry.
1-219	03-28-94	Groves	Env. exp., CIC.	20 kft. mil. pwr. PST entry.
1-220	03-31-94	Schneider	CICP.	Edward T. "Ed" Schneider, NASA DFRC pilot.
1-221	03-31-94	Knox	Env. exp.	265 kt & mil power PST entry.
1-222	03-31-94	Schneider	CICP.	
1-223	03-31-94	Kim	CICP	
1-224	04-07-94	Knox	Env. exp.	Mil. pwr. entry, 15 kft. 185 & 225 kts. Right LEF failed twice during PFB.
1-225	04-12-94	Kim	CICP.	Pt. Mugu F-18 adversary pilot.
1-226	04-12-94	Smith	CICP.	Pt. Mugu F-14 adversary pilot.
1-227	04-12-94	Kim	CICP.	Pt. Mugu F-18 adversary pilot.

1-228	04-12-94	Smith	CICP.	Pt. Mugu F-14 adversary pilot.
1-229	04-13-94	Knox	CICP.	Pt. Mugu F-14 adversary pilot. Failed right LEF deflection during PFB.
1-230	04-13-94	Kim	CICP	Pt. Mugu F-14 adversary pilot.
1-231	04-21-94	Smith	Env. exp.	265 kt. PST entry, 20 kft., 30 kft., 15 kft.
1-232	04-21-94	Groves	Env. exp.	265 kt. PST entry, 20 kft., 30 kft., 15 kft. Failed left LEF PFB once.
1-233	04-21-94	Lang	Env. exp.	265 kt. PST entry, 20 kft., 30 kft., 15 kft. Started engine during PFB, hyd. depres.
1-234	05-10-94	Lang	CIC.	119A retest & 265 kt. PST entry CIC. Generator off during start.
1-235	05-10-94	Kim	CIC.	Generator off during start.
1-236	05-10-94	Lang	CIC.	Generator off during start.
1-237	05-10-94	Kim	CIC.	Generator off during start.
CST	06-27-94		Engine run.	Production engine installed.
1-238	08-02-94	Smith	Warfight demo.	119C retest.
1-239	08-02-94	Lang	STEMS.	Gross acquisition.
1-240	08-04-94	Smith	STEMS.	Fine tracking.
1-241	08-04-94	Lang	STEMS.	Fine tracking.
1-242	08-09-94	Smith	CIC.	30° OBS missile.
1-243	08-09-94	Smith	CIC.	30° OBS missile.
1-244	08-09-94	Lang	CIC.	30° OBS missile.
1-245	08-09-94	Lang	CIC.	Nose gear light on handle did not come on during landing. 4 burnt out bulbs.
1-246	08-18-94	Smith	STEMS.	Fine tracking.
1-247	08-18-94	Lang	STEMS.	Fine tracking. Trigger did not activate FTB, so STEMS was performed instead.
1-248	08-18-94	Smith	STEMS.	Fine tracking.
1-249	08-30-94	Knox	FTB PID.	AIL servoamp failure during landing. FTB with DLR card, ICMD'RAMERR.
1-250	08-30-94	C. J. Loria	Familiarization.	USMC pilot "Gus" Loria's first flight.
1-251	08-30-94	Lang	FTB PID.	FTB with DLR card.
1-252	08-30-94	Loria	Familiarization.	
1-253	09-06-94	Knox	FTB PID.	FTB with DLR card. CAUT/WRN C/B popped during PA.
1-254	09-13-94	Lang	CICP.	Against F-15 from Nellis.
1-255	09-13-94	Smith	CICP.	Against F-16 from Nellis.
1-256	09-13-94	Kim	CICP.	Against F-15 from Nellis.

1-257	09-14-94	Lang	CIC.	Against F-16 from Nellis.
1-258	09-14-94	Smith	CIC.	Against F-15 from Nellis.
1-259	09-14-94	Kim	CIC.	Against F-16 from Nellis.
1-260	09-21-94	Lang	TV calibration.	Low power calibration for QT.
1-261	10-06-94	Kim	STEMS.	Fine tracking.
1-262	10-06-94	Lang	STEMS.	Fine tracking.
1-263	10-13-94	Loria	STEMS.	Fine tracking.
1-264	10-13-94	Smith	STEMS.	Fine tracking.
1-265	10-13-94	Loria	STEMS.	Fine tracking.
1-266	10-17-94	Kim	CIC.	Demo. for Lt. Gen. John (GMD).
1-267	10-17-94	Smith	STEMS.	Fine tracking.
1-268	10-18-94	Knox	STEMS.	HSLA for Congressman "Buck" McKeon.
1-269	12-13-94	Knox	QT 4 kft., 170 kt.	500th flight, 1st flight with TVV actuator mod. for less damped bypass.
1-270	12-13-94	Smith	QT 4 kft., 220 kt.	
1-271	12-13-94	Lang	QT 4 kft., 56° PLA.	
1-272	12-13-94	Knox	Precision approach.	ATLAS.
1-273	12-15-94	Smith	QT low altitude.	No altitude restriction for TV & QT.
1-274	12-15-94	Loria	QT low altitude.	
1-275	12-15-94	Knox	QT, precision approach.	
1-276	12-15-94	Loria	QT, precision approach.	
1-277	12-20-94	Loria	QT, precision approach.	
1-278	12-20-94	Knox	QT, precision approach.	Tech. bumped restore switch.
1-279	12-20-94	Loria	QT, precision approach.	
1-280	12-20-94	Smith	QT cruise.	F-18 chase low oil pres. emergency land.
1-281	12-22-94	Lang	QT cruise.	Beta fail during QT roll 425 KCAS, 8 kft Landing in R-2.
1-282	12-22-94	Kim	Air-to-Ground (A/G).	Replaced nose boom.
1-283	12-22-94	Lang	QT cruise.	Speedbrake close switch broken. Failed right LEF twice during PFB.
1-284	01-06-95	Smith	QT.	
1-285	01-06-95	Lang	QT.	
1-286	01-18-95	Lang	QT A/G.	
1-287	01-18-95	Loria	QT A/G.	
1-288	01-18-95	Kim	QT A/G.	
1-289	01-18-95	Lang	PID.	FTB with 120b software thresholds.
1-290	01-19-95	Loria	QT A/G.	
1-291	01-19-95	Kim	QT A/G.	

1-292	01-19-95	Lang	PID.	X-31 crashed. Pilot ejected. LTEI failure prior to ejection.

X-31 #2 (Bu.No. 164585)

Flight No.	Date	Pilot	Purpose	Comments
Taxi	12-23-90	Seeck	Shakedown.	Q_s & P_s sensors wired backwards. Divergent oscillation in pitch axis.
Taxi	01-14-91	Seeck	Flight readiness evaluation.	Pilot inadvertently shut down engine while attempting ground idle.
2-001	01-19-91	Seeck	Airspeed calib., subsys. check, FQ.	From Palmdale.
2-002	01-22-91	Seeck		Discovered canard actuator failure after flight.
2-003	01-23-91	Knox		
2-004	06-11-91	Seeck	Subsys. & FQ checks.	Actuators did not fade during engine start. Other problems.
2-005	06-25-91	Seeck	FQ & TV calibrations	Same problems during engine start. Blew left tire on landing, rudder fail IDs.
2-006	07-10-91	Lang	FQ	FCS reset button.
2-007	07-26-91	Lang	R-modes.	Slight canard buzz in R-1.
2-008	08-21-91	Dyson	FQ 30° AoA.	Problem with clock. Recovered data for 12.5 Hz only.
2-009	08-23-91	---	FQ 30° AoA.	No record of pilot.
2-010	08-23-91	Knox	Flight aborted.	Bad TM.
2-011	08-28-91	Seeck	30° AOA, 40 kft.	
2-012	08-28-91	Dyson	ADC calibration.	TV2 pressure too high, plume tracking canceled. Pacer points instead.
2-013	09-05-91	Knox	Plume line, 11K & 20 kft.	
2-014	09-06-91	Seeck	TV closed-loop, 20 kft.	TV closed-loop 1st time on A/C2. F-15 chase aircraft (Dyson) landed at China Lake.
2-015	09-11-91	Dyson	TV closed-loop.	F-8 (Lang) landed at Edwards. Anti-skid problem during T/O.
2-016	09-11-91	Lang	TV closed-loop to 30°.	30° FQ 1st time. R3 Hi/Lo switch FID. T-38 chase.
2-017	09-13-91	Seeck	TV closed-loop to 30°.	TV DIS switch failure when TV engaged.
2-018	09-16-91	Knox	TV closed-loop.	
2-019	09-16-91	Dyson	TV closed-loop.	R1 request during right full stick roll three times.
2-020	09-19-91	Seeck	FQ & PID with FTB.	
2-021	09-19-91	Knox	FQ, PID & TV calib.	Vane failures during TV calib., vane commanded to +27°. (+25.6° act. stop).

2-022	09-19-91	Dyson	FQ, FID.	Right roll during landing.
2-023	09-24-91	Lang	FQ, 40 kft., TV on/off.	
2-024	09-24-91	Seeck	PID, TV calibration.	TV1 failure @ 20 kft.OK at 30 & 40 kft. Rt. WoW switch failure during landing.
2-025	09-26-91	Knox	R3 FQ, tracking.	
2-026	09-26-91	Dyson	R2 FQ.	CCDL failure, reset during flight left LEF brake was set.
2-027	10-01-91	Lang	FQ, R2.	R1 request during right full stick roll.
2-028	10-01-91	Seeck	PID, tracking.	
2-029	10-04-91	Knox	Ferry from Palmdale.	Static display at EAFB.
2-030	10-08-91	Lang	Ferry.	Return from EAFB, flight OK. TM room problem.
2-031	11-11-91	Knox	FQ.	WoW failure during landing.
2-032	11-11-91	Dyson	FQ, R3.	N_y failures during WUT, LEF brake set, FCC3 invalid. Left WoW. Bird strike.
2-033	11-13-91	Lang	FQ.	
2-034	11-15-91	Knox	Loads.	
2-035	11-19-91	Lang	Loads.	N_y failure during WUT to 30°.
2-036	11-19-91	Knox	PST.	First PST flight. Command output failure at 40° AoA.
2-037	11-22-91	Seeck	PST.	Porpoising at 45° AoA.
2-038	11-22-91	Dyson	PST.	N_y during WUT. LEF brakes set. FCC 3 invalid 160 msec during WUT.
2-039	12-11-91	Groves	GPE.	
2-040	12-11-91	Dyson	PST.	
2-041	12-13-91	Groves	GPE.	Left WoW on landing.
2-042	01-20-92	Knox	Ferry & PST.	To Edwards AFB. TEF & N_y failures, aero buffet @ 39° AoA.
2-043	04-23-92	Lang	Sys. C/O, FQ, R1, R2, R3.	First ITO flight (from Dryden). Anti-skid not working.
2-044	05-07-92	Seeck	PST to 45°.	FCC #2 down during RTB, can't reset.
2-045	06-04-92	Knox	FQ, PST.	
2-046	06-04-92	Smith	Pilot checkout.	
2-047	06-09-92	Lang	FQ, PST 50°.	
2-048	06-09-92	Seeck	FQ, PST 50°.	
2-049	06-09-92	Wisneski	Pilot checkout.	
2-050	06-11-92	Smith	2nd flight.	
2-051	06-11-92	Wisneski	2nd flight.	
2-052	06-11-92	Knox	FQ, PST 50°.	
2-053	06-16-92	Lang	FQ, rolled to inverted position, A/B.	
2-054	06-16-92	Seeck	FQ, A/B, inverted.	

2-055	06-16-92	Smith	FQ, A/B, inverted.	
2-056	10-09-92	Lang	70° AoA, 360° rolls, 35 kft.	TV2 fail code twice during flight (trim resistor).
2-057	10-15-92	Smith	50° AoA, 360° rolls, 35 kft.	TV2 fail code after engine start.
2-058	10-16-92	Seeck	70° AoA, 360° rolls, 35 kft.	I0C power-up fail (sevro amp test). Asymmetry at 50° & 55°.
2-059	10-29-92	Knox	70° AoA, 35 kft.	IOC frame overrun during power-up TV2 fail code five times during flight.
2-060	10-29-92	Wisneski	70° AoA, 35 kft.	1st flight with grit on noseboom & radome.[1] TV2 fail code two times during flight.
2-061	11-03-92	Groves	50° AoA, 360°, rolls 35 kft.	TV servo amp faile codes twice on ground, once in flight.
2-062	11-03-92	Lang	60° AoA, 360° rolls, 35 kft.	Continued post-stall envelope expansion with grit strips.
2-063	11-03-92	Seeck	50° AoA, 360° rolls, 23 kft.	Post-stall testing.
2-064	11-05-92	Knox	PST env. exp. at 35 kft.	Flight abort, FTI problem.
2-065	11-05-92	Wisneski	65° AoA, 35 kft.	
2-066	11-06-92	Groves	65° AoA, B-B rolls, 23 kft.	
2-067	11-06-92	Seeck	70° AoA, B-B rolls, 23 kft.	
2-068	11-06-92	Knox	70° AoA, 360° rolls, 23 kft.	
2-069	11-06-92	Wisneski	70° AoA, 360° rolls, 23 kft.	
2-070	11-10-92	Smith	70° AoA, 360 rolls, 23 kft.	Grit on beta vane.
2-071	11-10-92	Seeck	70° AoA, 360° rolls, 23 kft.	Photo session. Grit removed from beta vane.
2-072	11-10-92	Knox	70° AoA, 360° rolls, 23 kft.	
2-073	11-30-92	Wisneski	2-G PST entry.	Departure during PST entry.
2-074	12-15-92	Smith	Pitch pulls, 30 kft.	TV servo amp fail codes 5 times.
2-075	12-15-92	Seeck	WUTs 30 kft.	TV servo amp fail codes once.
2-076	12-17-92	Wisneski	WUTs 20 kft.	TV servo amp fail codes twice
2-077	12-17-92	Smith	WUTs 20 kft.	
2-078	01-19-93	Lang	35 kft. 40° AoA.	Rt. WoW during landing.
2-079	01-19-93	Smith	35 kft. 50° AoA.	IOC frame overrun during power-up.
2-080	01-19-93	Knox	35 kft. 60° AoA.	TV servo amp fail codes 8 times during RTB.
2-081	01-21-93	Groves	35 kft. 60° AoA.	TM loss for few sec. during flight.
2-082	03-02-93	Wisneski	35 kft. 60° AoA.	Left WoW during landing.
2-083	03-02-93	Smith	35 kft. 60° AoA.	

[1] When the pilots started flying above 50 degrees AoA, they encountered kicks from the side that they called lurches. The international team added narrow 1/4-inch-wide strips of grit to the aircraft's noseboom and radome to change the vortices flowing from them. The grit strips reduced the randomness of the lurches caused by the vortices, enabling the pilots to finish envelope expansion to the design AoA-limit of 70 degrees at 1 g of acceleration.

2-084	03-02-93	Knox	35 kft. 60° AoA.	
2-085	03-02-93	Lang	35 kft. 60° AOA.	
2-086	03-30-93	Smith	Beta & roll stick checks.	1st flight with beta vane wedge to eliminate oscillations @ 60-65° AoA.
2-087	03-30-93	Wisneski	23 kft 50° AoA.	
2-088	03-30-93	Knox	23 kft. 70° AoA..	TV servo amp fail codes (2127) once prior to takeoff.
2-089	04-01-93	Knox	35 kft. 70° AoA WUT.	ECS problems.
2-090	04-02-93	Wisneski	30 kft. 70° AoA WUT.	ECS problems.
2-091	04-08-93	Lang	Tower flyby (Kiel probe).	1st flight with Kiel probe,[2] ECS problems.
2-092	04-15-93	Knox	35 kft 40°, 50° AoA rolls.	Extended nose strakes.
2-093	04-16-93	Smith	30 kft, M 0.4 45° split-S.	
2-094	04-16-93	Wisneski	30 kft, M 0.5 60° split-S.	2127 during RTB.
2-095	04-21-93	Knox	30 kft 70° diagonal pulls.	Extended nose strake removed. Grit added from strake to canard.
2-096	04-21-93	Wisneski	30 kft, M 0.4 40° split-S.	
2-097	04-22-93	Knox	30 kft, M 0.5 60° pulls.	
2-098	04-22-93	Knox	30 kft, M 0.5 50° split-S.	
2-099	04-23-93	Knox	30 kft, M 0.5 50° split S.	
2-100	04-23-93	Knox	23 kft, M 0.4 WUT.	AoA failure, broken contracts.
2-101	04-29-93	Groves	20 kft, M 0.4 60° pulls.	
2-102	04-29-93	Lang	20 kft, M 0.4 70° pull downs.	Milestone 4 test.
2-103	04-29-93	Smith	30 kft, M 0.6 70°.	
2-104	05-05-93	Groves	20 kft, 70°, J-turn.	Photo session with Lear Jet.
2-105	05-05-93	Groves	20 kft, 70°, J-turn.	Photo session with Lear Jet.
2-106	05-13-93	Wisneski	70° AoA.	Test for Kiel probe calibration.
2-107	05-13-93	Smith	Wake encounters.	
2-108	05-13-93	Lang	20 kft, pitch captures 30-60°.	
2-109	05-27-93	Knox	30 kft, 70° AoA.	Software version 117 installed, retest of aircraft.
2-110	05-27-93	Wisneski	30 kft, 70° AoA.	117 retest. Rt. WoW during landing.
2-111	05-27-93	Lang	30 kft, 70° AoA.	117 retest.
2-112	06-01-93	Smith	Wake encounters.	117 retest.
2-113	06-01-93	Wisneski	23 kft, split-S, scissors.	117 retest, software commanded acrft. out of PST.

[2] The Kiel pitot-static probe with a 10° downward cant was installed to solve the problem than when pilots flew for extended periods above 30 degrees AoA, the inertial navigation unit began calculating large but fictitious values of sideslip as a result of changes in wind direction and magnitude.

2-114	06-01-93	Lang	23 kft, J-turn.	117 retest.
2-115	06-01-93	Smith	23 kft, 720° rolls.	117 retest.
2-116	06-03-93	Wisneski	25 kft, lo. spd., J-turn.	Low speed expansion.
2-117	06-03-93	Lang	16 kft, J-turns.	Low speed expansion.
2-118	06-03-93	Smith	23 kft, lo. spd., J-turn.	Low speed expansion.
2-119	06-03-93	Wisneski	16 kft, lo. spd., J-turn.	Low speed expansion.
2-120	06-10-93	Groves	30 kft, 0.6M.	AoA Ch. 3 fail –84.71°. Broken wire, resoldered.
2-121	06-10-93	Lang	23 kft, 225 Kt. WUT, Split-S.	
2-122	06-10-93	Smith	Tactical.	
2-123	06-10-93	Wisneski	Tactical.	
2-124	10-05-93	Lang	Software version 118 C/O.	1st flight with load 118.
2-125	10-05-93	Smith	118 C/O.	
2-126	10-07-93	Knox	118 C/O.	
2-127	10-07-93	Wisneski	118 C/O.	
2-128	10-07-93	Lang	118 C/O.	
2-129	10-07-93	Smith	118 C/O.	
2-130	10-28-93	Knox	Pullups to 70°, pushovers. to 15°.	
2-131	12-16-93	Smith	Various maneuvers.	1st HMD Viper helmet flight.
2-132	12-16-93	Kim	Various maneuvers, HMD.	
2-133	12-20-93	Lang	Maneuvers, HMD.	Forebody strake change. FID during engine start.
2-134	12-20-93	Wisneski	Maneuvers, HMD.	FCC 1 failed to power-up 1st time. Forebody strake change.
2-135	01-06-94	Knox	Envelope expansion.	Blunted nose.
2-136	01-06-94	Kim	CIC w/o PST.	
2-137	01-06-94	Lang	Abrupt pulls.	Yaw rate failed during taxi after landing.
2-138	01-06-94	Wisneski	CIC w/o PST.	
2-139	01-11-94	Smith	CIC w/o PST.	Main gear wheel damage & blowout after landing.
2-140	01-11-94	Wisneski	CIC w/o PST.	Replaced both main gear wheels & tires.
2-141	01-11-94	Kim	CIC w/o PST.	
2-142	01-12-94	Lang	CIC w/o PST.	
2-143	01-12-94	Knox		FCC 2 failed during climbout; RTB.
2-144	01-20-94	Knox	Envelope expansion.	Replace main gear tires prior to flight. Lt. WoW during landing.
2-145	01-20-94	Smith	Envelope expansion.	
2-146	01-20-94	Lang	Envelope expansion.	
2-147	01-20-94	Knox	Envelope expansion.	
2-148	01-21-94	Smith	Envelope expansion.	

2-149	01-25-94	Groves	1st dedicated HMD flight, CICP.	Completed required sorties for all X-31 pilots using HMD before performing PST CIC flights.
2-150	01-26-94	Lange	CIC	Air intake lip servo amp FID, successfully reset during flight.
2-151	01-26-94	Groves	CIC.	
2-152	02-02-94	Lang	CIC.	HMD did not function due to loose connector.
2-153	02-02-94	Groves	CIC 15 kft.	Cloud cover.
2-154	02-08-94	Lang	CIC.	Crosswinds marginal during landing.
2-155	02-10-94	Lang	CIC.	TM time code problem.
2-156	02-10-94	Wisneski	CIC.	Lakebed runways wet.
2-157	02-10-94	Lang	CIC.	
2-158	03-15-94	Knox	Retest of X-31 with 119.	Software load 119 installed before flight.
2-159	03-15-94	Kim	Photo flight.	X-31, F-18 HARV, F-16 MATV.
2-160	03-15-94	Lang	Photo flight.	F-18 chase and X-31.
2-161	03-17-94	Knox	Envelope expansion.	265 Kt. PST, 30 kft., Spilt-S 45-70°. No HMD transmission to control rm.; circuit-breaker problem.
2-162	03-17-94	Kim	CIC.	45° AoA limit to investigate variable AoA CIC maneuvers.
2-163	03-29-94	Schneider	Various maneuvers.	Pilot familiarization.
2-164	03-29-94	Groves	CIC. Envelope expansion.	SSLA 45° AoA, mil. pwr. PST entry. Intermittant HMD operation.
2-165	03-29-94	Schneider	Aerobatic maneuvers. CCIP.	HMD tracker not working.
2-166	04-06-94	Lang	Envelope expansion.	265 Kt. PST entry, 15 kft.
2-167	04-14-94	Groves	CICP.	F-14 adversary.
2-168	04-14-94	Knox	CICP.	F-18 adversary.
2-169	04-14-94	Groves	CICP.	F-14 adversary.
2-170	04-14-94	Schmidt	CICP.	F-18 adversary, 400th X-31 flight.
2-171	04-19-94	Lang	Demo practice maneuvers.	Practice for Media Day.
2-172	05-18-94	Kim	Phasing maneuvers.	RTB because of engine problem.
2-173	06-08-94	Kim	CICP practice sorties.	Practice for visit of GAF chief of staff.
2-174	06-08-94	Knox	PID.	PID points for Langley.
2-175	06-08-94	Lang	PID	PID points for Langley.
2-176	06-10-94	Kim	CICP.	Demo for Lt. Gen. Hans-Jorg Kuebart (GAF chief-of-staff).
2-177	06-10-94	Knox	PID.	PID points for Langley.
2-178	06-15-94	Lang	CICP.	F-18 adversary.
2-179	06-15-94	Kim	CICP.	F-18 adversary.
2-180	06-15-94	Smith	PID.	PID points for Langley.
2-181	06-17-94	Lang	STEMS.	Longitudinal gross acquisition maneuvers (LGAM).

2-182	06-17-94	Knox	PID.	PID points for Langley.
2-183	06-17-94	Lang	LGAM.	Maneuvers to “acquire” the target F-18 aircraft.
2-184	06-17-94	Knox	STEMS.	LGAM.
2-185	06-22-94	Smith	CICP, 45° AoA limit.	SSLA and HSLA setups with F-18 adversary.
2-186	06-22-94	Kim	CIC.	F-18 adversary.
2-187	06-22-94	Smith	CIC, 45° AoA limit.	F-18 adversary.
2-188	06-22-94	Lang	CIC.	F-18 adversary.
2-189	06-24-94	Kim	CIC.	F-18 adversary.
2-190	06-24-94	Lang	CIC. Unlimited AoA.	F-18 adversary.
2-191	06-24-94	Kim	CIC.	F-18 adversary.
2-192	06-29-94	Smith	CIC. Unlimited AoA.	F-18 adversary.
2-193	06-29-94	Lang	CIC. Unlimited AoA.	F-18 adversary.
2-194	06-29-94	Kim	CIC. Unlimited AoA.	F-18 adversary.
2-195	07-11-94	Smith	Virtual warfight.	Practice for virtual warfight demo.
2-196	07-11-94	Lang	Virtual warfight.	QT engaged. Fixed point overflow FCS code necessitated RTB.
2-197	07-15-94	Smith	Virtual warfight practice.	QT, sim. bomb drop, missile evasion.
2-198	07-15-94	Lang	Virtual warfight planned.	Fixed point overflow FCS code necessitated RTB.
2-199	08-25-94	Lang	FCF, CICP/w HMD.	Left WoW indication ON during flight. Stuck WoW switch replaced after flight.
2-200	08-26-94	Lang	CICP w/HMD.	High off-boresight missile.
2-201	08-26-94	Kim	CICP w/HMD.	High off-boresight missile.
2-202	08-26-94	Kim	CICP w/HMD.	HMD failed and turned off.
2-203	09-01-94	Knox	QT power approach.	First QT subsonic flight.
2-204	09-01-94	Lang	QT power approach.	
2-205	09-01-94	Smith	CIC w/helmet.	F-18 adversary.
2-206	09-13-94	Kim	CIC with 422 TES.	Against F-16 from Nellis.
2-207	09-15-94	Smith	CIC with 422 TES.	F-16 from Nellis.
2-208	09-15-94	Lang	CIC with 422 TES.	F-15 from Nellis.
2-209	09-15-94	Kim	CIC with 422 TES.	F-15 from Nellis.
2-210	09-15-94	Loria	CIC with 422 TES.	F-15 from Nellis.
2-211	09-21-94	Kim	CIC w/HMD	Col. Tac Nix in 2nd F-18 chase airplane with Jim Smolka.
2-212	09-27-94	Knox		RTB because of ECS problem. Pressure relief valve replaced after flt.
2-213	09-27-94	Lang	PID and QT.	ECS problem recurred but flight not curtailed.
2-214	10-27-94	Lang	ECS/FCS checkout.	“Blue” fuel-control engine (GE F404-310) installed before flight.
2-215	10-27-94	Kim	STEMS.	Lateral gross (target) acquisition tracking tasks.

2-216	10-27-94	Smith	STEMS.	Longitudinal gross acquisition tasks.
2-217	10-27-94	Lang	STEMS.	Helicopter gun attack evaluation, F-18 adversary.
2-218	11-01-94	Kim	CIC w/HMD.	
2-219	11-01-94	Knox	CIC w/HMD.	Off-boresight missile capability.
2-220	11-08-94	Loria	FCF, QT testing.	QT PA.
2-221	11-08-94	Lang	QT in PA and cruise.	
2-222	11-08-94	Knox	QT in PA and cruise.	
2-223	11-17-94	Lang	QT, 8 kft. 170 & 220 KIAS.	TV plume boundary calibrations. Nose gear down-lock wire disconnected.
2-224	11-29-94	Loria	Practice approaches.	Simulated carrier operations.
2-225	11-29-94	Smith	QT in PA and cruise.	
2-226	11-29-94	Knox	Carrier suitability tasks.	
2-227	12-01-94	Smith	CIC w/HMD.	60° OBS-angle launchable missile.
2-228	12-01-94	Lang	QT, 8 kft.	Cruise configuration.
2-229	12-06-94	Knox	QT, 8 kft.	PA configuration, 170 & 220 KIAS.
2-230	12-06-94	Smith	QT, 220 KIAS.	9.5-7.5 kft . Rt. WoW during landing.
2-231	12-06-94	Lang	QT, 425 KIAS, 8 kft.	QT and BA also flown at 220 KIAS.
2-232	04-13-95	Knox	PST	1st flight after A/C #1 crash. Functional C/O.
2-233	04-13-95	Smith	PST	A/B ignition problem.
2-234	04-17-95	Knox	PST	Intermittent A/B problem.
2-235	04-17-95	Kim	PST	Intermittent A/B problem.
2-236	04-21-95	Kim	PST 13 kft.	Airdata observer fail during pull to 4.5 g's from 350 KIAS.
2-237	04-22-95	Kim	PST 13 kft., 5 kft. AGL.	
2-238	04-22-95	Smith	PST 13 kft., 5 kft. AGL.	
2-239	04-22-95	Smith	PST 13 kft., 5 kft. AGL.	
2-240	04-24-95	Knox	Airshow maneuvers @ 5 kft. AGL.	
2-241	04-24-95	Kim	Airshow maneu. @ 5 kft. AFL.	
2-242	04-26-95	Knox	Airshow maneu. @ 2500' AGL.	
2-243	04-26-95	Smith	Airshow maneu. @ 1500' AGL.	1st R3 landing.
2-244	04-26-95	Knox	Airshow maneu. @ 1000' AGL.	
2-245	04-28-95	Smith	Airshow maneu. @ 500' AGL.	1st A/B takeoff

2-246	04-28-95	Kim	Airshow maneu. @ 500' AGL.	
2-247	05-04-95	Knox	Airshow practice.	1st silent control room. 13° AoA landing.
2-248	05-04-95	Kim	Airshow practice.	Silent control room.
2-249	05-04-95	Knox	Airshow practice	Silent control room. 13k AoA landing.
2-250	05-05-95	Kim	Airshow practice.	
2-251	05-05-95	Knox	Airshow practice.	Drag chute not installed. Spin chute deactivated.
2-252	05-06-95	Kim	Airshow practice.	Drag chute not installed.
2-253	05-08-95	Knox	Airshow practice.	Drag chute not installed.
2-254	05-08-95	Kim	Airshow practice.	Drag chute not installed.
2-255	05-10-95	Knox	Airshow practice	Non-functional drag chute installed.
2-256	05-10-95	Kim	Airshow practice.	No chute installed.
2-257	05-10-95	Knox	Airshow practice.	Max. brake test at 100 KIAS.
2-258	05-12-95	Knox	Airshow practice.	1st drag chute deployment @ 130 KIAS. No problems.
2-259	05-12-95	Knox	Airshow practice.	Max. anti-skid braking @ 125 KIAS.
2-260	05-12-95	Kim	Airshow practice	Normal landing.
2-261	05-12-95	Knox	Airshow practice.	Drag chute deployment @ 155 KIAS.
2-262	05-13-95	Knox	Airshow practice.	Drag chute deployment @ 170 KIAS. Release @ 155 KIAS. Moderate braking.
2-263	05-13-95	Kim	Airshow practice.	Drag chute arming failed. Moderate braking.
2-264	05-13-95	Knox	Airshow practice.	Moderate braking.
2-265	05-15-95	Kim	Airshow practice.	Drag chute deployment jettison @ 110 KIAS.
2-266	05-16-95	Knox	Airshow practice.	Drag chute deployment.
2-267	05-16-95	Kim	Airshow practice.	Last flight at Dryden before airshow. Loaded on C-5A on 05-20-95.
2-268	05-29-95	Knox	A/C checkout & A/S practice.	1st flight at Manching, Germany. Problem with drag chute.
2-269	05-30-95	Kim	A/C checkout & A/S practice.	At Manching. Bad weather prevented more flights.
2-270	05-31-95	Kim	A/C checkout & A/S practice.	At Manching.
2-271	06-02-95	Knox	A/C checkout & A/S practice.	At Manching.
2-272	06-02-95	Kim	A/C checkout & A/S practice.	At Manching.
2-273	06-03-95	Knox		Ferry from Manching to Cologne. Stored heading INU for next flight.

2-274	06-03-95	Knox		Ferry from Cologne to Manching. Nose boom alpha failed after takeoff. R3 landing in Paris, France.
2-275	06-08-95	Knox	A/C checkout.	First flight at Paris. Noseboom replaced.
2-276	06-08-95	Kim	A/C checkout & A/S practice.	
2-277	06-08-95	Knox	A/C checkout & A/S practice.	
2-278	06-09-95	Kim	High show practice.	Both pilots now air show qualified.
2-279	06-10-95	Knox	High show demonstration.	Drag chute failed.
2-280	06-12-95	Kim	Demonstration flight.	Drag chute deployed successfully.
2-281	06-13-95	Knox	Low show demonstration.	
2-282	06-14-95	Kim	High show demonstration.	
2-283	06-15-95	Knox	High show demonstration.	
2-284	06-16-95	Kim	High show demonstration.	
2-285	06-17-95	Knox	Demonstration flight.	
2-286	06-18-95	Kim	Demonstration flight.	Last day of air show.
2-287	06-19-95	Kim	Ferry.	Paris to Cologne. Drag chute deployment.
2-288	06-19-95	Kim	Ferry.	Cologne to Manching. Last flight of aircraft in this project. 02-23-99 shipped to Boeing facility in Palmdale. Later it went to Patuxent River, MD, where it began a new flight test program in 2001 without Dryden involvement.

Sources: Flight logs compiled by the X-31 project office, supplemented by flight reports where available and information collected by Peter Merlin, Betty Love, and J. D. Hunley from various sources including project personnel.

X-31 Acronyms

2127	TV servo amp fail codes
a & b, a/b	Flight-control overlays a and b
A/B	Afterburner
A/C	Aircraft
ADC	Air data computer
A/G	Air-to-ground
AGL	Above ground level
AIL	Aileron
AoA	Angle of attack
ATLAS	Adaptable target lighting array system
BA	Basic airplane (no QT or TVV)
B-B	Bank-to-bank (rolls)
Beta	Angle of sideslip
BFM	Basic fighter maneuver
BIT	Built-in test
CAUT/WRN	Caution/warning
C/B	Circuit breaker
CCDL	Cross-channel data link
CIC	Close-in combat
CICP	Close-in combat practice
C.L.	Center line
C/O	Check out
DIS	Disabled
DLR	*Deutsche Forschungsanstalt f¸r Luft- und Raumfahrt* (German Research Institute for Air- and Spaceflight)
DVINU Qs	Inertial navigation, pitch rate selected
EAFB	Edwards Air Force Base
ECS	Environmental control system
EDV	Electrical depressurization value
FCC	Flight control check
FCS	Flight control system
FEST	Flight estimated thrust
FID	Fault instrumentation and detection
FQ	Flying qualities
FTB	Flutter test box—automated inputs for flutter testing
FTI	Flight test instrumentation
G	Acceleration equal to the force of gravity
GAF	German Air Force
GMD	German Ministry of Defense
GPE	Government preliminary evaluation
HARV	High Angle-of-attack Research Vehicle
HMD	Helmet mounted display
HSLA	High speed line-abreast

Hz	Hertz
ICMD'RAMERR	Ram error
IDs	Identifications
INU	Inertial navigation unit
IOC	Input and output controller
ITO	International Test Organization
KFT, kft	Altitude in thousands of feet
KIAS	Indicated airspeed in knots
Kt	Knots (probably always KIAS)
LCDR	Lieutenant commander
LEF	Leading edge flap
LFINS	Logic failure, inertial navigation system
LFTVS	Logic failure of the thrust vectoring system
LGA	Lateral or longitudinal gross acquisition (of target)
LGAM	Longitudinal gross acquisition maneuvers
LGRND	Weight on wheels, logical
L.O. PRES	Low oxygen pressure
LR1RQ	Logical for reversionary mode 1 request
LSLA	Low speed line abreast
LTEI	Left trailing edge inboard
LTEI & RLE	Left trailing edge inboard & right leading edge
M	Mach number (speed in relation to that of sound, Mach 1)
MATV	Multi-Axis Thrust Vectoring
MBB	Messerschmitt B^lkow Blohm
MIL PWR	Military power
N2S	Engine rotational measurement
N_x	Acceleration in the X axis
N_y	Lateral acceleration
OBS	Off boresight
PA	Power approach
PDM	Pulse duration modulation
PFB	Pre-flight bit
PID	Parameter identification
PLA	Power lever angle
PST	Post stall
q	Dynamic pressure
Qs & Ps	Impact pressure & static pressure
QT	Quasi tailless
R1, R2, R3	Flight control reversionary modes 1 (INU), 2 (angle of attack and sideslip), and 3 (airdata)
RADM	Rear admiral
RLE	Right leading edge
RTB	Return to base
RTE	Right trailing edge
SA	Solenoid actuator
SB	Synch. bit
SPDBRK	Speedbrake

SSLA MOD	Slow speed line abreast modified
STEMS	Standard test evaluation maneuver set
TB	Tie breaker
TEF	Trailing edge flap
TEO	Trailing edge outboard
TES	Test and Evaluation Squadron
TM	Telemetry
T/O	Takeoff
T/R	Transformer/rectifier
TV	Thrust vectoring
TVVs	Thrust vectoring vanes
USMC	United States Marine Corps
USN	United States Navy
WoW Sw	Landing-gear weight-on-wheel sensor
WTD-61	*Wehrtechnische Dienststelle-61* (roughly, Federal German Armed Forces Engineering Center for Aircraft-61)
WUT	Wind-up turn

Appendix Y

SR-71 Program Flight Chronology, 1991-1999

NASA crews flew four Lockheed SR-71 airplanes between July 1991 and October1999. They were used for research and to support the U.S. Air Force reactivation of the SR-71 for reconnaissance missions. The Air Force had retired the Blackbirds in 1990, but Congress reinstated funding for additional flights. Lockheed SR-71A (61-7980 / NASA 844) arrived at NASA Dryden Flight Research Center on 15 February 1990. It was placed into storage until 1992. It served as a research platform until October 1999. SR-71A (61-7971 / NASA 832) arrived at Dryden on 19 March 1990. It departed to Lockheed Palmdale on 12 January 1995, having never been flown by NASA crews at the Center. It was flown by NASA crews at Palmdale in support of the SR-71 Reactivation program. Another SR-71A (61-7967) was flown by NASA crews only at Palmdale in 1995 and 1996 in support of Reactivation. Steve Ishmael and Rod Dyckman flew two functional check flights in Lockheed SR-71B (61-7956 / NASA 831) at Palmdale in early July 1991 before delivering the aircraft to Dryden on 25 July. The SR-71B trainer served as a research platform and for crew training and proficiency until October 1997.

For the purpose of this chronology, flight numbers begin with the airplane's NASA number or last three digits of its Air Force serial number followed by the flight number for that particular aircraft. For the SR-71A, the name of the pilot is listed first, followed by the Research Systems Operator (RSO). For the SR-71B, the pilot or student pilot is first, followed by the instructor pilot (for training flights) or RSO (for research or RSO training flights).

SR-71 Flights

Flight No.	Date	Pilot/Research Systems Operator or Instructor Pilot	Max Altitude (Ft.)	Max Speed (Mach)	Remarks
831-1	1 Jul. 1991	Steve Ishmael/W. Rod Dyckman	71,400	3.10	Functional check flight (FCF) at Palmdale
831-2	10 Jul.	Dyckman/Ishmael	80,800	3.23	FCF at Palmdale
831-3	25 Jul.	Dyckman/Ishmael	74,500	3.09	Delivery to NASA DFRC
831-4	14 Aug.	Rogers Smith/Ishmael	75,500	3.05	First all-NASA crew
831-5	26 Aug.	Smith/Ishmael	79,800	3.22	First in-flight refueling
831-6	23 Sep.	Smith/Ishmael	78,700	3.19	Last training flight for Smith
831-7	3 Oct.	Ishmael/Marta Bohn-Meyer	81,450	3.23	RSO checkout, first female SR-71 crewmember
831-8	9 Oct.	Smith/Bob Meyer	81,450	3.23	RSO checkout
831-9	29 Oct.	Ishmael/Bohn-Meyer	79,500	3.20	Operational experience mission
831-10	1 Nov.	Smith/Meyer	81,500	3.25	Operational experience mission
831-11	7 Nov.	Ishmael/Bohn-Meyer	81,500	3.21	Operational experience mission
831-12	14 Nov	Ishmael/Meyer	81,800	3.21	Operational experience mission
831-13	28 Jan. 1992	Ishmael/Meyer	77,400	3.26	FCF, operational experience mission
831-14	26 Feb.	Smith/Bohn-Meyer	81,100	3.26	Operational experience mission
831-15	10 Mar.	Ishmael/Meyer	78,000	3.12	Level accelerations

831-16	22 Apr.	Smith/Ishmael	81,800	3.22	Operational experience mission
831-17	20 May	Smith/Ishmael	83,500	3.21	Crew proficiency
831-18	4 Jun.	Tom McMurtry/Ishmael	83,000	3.23	Operational and pilot check out
831-19	26 Jun.	Dr. Robert Barthelemy/Ishmael	82,160	3.23	Orientation flight for director of National AeroSpace Plane (NASP) Joint Program Office
844-1	24 Sep.	Ishmael/Bohn-Meyer	78,680	3.25	FCF, Orbital Sciences Corporation (OSC) frequency scanning experiment
844-2	6 Oct.	Smith/Meyer	80,000	3.26	FCF, OSC F3SAT back-up satellite carried in passive mode to check prelaunch operations
831-20	24 Oct.	Ishmael/Bohn-Meyer	78,600	3.20	FCF, Dryden Family Day fly-by
831-21	24 Nov.	Smith/Meyer	80,500	3.26	Crew proficiency with tanker
831-22	8 Dec.	Ishmael/Bohn-Meyer	76,500	3.09	Refueled from KC-10 tanker, Navy Radar Surveillance Technology Experimental Radar (RSTER) test over Atlantic Ocean. Aircraft flew almost 6,000 NM in less than 5 hours
831-23	20 Jan. 1993	Smith/Meyer	80,000	3.22	Inlet test, crew proficiency
844-3	9 Mar.	Ishmael/Bohn-Meyer	82,350	3.17	Ultraviolet Charge Coupled Device (UV CCD) astronomy experiment, Southwest University UV Imaging System (SWUIS) experiment
844-4	16 Mar.	Smith/Meyer	84,050	3.24	UV CCD, SWUIS
831-24	14 Apr.	Smith/Bohn-Meyer	81,500	3.23	FCF, crew proficiency
844-5	15 Jul.	Ishmael/Meyer	81,800	3.23	Near-Ultraviolet Spectrometer (NUVS) experiment, took on JP-8 fuel from tanker
844-6	28 Jul.	Smith/Bohn-Meyer	48,500	1.85	NUVS experiment, sonic boom test with F-16XL
844-7	3 Aug.	Ishmael/Meyer	83,950	3.23	NUVS experiment, handling qualities
844-8	17 Sep.	Smith/Ishmael	76,070	3.00	Optical Air Data System (OADS), handling qualities experiment
844-9	1 Oct.	Ishmael/Bohn-Meyer	76,500	3.17	OADS and NUVS experiments
844-10	6 Oct.	Smith/Meyer	73,025	3.03	NUVS experiment, handling qualities
844-11	13 Oct.	Ishmael/Bohn-Meyer	83,700	3.23	NUVS experiment, handling qualities
844-12	20 Oct.	Smith/Meyer	75,635	3.05	NUVS experiment, handling qualities
831-25	9 Nov.	Ishmael/Bohn-Meyer	75,850	3.07	FCF, crew proficiency
844-13	8 Dec.	Ishmael/Meyer	77,375	3.21	NUVS experiment
844-14	22 Dec.	Ishmael/Bohn-Meyer	76,000	3.11	NUVS, Low Earth Orbit Experiment (LEOEX)
844-15	22 Dec.	Ishmael/Bohn-Meyer	76,150	3.11	NUVS and LEOEX experiments
844-16	25 Jan. 1994	Smith/Meyer	77,600	3.04	NUVS, LEOEX, hot refuel and turnaround
844-17	25 Jan.	Smith/Meyer	77,600	3.04	NUVS and LEOEX experiments

831-26	18 Feb.	Ishmael/Meyer	81,200	3.20	FCF, crew proficiency
831-27	25 Feb.	Gen. John R. Dailey/Ishmael	82,100	3.17	FCF, orientation flight for NASA Associate Deputy Administrator
831-28	4 Mar.	Smith/Bohn-Meyer	80,800	3.23	FCF
844-18	7 Jul.	Ishmael/Bohn-Meyer	77,750	3.16	Dynamic Auroral Viewing Experiment (DAVE)
844-19	13 Jul.	Smith/Meyer	77,700	3.18	Handling qualities, DAVE
844-20	21 Jul.	Ishmael/Bohn-Meyer	80,300	3.19	NUVS, DAVE
831-29	24 Aug.	Smith/Bohn-Meyer	79,300	3.23	FCF
844-21	31 Aug.	Smith/Bohn-Meyer	75,700	3.05	NUVS, DAVE
831-30	6 Oct.	Ishmael/Meyer	83,600	3.22	Crew proficiency
831-31	18 Oct.	Ed Schneider/Ishmael	84,700	3.22	Pilot checkout, highest NASA SR-71 flight
844-22	25 Oct.	Smith/Bohn-Meyer	80,000	3.21	NUVS, DAVE, sonic boom test
831-32	17 Nov.	C. Gordon Fullerton/Ishmael	82,300	3.23	Pilot checkout
831-33	8 Dec.	Ishmael/Smith	82,500	3.23	Crew proficiency
832-1	12 Jan. 1995	Ishmael/Bohn-Meyer	3,200	0.39	Ferry flight to Palmdale
831-34	20 Jan.	Schneider/Ishmael	75,400	3.05	Crew proficiency
844-23	15 Feb.	Smith/Meyer	31,500	1.27	Sonic boom test with F-16XL
831-35	22 Feb.	Schneider/Smith	77,000	3.11	Crew proficiency
844-24	16 Mar.	Schneider/Bohn-Meyer	34,100	1.26	Sonic boom test with F-16XL
844-25	22 Mar.	Ishmael/Meyer	33,000	1.28	Sonic boom test with F-16XL and YO-3A, handling qualities
844-26	24 Mar.	Schneider/Bohn-Meyer	48,000	1.63	Sonic boom test with F-16XL and YO-3A
844-27	29 Mar.	Schneider/Meyer	48,200	1.56	Sonic boom test with F-16XL, YO-3A, and F-18
844-28	5 Apr.	Smith/Bohn-Meyer	47,600	1.54	Sonic boom test with F-16XL, YO-3A, and F-18
844-29	12 Apr.	Ishmael/Bohn-Meyer	44,300	1.28	Sonic boom test with F-16XL, YO-3A, and F-18
844-30	20 Apr.	Schneider/Meyer	44,300	1.35	Sonic boom test with F-16XL, YO-3A, and F-18
832-2	26 Apr.	Schneider/Bohn-Meyer	2,900	0.96	FCF for Air Force SR-71 reactivation
831-36	18 May	Schneider/Meyer	80,800	3.22	Crew proficiency
832-3	23 May	Schneider/Meyer	81,400	3.23	FCF for Air Force SR-71 reactivation
844-31	25 May	Smith/Bohn-Meyer	49,000	1.92	Sonic boom test with YO-3A, handling qualities, ferry to Palmdale for Linear Aerospike SR-71 Experiment (LASRE) modifications
831-37	31 May	Schneider/Bohn-Meyer	81,290	3.22	Crew proficiency
832-4	2 Jun.	Schneider/Bohn-Meyer	81,290	3.15	FCF for Air Force SR-71 reactivation
832-5	26 Jun.	Schneider/Meyer	72,800	3.21	FCF for Air Force SR-71 reactivation
831-38	27 Jun.	Lt. Col. Gil Luloff/Smith	78,245	3.23	Air Force training flight

832-6	6 Jul.	Schneider/Bohn-Meyer	79,100	3.20	FCF for Air Force SR-71 reactivation
832-7	12 Jul.	Schneider/Meyer	77,500	3.06	FCF for Air Force SR-71 reactivation
831-39	15 Jul.	Luloff/Smith	75,100	3.03	Air Force training flight
831-40	20 Jul.	Smith/Schneider	76,200	3.04	Instructor Pilot (IP) flight for Ed Schneider
831-41	28 Jul.	Lt. Col. Tom McCleary/Smith	76,650	3.15	Air Force training flight
831-42	2 Aug.	McCleary/Smith	77,450	3.05	Air Force training flight
831-43	11 Aug.	McCleary/Schneider	78,000	3.15	Air Force training flight
831-44	17 Aug.	Schneider/Bohn-Meyer	80,500	3.20	Crew proficiency
831-45	24 Aug.	Luloff/Smith	75,850	3.04	Air Force night training flight
967-1	28 Aug.	Schneider/Meyer	26,000	0.95	FCF for Air Force SR-71 reactivation
831-46	29 Aug.	McCleary/Lt. Col. Blair Bozek	76,500	3.15	Air Force crew requalification
831-47	31 Aug.	Smith/Meyer	79,800	3.24	Crew proficiency
967-2	6 Sep.	Smith/Bohn-Meyer	24,000	0.92	FCF for Air Force SR-71 reactivation
967-3	13 Sep.	Smith/Bohn-Meyer	23,500	0.93	FCF for Air Force SR-71 reactivation
831-48	27 Sep.	Smith/Schneider	80,300	3.21	FCF and crew proficiency
967-4	6 Oct.	Smith/Meyer	70,200	3.01	FCF for Air Force SR-71 reactivation
832-8	17 Oct.	Schneider/Bohn-Meyer	76,000	3.05	Exercise for Air Force SR-71 reactivation
967-5	25 Oct.	Schneider/Meyer	79,000	3.23	FCF for Air Force SR-71 activation, emergency landing at Nellis AFB, Nevada
967-6	27 Oct.	Schneider/Meyer	28,000	0.88	Ferry flight from Nellis AFB to Palmdale
831-49	30 Oct.	McCleary/Schneider	74,500	3.05	Crew proficiency, AF SR-71 reactivation
831-50	9 Nov.	Luloff/Smith	80,000	3.18	Crew proficiency, AF SR-71 reactivation
831-51	16 Nov.	McCleary/Schneider	77,050	3.02	Air Force night training flight
831-52	22 Nov.	Lt. Col. Don Watkins/Schneider	78,950	3.22	Air Force training flight
831-53	11 Dec.	Watkins/Schneider	80,250	3.21	Air Force training flight
831-54	14 Dec.	Schneider/BGen. Richard Engel	80,400	3.27	Orientation flight for commander of the Air Force Flight Test Center
967-7	15 Dec.	Smith/Bohn-Meyer	74,100	3.15	FCF for Air Force SR-71 reactivation
967-8	9 Jan. 1996	Smith/Meyer	72,800	3.06	FCF for Air Force SR-71 reactivation
831-55	11 Jan.	Watkins/Smith	77,400	3.23	Air Force training flight
967-9	12 Jan.	Smith/Bohn-Meyer	78,300	3.22	FCF, airplane released to Air Force
831-56	18 Jan.	McCleary/Schneider	77,700	3.02	Air Force requalification flight
831-57	1 Feb.	Schneider/Meyer	80,150	3.21	Crew proficiency
831-58	6 Feb.	Luloff/Smith	75,300	3.17	Air Force crew proficiency flight
831-59	16 Feb.	Schneider/Bohn-Meyer	48,000	1.50	Crew proficiency
844-32	14 Mar.	Schneider/Bohn-Meyer	26,650	0.98	FCF, ferry flight from Palmdale to DFRC

831-60	20 Mar.	Schneider/Bohn-Meyer	79,000	3.09	FCF, no-chute landing
844-33	22 Mar.	Schneider/Bohn-Meyer	80,400	3.22	FCF, engine up-trim tests, no-chute landing
831-61	29 Mar.	Smith/Meyer	80,000	3.22	Crew proficiency, no-chute landing
831-62	4 Apr.	Schneider/Bohn-Meyer	76,400	3.04	Crew proficiency, heavy-weight takeoff, no-chute landing
831-63	10 Apr.	Smith/Meyer	80,100	3.22	Crew proficiency, heavy-weight takeoff, no-chute landing
831-64	25 Apr.	Schneider/Meyer	82,200	3.26	Crew proficiency, tanker training
831-65	16 May	Smith/Bohn-Meyer	74,000	3.01	Crew proficiency, tanker training
831-66	5 Jun.	Schneider/Meyer	78,600	3.04	Crew proficiency
832-9	14 Jun.	Schneider/Bozek	78,600	3.05	Advanced Synthetic Aperture Radar System (ASARS), defensive avionics, and data link tests
844-34	12 Jul.	Schneider/Meyer	60,000	2.15	Simulated LASRE mission, pod-off
831-67	22 Aug.	Schneider/Bohn-Meyer	80,800	3.23	Crew proficiency, heavy-weight takeoff
831-68	7 Sep.	Dana Purifoy/Schneider	84,300	3.24	Pilot checkout and proficiency
831-69	8 Oct.	Luloff/Smith	80,000	3.17	Air Force training flight
831-70	23 Oct.	Luloff/Schneider	78,800	3.20	Air Force training flight
831-71	1 Nov.	McCleary/Smith	78,050	3.15	Air Force training flight
831-72	6 Nov.	McCleary/Schneider	79,600	3.16	Air Force training flight
831-73	20 Nov.	Luloff/Schneider	77,900	3.04	Air Force night training flight
831-74	3 Dec.	McCleary/Smith	76,500	3.02	Air Force night training flight
844-35	4 Dec.	Smith/Bohn-Meyer	42,000	1.75	Simulated LASRE mission, pod off
831-75	13 Dec.	Schneider/Meyer	81,000	3.20	Crew proficiency
831-76	28 Jan. 1997	Schneider/Meyer	80,000	3.20	Crew proficiency
844-36	30 Jan.	Smith/Bohn-Meyer	62,750	2.27	Handling qualities, crew proficiency
844-37	14 Feb.	Schneider/Meyer	60,500	2.10	Handling qualities, stability and control (S&C)
831-77	20 Feb.	Smith/Bohn-Meyer	77,600	3.07	Crew proficiency
831-78	4 Mar.	Schneider/Mark Stucky	81,400	3.27	Pilot checkout
844-38	18 Mar.	Schneider/Bohn-Meyer	62,750	2.17	Crew proficiency
844-39	2 Apr.	Schneider/Meyer	78,400	3.04	Crew proficiency, "hot time" on aircraft
831-79	17 Apr.	Marty Knutson/Schneider	80,850	3.27	Crew proficiency, orientation flight for Marty
831-80	2 May	McCleary/Luloff	75,800	3.04	Air Force training flight, IP check for Gil Luloff
831-81	8 May	Smith/Luloff	74,900	3.03	IP check for Gil Luloff
844-40	23 May	Smith/Bohn-Meyer	80,700	3.22	Crew proficiency, aircraft checkout
844-41	30 May	Schneider/Meyer	80,500	3.22	Crew proficiency, aircraft checkout
831-82	24 Jun.	Maj. Bert Garrison/Luloff	74,000	3.04	Air Force training flight
844-42	26 Jun.	Smith/Bohn-Meyer	80,000	3.19	Crew proficiency, air data calibration, S&C
831-83	2 Jul.	Garrison/Luloff	75,700	3.15	Air Force training flight

831-84	15 Jul.	Garrison/Luloff	75,900	3.06	Air Force training flight
844-43	17 Jul.	Schneider/Meyer	55,000	2.20	Crew proficiency, simulated LASRE mission
831-85	23 Jul.	Garrison/Luloff	73,500	3.04	Air Force training flight
844-44	25 Jul.	Smith/Bohn-Meyer	60,200	2.20	Crew proficiency, simulated LASRE mission
831-86	29 Jul.	Garrison/Luloff	78,000	3.18	Air Force training flight
831-87	2 Aug.	Smith/Meyer	74,600	3.05	Oshkosh Air Show fly-by, emergency landing at Milwaukee, Wisconsin due to fuel leak
831-88	11 Aug.	Smith/Meyer	63,000	2.55	Ferry flight from Milwaukee to DFRC, trouble with left engine resulted in damage to exhaust ejector
831-89	18 Sep.	Smith/Bohn-Meyer	80,150	3.22	FCF due to replacement of left engine
831-90	24 Sep.	McCleary/Luloff	75,000	3.03	Air Force night IP check for Gil Luloff
831-91	14 Oct.	Schneider/Meyer	76,600	3.22	Air Force 50th Anniversary Aerial Review at Edwards AFB
831-92	18 Oct.	Schneider/Bohn-Meyer	75,050	3.03	Edwards AFB Air Show fly-by
831-93	19 Oct.	Schneider/Meyer	25,000	0.70	Edwards AFB Air Show fly-by
844-45	31 Oct.	Schneider/Meyer	33,000	1.19	First LASRE pod flight, aerodynamic data
844-46	19 Dec.	Smith/Bohn-Meyer	50,050	1.61	Second LASRE aerodynamic flight
844-47	4 Mar. 1998	Schneider/Meyer	41,300	1.58	First LASRE cold flow
844-48	19 Mar.	Smith/Bohn-Meyer	50,050	1.63	Second LASRE cold flow
844-49	15 Apr.	Schneider/Meyer	51,000	1.78	LASRE ignition test, LOX carry
844-50	23 Jul.	Smith/Bohn-Meyer	32,350	0.92	LASRE LOX cold flow test, pod purge evaluation, hot fire simulation, crew proficiency
844-51	29 Oct.	Schneider/Meyer	41,200	1.4	LASRE LOX cold flow test, pod purge evaluation. Final LASRE program flight
844-52	30 Jun. 1999	Schneider/Meyer	55,000	2.25	Handling qualities with LASRE model removed from flight test fixture, pod purge evaluation
844-53	15 Jul.	Smith/Bohn-Meyer	63,800	2.73	Handling qualities with LASRE model removed from flight test fixture, pod purge evaluation
844-54	16 Aug.	Schneider/Meyer	67,800	2.73	Handling qualities with LASRE model removed from flight test fixture, pod purge evaluation, boundary layer rakes installed on pod
844-55	27 Sep.	Smith/Bohn-Meyer	64,000	2.70	Handling qualities with LASRE model removed from flight test fixture, pod purge evaluation, boundary layer rakes
844-56	9 Oct.	Smith/Meyer	80,100	3.21	Edwards AFB Air Show fly-by, a serious fuel leak led to cancellation of a scheduled Air Show flight for the next day.

Sources: SR-71 program chronology by Mike Reljia, DFRC Flight Operations Office Daily Logs, Buddy Brown's Blackbird crew list, Peter W. Merlin's personal log, Monthly Aerospace Projects Update memoranda, Weekly Aerospace Projects Highlights on file in the DFRC Historical Reference Collection

Index

Printed in the United States
by Baker & Taylor Publisher Services

Printed in the United States
by Baker & Taylor Publisher Services